TELEVISION

PRODUCTION

TELEVISION

PRODUCTION

Alan Wurtzel

SECOND EDITION

McGRAW-HILL BOOK COMPANY
New York   St. Louis   San Francisco   Auckland   Bogotá
Hamburg   Johannesburg   London   Madrid   Mexico
Montreal   New Delhi   Panama   Paris   São Paulo
Singapore   Sydney   Tokyo   Toronto

TELEVISION PRODUCTION

34567890DOCDOC89876543

ISBN 0-07-072131-9

**Library of Congress Cataloging
in Publication Data**

Wurtzel, Alan.
    Television production.

    Bibliography: p.
    Includes index.
    1. Television—Production and direction.
2. Television—Apparatus and supplies.   I. Title.
PN1992.75.W8 1983   791.45'0232   82-16209
ISBN 0-07-072131-9

This book was set in Helvetica Light by Black Dot, Inc. (ECU).
The editors were Marian D. Provenzano and James R. Belser;
the designer was Joan E. O'Connor;
the production supervisor was Phil Galea.
R. R. Donnelley & Sons Company was printer and binder.

For Susan

CONTENTS

PREFACE

TO THE SECOND EDITION

The television production industry has always been characterized by its rapid growth and development. Over the few years since the first edition of *Television Production* appeared, however, television production has undergone some of the most important and significant changes in its history:

- The capability of production equipment continues to grow as the cost of the equipment decreases. The result is greater production capability at every production level. In fact, even the most modest production facility today boasts a capability which was unmatched by the most sophisticated facility of a decade ago. The importance of all this is not simply that the equipment delivers more power and higher quality for less cost. The production opportunities which modern equipment offers are limited only by a producer or director's creative skills and imagination.

- New distribution systems such as cable, satellites, and home video, combined with the expanding use of video in such areas as industrial communication and educational media, have increased the need for television production. More original video production is being produced today for all these diverse applications than at any time during television's history.

- Remote production is now standard practice at virtually every production facility. The limitations of the studio walls are no longer a barrier as the production unit can go anywhere to take advantage of the unique elements which only location shooting can offer.

To reflect these and other important changes, the second edition of *Television Production* is not simply a revision; it is an entirely new book. Television production is not done today the way it was done in the past. The tools have changed, the approach has changed, and, most importantly, the many production limitations which existed in the past have all but disappeared. Today video is more flexible than motion picture film and, by incorporating the most up-to-date electronic technology, has become an incredibly powerful creative communications medium. Just as the original edition of *Television Production* was written to describe contemporary production equipment and techniques at the start of the 1980s, the second edition has been written to reflect television production throughout the remainder of the decade.

A number of changes are obvious. The addition of an entirely new chapter devoted to electronic news gathering reflects the increased importance which this production approach has attained. The expansion of the chapter on remote production indicates the importance of location shooting for today's television production. The chapters on video recording, graphics, and the switcher and special electronic effects were completely rewritten to reflect the many significant developments in these areas. In addition to these major changes, every chapter was revised and rewritten to incorporate the various changes and developments in both equipment and production techniques.

Readers familiar with the first edition will also notice a change in the sequencing of the chapters. I did this to provide the reader with the basics of production equipment and techniques early in the book. Secondary production aspects, such as graphics and set design, are covered later. Of course, many instructors have used the chapters in their own sequence to follow their particular classroom approach, and I have continued to follow the practice of writing each chapter to stand alone as an independent unit to facilitate each instructor's ability to tailor the readings to his or her classroom requirements.

I have also attempted to incorporate many production examples for such nonbroadcast situations as educational production and industrial video to reflect the growth and increased importance of these areas in the television production field.

One major element which has not changed in this edition is the emphasis on learning the capabilities and limitations of equipment and related production techniques to enable the production professional to make realistic, optimal production judgments. The capability of modern production equipment is dazzling, but even the most sophisticated equipment requires a creative and intelligent user in order to make the most effective use of the hardware's potential power and flexibility.

While this edition clearly reflects many important changes, something else has not changed at all: the assistance and cooperation I needed—and received—from so many professionals. Once again, I would like to thank all the individuals who helped me write the first edition, which served as a solid foundation for this revision. As I reread the book many times in preparation for the second edition, I was continually struck by how important their help and advice were to the success of the book. There is no doubt that their contributions remain significant in this second edition.

There are a number of individuals who were especially helpful in the preparation of the second edition and whom I would like to acknowledge. Among the industry professionals who graciously offered me their time, advice, and access to their staffs and facilities were Emil Neroda, The Sound Shop, and R. L. Pointer, Achille Raspantini, and Max Berry of ABC-TV. I am also grateful to the many manufacturers who provided me with photographs and with valuable information.

I would like to thank the various individuals who took the time to read and comment on various parts of the manuscript. They include Thomas McCain, Ohio State University; Dom Caristi, St. Mary's College; Joel Fowler, University of Texas at Austin; Barry Sherman, University of Georgia; Steve Ryan, Pepperdine University; and Richard D. Settle, University of North Carolina, Chapel Hill.

I am also very grateful to the many instructors, students, and industry professionals who offered me comments on the first edition and suggestions for the second. It was their positive and enthusiastic reaction to the original *Television Production* which made it so successful.

Finally, I want to acknowledge the help and encouragement I received from my wife. After writing tens of thousands of words, I find I am still unable to find the right ones to express adequately how much her support and enormous patience meant to me in every way. For all that—and much more—this book is dedicated to Susan.

Alan Wurtzel

PREFACE

TO THE FIRST EDITION

In order to communicate effectively in any medium, it is essential first to understand the capabilities and limitations of both the medium and its tools and techniques. Before artists can begin to paint, they must be familiar with brush, palette, and canvas. In much the same way, television communicators require a knowledge of a vast array of production equipment before they can work effectively in television. While proficiency in equipment operation is a logical first step, it is only a means to an end. We must know how best to use the equipment to produce programming which will communicate messages and ideas to the viewing audience.

A technical medium, such as television, defines its capabilities and limitations in large part by the equipment which is used to produce the programming. For all its power and pervasiveness, television, as we know it, is little more than thirty years old. Yet in that short period of time, the tools and techniques of production have changed at a breathtaking pace. Not long ago production capabilities were found at the two extremes of the production spectrum. At one extreme were the largest national networks and stations with vast manpower, facilities, and equipment resources. The remainder of the industry was confined to the opposite end and had a limited choice of equipment and methods, which severely hampered creativity and production flexibility.

Over the past few years, however, this situation has changed dramatically. The same technology which has produced the pocket calculator and the home computer has also greatly narrowed the technology gap, so that today even the most modest production facility boasts equipment and a resulting production capability which would have astounded the most sophisticated producer or director of a decade earlier. The rapid advances in portable color cameras and videotape recorders, which have revolutionized television news, are only one obvious example. The fact that electronic news gathering equipment was pioneered at the small-station level and later adopted by larger production facilities is another indication of how the medium and its techniques are rapidly evolving. This momentum will continue as more sophisticated equipment becomes accessible to increasing numbers of smaller studios and stations. The results will be changes in many production principles and a greatly enhanced technical and creativity capability, which translates into more effective television programming.

For all its technical complexity, however, television is as much an art as it is a science. Every technical operation in television has a creative component which is realized only through the proper application of equipment and technique. While I have devoted a great deal of attention to the operation of the equipment, I have also included much information dealing with production principles and methods—in other words, first how to *operate* the equipment, then how best to *use* it.

Along with major changes in equipment and technique have come radical changes in the television industry itself. At one time "television" was synonymous with "broadcasting." Today television is a multifaceted communications medium, which is used in a variety of applications ranging from broadcasting to business, from colleges to corporations, from instruction to entertainment. Indeed, the fastest-growing sector of the production industry is not at broadcast stations or networks but rather at the many production facilities which produce special programming for specific applications. This book was written to reflect all these changes in the medium, in equipment, and in production technique. It is intended to show how contemporary television production is actually practiced in the industry.

To do this, I have relied on the advice and expertise of many industry practitioners working in all phases of production. Their comments, suggestions, and tips—developed through years of hands-on trial and error—should prove useful to you in a number of practical ways. In discussing the operation of equipment, I have purposely omitted most references to specific manufacturers and model numbers. Once you understand the basics of equipment operation it is not difficult to apply your knowledge and skill to all other similar items of equipment. I have also written each chapter to be read independently. This will permit instructors to assign readings in whatever sequence will best suit their classroom approach. Every chapter is organized to introduce first the equipment and its operation, then the production team member involved, basic production principles, and, finally, more advanced techniques. This will enable the reader to become familiar with the fundamentals before attempting more sophisticated methods. For those interested in additional information, the annotated bibliography at the end of most chapters will provide further sources for reference and more detailed information.

A final word: all too often beginning production students become so enchanted with the equipment that they forget sophisticated hardware is valuable only when it is used properly and with a purpose. It is not hard to master the operation of even the most complex equipment with some practice. What is harder—and far more important—is to learn how and when to use the equipment to communicate effectively to an audience. To use television well takes talent, creativity, and imagination, none of which are to be found inside a camera, microphone, or videotape recorder. They must be supplied by you through practice and hard work. I hope this book will serve as a first step in that direction.

ACKNOWLEDGMENTS

Like most readers I have often skimmed through the acknowledgments on my way to the first chapter of a new book. Only after I began writing did I come to realize how much an author depends on others for assistance, advice, and support. I want to take this opportunity to thank some of those people who have helped me in writing this book.

First, I am very appreciative of the many equipment manufacturers, stations, networks, production facilities, and production organizations which graciously provided illustrations, information, and access. Their names, too numerous to list here, are included with the photographs throughout the book.

I am especially grateful to the many television professionals who took time from busy schedules to offer me their help and cooperation. Among those to whom I owe a special debt are William Klages, Imero Fiorentino Associates, who greatly helped with the lighting chapters; Ted Nathanson and Michael Weisman, NBC Sports; Neil Kuvin and Chris Glass, WXIA-TV; Lee Baygan, director of makeup for NBC; the American Society of Television Cameramen; George Heinemann, NBC-TV; Dwight Hemion; Allan Stanley, Dolphin Productions; Otis Riggs, NBC-TV; and Keith Jackson, ABC-TV. Joel Spector, of NBC, was especially helpful in providing information and much useful criticism.

James Morgenthaler took many of the original photographs and did the majority of the fine darkroom work for the picture prints. I also want to thank Billy Sherrill, Anthony Tantillo, and Al Wise for their photographic assistance. Many friends and students helped me, including Susan Jordan, Larry Aldridge, John Kelly, Norm and Judy Ornstein, and Ruth Reinhold.

I would like to acknowledge the help of many colleagues who read and criticized various portions of the manuscript. They include Robert E. Davis, University of Texas; Robert Smith, Temple University; Peter Mayeux, University of Nebraska; Frank Kahn, Lehman College; James Fletcher and Ed Lynch, University of Georgia; and John B. Haney, Queens College.

I would like to express a special word of thanks to my friend and long-time associate, Gady Reinhold, without whose help in so many ways this book would not have been possible. I also want to acknowledge the unending support and encouragement I've always received from my parents.

Most of all I want to thank my wife, Susan. Over the past few years she has had to endure hearing more about television production than could reasonably be expected of anyone. For her help, support, patience, and encouragement, it is to her that I dedicate this book.

Alan Wurtzel

CHAPTER 1

INTRODUCTION TO TELEVISION

PRODUCTION

The large set in CBS Studio 41 on Manhattan's West Side is strangely quiet. It is the lunch break before taping will begin on a musical-variety show. Four color cameras, one mounted atop a large crane, are scattered across the huge studio. Downstage, seemingly endless strands of black cable clutter the studio floor. Empty coffee cups and pastry wrappers are scattered throughout the studio, the vestiges of hours of rehearsal. With the bank of overhead lights off, the air conditioning keeps the temperature in the studio at a chilly and almost uncomfortable level.

Crew members slowly return to the studio; the lunch break is almost over. The overhead lights are turned on, bathing the set in a bright glow. The microphone boom operator climbs atop the boom and adjusts the controls that suspend the microphone above the performers' heads. The crane camera with its crew of four sweeps high into the air, practicing a difficult movement.

Down the hall from the studio, in a large sound-proof room, about twenty musicians are noisily tuning up. They are located outside the television studio so that the sound of the orchestra can be mixed and blended together without interfering with in-studio operations. The final mixed sound of

FIGURE 1-1 **The Popular Children's Program** *Sesame Street* **in Production.**
(Courtesy: Children's Television Workshop)

the orchestra will be fed into the studio through large loudspeakers, which are wheeled to the center of the floor just outside camera range.

Directly next to the studio floor is the darkened control room with rows of television monitors and seemingly thousands of multicolored buttons and switches. The control room serves as the studio's command center, where the director supervises the actual production of the show. The program's director is meeting with the lighting director and the scenic designer. They are discussing some last-minute problems which occurred during camera rehearsal. One of the major difficulties concerns a specially designed lighting fixture which has not operated properly. The lighting director and scenic designer go off to see if they can solve the problem as the star of the show walks into the control room to speak with the producer and director. A change in one of the

musical numbers is worked out among the three. The director sends a production assistant out to the studio floor to inform the camera operators about the changes and to give them their newly revised shots.

Out on the studio floor, the set, which was almost deserted a few minutes ago, is now a flurry of activity. Dancers rehearse their moves, accompanied by a studio pianist. Production assistants gather cue cards while technicians and production staff members run back and forth, completing final arrangements, making notes, and taking care of the thousand and one small details which must be attended to before the taping session can begin. The makeup artist and wardrobe dresser put their last minute touch-ups on the star and her guests.

Inside the control room, the technical director is seated before the "switching console." The switcher is a multibutton device which permits the technical director to put any video source— cameras, film or slides, videotaped segments, or remote feeds—on the air. Each of these video sources is displayed on a row of black and white monitors facing the console where the technical director, program director, and other production personnel sit during the program. The technical director punches up each of the four cameras and checks their picture quality on the large color monitor labeled "line," which is adjacent to the row of black and white screens.

The audio engineer, seated behind the audio console in a small booth at the end of the control room, balances the sound of the orchestra as it rehearses the opening number. The brass section is too loud, and the engineer lowers its volume by manipulating various dials and controls.

The program's director sits in a chair next to the technical director and puts on a headphone which permits communication with all production crew members out in the studio. On the set, the floor manager, also wearing a communication headset, listens for the director's commands. At the instruction from the floor manager, "OK. Places, everyone. This will be a take," the cameras move to their assigned places while the technicians and production staff on the studio floor quickly move behind the cameras.

The program director checks with everyone in the control room. The technical director and audio engineer are ready. The video engineers, seated in their own booth at the other end of the control room, peer into their waveform monitors, make a last check on video levels and picture quality, and nod to the director. The assistant director pushes a button on the console and speaks into a small microphone to the videotape operator, who is located on the floor below. "Roll tape," says the AD (assistant director). Through the loudspeaker on the console comes the reply "VTR rolling and up to speed." The machines are ready to record the show.

On the studio floor, the floor manager calls out, "Quiet please. Tape is rolling," and holds a small slate, on which are written the program title, taping date, and scene and take numbers, in front of one of the cameras.

The program director reads the slate off the line

monitor in the control room, calls for the technical director to fade to black, and calls out over the headsets, "Stand by. Ready to hit the music, cue talent, and fade in on Camera 1. OK. Hit the music, cue talent, and fade in 1."

The floor manager, standing beside one of the cameras, with upraised arm, hears the director's cue and quickly points to the star as the orchestra's first notes blare out of the loudspeakers. Camera 1 starts to move in; the talent begins to sing.

The taping has begun.

Located behind the gymnasium on a college campus is a small remote production truck. The final championship game of the basketball season will be broadcast live over a regional cable system. Upstairs, in the crowded gym, the program director is speaking to the play-by-play announcer. They are going over the details of the program's opening. Engineers are working on the alignment of the cameras, which are positioned on large platforms above the playing court. The camera operators and the director meet for the last time before the broadcast will begin. The game will be broadcast live, and once it has begun, there will be no chance for changes or retakes. The director, satisfied that all know their jobs, leaves the gym as the camera operators take their positions behind the cameras.

The inside of the control truck is a miniature control room. It is five minutes to air time, and the assistant director is on the telephone talking to the master control engineer. A small monitor shows the program now being cablecast. The director enters and sits down, checking with the audio engineer to be sure that the theme music for the opening is cued and ready. Camera 1 zooms in on the opening title card while Camera 2 focuses on a wide shot of the gymnasium. The assistant director counts down to air time. "Two minutes to air. Stand by."

Tension builds in the truck as the second hand sweeps toward air time. At thirty seconds to air, the program director calls out over the headset,

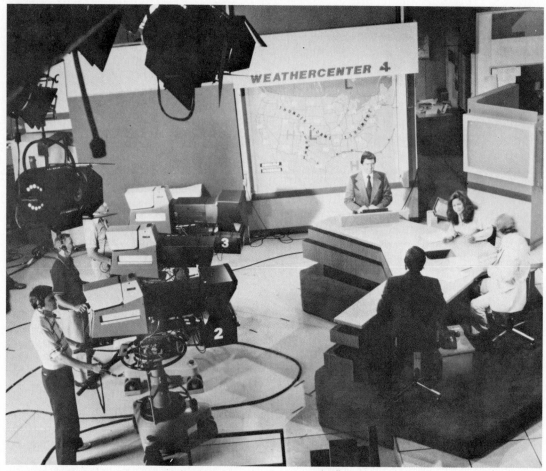

FIGURE 1-2 **Local Station Newscast in Production.** (*Courtesy: RCA*)

"Thirty seconds to air. Ready to hit music, fade up on Camera 2, insert the title on 1, and cue announcer." The assistant director watches the clock and calls out. "Ten seconds to air. Nine, eight, seven . . ."

The "off-air" monitor shows the end of the last program. At exactly "straight-up," on the hour, an engineer at the cable system master control room presses a button, and the feed from the remote truck replaces the ID slide on the off-air monitor. As the shot from Camera 2 appears on the screen, the director calls out, "Hit the music, insert the title, and cue announcer."

The live program is on the air.

It is eleven o'clock at night when the last customer leaves the fast-food restaurant. At the same time, the five-person television crew begins unloading equipment from their production van and starts to set it up inside the restaurant. The crew is about to work through the night to produce a series of training tapes which will be shown to all of the chain's employees about the proper food preparation and serving procedures which they must follow on the job. In addition to the technical crew, a producer-director is on hand to supervise the taping. The program's "talent" consists of experienced restaurant employees who will demonstrate how the various tasks must be accomplished.

A few days earlier, the producer-director discussed the script with each of the participants and explained to them exactly what they were to do during the production. Long before the taping date, the producer-director carefully worked out a detailed shooting schedule. To expedite the production, she has decided to shoot the footage out of sequence and so she now explains once again to both talent and the production crew how she expects to proceed throughout the shoot.

The first sequence that will be taped involves working the cash register at the front counter. The

producer-director walks the talent through each of their moves while the camera operator and the audio engineer watch the action closely. Once she is satisfied with their performance, she calls for a rehearsal with camera, lighting, and audio.

FIGURE 1-3 Studio Control Room in Production.
The program's *director* (seated in the middle) watches the pictures from the four camera monitors and selects the shots which appear on the larger line monitor located above. The *technical director* (seated to the left of the director) operates the switcher. The *assistant director* (seated to the right of the director) prepares upcoming camera shots and keeps track of the program timing. *(Courtesy: ABC-TV)*

The rehearsal proceeds smoothly, and the director decides to go for a take. The camera operator's assistant holds a slate in front of the camera while the director calls out, "Roll tape." The tape operator confirms that the videotape is rolling up to speed and nods to the director. The director says, "Action," and the talent begin their preplanned moves. In the middle of the sequence, the camera operator suddenly calls out, "Boom in the shot," indicating that the boom microphone, which is held over talent's heads in order to pick up the sound of their voices, has become visible in the camera viewfinder. "Cut," calls the director, and the sequence is repeated.

At the end of the take, the director calls for a videotape playback as the cast and crew gather around a small television monitor to watch their work. While the tape plays back, the audio engi-

neer monitors the sound, using a pair of headsets, while the lighting director examines the screen closely to see if there are any lighting problems. The director is satisfied with the last take, and there are no technical problems, so the crew moves on to tape the next sequence.

Since the production crew is using only one camera, every "cut" to a new shot in order to show the action from a different angle must be accomplished by physically moving the camera to a new position while one of the talent repeats his movements for the new shot. As the early morning shift arrives to open the restaurant for the breakfast rush, the production unit complete the last take and begin to strike their equipment and pack up the production van.

Although the taping is complete, the producer-director's job has only begun. Back at the production studio, she must screen the raw footage to decide which take to use for each sequence. Sometimes no single take is completely satisfactory, but by combining a piece from one take with a piece from another, she is able to assemble a perfect sequence. "Let's see that sequence,

FIGURE 1-4 **Remote Truck Control Room.**
The remote truck's control room is a miniature studio control room enclosed in a large mobile van. *(Courtesy: Echo Sciences Corp. and ABC-TV)*

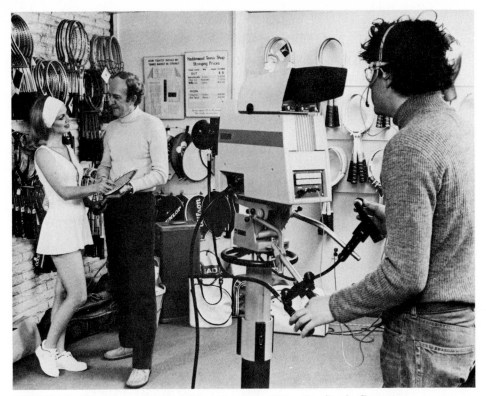

FIGURE 1-5 **Shooting a Commercial on Location Using the Single-Camera Videotape Production Approach.**

again, when his hand moves to the register," says the producer-director to the tape editor. The editor touches a few buttons, and the tape begins to show the performer's hand reaching for the register. "If we cut in a close-up of the hand here," says the tape editor, "we can eliminate the jump cut and pick up the rest of the action from Take 4." The director nods approval as the editor begins to edit the tape at precisely the point they decided upon.

Although it took only six or seven hours to videotape all of the segments, it takes the producer a full week of ten-hour days inside the edit room to screen the footage, select the takes she wants to use, and then edit the pieces together until a final rough-cut is complete.

Once the rough-cut version is completed, the producer-director turns her attention to the audio sound track. Sitting inside an audio studio, she works with the announcer, who reads the voice-over-narration. "Give me a little more energy on that line," she says, and the announcer reads the copy over again until the director is satisfied. Once the narration recording is finished, the director works closely with the audio engineer to select the music and sound effects that will be added to the final audio mix.

Having assembled all the audio elements, the producer-director returns to the postproduction room, and the delicate job of mixing the sound track begins. The original sound which was recorded during the location taping must be mixed with the narration, music, and sound effects. The process is laborious not only because the audio engineer must perfectly balance all the sound elements together, but also because the entire audio track must match the visual action on the television screen.

Finally, both the audio engineer and the producer-director are satisfied with the audio mix, but the production is still not completed. The final step is the integration of the master videotape footage which was used to produce the rough-cut workprint, with the audio which was mixed into a composite track. To do this, the producer-director and the tape editor return to the postproduction room, where a computer-assisted editing system coordinates the sound and pictures into the completed master tape.

Over a hundred hours were required to shoot, edit, and postproduce the thirty-minute training tape. The master version is used to make dozens of videocassette copies of the program, which are shipped to each of the restaurant chain's regional offices. Exactly one month after the production crew taped the sequences inside the restaurant, the first group of employees sits before a television monitor to watch a program which shows them, in an interesting and informative way, how to do their jobs better. Although the program may not have any Hollywood stars or the excitement of a basketball play-off game and although it will be seen by an audience of only a few hundred, it is another example of the effective use of television production to communicate a message to a viewing audience.

INTRODUCTION TO TELEVISION PRODUCTION

A network musical-variety special, a cable system's sports remote, and a corporate training tape—these are only a few of the many production situations which make up television. If you were to ask ten different people to define "television," you would probably get ten different answers. For some, television is entertainment: dramatic programs, musical shows, or comedy. For others, television is news: international, national, and local news, and special events. Others think of television in terms of sports: from coverage of Olympic games in exotic locations to a local high school football game. For still others, television is

closed-circuit systems used for internal communication in schools, hospitals, and industry.

Of course, television is all these things and more. Regardless of the size and complexity of a program, how it is produced and transmitted, or what its content is, all television has one thing in common: the use of the television medium to communicate messages and ideas to a viewing audience. Since television production encompasses such a vast array of equipment and technique, we will use this chapter to provide you with a brief overview of the production process and to introduce you to some of the basic concepts which will be used throughout the book. First, we will introduce the principal members of the television production unit, or team, and orient you to the studio and control room, where the production takes place. Then we will discuss the four stages of television production and describe how the technical operation of the television system works to reproduce an image. Finally, we will introduce an important concept called "videospace," which plays a crucial role in almost every production decision.

THE TELEVISION TEAM

Television is a hybrid medium which has borrowed and adapted many techniques from the theatrical stage, from motion pictures, and from radio. Early television productions were often produced on a trial-and-error basis. Creative production staffs and technical crews worked together to explore the capabilities of the new medium and, as they worked, developed many of the guidelines and practices which are used today. As producers and directors became more experienced, they constantly challenged engineers to develop more sensitive cameras, better lenses, newer and more flexible electronic effects, and more elaborate audio systems to permit them to expand and perfect their productions. The engineers quickly rose to the challenge and, within a few short years, developed color television, videotape, and countless sophisticated electronic effects which have radically altered the way in which programs are produced.

Many changes have occurred in television production over the past thirty-five years, but one constant is the nature of television as a team operation which requires the skills and abilities of a variety of artists and craftspeople to produce the programming successfully. From the simplest program to the most complex, a television production is the sum total of the coordinated efforts of dozens of skilled individuals who make up the *television team*.

Perhaps Bob LaHendro, a talented director of such successful programs as *All in the Family*, put it best when he said:

> I started in this business pulling cue cards and spent five years until I became a floor manager, assistant director, and then director. And while working at these various levels I was able to really appreciate every job that everyone was doing to make the show successful. No matter what your pay scale is or what your job is, you take a certain amount of pride in doing that job well. Television is a team effort and the entire team can make the director look good or make the director look bad. All it takes is one misplaced cue, a line not correctly picked up on audio, or one missed shot to ruin what could be a beautiful moment in a show. No matter how good a person may be in his job, unless everyone else puts out 100% effort, you won't look good and the program can't possibly be a success. If I've learned one thing about television, it's that the team approach is probably the single most important element in producing the kind of programs people will want to watch and that everyone can be proud to say they worked on.

A smoothly operating television team requires the integration of many different jobs, all performed and coordinated perfectly. In order to give you some understanding of the composition of the television team and to let you see how all the elements fit together in the production of a show, we provide here a brief description of some of the principal team members and their primary responsibilities.

The television team can be roughly divided into two groups: *production staff* members and *production crew* members. The production staff is composed of those whose jobs are considered to

be "creative": the producer, director, writer, and a variety of production assistants. Sometimes in terms of budgeting, these are called "above the line" positions. The production crew is made up of individuals who work primarily with the production hardware and equipment: the technical director, audio engineer, camera operators, floor managers, and a variety of crew assistants and technicians. These jobs are sometimes referred to as "below the line" since they appear on a production budget in a separate category from the production staff. To a certain extent, this division between staff and crew is rather arbitrary since many jobs will invariably cross lines. For example, the director is considered to be a member of the production staff. Although he or she may not actually operate any equipment, a good director must be as knowledgeable about the capabilities and limitations of the production hardware— cameras, microphones, lights, and the like—as with directing performers or selecting camera angles. Similarly, the production crew, although not technically considered to be "creative," requires members with a great deal of imagination and ingenuity to do a good job. A good camera operator must not only understand the technical operation of the camera but must also display a good sense of picture composition and creativity to frame and compose shots for the director. There is little question that members of the production crew lend a production technical expertise, skillful artistry, creativity, and imagination that go a long way toward making an entire show a success.

The Production Staff

THE PRODUCER The producer is responsible for the entire television production. He or she is the ultimate authority in charge of all production aspects from the planning and writing of the script to the final production and editing. Since the producer must also be concerned with program budgets and organizational matters, as well as

FIGURE 1-6 **Television Team—Production Staff.**

Position	Responsibilities During Four Production Stages			
	Preproduction	Setup and Rehearsal	Production	Postproduction
Producer	Develop program concept. Develop production budget. Assign program's director. Work with writer on script. Approve director's approach, light design, and set design. Supervise and coordinate all preproduction planning.	Supervise overall production activities. Watch rehearsals as surrogate audience and make notes for changes or improvements. Keep production moving on time and within budget. Approve last-minute changes as they arise.	On live shows help director as needed. On taped shows work with director on which takes are usable.	Approve final edited version. Coordinate with station for promotion/publicity. Evaluate program to see if it met objectives.
Director	Participate in all preproduction meetings. Work with producer and writer in script development. Establish production approach in consultation with producer. Consult with lighting director, set designer, audio engineer and approve their various designs and approaches. Cast performers. Work out camera shots.	Rehearse performers. Rehearse camera shots in studio. Integrate all production elements into a coordinated show.	Execute production.	Supervise editing.
Writer	Work with producer and director in developing script or format. Revise script until approved.	Available for rewriting if necessary		
Assistant Director	Help director in planning production approach	Assist director during out-of-studio rehearsal. Ready camera shots and other cues during studio rehearsal.	Assist director by readying camera shots and other cues. Keep track of program timing. Roll in film or videotape segments.	Help director during editing. Keep track of timing during editing.

with aesthetic decisions, in many large production studios the producer may have a number of associate producers assisting.

THE DIRECTOR   The director is responsible for creating the look and sound of the production. To do this, the director oversees the performance of the on-air talent and coordinates the operation of the technical crew. A director's job is very complex and demanding as it requires the ability to coordinate a tremendous number of different operations, often simultaneously. The director must watch several different camera shots; select the shot which will be sent over the air; direct the camera operators for their upcoming shots; listen to and cue the program audio; direct all production elements, including talent, cameras, audio, lighting, and so on; approve all art, graphic, and lighting designs; and make certain that the entire effect created is consistent with the producer's original concept of the production.

ASSISTANT DIRECTOR   The assistant director (AD) helps the program director by readying talent, cameras, and film or tape roll cues and by alerting other members of the production team to upcoming events. The AD must also keep careful track of the time for each program segment and for the overall production, making sure that the program begins and ends on time.

PRODUCTION ASSISTANTS   The job of the production assistant (PA) often varies from program to program, but basically a PA's primary responsibility is to assist the producer, director, and other members of the production team. Sometimes this means that the PA is a "gofer" who takes care of whatever details require immediate attention (including such mundane tasks as going for coffee or sandwiches). Other times, the PA will work on the studio floor, holding cue cards for talent or keeping track of the production by following the script and prompting performers should they forget their lines during rehearsal. Usually, the PA works in the control room and helps the producer and director by taking notes; making necessary changes on all scripts when they occur; assisting the AD in timing the show; publishing scripts, script changes, and production forms; and keeping track of different program material such as films, tapes, and slides.

The Production Crew

TECHNICAL DIRECTOR (SWITCHER)   The technical director (TD) sits next to the program director in front of a large bank of buttons and controls called the "switcher." The TD operates the switcher on the director's command, "punching up" whatever video source is called for onto the air. In some studios, the TD is also responsible for supervising the activities of the technical crew. In those studios where the TD's function is only to switch for the production, the position is sometimes referred to simply as the "switcher."

AUDIO ENGINEER   The audio engineer is responsible for the sound of a television production. During a show, the audio engineer sits at a sound-control console mixing the various audio inputs from studio microphones, tape recorders, record turntables, film and videotape tracks, and remote feeds from outside the production studio. The audio engineer must balance all of the sound inputs together to create the mixed sound of the program.

The audio engineer is also responsible for planning the audio pickup after consulting with the program's director. The audio engineer supervises the operation of the audio crew and coordinates the setup of all audio equipment.

LIGHTING DIRECTOR   The lighting director plans and executes the lighting for a production. Much of the lighting director's job takes place before the production ever enters the studio. The lighting director consults with the director and plans a lighting approach which will complement the director's concept of the show. Lighting is a crucial production element, not only because it provides the necessary illumination for the operation of the cameras, but because, as a means of setting the mood and tone of the program, it is one important way of presenting the performers as effectively as possible on screen.

The lighting director supervises the activities of the lighting crew in hanging and focusing the

FIGURE 1-7   **Television Team—Production Crew.**

Position	Responsibilities During Four Production Stages			
	Preproduction	*Setup and Rehearsal*	*Production*	*Postproduction*
Technical Director (Switcher)	Consult with director and producer on necessary technical facilities.	Responsible for overall technical quality (if acting as technical director). Operate production switcher during studio camera rehearsals.	Operate production switcher.	Operate switcher during postproduction.
Audio Engineer	Consult with director and other key team members on production approach and necessary audio. Plan audio approach and necessary audio facilities. Prepare necessary audiotapes in advance.	Supervise audio crew in studio and control room preparation. Prepare audio control console. Check all microphones and balance audio sources.	Mix program audio.	Operate audio console during postproduction audio sweetening.
Lighting Director	Consult with director, producer and scenic designer on overall design approach. Develop lighting approach. Prepare lighting plot for production.	Supervise hanging and focusing of lighting instruments. Balance all instruments until proper illumination and effect are achieved. Make whatever changes are necessary as problems develop during studio rehearsal.	Coordinate all lighting cues. Operate lighting dimmer board.	
Scenic Designer	Consult with director, producer, and lighting director on overall design. Develop set design approach and design settings.	Supervise set construction. Supervise activities of stagehands as set is erected in studio. Make necessary changes as problems develop during rehearsal.		

12

Responsibilities During Four Production Stages

Position	Preproduction	Setup and Rehearsal	Production	Postproduction
Floor Manager		Responsible for all activities on studio floor. Serve as director's "eyes and ears" on floor during rehearsal and production. Responsible for props and costumes during rehearsal and production. Relay cues to talent as they come from director.	Relay all cues to talent.	
Camera Operators		Prepare cameras for production. Operate cameras during camera rehearsals.	Operate cameras during production.	
Video Engineers		Set up and align cameras for best picture. "Shade" cameras to control for variations in scene brightness. Help director to achieve special visual effects as necessary. Consult with lighting director should illumination problems arise which affect camera operation.	Shade cameras during production.	

lighting instruments. During the production, the lighting director is responsible for coordinating all lighting cues which may be necessary.

SCENIC DESIGNER The scenic director—sometimes called the "art director," or "set designer"—is responsible for devising the physical setting for a program. The scenic designer works closely with the program's director and with the lighting director in planning and executing the program's overall design. The designer also supervises the stagehands and crew members who erect the set on the studio floor.

FLOOR MANAGER (STAGE MANAGER) The floor manager (FM), sometimes called a "stage manager," is responsible for all operations on the studio floor. Since the program director operates from the control room, usually out of direct visual communication with the studio floor, the FM acts as the director's eyes, ears, and voice. He or she is responsible for seeing that everything on the floor goes smoothly and for cuing the performers by relaying the director's commands, which come over the headset.

CAMERA OPERATORS The camera operator controls the television camera during a production. Studio cameras are mounted on pedestals, which are wheeled around the floor to set up different shots and angles. The camera is also equipped with a lens system, which the camera operator uses to compose and frame a shot. Although the camera operator receives shot instructions from the director via the headset, a good camera operator who exhibits a strong sense of composition and visualization is a valuable asset to a production.

VIDEO ENGINEERS The video engineer is responsible for the technical quality of the camera picture. Each studio camera has its own control unit, which enables the video engineer to "ride levels," controlling for variations in scene brightness, contrast, color balance, and registration.

Although the video engineer concentrates primarily on the technical aspects of the picture, a good video engineer can make an important contribution to a production by helping to achieve the desired visual effect through the manipulation of various "shading" controls.

Other Members of the Television Team

The list of jobs just described is by no means complete. We have purposely omitted many important members such as the writers, graphic artists, makeup and wardrobe personnel, and a host of technicians who operate videotape recorders, and film and slide projectors and who repair and maintain the sophisticated and delicate production equipment. In order to keep this introduction from becoming too complicated, we will postpone describing these other functions until we have had a chance to discuss in detail the equipment and operations employed by the team members already mentioned.

Of course, production jobs require special skills and talents, and in most large studios they are performed by specialists. In many smaller operations, however, one person may be assigned to a number of different responsibilities. For example, it is not uncommon for the scene designer to handle lighting also or for the director to operate the switcher. However, for the sake of clarity, we will discuss each production position as a particular "role" with the understanding that it is possible and quite common for one individual to undertake a number of responsibilities during a production.

THE TELEVISION STUDIO

Although a television program can be produced almost anywhere today, from a remote news location to a large athletic stadium, most programs are still produced within a television studio. Studios vary in size and complexity from the immense, barnlike network complexes, which are as large as a full city block, to the small, unpretentious studios found at many closed-circuit installations. Regardless of their size and sophistication, all studios are made up of two areas: (1) the *control room*, which is the operational nerve cen-

Line monitor

Preview monitor

Camera monitors

Associate director's position

Video switcher

Director's position

FIGURE 1-8 **Studio Control Room.** *(Courtesy: ABC-TV)*

ter for the production, and (2) the *studio floor*, where the production takes place.

Control Room

The control room is where the program's director, assistant director, technical director, audio engineer, and video engineer work. Producers and production assistants also operate from the control room during rehearsal and production.

As you walk into a control room, you will see a wall of monitors, each one displaying the video output of a studio camera. If there are three cameras in the studio, there will be three monitors, each showing the picture from one of the cameras. Additional monitors show the video output of film or slide projectors, videotape machines, electronic character generators, and remote video feeds. Many broadcast control rooms also include an off-air monitor, which shows what the station is broadcasting over the air at the time.

Adjacent to the bank of black and white monitors are two large monitors—color if the studio is equipped to produce in color—one labeled "Preview" and the other labeled either "Line" or "Program." The line or program monitor shows the actual picture which is leaving the control room to be broadcast live or to be fed to videotape for recording. The preview monitor is used to check any picture or special video effect before it

is actually sent out over the line. (See Figure 1-8.)

In front of the monitor bank is a long table called the *production console*. This is where the director, technical director, assistant director, and production assistants sit during a production. The director usually sits in the middle of the console because all the monitors are clearly visible from there. Seated next to the director is the technical director, who sits at the production switcher. The technical director controls the video picture, which is on the line monitor, by operating the switcher according to the director's commands.

Seated on the side of the director opposite the technical director is the assistant director. Both the director and the AD wear intercom headsets so they can talk to various members of the production team who are out on the studio floor. The director and the AD have another intercom which permits them to talk with videotape operators qr to film or slide projector technicians, who are usually located in areas outside the production control room.

Some studio control rooms are built on two levels with a second long desk arranged one step up behind the production console. This is for the producer and various production assistants, and it gives them a good view of the monitor bank and the production operations without interfering with the director who is seated below.

The audio engineer and the audio control console are usually located to the side of the production console area, often isolated behind a glass partition so the audio engineer can listen to the audio mix or preview audio sources without interfering with the activities in the control room. Off to the side or to the rear of the control room are the camera control units which are used by the video operators to regulate the camera pictures.

Although the above description is fairly typical, the exact configuration of a studio control room will vary from one facility to another. For example, in some operations the video engineers are located in an entirely separate area of the facility and communicate with the production team in the control room via an intercom system. While the exact layout will be different, the individual components found within a control room are common to all productions; and once you are familiar with their use and application, it will not be hard to orient to a slightly different control room.

Studio Floor

The studio floor is where the production actually occurs. During the days of radio it was important for directors and audio engineers located inside the control booth to have a direct view of the performers in the studio. This enabled the director to send hand signals to the performers while a production was in progress. Since the television director is concerned only with the pictures on the monitors, however, many newer studio facilities are built without a glass window separating the studio floor and the control room. In fact, some control rooms are located a floor above or below the actual studio without any line of sight possible between the two areas.

The studio floor is an open area, which contains the television cameras, microphones, lighting equipment, sets, and, of course, performers and crew. The size of the studio floor will usually determine the complexity of the programming which is possible. The larger the studio, the more space for sets, performers, and equipment and the more flexible the operation. Smaller studios restrict equipment, technicians, and performers and tend to limit the size of the set and the number of performers which are practical.

Located around the walls of the studio are various connector boxes to which cameras, microphones, and lighting equipment are connected. With so much activity taking place on the floor simultaneously, it is important to keep cables and wires to a bare minimum. (See Figure 1-9.)

The lighting equipment, which is hung above the studio floor, generates a considerable amount of heat, which is why most studios are equipped with powerful air conditioning to keep temperature levels down. The temperature is important both for personnel comfort and especially for the operation of delicate electronic equipment, which requires a fairly stable environment to function properly.

A well-planned studio facility will provide a large area adjacent to the studio floor for storing props, sets, and equipment. Otherwise, a portion of the studio floor must be used for storage, and this limits the amount of room available for the production. Oversized doors permit equipment and sets to be moved between the studio floor and the storage area.

Most studios employ large, soundproof doors, or "sound locks," which prevent extraneous noise from entering the studio, where it might be picked up by a sensitive microphone. A sign above the studio doors automatically lights wherever a microphone is "live" to warn that the studio is in operation.

Orson Welles, the actor and film director, once said, "A movie studio is the greatest toy a boy can have," and much the same can be said about a television studio. While there is no question that the studio and control room are fascinating places, they are also potentially very dangerous. In the excitement of a production it is sometimes easy to forget that heavy lights are suspended overhead, high voltage runs through equipment, and cables and wires are present everywhere. More than anything, it is important to treat the studio and equipment with a healthy measure of respect and to take care that performers, crew members, and production personnel are safeguarded at all times.

FOUR STAGES OF TELEVISION PRODUCTION

Although the actual production of a show takes place inside the studio complex, this is only a small part of the overall production process. Long before anyone enters the studio or control room, the program must be carefully planned and many detailed preparations completed. This involves work in a number of different steps which can be divided into four separate stages: (1) preproduction planning, (2) setup and rehearsal, (3) production, and (4) postproduction. Of course, not every production will require work in each stage, nor is the same emphasis always given to each. For example, a live, daily news program will probably require little setup and rehearsal and no postproduction, while a complicated dramatic special, which is videotaped in segments and assembled later through tape editing, would operate in all four stages. Throughout the book we will discuss the operation of equipment and production personnel as they apply to each phase of the production process.

Preproduction Planning

The preproduction planning for a program may begin days, weeks or even months before the actual production date. The more complicated and involved the production, the more preproduction time is necessary. During this stage, the producer and director work with the writer to complete the script and to develop the overall production approach. The key members of the television team—producer, director, TD, audio engineer, lighting director, scenic designer—meet to discuss the program and the part each will contribute.

Preproduction planning is essential for a successful show. Enormous difficulties can be avoided if the production has been planned out carefully in advance, with all key members of the team thoroughly aware of their contributions and areas

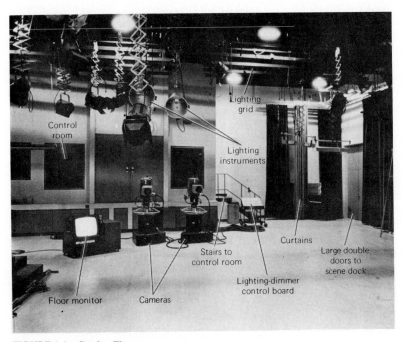

FIGURE 1-9 **Studio Floor.**

of responsibility. It is much easier to correct a problem on paper during the preproduction stage than later when unanticipated difficulties can halt work in the studio, delay the production, and increase costs. Remember Murphy's law, which always seems to operate overtime in television: "If anything can go wrong, it will." There will always be problems even in the most carefully planned production, but producing a program without adequate preproduction is an invitation to disaster.

Setup and Rehearsal

SETUP Just prior to the actual production, the studio and control room must be prepared for the program. The amount of time allotted for setup will be determined during the preproduction and usually depends upon the program's complexity and the size of its budget. In order to maximize the amount of time available, all key members of the production team must know exactly what will be required and must carefully supervise their crews to accomplish the task. The various crews should try to work simultaneously, whenever possible. It is inefficient to have to wait until a set is completely finished and dressed by the stagehands before even beginning to hang lights or to wait until the lighting is completed before setting up microphones.

While the studio floor is being set up for the production, a similar operation is occurring in the control room. The technical director must check on all external feeds from tape machines, film/slide projectors, electronic character generators, or remote locations. The audio engineer has to prepare the audio console by patching in microphones and sound from videotape and film tracks, setting levels, and checking remote audio feeds and communication lines. The video engineers must align the cameras to provide the best possible picture. Usually, most of the setup in the control room can proceed while the studio setup takes place on the studio floor.

A smooth and efficient setup is a crucial factor in the success or failure of a production. If too many details go unattended during the setup, they will have to be taken care of during rehearsal. Not only is this wasteful and inefficient, but it takes away valuable rehearsal time which is better spent in perfecting the performance and coordinating the various production elements into a smoothly operating team.

REHEARSAL Once the studio and control room are properly setup and prepared for the production, rehearsals begin. Since the amount of rehearsal time is always limited, even on the largest shows, the director must carefully plan the use of rehearsal time.

It is during the rehearsal that all the production elements should finally come together: the set, costumes, and lighting; the music and sound effects; the camera shots; filmed and taped inserts; and so on. While the director works on perfecting performances and camera shots, the other key production team members must watch the operation of their individual areas and correct whatever problems may arise. The audio engineer may reposition a microphone; the lighting director might add another instrument to illuminate a dark area; the scenic designer may decide to move some furniture to help out a difficult camera shot.

During this period, the producer carefully watches the program monitor, acting as a surrogate audience and making notes for suggested changes to improve both the aesthetic and technical quality of the production. During a break, these notes are discussed with the director, performers, and production crew.

Production

In the early days of television, before the development of videotape, all programming was produced live. The show started at the appropriate time and ran straight through until the end with no possibility to stop or correct any mistakes. This era of broadcasting was responsible for some classic stories such as the "dead" body which crawled offstage while still on camera or the actor who, after blowing his lines throughout the entire program, turned to another "passenger" on an airplane set and said, in frustration, "This is where

I get off." Whereupon he promptly walked off the airplane, ostensibly in midair.

The development of videotape has changed this and allows the producer and director far more flexibility. With the advent of tape and tape-editing techniques, shows can be produced in a number of different ways. We have divided them into two groups: live and videotaped for editing.

LIVE For a program which is broadcast live, the production stage is the final phase in the production process. Most news programs, sports coverage, and other such types of "immediate" programming are produced live.

Sometimes a program may be produced "live on tape." While this may seem to be a contradic-

tion in terms, it means that the program is produced in real time, as though it were live, but instead of being broadcast during production the show is recorded on videotape for later broadcast. Tape is used essentially as a storage medium and permits the show to be produced at a more convenient time than the actual broadcast time. Only in the case of a serious blunder is the tape stopped and edited. Most talk programs and game shows are produced live on tape.

FIGURE 1-10 **Four Stages of Television Production.**

PREPRODUCTION

- Develop concept
- Establish objectives and production approach
- Write program script/format
- Production meetings with key team members
  (producer, director, lighting director,
  scenic designer, technical director, audio)

SET UP AND REHEARSAL

Setup
- Set construction
- Hang and focus lights
- Audio setup
- Prepare videotape and film playbacks

Rehearsal*
- Dry rehearsal outside studio (for fully scripted shows)
- Camera blocking
- Run-throughs (problems are corrected as they arise)
- Dress rehearsal

*Varies depending on production complexity and whether or not program is fully or partially scripted

POSTPRODUCTION

- Studio strike
- Videotape editing
- Audio sweetening
- Evaluation of program to see if it met objectives

PRODUCTION

Live
- Program starts and ends according to preplanned airtime

Videotape
- *Live on tape:* Production proceeds straight through without stops—editing only in special cases.
- *Taped in segments:* Production is taped in short segments with multiple cameras. Shooting schedule is planned for maximum convenience and efficiency.
- *Single camera:* Production is taped with a single camera and multiple set ups for each scene—allows for maximum creative control.
- *Multiple camera/multiple: VTRs:* Production proceeds straight through or in segments with each camera feeding a separate videotape machine. All editing is done in postproduction.

VIDEOTAPING FOR EDITING The development of sophisticated videotape-editing techniques permits the use of tape not only as a storage medium but as a primary production tool like a camera or a microphone. Complex productions can be taped in short segments, which can be assembled later through videotape editing. Sometimes, segments may not be taped in the exact order in which they will appear on the completed show but rather for the convenience of the production. On other occasions, a program may be taped in segments because some performers may not be available during the entire production period. In this case, all scenes involving a particular performer are taped at one time, and, once they are completed to the director's satisfaction, the performer is released.

Taping in segments for later editing provides the program director with a high degree of creative control. Instead of the director and crew having to worry about the entire production, they can concentrate on each segment and, once that portion of the show is recorded and "in the can," move on to the next. Some productions are even taped with only one camera rather than with the more conventional multiple camera technique. Shooting with a single camera, much like motion pictures, allows the director to concentrate on a very small part of the performance at any one time and permits lighting, audio, and other production elements to be perfected and carefully controlled. Of course, this is a very costly and time-consuming technique, but some programs justify the additional expense involved.

Postproduction

Programs which are produced on videotape for later editing require a postproduction phase. At this time the director supervises the tape editing, selecting those takes or segments which are to be included in the final, edited version. Postproduction can be a simple matter of assembling a bunch of completed segments or a highly complicated procedure which employs computers to help the director and tape editor combine hundreds of individual shots into the completed program.

One of the major advantages of postproduction is the creative control it offers the director in the selection of both shots and performances. It is possible to select the very best performance of both cast and crew from a number of different takes, literally building the show by assembling the best scene, shot, or even a performer's delivery of a single line. In many instructional or industrial productions, postproduction permits the addition of visual elements, such as graphics, film, or tape, to produce an even more effective presentation. Additional audio can also be added during postproduction to enhance or modify the existing sound track.

VIDEOSPACE

Up to this point, we have discussed the technical and production aspects of television. But these are really only a means to an end, the tools which enable us to communicate through the television medium. Now we must consider how to make this communication effective; in other words, how to create and present visual and aural messages which will convey to the audience the impression we are trying to make in their minds.

When most viewers have an opportunity to visit the set of a favorite television program, they are invariably disappointed: the living room set which looks so large on screen appears much smaller in real life; the elegant furniture may be chipped or cracked; the walls and doors appear flimsy and frail; the colors which look so vibrant and alive on television seem faded and washed out. Usually, their faith is restored when they see the same set on a television monitor. Magically, the room appears to be its "normal" size, the walls look strong and sturdy, and the colors are alive and vital.

This experience illustrates an essential concept in television production which we call *videospace*. Videospace means that the only measure of reality for television viewers is what they see and hear through the television receiver. What exists in real life does not exist for viewers until that reality is translated through the television medium and onto their television sets.

Just as the proscenium arch on stage defines the boundaries of reality for the theater audience, videospace is the measure by which the viewers judge what exists on television. All the spatial relationships, all the sound cues, the appearance of the performers, the total environment in which the program takes place must be created and conveyed through the videospace. Videospace and its aural counterpart, "audiospace," really consist of two interrelated components: the technical aspect of the production and the aesthetic or creative elements. Together these create a reality for the viewing audience.

Media consultant Tony Schwartz discusses this idea in his book, *The Responsive Chord*. As Schwartz puts it, "If we seek to communicate a situation or event, our problem is not to capture the reality of that situation but to record or create stimuli that will affect the viewer in a manner similar to a viewer's experience in the real situation. What counts is not reality, as a scientist might measure it, but the ability to communicate the situation in a believable way."[1]

It may seem paradoxical, but sometimes we must modify what actually exists in the television studio in order for the viewer to perceive it as "real." For example, an actor wearing color television makeup may look rather odd in person but perfectly normal when seen on camera. Were he not wearing any makeup at all, his in-studio appearance would be quite normal but his image on the screen would not look attractive or "natural." Similarly, if we photograph a performer outside the studio on location under a bright, sunny sky, we may have to add artificial light to make her look natural on camera. The use of artifical illumination may seem superfluous, but the additional light is sometimes necessary to fill in the shadows on the performer's face and make her appear normal on screen.

These examples illustrate the application of the videospace concept to the television medium's technical capabilities and limitations. Quite often, because of the way in which studio equipment

[1]Tony Schwartz, *The Responsive Chord*, Anchor Books, New York, 1974, p. 31.

operates, we must vary in-studio reality so it will appear normal onscreen. The makeup and lighting we referred to earlier are necessary because of the technical limitations of the television camera.

The other applicaton of the videospace concept refers to its use to create an entirely new "reality" for the viewing audience. Sometimes we can use certain techniques to produce a reality in the videospace which actually does not exist. One commonly used technique is an electronic effect called "chroma key" in which two different video sources are electronically combined into a composite shot. Thus, a sportscaster inside the studio can be electronically inserted into a wide shot of a football stadium. The illusion produced is as though the sportscaster is actually present at the stadium when, in fact, that "reality" exists only in the videospace. The use of depth perspective in planning sets, lighting, and camera shots can also modify the viewer's perception of reality. A studio can be made to appear larger or smaller than it actually is through the manipulation of the videospace.

Audiospace is the aural counterpart of videospace and refers to the sound portion of a program. We can manipulate audio through various mechanical and electronic methods to modify and enhance the original sound. Natural sound can be filtered, echo can be added in varying amounts, and "equalization" permits us to accentuate or to minimize certain sound frequencies which, in turn, effect the audience's perception of aural realtiy.

Videospace and audiospace must always be a major consideration when planning and producing a program. The idea is to learn how to use them to convey the intended impression, mood, and overall message to the viewer. Everything you do in television should be undertaken with the videospace concept in mind. You must learn to translate what you see and hear inside the studio into how it will appear to, and be heard by, the viewing audience. It does not matter how it looks or sounds in the studio since the viewer's only measure of reality is what is seen and heard through the television receiver. If the videospace and audiospace you achieve for the viewer create the proper mood and impression, establish the intended environment in which the production takes place, and complement the overall program objectives, then the production has moved a long way toward satisfying the basic goal of any program: to communicate ideas and messages to the audience successfully. The effective use of both videospace and audiospace will enable you to take full advantage of the many capabilities which the television medium has to offer.

SUMMARY

Television production is a team operation which requires the combined effort of dozens of skilled individuals to produce a program. The television team can be divided into two major groups: (1) the production staff, including the producer, director, assistant director, lighting director, scenic designer, and production assistants, who are concerned primarily with the "creative" side of the production, and (2) the production crew, including the technical director or switcher, the audio engineer, video engineers, floor manager, camera operators, and the rest of the technical crew, who are primarily concerned with equipment operation.

The television studio complex is composed of two main areas: (1) the control room, which is the operational command center where program elements are directed and coordinated, and (2) the studio floor, where the production actually takes place.

Television production operates in four separate stages: (1) preproduction planning, when the overall production concept and approach are developed and organized; (2) setup and rehearsal, when the studio and control room are prepared for the production and the program elements are rehearsed and coordinated; (3) production, which can be either live or on videotape; and (4) postproduction, when videotape is edited and additional video and audio material can be added to the edited master tape.

A primary consideration in any production decision is the effective use of videospace and its sound counterpart, audiospace. Videospace means that the only measure of reality for the viewer is what is seen and heard through the television receiver. It is an essential factor when planning and producing any television program since the videospace will ultimately determine the context in which the audience views the program and the way in which the aural and visual messages are perceived.

CHAPTER 2
THE TELEVISION CAMERA

The camera is the primary instrument of television communication. As it is the basic production tool, most production decisions and techniques depend upon the camera's capabilities and limitations. In this chapter we will cover the basic characteristics of the camera: how it works, its operation in production, and most importantly, what it can and cannot do. Before you can hope to use the camera effectively, you must understand its characteristics and its operation.

SCANNING AND REPRODUCTION

Television is an electronic process which converts light energy into electrical signals. Both television and motion picture film operate by breaking down a moving image into a series of rapidly changing frames. Unlike film, which can capture an entire frame in one instant, television is an electronic medium, and the electronic circuitry can deal with only one piece of information at a time. This means the television camera must dissect the entire image into a series of picture elements. It is as though your eye read a page of copy by looking at each individual letter as it scans across every line. Once the complete picture has been disassembled into this electronic mosaic, the electronic impulses are processed and ultimately sent to a television receiver, which decodes the signals and recreates the original image.

The way in which the television system works is, to say the least, a highly complex procedure. Our purpose in explaining this operation in a simplified version is not to make you into a television technician, but to give you a basic understanding of the process since many production decisions are based upon the technical capabilities and limitations of the scanning and reproduction system.

Pickup Tube

The heart of the television scanning system is the *pickup tube* which is housed inside the camera. The tube is an optical-video transducer, which simply means it is a device designed to convert light energy into electrical impulses.

Although there are a number of different types

of pickup tubes used in television cameras, for now it is enough to know that they all work along the same basic principles. The entire tube is encased inside a glass envelope, which protects the interior components and provides a vacuum which is necessary for the tube's operation.

Light reflected from a subject is gathered by the camera lens and focused onto a light-sensitive element at the front of the pickup tube. When the light strikes the electrically charged element, it creates an electrical reaction, which corresponds to the brightness of the original image. The brighter the image, the stronger the electrical reaction. Because the process is electronic, however, the tube must break down the entire image into thousands of tiny dots, which can be processed individually and later reconstructed in the proper order to re-create the original image on the television receiver screen. (See Figure 2-1.)

The way the tube encodes or processes this information is by using an *electron gun* with an *electron scanning beam*. The gun is located at the rear of the tube and shoots out a stream of electrons which "reads" the electrical information dot by dot as it appears on the rear of the photosensitive element at the front of the pickup tube. The gun focuses the beam to begin reading the lower left-hand corner of the photosensitive element (as viewed from the front lens side of the camera) and sweeps the beam across until it reaches the lower right-hand corner. Then the beam retraces to the left, one line up, to begin reading the next line, much as a typewriter carriage returns to the margin after completing a line. The reason the beam starts in the lower left-hand corner is because the image is inverted by the lens. In order to reproduce an upside-down image correctly, the beam must start at the lower left corner and work its way upward.

This process is called *scanning*, and it continues until the beam has read the entire light-sensitive area line by line. Once the beam reaches the last line, it immediately returns to the starting point and begins its sweep all over again.

If the scanning beam were to read each line sequentially from top to bottom, the image produced would have an annoying flicker. To correct for this flickering effect, the electron scanning beam scans *every other line* before returning to

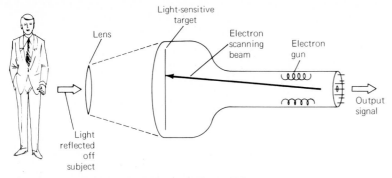

FIGURE 2-1 Operation of Television Pickup Tube.
Light reflected from the subject is gathered by the lens system and focused on the light-sensitive surface of the camera's pickup tube. An electron scanning beam produced by the electron gun is the rear of the tube scans across the target area and "reads" the light intensity of every point along 525 lines. A return electron beam signal is amplified by the camera circuitry and enters the video system.

the bottom of the target to scan the lines which were missed the first time around. In other words, the beam would first scan every odd line: 1,3,5, and so on and then return to scan every even line: 2, 4, 6, 8, and so forth. This process of scanning every other line is called *interlace scanning* and is used on all television reproduction systems. (See Figure 2-2.)

Each time the scanning beam reads one complete set of lines (that is, all the odd lines) it completes a single *field*. It takes the scanning beam one-sixtieth of a second to complete one field. Because it requires two fields to scan the entire target one time, the beam must scan the target twice to complete one *frame*. If each field takes one-sixtieth of a second and two fields are required to complete a frame, then the frame rate of television is one-sixtieth (all odd lines) plus one-sixtieth (all even lines), which equals one-thirtieth of a second, or there are *thirty frames per second* in television reproduction.

When we watch television, what we are really seeing are thirty individual still frames reproduced each second. If you recall that motion picture film reproduces twenty-four still frames per second, you can see that the idea behind both processes is quite similar. In fact, they both take advantage of *persistence of vision*, which allows us to perceive continuous movement even though we are really watching a series of rapidly changing individual still frames.

Cathode Ray Tube

The cathode ray tube (CRT) serves as the reproduction counterpart to the camera's pickup tube. Like the pickup tube, the CRT is encased inside a glass envelope to protect its components and to supply the vacuum which is necessary for its operation.

The inside of the tube face is coated with a phosphorescent material which glows after being struck by an electron beam. The intensity of the glow is directly proportional to the intensity of the original signal which was produced during scanning. A strong signal causes a very bright glow while a weaker signal causes a less intense, dimmer glow or no glow at all. The electron beam is supplied by an electron gun, which is situated at the rear of the CRT inside the narrow neck. Its operation is the exact replica of the operation of the electron scanning beam inside the pickup tube, but instead of reading the information, it sends out a signal which is "modulated" or controlled by the original picture signal produced during the scanning process.

The beam scans the inside face of the picture tube creating 30 frames per second. Because of the afterglow of the phosphorescent dots and the persistence of vision phenomenon, we perceive the glowing dots as a composite image—the replica of the original subject which was photographed by the television camera. (See Figure 2-3.)

Sync Generator

By now you may have realized that we need some way to keep the picture tube's scanning beam and the CRT's electron beam in perfect step to reproduce the image. This is done with a device known as the *sync generator*—short for "synchronizing generator"—which produces a series of electrical timing pulses, which control the entire scanning and reproduction operation.

Sync pulses, produced by the sync generator,

ensure that all the electronic reproduction equipment used to produce television pictures operates in unison. Unless there is perfect synchronization among various cameras, videotape recorders, and television receivers (to mention only a few items of equipment which rely on the sync generator), no television pictures are possible.

Resolution

One of the most important characteristics of any picture-reproduction system is its ability to reproduce, in clear and sharp detail, all the elements which are in the original subject. In television, this is known as picture *resolution*, and it can be

FIGURE 2-2 **Interlace Scanning.**
To reduce picture flicker, the scanning beam actually scans every other line in an "interlace" fashion. In the first frame, the beam scans every odd line (1, 3, 5, 7, . . . to 525) to produce one field. The next scanning cycle covers all even-numbered lines (2, 4, 6, 8, . . . to 524). Together the two fields equal one complete frame thirty times a second.

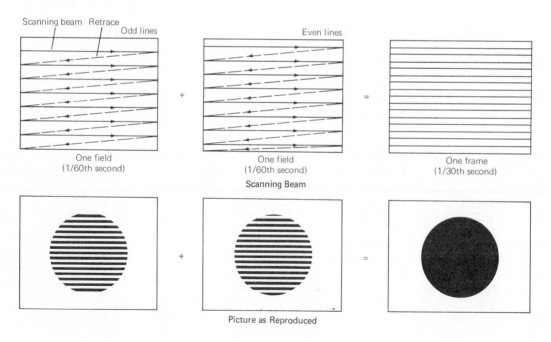

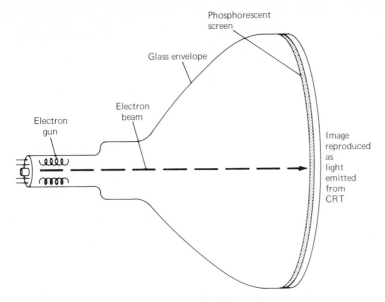

FIGURE 2-3 Operation of Cathode Ray Tube (CRT).
The electron gun of the CRT emits an electron beam which is an exact duplicate of the original scanning beam in the camera pickup tube. The beam scans across the phosphorescent face of the CRT, which emits light whenever it is struck by the scanning beam. The more reflected light which enters the pickup tube, the greater the intensity of the electron beam in the CRT, and the brighter the phosphorescent dot glows. Our eyes perceive the rapidly changing mosaic as a complete image.

measured with test charts. A picture with high resolution is capable of distinguishing fine detail in the subject. A picture with low resolution is incapable of differentiating and reproducing minute detail.

The resolution depends upon the number of lines which are used to reproduce the image. We are all familiar with newspaper pictures which use tiny dots of various gray tones to reproduce the image. The more dots used, the more detailed the image since each dot must represent less total information, which permits the entire reproduction process to discriminate subtle differences in the subject.

Since television operates by using time phosphor dots, or "pixels", arranged in lines, it too depends upon the number of lines, among other things, to determine the picture resolution. When American television developed, 525 lines was the

most practical number which could be utilized. By the time it was possible to increase the number of lines and improve resolution, all the stations in the country, as well as all the home receivers, were already designed for 525-line operation. To change the number of lines would have required extensive retooling of all television studio hardware and home receivers. Since the investment in equipment was too large to consider such extensive modifications, the 525-line system has remained the United States standard. Those countries which introduced television later were able to take advantage of newer technology; television systems in Europe use a greater number of lines and can reproduce pictures with much better resolution and clarity than possible with our 525-line system.

Because of the relatively small size of receivers and to the resolution limitations of the broadcast picture, television has had to resort to using very tight shots to overcome resolution problems. This is one of the reasons why television is often called a "close-up medium."

High-Definition Television

Recent technological advances have resulted in the development of a prototype high-definition television (HDTV) system, which is currently being used on an experimental basis in Japan and in the United States. The system uses 1,125 lines to create the video image as compared with the 525 or 625 lines currently in use on conventional broadcast systems. Doubling the number of scanning lines produces an image which is equivalent in quality to the picture we currently obtain from 35-mm motion picture film (See Color Plate 6.)

Unfortunately, such a high-definition system cannot use existing hardware; and new cameras, pickup tubes, videotape recorders, and auxiliary equipment must be developed before such a system can become truly viable. Nevertheless, the possibilities of a true HDTV system are virtually limitless. Even if a home delivery system were developed sometime in the future, the HDTV system could be used now to produce a high-resolution master tape, which would eliminate the cost and time problems associated with 35-mm motion picture film. The HDTV master tape could then be used to produce conventional videotape dubs or for distribution via video cassette or video disc.

Some motion picture directors and producers, such as Francis Ford Coppola, are excited about the possibility of using HDTV as a substitute for motion picture film in the production of theatrical movies. In this case, the production would be produced entirely with electronic video technology and edited on videotape. Once the master HDTV tape is completed, the production can be transferred onto 35-mm film by means of a laser recording system for distribution to movie theaters. Using this method, motion picture theaters would simply receive a conventional film release

print, and no new equipment would be necessary. Some, however, envision movie theaters as going to the next step of installing HDTV large projection screen units in which an actual videotape would be projected onto the large screen of the theater and produce a picture with the same quality as 35-mm film.

Since the high-definition video signal requires the processing of much more electronic information than we need to use for conventional broadcast pictures in 525 or 625 lines, we cannot use conventional broadcasting methods of transmitting the picture to the home receiver. One possibility would be the use of direct broadcast satellites, which have the necessary capabilities of transmitting the more complex HDTV signal direct from the satellite orbiting in space to the viewer's home via a small rooftop receiving dish. Another possibility for delivering a high definition signal would be via a sophisticated cable system, and a third would be through the use of high definition videocassette or video disc machines hooked up to an HDTV receiver.

Since home delivery of HDTV requires the ability to solve a number of complex technological and policy questions, it seems likely that HDTV will first be used as a primary production medium in the development of a top-quality HDTV master tape, which is used to strike copies of the production onto conventional videotapes. Even though the videotape copy is limited by the existing number of lines in our conventional television system, the ability to produce programs electronically which have the same quality as 35-mm motion picture film offers tremendous savings in production cost and greatly expanded flexibility since all of the various electronic methods of processing and manipulating the video signal which we currently employ in television production can be applied to HDTV production as well.

COLOR REPRODUCTION

Up to this point, we have discussed scanning and reproduction only as they apply to monochrome, or black and white, television. Although color television uses essentially the same scanning process, the reproduction of color information requires some additional elements in both the scanning and reproduction systems.

The Nature of Color

Before we can talk about the operation of the color system, we must spend a little time discussing the basic properties of color. Color light is a portion of the visible light within the electromagnetic spectrum. Our eyes perceive visible light as "white light" although we are all familiar with the operation of a prism which refracts white light and separates it into its component colors, creating the spectral range of colors from red to violet.

Color light operates under the additive principle, which means that mixing the three primary colors of *red, blue,* and *green* in varying proportions can create every color in the visible spectrum.

To illustrate this, imagine three spotlights, each of a primary color, which are positioned to overlap slightly along their edges. (See Color Plate 1.) Where the three colors overlap equally, we produce *white* light. Where two primary colors overlap, a *complementary* color is formed. Red and green produce the complementary color *yellow*. Green and blue produce *cyan*. Red and blue produce *magenta*. An absence of all three primary colors produces no light, or *black*. Remember that we are talking about the additive property of light, and this is not the same thing as mixing paint pigments, which uses a subtractive process and results in different color combinations.

But the color of light is only one of three components which interact to form the colors we see. The three components are (1) hue, (2) saturation, and (3) luminance.

HUE Hue is the tint of color. In effect, it is the color which we see. In the example of the three spotlights, we produced different hues by combining the three primary colors of light.

SATURATION Saturation, usually referred to in television as "chroma," is the degree of color strength or purity, or the amount of dilution of the color through the addition of white light. A 100 percent saturation represents the pure hue with no white light added. For example, 100 percent red is highly saturated, and appears very strong and vibrant. Diluting the saturated red with white light would produce a weaker, washed-out pink.

LUMINANCE The brightness component of the

color is called "luminance" and depends upon the amount of light which the color reflects. A color with high luminance reflects much light and appears bright. A low-luminance color reflects little light and appears dark.

By this time it should be obvious that by manipulating the three primary colors in different intensities and with different quantities of luminance and saturation, we should be able to produce any color in the visible spectrum. In fact, this is the basic idea behind color television reproduction. (See Color Plate 2.)

Color Camera

The color reproduction system used in the United States is often referred to as the "NTSC color system." NTSC refers to the National Television Systems Committee, which was a panel of experts organized to research and develop a compatible color system that would produce color pictures on specially designed receivers while permitting conventional monochrome sets to reproduce the same signals in black and white.

The color camera contains four basic components: (1) the *optical system,* which splits the reflected light into the three primary colors, (2) the *chrominance channels,* which transform the light into electrical signals, (3) the *luminance channel,* which provides brightness information and (4) the *encoder,* which processes the color and brightness information for transmission through the system. (See Color Plate 3.)

COLOR CAMERA OPTICAL SYSTEM Top quality production cameras use a *dichroic mirror* assembly or a *prism beam-splitter* to dissect the reflected light from the subject into the three primary colors. The optical system is designed to permit only one primary color to enter each of the chrominance channels. As the multicolored white light passes through the dichroic mirrors or beam-splitter, only the red light is permitted through to the red pickup tube, only green light passes through to the green tube, and only blue light passes to the blue tube.

CHROMINANCE CHANNELS The highest quality production cameras utilize three separate pickup tubes, each for a primary color. When the reflected light from the subject enters the lens and is split by the optical system, each of the three tubes receives a part of the light in direct proportion to the amount of the primary color which is present in the subject. For example, if the camera were focused on a pure red card, red light would enter the red gun, but since there is no blue or green light present, no light would pass through to the other two guns. Now, if we focus the camera on a yellow card (which is actually a mixture of red and green light), then about half the light would reach the red tube, and half the light would reach the green tube. Since there is no blue light present, no light would enter the blue tube. A pure white card, which is, theoretically at least, a combination of red, blue, and green light, would activate each pickup tube about equally since the beam-splitter would permit each primary color to pass through to its respective pickup tube.

Once the dissected light reaches the pickup tubes, the scanning operation is exactly the same as in black and white reproduction. In fact, you might think of color television as three black and white "cameras", each receiving a portion of the white light and operating simultaneously in perfect synchronization. (See Color Plate 4.)

LUMINANCE CHANNEL Although the three pickup tubes process the "chroma" or color information, we still need a reference to form the luminance or brightness component of the picture. This information is used in three ways: (1) to provide brightness information for the color reproduction, (2) to outline and separate colors in the picture and provide more sharpness and detail, and (3) to produce a monochrome signal for black and white receivers.

Some early color cameras used a fourth tube, which supplied luminance information and worked much like a coloring book, the luminance channel supplying an outline which was filled in by the color channels. A better method has been devised using one of the primary color channels, usually green, to provide contouring and luminance information. This is the most commonly used approach in modern studio production cameras.

ENCODER Once the color information has been separated into the three tubes and the luminance information provided, we need some means of processing and transmitting the separate color signals simultaneously while still keeping their information separate from each other. This is the function of the *encoder*, which is a device designed to combine the three color signals (or chrominance information) with the brightness (luminance information) into a single, composite signal. Encoders which process signals along the guidelines developed by the NTSC are called "NTSC encoders" and produce a broadcast-quality signal.

Color Cathode Ray Tube

Instead of using a single electron gun as in monochrome, the color tube has three, one for each of the primary colors. The inside face of the color CRT is coated with phosphorescent elements, but instead of one dot per element there are three, again one for each primary color. When struck by electrons from the electron beam, the red, blue, and green phosphorescent elements glow their respective colors. Approximately one million color dots are arranged in triad groups across the face of the tube. If you look closely at a color CRT in operation, you can see the many dots which comprise the overall picture. However, at normal viewing distance, the closely spaced dots produce a mosaic, which displays a composite color image. (See Color Plate 5.)

It is the glow of the phosphorescent dots which actually generates the color in a picture. When all three dots in a triad glow simultaneously and with equal strength, the triad appears to be white. When only a single dot in a triad glows, the primary color is generated. If two dots glow together in combination, a complementary color is formed. In this way, the combination of dots glowing in varying proportions of hue, saturation, and brightness produces the range of colors. It is the combination of many triads glowing together which forms the complete color image we see. The system is "compatible" since monochrome receivers will simply ignore the color information and will use the luminance signal to reproduce the picture in black and white.

Alternative Color Systems

The three-gun method of color reproduction produces the highest quality picture in terms of both color and resolution. This is why it is the method used in all top-level studio production cameras. However, the three-tube system does have a number of drawbacks. The camera is larger, heavier, and more complex in order to accommodate three tubes and the necessary electronic circuitry. Secondly, the camera must be carefully aligned by a skilled technician daily to ensure that the three images overlap perfectly and result in a sharp and clearly defined composite image. Finally, the three-gun camera is expensive to purchase, operate, and maintain.

A number of alternative color systems have been developed which are used on lighter and less expensive color cameras. Although these systems cannot compete with the picture quality available from the three-gun method, they do offer some advantages in terms of decreased cost, lighter weight, and increased portability and less technical maintenance.

SINGLE-GUN SYSTEMS The single-gun system uses only one pickup tube to provide both chrominance and luminance information. In the single-gun design, a special color filter made of fine, crisscrossed color stripes is positioned over the face of the pickup tube. The tube views the subject through the mosaic of color stripes. Usually the striped filter contains only two primary colors since the third color can be derived electronically by subtracting the known colors from white. Special electronic circuits separate the color information into three discrete color signals, which are encoded and sent through the system.

Since there is only one pickup tube, there is never a registration or balancing problem, and this is a primary advantage of the single-gun system. Also, fewer pickup tubes require less electrical power, and this is a significant advan-

tage when using battery-powered cameras on location in the field.

We must pay something for this relative simplicity, however, by producing pictures with somewhat less resolution and color quality than is possible using the three-gun camera design. Some manufacturers have developed a new line of pickup tubes, especially for single-gun cameras, and these tubes can produce remarkably good pictures particularly when they are used in less-demanding production situations such as for news gathering or for nonbroadcast closed circuit application. (See Figure 2-4.)

SOLID STATE CAMERAS

All television cameras currently use a vacuum pickup tube to convert light into electrical energy. Although these tubes produce top-quality pictures, they have a number of serious operational drawbacks. First, they tend to be very fragile and don't operate well in extremely hot or cold temperatures. They also require a lot of electrical power in order to operate, and their physical size and shape result in larger and heavier cameras.

Engineers are currently working on an alternative to the pickup tube which employs a solid state charge-coupled device, or CCD, which will replace the conventional pickup tubes in certain cameras in the near future. The CCD is a light-

FIGURE 2-4 Single Gun Color System.
In the single-gun system, only one pickup tube takes care of both color and luminance processing, using a color-stripe filter over the face of the tube.

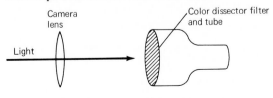

Single-gun Color System

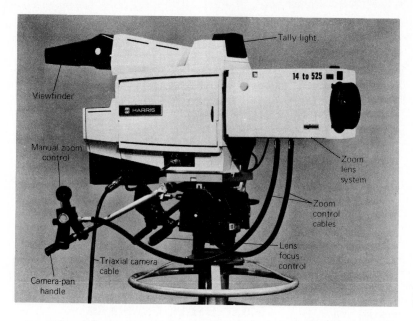

Tally light

Viewfinder

14 to 525

Manual zoom control

Zoom lens system

Zoom control cables

Lens focus control

Triaxial camera cable

Camera-pan handle

FIGURE 2-5 Television Camera Head.

sensitive electronic chip which converts the reflected light of a subject into electrical energy much as the conventional pickup tube does. However, the CCD offers a number of important advantages over tubes: It is very small, rugged, and lightweight and uses relatively little electrical power for its operation. These characteristics make the CCD important in the design and development of small, lightweight portable cameras for use in news production and similar remote situations.

CAMERA CHAIN

The camera chain consists of (1) the *camera head,* which contains the optical and electronic systems required to convert light energy into electrical signals, and (2) the *camera control unit* (CCU), which contains the power supply and controls necessary for regulating the technical quality of the camera's picture.

Camera Head

The camera head consists of five basic systems: (1) the *lens system,* (2) *the internal optical system* (found only on color cameras), (3) the *electron*

pickup tube (or multiple tubes if a color camera) and associated electronic equipment, (4) the *viewfinder,* and (5) the *camera communication system.*

LENS SYSTEM All television cameras are equipped with a lens system which gathers the reflected light from a subject and focuses the light rays onto the camera's pickup tubes. All color cameras utilize a zoom lens with continuously variable field of view. The zoom lens is operated by a control situated at the rear of the camera.

Although most monochrome cameras are now fitted with zoom lenses because of the increased production flexibility which the lens provides, some still utilize a complement of fixed focal length lenses mounted on a camera turret. We will discuss lenses in greater detail in the following chapter.

INTERNAL OPTICAL SYSTEM All color cameras use an internal optical system, which dissects the composite light from the image into the three primary colors of light. The most commonly used optical system is the "prism beam-splitter," which provides maximum light output with a minimum of light loss or optical distortion. Some color cameras use a "dichroic mirror" optical system, which does the same thing but has a greater light loss between the lens and the pickup tubes. Since monochrome cameras do not require light-splitting, they have no internal optical system.

PICKUP TUBES   The electronic pickup tube, which converts the optical image into an electronic signal, defines the camera's operating characteristics and determines its picture-reproduction capability. The most commonly used tubes today are a family of vidicon-type tubes which are known under such trade names as Plumbicon, Saticon, Chalnicon, and Leddicon. These tubes are capable of producing high-quality color images over a wide range of production situations.

Originally, the Plumbicon and similar tubes

were made for studio-type cameras and were available in two sizes: a 1¼ inch (30-mm) diameter format and a 1-inch (25-mm) diameter format. The format refers to the size of the photoconductive face of the tube; and, all things being equal, the larger the tube face, the better quality the image. With the introduction of smaller studio and miniature portable cameras, however, tube manufacturers were able to produce high-quality pickup tubes in smaller ⅔-inch (18-mm) and ½-inch (12-mm) formats which are capable of producing excellent pictures. The smaller physical size of the tube enables cameras to be made smaller and more lightweight, and to require less electrical power to operate. (See Figure 2-6.)

VIEWFINDER SYSTEM   All television cameras are equipped with an electronic viewfinder, which continuously shows whatever the camera photographs. The viewfinder is simply a miniature television CRT ranging in size from about 3 inches to 9 inches in diameter. Many viewfinders can be tilted, affording the camera operator a better viewing angle; and others can also be rotated, making it convenient for the camera operator to manipulate the camera and see the viewfinder while shooting from the most favorable angle.

The camera operator uses the viewfinder to frame, compose, and focus the camera shot. Some camera viewfinders also display an indication of the focal-length setting of the zoom lens.

On most cameras, the viewfinder can be switched from the camera's picture to show an external feed from the video switcher in the control room. This valuable feature enables the camera operator to see how his camera shot will combine with the shot from another camera for certain composite special effects. For example, if you had to follow a baseball runner on first base in the upper right-hand corner of the shot, the external viewfinder feed would help you frame the subject accurately within the small corner square.

CAMERA COMMUNICATION SYSTEMS   The camera operator is always in direct contact with

FIGURE 2-6   **Television Pickup Tubes.**
Three lead-oxide tubes: (left to right) 1¼-inch (30mm) tube used in large studio cameras; 1-inch (25mm) tube used in studio cameras; ⅔-inch (18mm) used in small studio cameras and in most portable ENG/EFP cameras. *(Courtesy: Amperex Electronic Corp.)*

the program's director and other members of the production team through a headphone intercom system, sometimes called a "private line" or PL. Throughout the production, the camera operator sets up each shot according to the director's commands over the PL. The technical director, video engineers, and assistany director can also talk to the camera operator over the PL. The camera operator can talk back through a small mouthpiece on the headset. Some camera communication systems use a dual headset in which one earpiece transmits the production commands and the other transmits program audio. This is particularly valuable on such unscripted productions as a sports remote, where the camera operator may have to follow the commentator's words quickly in order to cover the event.

Because television production uses a number of cameras operating simultaneously to pick up different shots from a variety of angles, we need some means of notifying the camera operator and performers which camera has been selected by the director and punched up on the air at any particular time. This is accomplished silently with *tally lights,* the large red light or lights atop the studio camera which light up to signal the crew and performers that the camera's shot is on the air. The camera operator has another tally light located directly next to the viewfinder, where it can be easily seen as the operator watches the viewfinder picture. Both tally lights operate automatically when the camera's picture is punched up on the air.

Camera Control Unit

The second half of the camera chain is the camera control unit which is located in the studio control room and includes all the controls necessary to register and regulate the picture's exposure during camera operation. (See Figure 2-7.)

All·cameras must be registered and aligned prior to use. This is especially important with color cameras since their three separate picture tubes must be precisely aligned with each other to

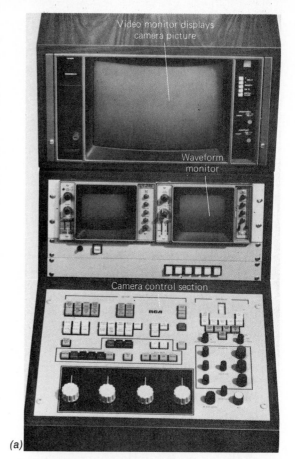

(a)

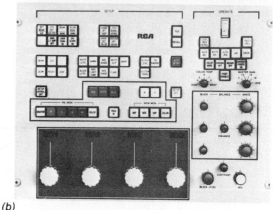

(b)

FIGURE 2-7 Camera Control Unit (CCU).
The CCU contains all of the necessary controls to set up and align cameras and to shade them during operation. *(a)* The entire CCU with video monitor, waveform monitors, and operational controls. *(b)* A close-up of the control panel.

produce a clear, sharp, and color-true image. Usually, the video engineer will register and align the camera before you enter the studio for rehearsal or production. Although it may seem frustrating to wait until the alignment procedure has been completed before starting your production, unless the cameras have been properly registered, they cannot possibly provide the best quality pictures. In fact, at many studios cameras are aligned early in the day and then, just before taping or production, the video engineers touch up the alignments, which naturally change or "drift" over a period of time. Once a camera has been set up and aligned, it should never be physically moved off its pedestal or tripod, since this will tend to upset the alignment and deteriorate the picture quality.

While a production or rehearsal is under way, the video engineers must adjust certain controls to compensate for variations in scene brightness and to keep the exposure of each camera within the proper technical limits. On color cameras, the shading is accomplished with two controls: iris and master black.

The *iris* control operates the diaphragm or f-stop of the camera lens through a remotely controlled electrical motor system. Opening the iris allows more light to enter the camera and increases the intensity of the whites in the picture. The *master black* level controls the black reference or "pedestal" and affects the dark portions of the picture. The video engineer manipulates both controls while watching both the picture monitor as well as a *waveform monitor*, which graphically displays the video signal on an oscilloscope. As the camera focuses on a new shot or scene, the shader "rides levels" to keep the exposure constant and within the necessary technical limits for proper video reproduction. (See Figure 2-8.)

Many color cameras have automatic white and black "balance" control circuits which enable the camera operator or shader to balance the camera

FIGURE 2-8 Waveform.
The waveform monitor displays an oscilloscope's graphic representation of the video picture. The whites in the picture should be adjusted to peak no higher than 100 percent at the upper end of the waveform scale. Blacks should just touch about 7 percent at the lower extreme. Compressing the blacks results in a dark and muddy picture with little or no gray shade definition. Clipping the white level by exceeding the 100 percent level merges light gray tones into white. The operations of the white and black levels are related: adjusting one affects the other and vice versa.

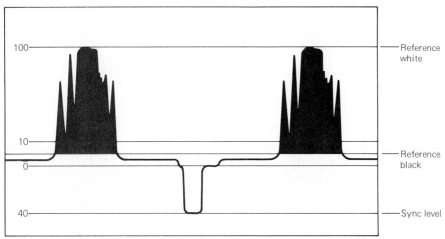

for optimum color reproduction quickly and accurately The white balance compensates for changes in color temperature—the amount of red or blue color quality of the light—and is quickly accomplished by focusing the camera on a white card and depressing the white balance button. On most modern color cameras, "capping" the camera—shutting off the operation of the pickup tubes—automatically balances the black level so the camera will "track" or properly reproduce colors across the entire brightness range. Balancing the white and black levels ensures good color reproduction just as taking a preliminary audio level before making an audio recording assures accurate sound.

Another feature commonly found on newer cameras is a *gain control*, which enables the video operator to increase the output of the camera by electroncially amplifying the video signals. This permits you to operate the camera under less than ideal, low-light conditions and is quite valuable on remote productions where lighting cannot always be controlled. Since the amplification results in an increase in picture noise and a slight deterioration in overall quality, additional amplification should be used only when necessary.

DIGITALLY CONTROLLED CAMERAS

One of the most exciting advances in recent camera design has been the development of the "digitally controlled camera." Unlike conventional cameras, which require large, multiconductor cables to send and receive picture information and operating commands between the camera head and the CCU, digital camera systems transform all "analog" signals into computerlike digital impulses or code signals. These digital signals are not only more precise than conventional analog signals, but they are also impervious to electrical interference over long cable distances. This provides the digital cameras with a number of signifi-

cant advantages over conventional cameras.

1 Since all electronic information to and from the camera head is digitized, it can be processed and handled more accurately than conventional signals. Digitizing the electronic signals permits a number of different commands and operations to be "multiplexed" or combined to travel together along a single conductor cable. Digital cameras use "triaxial cable," which is lighter and narrower than the multiconductor cable which conventional cameras use. Since the triaxial cable is one-fifth the size and weight of multiconductor cable, more cable can be carried on remote trucks and run to cameras in a much shorter time by fewer crew members. In the studio, the smaller cable results in easier camera operation and fewer cable problems than with the heavier, more rigid multicore cable.

2 Digital signals are extremely flexible and, because they are not affected by long transmission distances, permit a much greater operating distance between the camera head and the control unit. Digital cameras are capable of operating distances of up to 1 mile from the CCU, and the signals can even be transmitted via public telephone lines, radio-telephone systems, or radio-frequency (RF) links.

3 Since digital techniques utilize the same operating principles as used with computers, the camera head includes an internal memory which automatically retains the correct technical camera settings. Once the memory is programmed and the proper technical values established, the camera automatically checks to ensure that it is always operating within its proper technical limits. Any deviations from these values are sensed by the camera and immediately corrected. This ensures accurate timing and synchronization between the camera and the base station, or CCU, at all times. The cameras also contain rechargeable batteries which automatically switch on after all ac (alternating current) power is shut off, providing a small current to keep the memory system operative. This means that a digital camera can be shut off, turned on the next day, and will still "remember" its proper tech-

nical values and settings, automatically correcting the camera to match them if necessary.

4 The tedious and time-consuming job of registering and aligning the three separate pickup tubes to produce a sharp and color-perfect image can be done automatically with a digital camera by using a microprocessor. Registration and alignment operations, which take a considerable amount of a video engineer's time, can be accomplished in literally seconds, and the results are consistent for each camera because of the ability of a computer to perform highly complex functions with perfect accuracy.

In digital systems the complex camera control unit is called a *base station*. The base station's primary function is to transform all electrical signals—camera controls, synchronizing pulses, audio channels, program intercom channels, and remote viewfinder feeds—into digital pulses which the camera can understand and process. The base station also continuously sends the camera information concerning the camera's technical operation and automatically corrects any deviations from the memory's preset values.

Of course, all manual operations such as shading "iris" and "black level" are still performed by the video engineer. These signals are also digitized and sent to the camera head over the triaxial cable.

The advantages of the digital cameras are obvious. Maximum control and operational flexibility, the ability to operate the camera head at long distances from the base station without interference of the camera signals or command data, the choice of using lightweight triaxial cable or alternative transmission systems such as telephone lines, or radio-frequency links, the camera's internal memory system, which simplifies setups and ensures precision operation of all technical functions.

The disadvantages of the digital camera systems are primarily in terms of its cost and its extremely sophisticated construction and design. The highly advanced electronics which control the camera's operation are borrowed from computer and integrated circuit technology. This makes the digital camera extremely expensive to

purchase and requires sophisticated technical maintenance to keep it working properly. However, in the long run, the camera's inherent advantages far outweigh its disadvantages, and it is quite conceivable that digital cameras will ultimately replace conventional cameras within a few years as the older cameras become outmoded and the price of the digital camera system is reduced.

OPERATING CHARACTERISTICS OF THE TELEVISION CAMERA

If a television camera were capable of reproducing an image exactly as our eyes see it, many production problems would be made much simpler. We would know immediately that whatever we see with our eyes is what will appear on the television screen. Unfortunately, this is not the case as no television camera yet devised is as sensitive and discriminating as the human eye. Since the camera is not nearly as responsive as our eyes, we must continually take into account its operating abilities—what it can and cannot do. The most important of these characteristics for both monochrome as well as color cameras are (1) the operating light level, (2) the contrast range, and (3) the picture's resolution. Taken together, these characteristics outline the television camera's operating capabilities and limitations.

Operating Light Level

All cameras require a mimimum light level to function properly. The primary determinant of the minimum operating light level for any camera is its "signal-to-noise" ratio, which is the ratio of video signal strength to electrical "noise" or interference. All electronic devices generate a certain amount of "noise" during normal operation. However, the strength of the signal or, in this case, picture information is usually strong enough to overcome the interfering noise. To illustrate this, imagine a television receiver which is unhooked from its antenna. The screen is covered with grainy snow, which is "noise." Because the signals in the air are too weak to overcome the noise and mask it, the picture is very blurry. Once the antenna is connected to the receiver, the signal strength increases sufficiently to overpower the noise, essentially removing it from the picture.

Television cameras work in much the same way. Since modern cameras have a much higher signal-to-noise ratio than was possible a few years ago, they can operate under much lower light levels and still deliver a crisp, high-quality picture. We measure light level or light intensity in *footcandles* (fc). The signal-to-noise ratio and the minimum number of footcandles necessary to overcome the inherent noise in the system determine the camera's basic operating light level.

Modern pickup tubes have a high signal-to-noise ratio and the ability to reproduce pictures even under low light conditions. In the studio, most of the standard pickup tubes can operate quite comfortably at light levels between 75 and 250 fc. On remote locations, however, where lighting can not be controlled as easily, the camera is still capable of producing quality color pictures with light as low as 5 or 10 fc.

Additional Light Level Characteristics

There are a number of additional operating characteristics which are directly related to the camera's operating light level and to lighting conditions. They are lag, burn-in, blooming, and comet-tailing.

LAG   Under low light conditions, Plumbicon and similar video-type tubes are susceptible to image smearing or "lag" either as the subject moves across the camera's range of view or as the camera pans or tilts across the scene. As the pickup tube ages, the appearance of lag tends to increase, but the problem is compounded by decreased light levels. In an attempt to reduce lag, many color cameras are equipped with a *bias light* inside the camera head. The bias light is a tiny bulb which floods the faceplate of the pickup tube with a soft, uniform illumination. This keeps a small, constant current flowing through the tube and decreases its tendency to lag or smear. The bias light can be switched on when the camera must operate under low light conditions and switched off for normal operation. The bias light is a valuable accessory, particularly for those cameras which must operate on remote locations.

BURN-IN   When a television camera is trained on a scene with high contrast, if it has been focused on a static picture for an extended period of time or has been accidentally pointed into a light, the picture tube tends to retain an "after-ghost" of the original image, which "burns" into the tube. Usually the burn disappears quickly by itself, but some severe burns must be removed by panning the camera across a neutral area such as the studio floor or a lit cyclorama. Burn-in tends to increase as the tube ages and can be an annoying problem if it appears too frequently.

Most modern generation pickup tubes are virtually insensitive to burn-ins and, in fact, can be focused directly into studio lights or onto highly reflective surfaces without burning or damaging the tube. This ability to reject burn-ins offers the director some dramatic production possibilities. Two exceptions to the rule, however, are direct sunlight and an electronic flash from a still camera. Either can seriously damage the camera tube beyond repair.

BLOOMING   Excessively bright lights or highly specular reflections from metal or jewelry can cause a portion of the picture to "bloom." On color Plumbicon and vidicon tubes, blooming appears as a multicolored highlight or smearing. If the subject or camera moves, this blooming is sometimes called "comet-tailing" as the resulting image looks like a comet with a reddish tail. It is quite common in coverage of football games when the sunlight reflecting from a player's helmet produces the comet-tail effect. Although blooming and comet-tailing can be purposely induced for a special effect—when shooting a musical group, for instance, or for a fantasy effect—it is usually a distracting annoyance, and there are a number of ways to reduce its appearance in a shot.

Some new cameras include an "anticomet-tailing" tube and circuitry, which enables the camera to handle a highly exposed portion of a scene without washing out detail or blooming. Another solution is to coat the offending surface with a dulling spray which will minimize its reflection and the resulting comet-tailing.

Contrast Range

The contrast range refers to the ability of the television camera to pick up and faithfully reproduce the variations in brightness within a scene. The contrast range is usually expressed in terms of a contrast ratio, which generally should not exceed 30:1. In other words, the brightest portion of any scene should be no more than thirty times lighter in footcandles than the darkest area included in the shot. If you exceed this contrast range, the camera has difficulty reproducing both the light and dark ends of the picture, and you must sacrifice brightness detail at either one end or the other. The video engineer can shade for the excessive white values in the picture by "crushing" the dark grays and shadows completely to black. If you want to capture the detail in the dark end, the video engineer must "stretch" the blacks, merging the light grays and highlights completely into white.

It is important to understand that the contrast range is a *relative* concept, which is a separate consideration from the basic operating light level. Regardless of how much light is present on the scene—50 fc or 500 fc—if the difference between the brightest and darkest portions of the picture is greater than a 30:1 ratio, the picture quality may suffer since you have exceeded the contrast range. This does not mean that you should design your setting without any contrast, however. Including some contrast is important—both technically and aesthetically. A picture which consists only of medium brightness values will look tired and washed out, without the punch and snap of a contrasty scene. The camera's contrast range is one of those operating characteristics which must always be approached from an aesthetic, as well as a technical, standpoint. The most important consideration is to be certain that the principal portion of the picture—which is generally the subject's face—falls well within the contrast

range. Also, avoid having excessively bright and dark areas overlap, which may create additional shading problems.

Sometimes, particularly on out-of-studio productions, it is impossible to control for excessively wide contrast ratios. For example, many outdoor athletic stadiums develop a deep shadow over a portion of the playing field during late afternoon. Often a player will start in bright sunlight and run into the shadow area. As the camera follows the player the extremely wide difference in contrast becomes too much for the camera to handle and the picture either blooms, overexposing the lighter portions, or completely darkens in the blacks, losing all definition and detail. To avoid this, many top-level cameras are equipped with a special "contrast compression" device, which electronically increases the cameras's contrast handling capability and permits the camera to capture picture detail in the shadow and dark areas without overexposing the white highlights.

The contrast range is of particular concern to the lighting director and the scenic designer since they must plan the lighting and sets around the contrast capabilties of the television camera. We will consider contrast range and its effects on lighting and set design in later chapters.

Resolution

The *resolution,* or clarity of the picture is determined by a number of interacting factors, including the contrast range, the lens system, and most importantly, the camera's pickup tube. Most medium and top-level cameras employ an *image enhancer,* which is an electronic device that improves the picture's resolution and detail. The enhancers are usually contained inside each camera chain's CCU although some studios utilize a single enhancer, which is attached to the output of the program switcher and enhances the final video signal as it leaves the control room.

Even with modern pickup tubes and image enhancers, a conventional, 525-line television picture cannot compare in sharpness and detail with 35 mm motion picture film. You should always keep the resolution limitations of the television medium in mind, particularly when approaching such production decisions as the shot size and the design and use of graphics, lighting, and sets.

TYPES OF TELEVISION CAMERAS

Only a few years ago, comparatively few camera models were manufactured, and those were either top-quality broadcast models or "industrial" cameras which were developed for nonbroadcast, closed-circuit use. Today there is a wide variety of cameras for every price and performance range, and what is even more significant is that every one produces an excellent picture within its respective price level. Although there is little question that more expensive cameras produce technically superior pictures, the remarkable fact is that even the most inexpensive professional television camera available today delivers a picture quality which is equal and, in some cases, superior to images produced by the most expensive broadcast cameras a decade ago.

The basic differences among the cameras of various price ranges are the additional features which offer you greater production flexibility and, of course, better picture quality. You will quickly discover that all cameras, regardless of model or manufacturer, operate pretty much the same way. Once you have learned how to use one camera, learning to operate another is easy. It usually takes only a brief orientation to a new camera before you feel perfectly comfortable operating it.

With so many cameras available, we have divided them into two broad categories: (1) studio cameras and (2) portable cameras.

Studio Cameras

The studio camera is the workhorse of the television industry. Although it is most commonly found inside the studio, it is also used outside in the field, where it is mounted in a relatively fixed position.

All studio cameras utilize a camera-control unit where the video levels are regulated by a video engineer during production. Studio cameras also

FIGURE 2-9 **Studio Camera —Low End.**
Hitachi 1212 is a three-tube color camera, which is inexpensive and produces high-quality pictures. *(Courtesy: Hitachi Denshi America, Ltd.)*

require conventional ac power for operation and are generally mounted atop pedestal devices which enable the operator to wheel them around the studio floor to set up various shots during a show.

LOW-END STUDIO CAMERAS   Low-end studio cameras are the Instamatics of the camera line, without frills or many features. Nevertheless, pound for pound and dollar for dollar they produce excellent pictures although they require optimum lighting and production conditions in order to deliver consistently good quality images. Low-end cameras utilize single-gun systems as well as three-gun systems and can take advantage of many of the most advanced pickup tubes such as the Saticon and Plumbicon for superior picture performance. These cameras cost between $3,500 and $20,000 and can be found in many closed-circuit facilities as well as at smaller production studios and broadcast stations.

MIDDLE-RANGE STUDIO CAMERAS   Middle-range studio cameras are heavier and more sophisticated than less expensive models and often incorporate some of the advanced electronic features found on the top-line camera models. Cameras in the middle-range have a much broader operational capability than low-end cameras and can tolerate much lower light levels and more extreme production conditions without sacrificing picture quality.

These cameras are used in many medium-to-large closed-circuit facilities and by broadcast stations and production houses. They generally employ three-gun pickup tube systems, and their overall picture quality is close to that provided by the most expensive camera models. They generally cost between $20,000 and $50,000 depending upon the camera model and the electronic and optical features included.

HIGH-END STUDIO CAMERAS   High-end studio cameras are the finest cameras available and can cost anywhere from $50,000 on up. The advanced electronic and optical systems contained in these cameras produce exceptionally beautiful pictures across a very wide range of production conditions.

Most cameras in this price range utilize digital control technology and offer microprocessor-assisted setup and registration. Cameras of this quality are found in most broadcast studio operations as well as in those sophisticated production houses and closed-circuit systems which require consistently excellent picture reproduction, true color fidelity, and high resolution.

Portable Cameras

The development of a lightweight, portable television camera has literally revolutionized television production. Portable cameras are used so often today that it is hard to realize that only a few years ago most television productions were confined inside the walls of a studio because it was too expensive and time-consuming to haul tons of heavy studio cameras and support equipment into the field for a location shoot. Not only have portable cameras completely changed the way we cover news, but they have also affected almost every kind of television production by providing the producer and director with the ability to leave the studio and shoot in the field when the production warrants it.

Although there are a variety of portable cameras currently available, they all have certain operational characteristics in common. They are powered by a battery pack for complete mobility, they do not require a camera control unit but feed the video signal directly to a videocassette or videotape recorder, and most have automatic gain controls which enable the camera operator to cover the action without riding video levels and still produce a technically acceptable picture. Since they are commonly used with a videotape or videocassette recorder, most portable cameras have controls for stopping and starting the tape machine and the ability to view previously recorded material through the camera's electronic viewfinder.

We will discuss portable cameras by dividing them into four categories: (1) ENG and EFP cameras, (2) convertible cameras, (3) video-

(a)

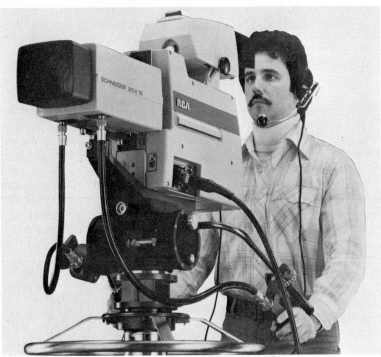

(b)

FIGURE 2-10 **Studio Camera—Middle Range.** Hitachi SK-96 *(a)* and the RCA TK-761 *(b)* are examples of midrange cameras which are capable of producing excellent quality pictures. *(Courtesy: Hitachi Denshi America, Ltd. and RCA Corp.)*

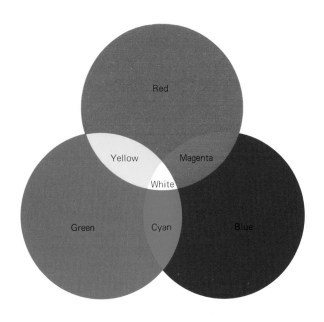

Color Plate 1 Additive Property of Color Light.
When two of the three primary colors of light are combined, they produce a complementary color. Where all three primary colors overlap equally, they produce white. Any color can be reproduced by combining the primary colors in varying degrees of hue, saturation, and brightness.

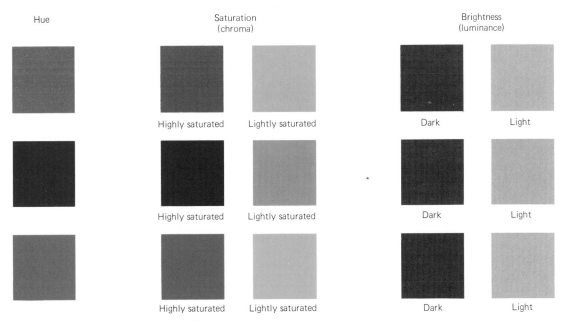

Color Plate 2 Hue, Saturation, and Brightness.
Hue refers to the tint of the color. Saturation (sometimes called "chroma") refers to how much or how little white light is mixed in. The square on the left is highly saturated with the pure hue and no white. The lightly saturated square shows how the same hue would appear when mixed with white. Brightness (or luminance) values show how the colors in the second column will reproduce on a black and white receiver.

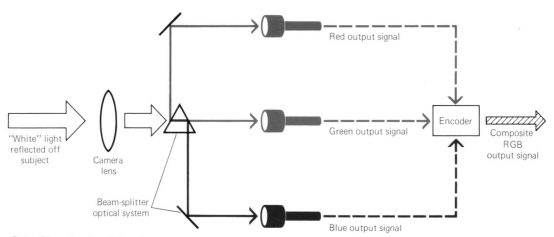

Color Plate 3   The Color Camera.
The light reflected from a subject contains many colors in varying combinations of hue, saturation, and brightness. The light is collected by the camera lens and focused onto a beam-splitter optical system, which dissects the "white" light into the three primary colors and sends each color to its respective pickup tube. The electronic output of each tube is fed into an encoder, which combines the three color (chrominance) signals with a brightness (luminance) signal to produce a composite video signal.

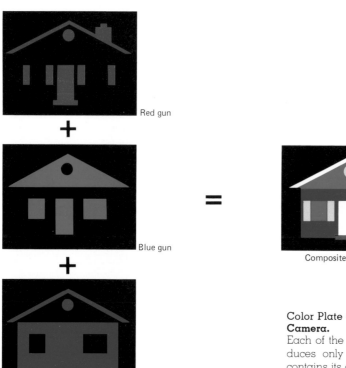

Composite color image

Color Plate 4   Operation of Three-Gun Color Camera.
Each of the three primary color pickup tubes reproduces only that part of the actual picture which contains its color. In this illustration, the images from the red gun, blue gun, and green gun are combined to produce the composite, full-color image.

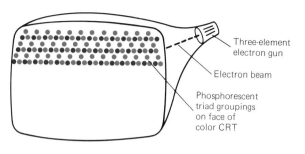

Color CRT operation

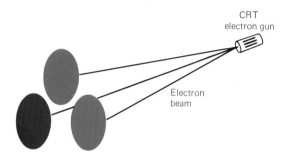

Three-element electron gun

Electron beam

Phosphorescent triad groupings on face of color CRT

CRT electron gun

Electron beam

Close-up of single phosphorescent color triad group

(a)

(b)

Color Plate 5 Color Cathode Ray Tube.
The first illustration shows how a single triad group of red, blue, and green phosphors are made to glow when struck by the CRT electron beam. The second diagram shows the configuration of triad groups as they appear on the face of the CRT.

Color Plate 6 High Definition Television (HDTV).
The photo (a) shot off a 525-line monitor shows limited resolution and discernible scanning lines. The photo (b) shot off a high definition monitor of the same subject using 1,125 scanning lines shows much higher resolution and sharper color. *(Courtesy: NHK)*

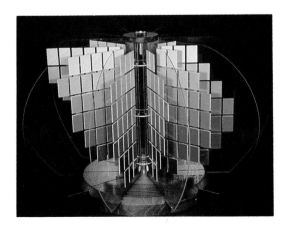

Color Plate 7 Munsell System.
The Munsell system is a convenient way for the designer to determine how variations in brightness and saturation will affect a particular hue. It is also useful when attempting to produce a color mix with sufficient contrast so that the hues are easily distinguishable on a monochrome receiver. The central axis is divided into nine brightness levels from black to white. The various hues revolve around the brightness axis; the farther from the axis a square lies, the more highly saturated the color. *(Courtesy: The Munsell Color Co.)*

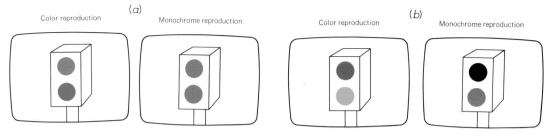

Color reproduction   (a)   Monochrome reproduction   Color reproduction   (b)   Monochrome reproduction

Color Plate 8   Color Contrast and Monochrome Reproduction.
In figure (a), the reproduction of a traffic light in color appears normal, but the same picture on a monochrome receiver shows no difference between red and green since the brightness values of the two colors are almost identical. Figure (b) shows the same light, but with a slightly different shade of green. Although the red and green appear normal in color, the difference in their respective brightness, as reproduced on a black and white monitor, is much more obvious to the viewer.

Color Plate 9   Color and Graphics.
The effective use of color can greatly enhance the impact and meaning of a graphic. *(Courtesy: Phyllis Essex–WPLG-TV and Ken Dyball)*

Color Plate 10   Digital Video Art.
Both these pictures were created entirely electronically using a digital video art system. The speed and flexibility of a digital video art system offer the graphic artist enormous creative capability. *(Courtesy: MCI/Quantel)*

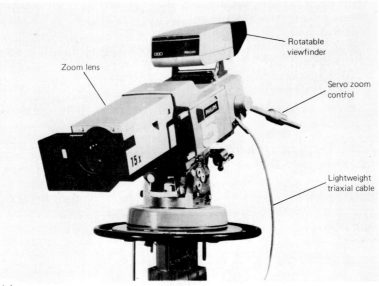

Zoom lens

Rotatable
viewfinder

Servo zoom
control

Lightweight
triaxial cable

(a)

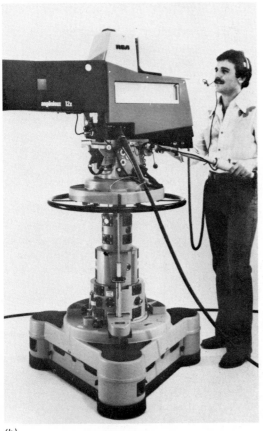

(b)

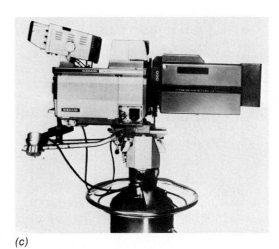

(c)

FIGURE 2-11 **Studio Camera—High End.**
The Phillips LDK-5 *(a)* RCA TK-47 *(b)* and Ikegami
HK-322 *(c)* are examples of state-of-the-art cameras.
(Courtesy: Phillips Corp., RCA Corp., and Ikegami.)

recording cameras, and (4) electronic cinematography cameras.

ENG AND EFP CAMERAS   The primary reason for the development of the portable camera was to replace 16-mm motion picture film for television news production. Electronic news gathering (ENG) employs a battery-powered camera and videocassette recorder to record both sound and picture in the field. There are generally dozens of ENG cameras currently available, and they range in price from relatively inexpensive single-gun cameras to sophisticated three-gun cameras which can produce a picture that rivals the most expensive studio camera. Most ENG cameras offer automatic operation to enable the camera operator to maintain his or her attention on covering the story without having to worry about the camera's video level. Automatic white balance, black balance, and iris controls are standard on most ENG cameras.

Electronic field production (EFP) is really an extension of ENG operation, but EFP work is usually considered any remote production with a single camera which does not include straight news. For example, producing training tapes, commercials, segments which will be inserted within a studio produced production, and documentaries or magazine-type feature programs are all examples of EFP. Although EFP often uses the same cameras as ENG, some production facilities utilize higher-quality cameras for EFP in order to produce a better quality image, which may be

(a)

(b)

(c)

FIGURE 2-12   **ENG and EFP Cameras.**
The Hitachi FP-10 (a) is a single-gun color camera. The Ikegami (b) and RCA TK-86 (c) are three-gun cameras commonly used for both ENG and EFP production applications. (Courtesy: Hitachi Denshi America, Ltd., Ikegami, and RCA Corp.)

combined with footage shot by top-quality studio cameras. Some EFP cameras are capable of being connected to a camera control unit which offers manual shading of the video level and can produce a better quality image than is possible with automatic gain controls.

ENG and EFP cameras generally weigh between 6 and 20 pounds, depending upon the number of pickup tubes. Although they are commonly powered off of a battery belt, they can also be run off AC power by using an adapter.

CONVERTIBLE CAMERAS A number of high-end ENG/EFP cameras are capable of being converted from a portable configuration into a studio-type camera head complete with a large viewfinder and sophisticated zoom lens system. The advantage of the convertible camera is basically getting two cameras for the price of one. On a remote production, the camera can be used in its portable configuration, and back in the studio it can be converted into a studio camera, hooked up to its camera-control unit and operated like any studio camera. Many production facilities

which do a great many remotes requiring multiple cameras, such as covering sports events or special entertainment events, use the convertible camera in its studio configuration. The convertible camera is much lighter in weight than a studio camera, which makes setting and striking a number of cameras much easier.

VIDEO RECORDER CAMERA The video recorder camera (VRC) is actually an ENG camera with a built-in videocassette recorder. Conventional ENG cameras must be connected to the videocassette recorder by a cable which limits the camera operator's mobility and could short out after being subjected to the stress of a camera operator's pulling against the VCR cable connection. With a single, integrated unit, there are no external cables, and a single operator can control both camera and video-recording operations.

The camera operates like a conventional ENG camera and utilizes ½-inch Plumbicon or Saticon tubes to reduce the size and weight of the unit. The VCR uses ½-inch videocassettes, but these are not compatible with either conventional VHS or Beta VCR units and must be played and edited on special videocassette recorders. The combination of videocassette recorder and camera brings the weight of the total package to about 22 pounds. (See Figure 2-14.)

ELECTRONIC CINEMATOGRAPHY CAMERA The electronic cinematography (EC) camera is a special purpose camera designed to replace 35 mm motion picture cameras for use in film-style video production. The camera looks more like a film camera than a video camera and is, in fact, designed to produce a color image with characteristics more like motion picture film than television. Of course, the resolution limitations of the 525-line video system do not make the television picture as sharp as one produced with 35 mm film, but the production flexibility and cost savings possible with electronic production and postproduction techniques are significant.

FIGURE 2-13 **Convertible Camera.**
This Ampex BCC-20 camera can be converted from a portable ENG/EFP camera *(top)* to a full-sized studio camera. *(bottom) (Courtesy: Ampex Corp.)*

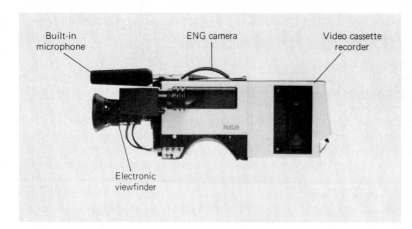

Built-in microphone

ENG camera

Video cassette recorder

Electronic viewfinder

FIGURE 2-14 **Video Recorder/Camera.** The VRC integrates an ENG camera with a lightweight videocassette recorder which eliminates the needs for an external VCR and permits a single operator to control both camera and recording. *(Courtesy: RCA Corp.)*

FIGURE 2-15 **Electronic Cinematography Camera.** The camera is designed to operate much like a motion picture camera and produces a picture with image characteristics which look much like that obtained with film. *(Courtesy: Ikegami.)*

The EC camera is designed to operate much like a motion picture camera. The camera accepts movie-type fixed focal length lenses, which can be pre-set to a particular f-stop to enable the camera operator to establish the appropriate depth of field for each shot. Although the EC camera is designed for very special purpose applications and will not be found in most television production facilities, its development is a clear indication of the movement toward replacing motion picture film with electronic video production techniques. (See Figure 2-15.)

CAMERA MOUNTING EQUIPMENT

During a production, the camera operator must be able to move the camera with relative ease and with maximum flexibility. Even though a large studio camera may weigh well over 100 pounds, it must be moved, panned, and tilted quickly and smoothly without jerking or shaking. This is made possible by attaching the camera to a special camera head and mounting device which permits smooth and flexible operation.

Camera Mounting Head

The camera itself rests on a *mounting head,* which is attached to a tripod or pedestal. The mounting head's function is to hold the camera securely in place while permitting smooth rotation horizontally (panning) and vertically (tilting).

Some heads use a *wedge mount,* which simplifies the mounting operation of the camera to the head. A plate with the male wedge is attached permanently to the underside of the camera. The camera and wedge are then inserted into the female slot, which is bolted to the top of the pedestal or tripod. This is a particularly convenient system if your cameras will often be used on remote productions where frequent mounting and

Head lock

Tilt control

Pan control

(a)

FIGURE 2-16 **Camera Wedge Mount.**
The camera wedge *(on left)* is attached to the camera. The wedge mount receptable (on right) is attached to the tripod or pedestal. The camera and wedge mount can then be easily and rapidly mounted or dismounted as necessary. *(Courtesy: Innovative Television Equipment Corp.)*

(b)

FIGURE 2-17 **Friction and Fluid Heads.**
(a) Friction-head camera mount which uses a strong spring to control movement. *(b)* Fluid-head mount which uses a chamber filled with a viscous fluid to control movement. *(Courtesy: Innovative Television Equipment Corp. and Listec Television Equipment Corp.)*

dismounting must be done quickly and safely. (See Figure 2-16.)

FRICTION AND FLUID HEADS   The simplest and most basic mounting is the *friction head,* which uses a heavy spring to control the camera's movement. Friction heads are not recommended for most studio cameras since they offer little control over such heavy equipment and have a tendency to overbalance when the camera is tilted at extreme angles. *Fluid head* tripods offer more control and result in smoother camera movements. However, they are also subject to overbalancing when used with heavy cameras. Friction and fluid heads both have two controls which enable the camera operator to set the amount of friction on the pan and tilt for the smoothest movement. (See Figure 2-17.)

CRADLE HEAD   The cradle head is designed to balance a heavy studio camera automatically

even when it is tilted at the most extreme vertical angle. Two controls on the side of the cradle head permit the camera operator to adjust the amount of "drag" or friction over the pan and tilt. A separate control locks the cradle into any preset position. (See Figure 2-18.)

CAM HEAD   The cam head utilizes a number of "cams" or cylinders to control the camera's pan and tilt movements. These heads are especially designed for heavy studio/field cameras and offer extremely smooth, effortless pan and tilt control. The cam head is equipped with both pan and tilt drag controls to adjust the friction and with brake controls to lock the camera securely into any position. (See Figure 2-19.)

Camera Tripod

The simplest camera mounting device is the tripod. Tripods are generally lightweight and collapsible, which makes them ideal for use on remote productions. However, the operating height of the camera cannot be varied during the production, which is a serious limitation. Tripods with wheels attached to the legs are more flexible

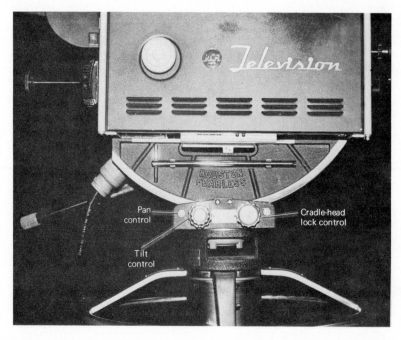

FIGURE 2-18   Cradle Head.

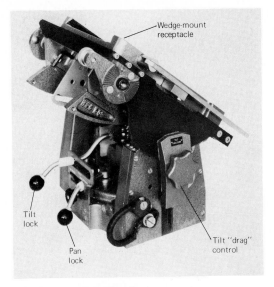

FIGURE 2-19 **Cam Head.**
(Courtesy: Listec Television Equipment Corp.)

than simple tripods since the camera can be moved around the floor. If stationary operation is desired, the wheels can be locked to prevent the tripod from moving. (See Figure 2-20.)

Camera Pedestals

Camera pedestals are the most commonly used camera mounting devices and are found in almost all production studios. The pedestals are designed to move smoothly across the studio floor and to permit the camera operator to easily change the camera's height, even while on the air.

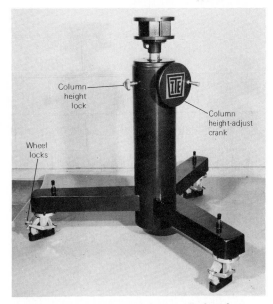

FIGURE 2-21 **Studio/Field Camera Pedestal.**
(Courtesy: Innovative Television Equipment Corp.)

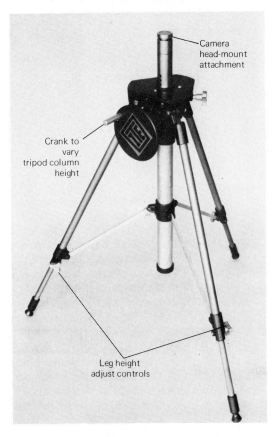

FIGURE 2-20 **Camera Tripod.**
(Courtesy: Innovative Television Equipment Corp.)

Cradle
head

Column
lift/steering
ring

Column lock
ring

Weight system
travels up
and down

Counterweights

Steering mode controls
for "tracking" or "crabbing"
operation

Cable
connector

Cable
guard
skirts

FIGURE 2-22 Counter-
weight Pedestal.
With the cover opened, the
counterweights are visible.
Weights can be added or
subtracted to accommodate
specific cameras.

STUDIO/FIELD PEDESTAL The studio/field pedestal is a combination tripod and pedestal. It is sufficiently lightweight to be easily carried to remote locations, yet its oversized wheels permit the camera operator to move the camera and pedestal smoothly across a studio floor. The camera's height can be varied by pumping air or turning a crank to raise the vertical column, but the operation is jerky, and the camera's height cannot be changed while its shot is on the air. (See Figure 2-21.)

COUNTERWEIGHT PEDESTAL The counter-

weight pedestal is one of the most commonly used and most flexible mounting devices. The camera is mounted atop the pedestal's center column, which can be raised or lowered effortlessly by a slight pressure from the camera operator on the large center lifting ring. The pedestal is counterbalanced with weights and springs and is adjustable for each individual camera weight, enabling you to vary the camera height even while on the air.

The pedestal is mounted on three pairs of wheels, which can be adjusted to steer in two different modes. The most common method is to lock all three pairs so they steer in parallel. This is called "crabbing" and permits you to move the pedestal smoothly in any direction. However, if you must rotate the pedestal itself, then you have to change the steering to the "tracking" mode, in which the first wheel steers while the rear wheels remain locked in parallel, much like a child's tricycle. The selection of steering modes is accomplished by a foot switch located at the base of the pedestal. The camera operator steers the pedestal by turning the large center ring in the appropriate direction. The wheels of the pedestal are covered with cable guard skirts, which prevent the pedestal from running over cables on the studio floor. (See Figure 2-22.)

The pedestal's major advantages are its ease of operation and its even movement across the studio floor, resulting in smooth dollies and trucks. However, because the pedestal is large and heavy, coordinating camera movement and pedestal operation takes some practice. Also, its large size and extremely heavy weight make it impractical to use on remote locations.

PNEUMATIC PEDESTAL   A more sophisticated version of the counterweight pedestal is the pneumatic pedestal, which uses compressed air to operate the action of the telescoping center column to change the camera's height. The operation of the pedestal is very flexible. Some permit the camera to vary over 3 feet, from as low as 21 inches off the studio floor to a maximum height of 58 inches.

The operation of the pneumatic pedestal is identical to the counterweight pedestal except that the center column height is controlled through the use of compressed air, which is carried in an air tank. Since the pedestal uses air and not heavy counterweights, the pneumatic pedestal is much lighter than the counterweight design. This makes it easier for the camera operator to move and guide the camera and pedestal across the studio floor. (See Figure 2-23.)

Cranes and Dollies

Although the camera pedestal provides sufficient flexibility for most in-studio requirements, some productions require the camera to operate from angles which are inaccessible with the studio

pedestal. For these special purpose applications, cranes and dollies can be used. However, their increased cost, whether purchased or rented, and the additional manpower needed for their operation are serious disadvantages.

CRAB DOLLY   The crab dolly had been borrowed from the motion picture industry where it is the mainstay of film camera mounting devices. The camera is mounted atop a small arm, which is raised or lowered electrically and provides a wider range of operating height than with conven-

FIGURE 2-23   **Pneumatic Pedestal.**
(Courtesy: Listec Television Equipment Corp.)

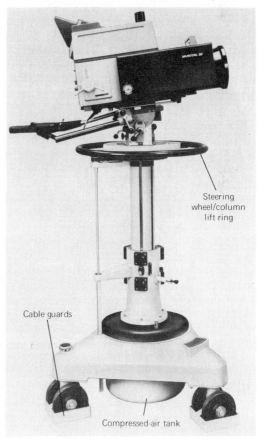

Steering wheel/column lift ring

Cable guards

Compressed-air tank

tional pedestals. Most crab dollies permit the camera to operate from a minimum height of 14 inches to a maximum height of 61 inches.

The camera operator sits on a small seat next to the camera and rides with the arm as it travels up and down. A second operator pushes and steers the crab dolly across the studio floor. The dolly's four wheels move in a parallel or "crabbing" fashion, hence its name.

The advantage of the crab dolly is increased camera flexibility and very smooth dolly movement, but the additional cost in manpower and the size of the equipment are major disadvantages.

CAMERA CRANE The camera crane is a huge mounting device with four pairs of wheels on a base and a large arm extending outward. The crane requires a very large studio with a high ceiling to make the most of its operating potential. It requires a crew of four: a camera operator who sits at the end of the crane arm, a crane operator who steers the crane, and two arm operators who

manually lift and pan the counterweighted crane arm. (See Figure 2-24.)

Special Purpose Mounting Equipment

Occasionally, a director will require extreme camera angles which cannot be achieved with conventional equipment. In this case, a variety of special mounting devices can be used to position the camera for the shot.

HIGH-HAT AND FLOOR SLED A *high-hat* is a small mount, only a few inches high. The camera and camera head are mounted atop the high-hat, and its extremely low height permits dramatic low-angle shots. In some cases, the high-hat is mounted atop a small dolly called a "floor sled," which permits the camera to be wheeled across the floor, increasing its shooting flexibility.

The high-hat and floor sled are relatively inexpensive to buy or to build. Once the camera is mounted on the equipment, it can shoot only from the extreme low angle. This reduces the number of cameras available for more conventional shots.

BODY MOUNTS In many ways, the portable

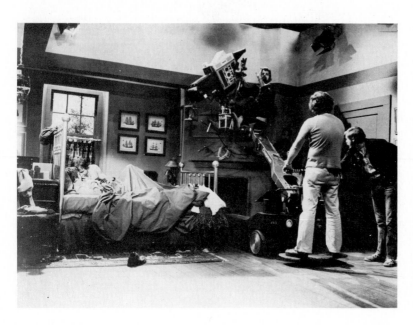

FIGURE 2-24 **Camera Crane.**
The camera crane enables the camera to photograph a scene from a variety of perspectives and angles. It also permits very fluid movement by coordinating the movement of the crane on its wheels and the vertical/horizontal movement of the crane arm. *(Courtesy: PBS)*

camera can take advantage of the best mounting device of all: the human body. When used by a skillful operator, hand-held cameras offer enormous versatility in shooting angles. Imaginatively used, the portable camera can substitute for such special mounting devices as the high-hat and even, with sufficient ingenuity, the studio crane.

The body mount attaches around the operator's waist, distributing the camera's weight and keeping it balanced. Even though a camera may be "portable," it is still heavy and gets heavier after only short periods of use. The body mount helps you keep the entire weight of the camera off your arms and shoulders and results in much smoother camera movement.

STABILIZERS Some special body mounts are equipped with a series of "servostabilizers," which absorb shake and jitter, maintaining a rock-steady picture even when the operator is in motion. The body stabilizer allows the camera operator to replace pedestals, dollies, and cranes while delivering dolly-smooth, jitter-free, hand-held moving shots. An added advantage is that the body stabilizer can be used to shoot from moving cars, trucks, trains, or helicopters, turning almost any vehicle on location into an instant camera platform. (See Firgure 2-25.)

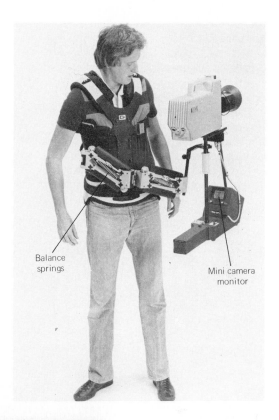

Balance springs

Mini camera monitor

FIGURE 2-25 **Camera Stabilizer.**
The Steadicam enables the camera operator to achieve rock-steady pictures even while moving or shooting from difficult positions. *(Courtesy: Cinema Products Corp.)*

SUMMARY

The camera is the basic instrument of television production. Because of its central role, almost every other production decision must revolve around its capabilities and limitations.

The camera chain consists of two major systems: (1) the camera head—including the lens, internal optics, electronic pickup tube and amplifiers, viewfinder, and camera communications systems and (2) the camera-control unit, or CCU—containing all the necessary equipment to set up and control the camera's operation.

The camera's pickup tube is its most important component. Modern production cameras use either the lead-oxide, Plumbicon tube, or a variety of vidicon-type tubes.

Three important operating characteristics define the camera's capabilities and limitations: (1) the operating light level, (2) the contrast range, and (3) the camera's picture resolution. Most modern cameras are capable of operating under widely varying light levels, some with light as low as 5 or 10 fc. For optimum picture reproduction, the contrast range—the difference between the brightest and darkest areas of the picture—should not exceed a 30:1 ratio. The camera's resolution—its ability to distinguish detail in the picture—is limited, even with electronic picture enhancement. Camera shots, angles, and design considerations must be planned to take this restriction into account.

Cameras are grouped into two major designs: (1) studio cameras, which operate on a tripod or pedestal, and (2) portable cameras, which are lighter in weight than studio cameras and are operated hand-held, shoulder-mounted, or in a convertible design as either a mounted studio/field camera or a portable shoulder-mounted camera. Most portable cameras operate on battery power.

The most common camera-mounting equipment is the pedestal, which, operated either with counterweights or with air pressure, enables the operator to vary the camera's operating height smoothly even while on the air. For portability on remote productions, tripods or studio/field pedestals are used. Such special mounting devices as camera cranes, dollies, and high-hats offer dramatic camera angles and flexible movement. For portable cameras, various body braces are available, some with built-in stabilizers which eliminate shake and jitter and produce rock-steady, hand-held or shoulder-mounted shots.

CHAPTER 3

TELEVISION LENSES

Before the television camera can begin to scan and reproduce the image, a picture must be formed by the camera lens. While the primary function of the lens is to produce a sharp image of the subject on the face of the camera tube, the lens actually does far more than that. The type of lens selected and the way it is used determine the magnification and size of the subject and the field of the view which is photographed and establish the viewer's visual perspective of the scene. Since the lens is a primary production tool, you should understand its physical properties and its optical characteristics.

PHYSICAL CHARACTERISTICS OF LENSES

A lens is made up of three basic component parts: (1) the optical elements, (2) a variable iris, and (3) the mounting system which is used to attach the lens to the camera head.

Elements

A lens contains a number of optical elements which are composed inside a hollow metal cylinder. The elements are specially coated glass components which collect the reflected light rays from the subject and focus the rays through the rear of the lens into the camera. The internal elements not only magnify and focus the image, but also correct for optical and color aberrations which occur whenever light rays are bent or "refracted." Whenever we speak about a single "lens," we are really referring to a complex optical system consisting of a number of elements designed and constructed to operate together to collect, magnify, and focus the light rays onto a photographic surface or, in the case of television, onto the face of the pickup tube.

As light travels through the various glass components, some of the rays are absorbed by the interior elements of the lens before they can reach the camera. This is why no lens made can transmit 100 percent of the light which enters it. Top-quality lenses used in television production are capable of transmitting about 75 to 80 percent of the original light, which means about 20 percent of the light which enters the lens is lost before it reaches the camera. Generally, the more complex the lens, the more internal optical elements and the less overall light which reaches the pickup tube. This is why some very sophisticated zoom lenses or high-powered telephoto lenses require higher light levels since they tend to absorb some of the original light from the scene.

Iris (Aperture)

As we walk between a bright area outdoors and a dark, indoor area, the pupils in our eyes enlarge or diminish in diameter to vary the amount of light which reaches the retina. This enables our eyes to function effectively across a variety of light level conditions. Lenses also have the ability to vary the amount of light which enters the camera to accommodate for different levels of illumination by changing the opening of the *iris*, or *aperture*. When the circular iris opening is completely wide open, the lens transmits the maximum amount of light into the camera. If we close or "stop down" the iris, diminishing the size of the aperture, we permit less light to enter the camera. (See Figure 3-2.)

The size of the iris opening has been calibrated as a series of numerical *f-stops*, which run from about f/1.4 to f/22. As the numerical f-stop *increases*, the amount of light which enters the camera *decreases* because we are closing the opening of the iris. Therefore, a lens set at f/2 would permit much more light to enter the camera than a lens set at f/16.

Lens Mounting Devices

Most modern color cameras utilize a zoom lens which has been specifically matched to the camera's internal optical/pickup tube system, and it is generally purchased along with the camera head. However, many color cameras and all monochrome cameras are also capable of being fitted

FIGURE 3-1 **Cross Section of a Zoom Lens.**
Each of the individual glass elements is designed to modify the image which enters the lens and to correct for optical distortion. Zooming in or out moves certain elements to vary the optical focal length and to change picture size and field of view.

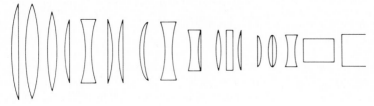

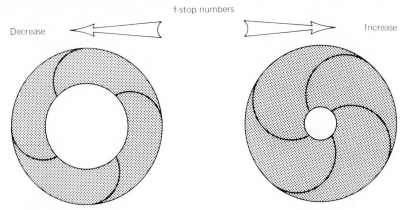

f-stop numbers

Decrease

Increase

FIGURE 3-2 Lens Iris.
This iris, or aperture *(on the left)* is open wide, permitting more light to enter the camera. The iris *(on the right)* is closed or "stopped down," reducing the amount of light which can reach the television camera pickup tube.

with a variety of different lenses, increasing their production flexibility. Two frequently used mounting systems you should be familiar with are (1) the *C-mount*, which is a screw-type mounting system used on most 16-mm film camera lenses and adapted for television use, and (2) the *bayonet lock mount*, which permits lenses to be quickly mounted and dismounted by fitting the rear of the lens into a keyway slot in the camera head, securing the lens in place with a half-twist.

CAMERA MOUNT ADAPTERS Although most production situations can be adequately covered with the usual complement of television camera lenses, there are times when certain visual effects call for the use of special lenses which are produced primarily for motion picture work. In this case, the lens must be coupled with a camera mount adapter which permits its use on a television camera.

Lens Formats

You should recall that different types of camera pickup tubes have different sized tube faces. For this reason it is important to match the image size produced by the lens to the corresponding format size of the camera tube. Otherwise, the image formed by the lens will be either too large or too small for the camera tube, resulting in an inferior

picture. Since the face sizes of 1¼-, 1-, ⅔-inch, and ½-inch Plumbicon and vidicon tubes closely resemble the size of a 16-mm film frame, cameras equipped with these pickup tubes all use "16-mm format" lenses. The larger 3- and 4½-inch I-O (Image-Orthicon) tube faces more closely match the size of a 35-mm film frame, and cameras equipped with these pickup tubes use "35-mm format" lenses. Sixteen-millimeter format lenses are smaller, lighter, and less costly than their 35-mm format counterparts, which is a primary advantage in using a camera that utilizes the smaller format lens.

OPTICAL CHARACTERISTICS OF LENSES

All lenses have three important optical characteristics which determine the size and magnification of the image, produce the horizontal view of the scene photographed, and establish the visual perspective of the shot. These three characteristics are (1) focal length, (2) f-stop, and (3) focus and depth of field.

Focal Length

The focal length is a basic lens quality which determines the amount of image magnification and the horizontal field of view which the lens photographs. The focal length is determined by measuring the distance from the "optical center" of the lens (which is not always its physical center) to the point at which the light rays converge at the rear of the lens to produce a perfectly focused image which, in a video camera, is the face of the pickup tube. The distance is measured in millimeters (25 mm=1 inch), so if a particular lens requires a distance of 2 inches to focus the image, it is a 50-mm lens. (See Figure 3-3.)

The shorter the focal length, the wider the horizontal field of view and the smaller the image. As the focal length increases, the horizontal field of view narrows, and subject magnification increases. We see less of the overall scene, but what is included in the shot is greatly enlarged.

Lenses are frequently described in terms of their focal length. A "normal angle" or "medium" lens reproduces a scene much as our eyes do. Its horizontal field of view and image magnification are about the same as if you were standing where the camera is. In the 16-mm lens format size, a normal or medium lens is about 25 mm. Lenses with a focal length shorter than 25 mm are called "wide-angle" or "short" lenses because the lens produces a wide horizontal field of view and decreases subject magnification. Lenses with focal lengths greater than 25 mm are called "narrow-angle" or "long" lenses since they provide a narrow horizontal field of view but magnify and enlarge the subject from long distances. Lenses with focal lengths of 100 mm and beyond are frequently called "telephoto" lenses. They are used to shoot subjects from great distances and to enlarge and magnify the subject, much as a telescope does.

HORIZONTAL FIELD OF VIEW The horizontal field of view of a lens tells you how wide a shot the lens will deliver. Knowing the horizontal field of view for any particular lens in advance is of great help to the director in preplanning shots and camera locations. You can determine the horizontal angle which any lens will produce by using the following simple formula:

$$\frac{676}{\text{Focal length in mm}} = \begin{array}{c}\text{Horizontal angle of the lens}\\ \text{(in 16-mm format)}\end{array}$$

If we were using a 25-mm lens, then 676/25=27°. This means that the horizontal angle of

FIGURE 3-3 **Focal Length.**
A lens with a long focal length requires a greater distance to focus the image on the focal plane, which is the face of the pickup tube. A short focal length lens requires a shorter distance to focus the light rays. Focal length is measured in millimeters.

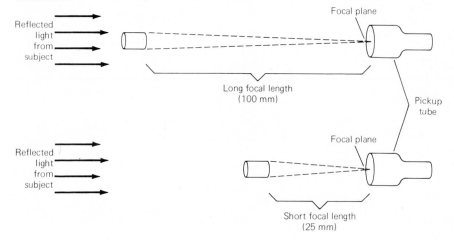

Iris ring
for setting
f-stop

Focusing
ring for
setting
distance

Depth of
field
indicator

FIGURE 3-4   Fixed Focal Length Lens.

"Take" lens
in front of
pickup tube

FIGURE 3-5   Lens Turret.

the lens is 27° or that the lens will include all objects within a 27° angle. Later on, we will discuss how to use this information in planning shots and camera angles with a studio floorplan.

Fixed Focal Length Lenses

A *fixed focal length* lens, or a "primary" lens, is designed to operate at a predetermined focal length, producing a fixed amount of image magnification and horizontal field of view. Primary lenses are available in a wide variety of focal lengths from extreme wide angle lenses with focal lengths as low as 6 mm to extreme telephoto lenses with focal lengths as long as 1,000 mm and beyond.

Older monochrome cameras are equipped with a lens *turret*, which holds a complement of four fixed focal length lenses to provide the camera operator with a variety of different camera shots. The camera operator simply flips the turret with a handle at the rear of the camera to position any one of the lenses in front of the pickup tube to achieve the desired magnification and field of view. The problem with the turret is that although the four lenses are selected to encompass a wide range of focal lengths—usually a wide-angle, medium, long and telephoto lens complement—simply changing to a new lens will rarely produce

a well-framed shot. The camera operator must also physically move the camera closer to or farther from the subject in order to compose the picture precisely. This is time-consuming and inconvenient and impossible on those remote or in-studio productions where the camera's movement capability is severely limited or not possible at all. (See Figure 3-5.)

For these reasons, fixed lenses have been replaced by the more flexible variable focal length or "zoom" lens. However, some special purpose fixed focal length lenses—such as the extreme wide-angle "fisheye" lens, or the extreme telephoto lens—are still used on studio or portable cameras to achieve a unique visual perspective.

Zoom Lenses

The most popular lens used in television production is the variable focal length or *zoom* lens. The zoom lens can be continuously varied throughout its entire focal length range from its widest horizontal angle to its narrowest, or longest focal length. You can start at any desired focal length and zoom in or out, varying the magnification of the subject and the size of the horizontal field of view at any desired speed. The large number of focal lengths which are instantly available with the

FIGURE 3-6 **Television Zoom Lens.**
*(a)*Exterior; *(b)*Interior with case removed. The optical elements are contained inside the tube. The circuit board and mechanical gears are part of the electroservo system which controls focal range, focus, and iris opening. *(Courtesy: Angenieux Optical.)*

(a)

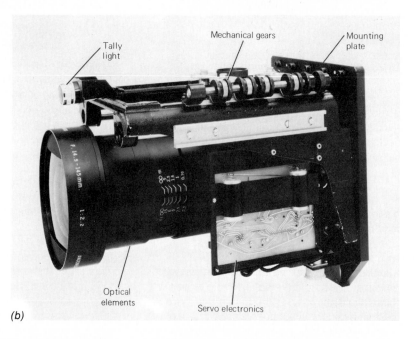

(b)

zoom lens makes it easy for the camera operator to frame a shot precisely without moving the camera. (See Figure 3-6.)

ZOOM RANGE A zoom lens is commonly described with a two-number figure such as "10×12" (pronounced "ten by twelve"). The first number, "10," is the *zoom range* and is the ratio of the shortest focal length of the lens to the longest focal length. Here it is a 10:1 ratio. The second number, "12," is the shortest focal length in

millimeters and is the zoom lens's widest angle. Knowing this information makes it easy to see that our 10×12 lens is capable of zooming from 12 mm (the widest possible focal length) to 120 mm (the longest possible focal length).

Most studio cameras are equipped with lenses with a zoom range of about 10:1 or 15:1. These lenses combine the necessary wide angle for studio work with a sufficiently long focal length. Since in-studio cameras most often require the widest possible angle at the short end of the range, studio zooms are usually made with the maximum wide angle as an important consideraton. However, if you will be shooting many remote productions where telephoto capability becomes much more important in order to magnify subjects from great distances and bring the action closer, you may wish to sacrifice the extreme wide angle for more magnification power at the longer end of the zoom range. A number of manufacturers have produced a "universal" zoom lens, which provides the camera operator with the best of both worlds—the short end of the lens has a sufficiently wide angle for in-studio use, and the same lens has a very powerful long range for telephoto field work. Some of these universal zooms have ratios as large as 42:1, with a wide-angle capability of 12.5 mm and a maximum long lens of 525 mm.

As you've undoubtedly realized by now, a zoom lens is capable of producing a virtually unlimited number of focal lengths within the extremes of its zoom range. Whenever we use the zoom at the wide-angle end of the range, we will produce pictures with all of the characteristics of a wide-angle lens. Similarly, if we use the zoom toward the telephoto end, we produce pictures with the characteristics of long lenses. Throughout the remainder of this chapter, whenever we refer to a "wide," "medium," or "long" lens, we are really referring to the characteristics of the image which are produced by the focal length we are using. Whether the shot is produced by a fixed focal length lens or, as is far more common today, by a zoom lens which is positioned at a particular focal point along its zoom range, the visual characteristics and the production applications remain identical.

RANGE EXTENDERS Range extenders, or "multipliers," are supplementary optical devices which are used to extend the range or focal length of a zoom lens. *Fixed extenders* have a predetermined power, which is rated in terms of the amount of additional magnification provided. For example, a "2×" (pronounced, "two times") extender on a normal 12-mm to 120-mm lens will double the focal length across the entire zoom range producing, in effect, a 24-mm to 240-mm lens. Some range extenders are *variable* rather than fixed and permit the camera operator to

change their multipying power continously. These "double zooms" give the camera operator the flexibility to select the exact amount of multiplier magnification power for every shot.

Using a range extender is always a compromise since the extender increases focal length at the expense of the lens's operating light level. A 2× range extender will reduce by four times the amount of light that would have entered the camera without the extender. This can be a serious problem on remote telecasts when light levels, which cannot be easily controlled, get critically low.

ZOOM LENS OPERATION Zoom lenses come in two varieties: (1) *manual zooms* in which the camera operator controls the angle and speed of the zoom by pushing a control rod or by turning a handle-crank, and (2) *servo zooms*, in which the angle and speed of the zoom shot are controlled by operating a switch which activates a small electronic motor inside the lens.

MANUAL OPERATION Manual zooms are operated by either a handle or rod control.

ZOOM HANDLE CONTROL Most manual zooms are controlled with a crank attached to the right-hand pan handle at the rear of the camera. Turning the crank varies the focal length of the lens. Many controls also have a two-gear switch which permits either "slow" or "fast" zooms by changing the number of turns necessary to cover the entire zoom range. Focus is controlled by rotating a handle-grip located on the left-hand pan bar. (See Figure 3-7.)

ZOOM ROD CONTROL Less expensive zoom lenses utilize a control rod which is pushed in or pulled out to vary the focal length of the zoom lens. You do not have as much control over the zoom's speed with a rod, and pushing and pulling the rod evenly to obtain a smooth, on-air zoom is often difficult. Focus is achieved by rotating the knob at the end of the zoom rod. (See Figure 3-8.)

Zoom control cables

Zoom crank

Zoom focus

FIGURE 3-7 **Zoom Handle Control.**

Zoom rod

Zoom-lens focus control

FIGURE 3-8 **Zoom Rod Control.**

SERVO ZOOM OPERATION Lenses equipped with a servo system utilize a tiny electric motor inside the lens to control both the focal length and the focus. The zoom range is controlled with a thumb switch, located on the right-hand pan bar, which is pushed either right or left to zoom in or out. The farther from the central position you push the switch, the faster the zoom's rate of speed. Focusing is also accomplished electronically with a control on the left-hand pan handle. Because the servo system is electronic, the zooming speed need not depend on the operator's manual touch, and servo zooms produce exceptionally smooth and precise zooms.

Servo zooms are generally used with a *shot box* which enables the camera operator to preset a number of specific focal lengths, much as you preset your car radio for a number of different station frequencies. During the production, the camera operator simply presses the appropriate button on the shot box, and the lens automatically produces the preset focal length. (See Figure 3-9.)

CALIBRATING THE ZOOM A zoom lens can maintain perfect focus throughout its entire range only if you have "calibrated" or "prefocused" the lens before each series of shots. As the calibration operation is slightly different for built-in zooms on color cameras and for zooms on monochrome cameras, we will discuss each separately.

CALIBRATION ON THE COLOR CAMERA If your camera has a built-in zoom as all color cameras have, the calibration or prefocus operation is relatively simple:

1   Zoom the lens all the way in to the longest possible focal length (even though the shot may be too tight to use) and focus on the principal subject.
2   Use your focus control to produce a sharply focused image. The picture will remain in perfect focus throughout the entire zoom range, as long as neither the camera nor the subject moves too far from its original position. If either subject or camera does move or if you must set up for a shot on another subject at a different position, you must recalibrate your focus as soon as the director switches to

another camera shot. Good camera operators automatically calibrate their zoom focus each time they set up a new shot to assure accurate focusing across the entire zoom range.

CALIBRATING ON A MONOCHROME CAMERA If you are using a monochrome camera, you have two focus controls to operate: the focus for the zoom lens, and the camera's focus control which is found on the right-hand side of the camera. In this case, the prefocus operation is slightly different from that of the color camera:

1   Zoom the lens to the *widest* possible focal length.

FIGURE 3-9 Servo Zoom Control.
The shot box enables the camera operator to preset six specific lens focal lengths.

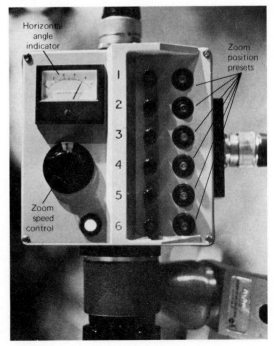

2 Using the *camera's* focus control, adjust for maximum focus.

3 Zoom the lens into its longest focal length, and use the *zoom's* focus control to focus the picture.

The zoom will now be in focus across the entire range provided camera and subject do not move too far away from their original positions. As with the color camera, each time you focus on a new shot, you should automatically prefocus your zoom lens to ensure accurate focus across the entire zoom range. The reason you have to first focus with the camera's focus control is that monochrome cameras require you to adjust for back-focus before focusing the lens. On color cameras, the back-focusing is preset when the lens is mounted on the camera and need not be adjusted during normal camera operation.

F-Stop

We mentioned earlier that all lenses are equipped with a variable iris which permits us to control the amount of light that enters the camera. The exact iris opening has been calibrated as a series of numerical f-stops. As the numerical f-stop *increases*, we close down the opening of the iris and *decrease* the amount of light which enters the camera.

F-stop numbers for television lenses run from about f/1.4 to f/22, and the lens is designed in

such a way so that changing from one f-stop to the next doubles or halves the amount of light entering the camera. For example, if we go from f/4 to the next highest f-stop, f/5.6, we decrease by one-half the amount of light which enters the camera. Similarly, opening the iris one full f-stop, from f/4 to the next largest f-stop—f/2.8, doubles the amount of light which enters the camera. (See Figure 3-10.)

Lenses are commonly rated in terms of their widest possible f-stop as this determines the maximum light-gathering capability of the lens. A lens with maximum iris aperture of f/1.8 or f/2.0 would be considered a "fast lens" since the lens is capable of gathering more light under low-light conditions.

You might wonder why all lenses are not simply designed to operate with the largest possible f-stop. The problem is that it requires a great deal of sophistication and complexity in the design and construction of a lens to increase the aperture size and still produce a sharp, distortion-free image. As the speed of a lens increases, its flexibility increases, but its cost rises dramatically. A fast lens is not crucial if it will be used exclusively inside the studio where more light can be added to the scene without much difficulty. However, if you expect to use your lens on many remote locations where extra light to increase exposure cannot always be easily added, a faster lens will provide you with greater shooting capability under varying light conditions.

REMOTE IRIS CONTROL Virtually all cameras are equipped with a remote iris control which permits the video engineer to control the f-stop from the CCU. In fact, once a color camera has been properly aligned, varying the f-stop from the

FIGURE 3-10 **Lens F-Stops and Iris Openings.**
As the numerical f-stop increases, the size of the iris opening decreases.

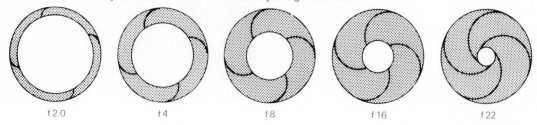

f 2.0   f 4   f 8   f 16   f 22

CCU is one of the primary methods used to shade the camera to compensate for exposure variations during the production.

Many cameras are also equipped with an *auto-iris* which automatically changes the size of the iris as the exposure varies on the scene being photographed. The problem with the auto-iris is similar to that found with many inexpensive still and movie film cameras which also use an automatic exposure control. The auto-iris reads the average amount of illumination in the scene but cannot discriminate between various elements within the camera shot. Therefore, it simply shades for the best average exposure without any regard for the most crucial elements within the shot. For instance, if the principal subject is standing in front of a bright background, the background area, which contributes the most illumination to the scene, will be perfectly exposed by the auto-iris, but the important foreground subject will probably appear too dark. This is why the auto-iris really cannot substitute for the human touch of a video engineer, who can take both aesthetic and technical considerations into account when shading the picture.

MANUAL IRIS CONTROL Cameras without a remote iris require the aperture on each camera lens to be preset manually by adjusting the f-stop ring on the barrel of the lens prior to the production. The video engineer will generally set the f-stop for the average illumination level on the set and then shade the camera electronically while the f-stop remains fixed. Of course, if the light levels change too drastically during the production, the lenses must be reset manually to provide the shader with more range of control at the CCU.

Focus and Depth of Field

FOCUS A camera shot is in focus when the light rays from the rear of the lens converge precisely on the face of the camera's pickup tube. Since this distance varies depending upon the focal length of the lens and the camera-to-subject distance, we must continually adjust the distance between the lens and the pickup tubes in order to maintain accurate focus. This is accomplished by turning the focus control, which is usually located on the left-hand pan handle.

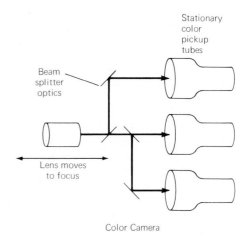

Color Camera

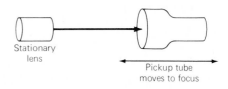

Monochrome Camera

FIGURE 3-11 **Camera Focusing.**
In all color cameras, the lens—or its internal elements—moves back and forth to focus the light rays on the camera's internal optical system. In monochrome cameras, the pickup tube travels back and forth while the lens remains stationary.

FOCUS AND FOCAL LENGTH The closer an object is to the camera lens, the greater the distance necessary between the lens and the pickup tube in order to focus the image. All lenses—both zoom and fixed focal length—have a minimum object-focusing distance, which is the shortest possible distance between the object and the lens where the picture can still be focused. In general, shorter focal length lenses can operate with objects closer to the lens than longer focal length lenses because the short focal length does not require a very large distance between the lens and the pickup tube in order to achieve sharp focus. As the focal length increases, however, you need greater distance between the lens and the pickup tube. Ultimately you may reach a point where the lens and pickup tube cannot move sufficiently apart to focus the image properly.

There may be situations in which you are asked to obtain an extremely tight shot of a very small object or a graphic, yet using a long focal length lens, or a zoom at its longest range, will not permit you to focus the picture. In this case, if you switch to a wider lens or zoom out to a shorter focal length and move the camera closer to the subject, you may find that you can fill the screen with a tight shot of the small object and still maintain focus. This is because the short focal length does not require a large lens-to-pickup tube distance to achieve a focused shot. You will usually have to get the camera within inches of the subject, however, and this may cause lighting problems or increase the possibility of including the close-up camera in another camera's shot. These problems can be solved by careful planning, coordination and adequate rehearsal prior to the telecast. (See Figure 3-12.)

CLOSE-UP LENSES If your production requires shooting a great many small objects extremely close, you might consider using an auxiliary close-up lens attached to the camera lens. This increases the camera lens's magnification and permits a zoom lens to vary its focal length within predetermined limits. A close-up lens will usually allow you to shoot small objects from greater distances than the wide-angle, short focal length technique we've just mentioned, which may solve some lighting and camera problems. However, when you are using a close-up lens, your normal lens capability is extremely limited, and the lens can focus only on objects located within a fairly small range.

DEPTH OF FIELD Whenever a lens is focused on a subject, there will be an area in front of the subject and behind the subject in which all objects are seen in focus. This area of acceptable

focus is called the *depth of field*, and it is an important optical characteristic of lenses which we can use as a creative tool. When there is a very large space surrounding the principal subject in which objects are still seen in focus, the lens is said to have a "wide" or "deep" depth of field. If the area surrounding the subject is not very large, the lens is said to have a "shallow" or "narrow" depth of field. (See Figure 3-13.)

Depth of field is important for both technical and aesthetic reasons. Technically, a shot with a wide depth of field makes it relatively easy to follow the action. A shallow depth of field requires you to continually change the focus—in effect, shifting the depth of field—as either the camera or

FIGURE 3-12 **Lens Focal Length and Focusing.**
Shooting extremely tight close-ups with a long focal length lens may result in the inability to achieve focus because the long focal length falls beyond the focusing limits of the camera and the lens. In this case, a short focal length lens, shooting close to the subject, may produce the close-up shot and still permit the camera and lens to focus the image owing to its shorter focal length.

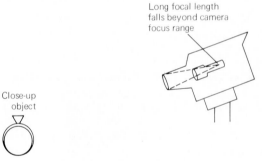

Long focal length falls beyond camera focus range

Close-up object

Long Focal-length Lens from Distance

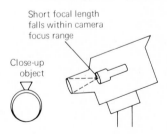

Short focal length falls within camera focus range

Close-up object

Short Focal-length Lens Shooting Close

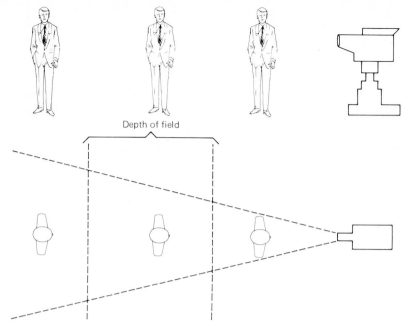

FIGURE 3-13 Depth of Field.
The depth of field is that area between the point closest to the camera and
farthest from the camera which is in perfect focus. In this illustration, the subject
in the middle is positioned within the lens's depth of field.

the subject moves. Aesthetically, the depth of
field plays an important role in creating the shot's
overall visual perspective.

Three factors determine the depth of field of a
lens: (1) the focal length, (2) the f-stop, and (3) the
camera-to-subject distance.

FOCAL LENGTH The shorter the focal length or
the wider the zoom lens position, the deeper the
depth of field. As the focal length increases, the
depth of field decreases.

F-STOP The smaller the lens opening (the larger
the f-stop number), the deeper the depth of field.
As we open the iris (decrease the numerical
f-stop), we decrease the depth of field.

CAMERA-TO-SUBJECT DISTANCE The larger
the distance between the subject and the camera, the greater the depth of field. The shorter the
camera-to-subject distance, the shallower the
depth of field. Also, depth of field is always
greater *behind* the principal subject than in front

of the subject, which means that you have more
range of focus with those objects positioned
behind the principal subject than in front of it.

Although each of the three factors will influence
the depth of field of a lens, the most practical
means of varying depth of field is to change either
the focal length of the lens or the camera-to-
subject distance. You can certainly light at high
illumination levels which will cause the lens to
stop down and increase the depth of field, but
lighting levels are often dictated by such other
considerations as the camera's minimum operating light level. Varying the overall light level is
usually the least versatile approach to controlling
the depth of field.

Using Depth of Field in Production

CAMERA MOVEMENT If a camera must move
around a set and still keep the picture in sharp
focus while on the air, the widest possible depth

of field will make camera operation easier and ensure a sharply focused picture. As the depth of field narrows, relatively short movements of either the camera or the subject will require you to refocus the picture constantly. A deep depth of field is also helpful when using a portable camera since you won't have to worry about refocusing the picture constantly as you move about.

ISOLATING IMPORTANT PICTURE ELEMENTS

Most camera shots are composed to isolate a principal subject, or subjects, which are of central importance to the viewer. In order to help make this distinction and to avoid distracting the viewer by other elements within the shot, use a slightly shallow depth of field—with foreground and background elements partially or completely out of focus—to direct audience attention to the most important elements within the frame. (See Figure 3-15.)

Fortunately, this is often automatically accomplished with the particular lens you would normally select to frame a shot. In a wide-angle shot, where we want to show the audience an establishing view of the entire scene. The short focal length which produces a wide horizontal field of view also provides a very deep depth of field in which almost all elements are in sharp focus. As we zoom to a tighter shot (or change turret lenses to a longer focal length), the depth of field becomes shallower, throwing background and foreground objects slightly out of focus and increasing the visual dominance of the principal subject. An extremely tight close-up, in which the subject is prominent in the shot, will produce a very shallow depth of field since the long focal length lens and/or the close subject-to-camera distance which is necessary to frame the shot also conveniently produces a shallow depth of field.

USING DEEP DEPTH OF FIELD A very wide or deep depth of field can be effective when we wish to show the viewer a number of different actions which occur on different depth planes within the same shot. One of the most effective uses of deep focus photography can be seen in Orson Welles'

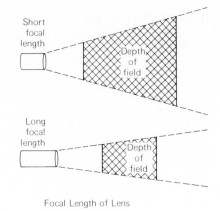

(a)   Focal Length of Lens

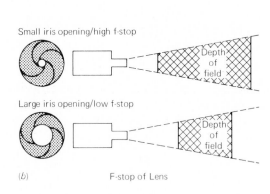

(b)   F-stop of Lens

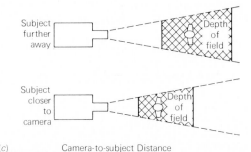

(c)   Camera-to-subject Distance

FIGURE 3-14 Factors Influencing Depth of Field.
(a) The shorter the focal length, the greater the depth of field; (b) the higher the numerical f-stop, the greater the depth of field; (c) the greater the camera-to-subject distance, the greater the depth of field.

film *Citizen Kane*. Welles frequently used a very deep depth of field to show one character in the foreground of a shot while showing the actions of another character in the background. In general,

(a)

(b)

(c)

FIGURE 3-15 **Isolating Picture Elements with Depth of Field.**
In figure (a) the depth of field includes subjects on all depth planes, and everything is in focus. In figure (b) the depth of field is shallower, and background subjects begin to go out of focus, lending more visual dominance to the foreground subjects, which remain sharp and clear. In figure (c) the shallow depth of field throws all but the foreground subject out of focus, adding to its visual dominance in the frame. (Photo: Jim Morgenthaler.)

a deep depth of field can establish a strong visual relationship between various elements within a shot and often eliminates the need to cut between different cameras.

USING A SHALLOW DEPTH OF FIELD In general, a shallow depth of field is desirable when we wish to isolate the principal subject against a nondistracting background. This approach is commonly used in shooting a sponsor's product: the product is seen against a blurry background which does not compete for the viewer's attention.

Shallow depth of field can also work to your advantage when you must shoot through foreground elements in order to obtain a shot. For example, the camera positioned behind home plate in a baseball game invariably must shoot through a protective wire screen or fence. Since you must usually use a long focal-length lens with a very shallow depth of field, the screen in the foreground is so out of focus that it essentially becomes invisible in the camera shot.

A shallow depth of field can also be used to your advantage when you lack an effective background set. For example, a director needed to show a character seated inside a car during a rain storm. The production was shooting on location with a portable camera and videotape recorder, but it was inconvenient and too expensive to wait for a rainy day to shoot the scene. Instead, the director used a long, telephoto lens on the camera to shoot a close-up of the subject seated inside the car and had a stagehand spray the car with a garden hose to simulate rain. Since the long lens produced an extremely shallow depth of field, the character was seen in perfect focus while the background area through the rain-streaked windows was blurry and out of focus. Through the effective use of a shallow depth of field the director was able to maintain the illusion of a rainy day with a convincing background.

RACK FOCUS A variation on the use of a shallow depth of field is to vary the focus within a shot selectively, drawing the viewer's attention to a number of different objects which lie on various depth planes within the frame. This is accomplished by using a lens with a narrow depth of field and instructing the camera operator to "rack" or "throw" the focus to different subjects in the shot. For example, you might have a character who is waiting for an important telephone call. The director could set up the shot with the telephone in the foreground and the character seated in the background. Using a long focal length lens, the camera operator would first focus on the telephone in the foreground, throwing the background slightly out of focus. As the phone rings, the camera operator racks focus to the back-

ground, showing the viewer the character's reaction to the ringing phone and directing the audience's attention from the telephone to the performer.

Another use of rack focus is to draw graphic material on a clear plastic or plexiglass plate which is positioned close to the camera in front of the set. When the focus is trained on the graphic, the background appears blurry and out of focus. As the focus is slowly racked to the background, the foreground graphic visually "disappears" as it is thrown far out of focus, and the audience's attention is now directed to the background area of the shot. (See Figure 3-16.)

PERSPECTIVE

By now it should be obvious to you that a long focal length lens and a short focal length lens produce two very different views of the same scene. The long lens has a greater magnification, a narrower field of view, and a shallower depth of field than the short lens. But there is one more crucial difference between long and short focal length lenses. It is the way the lenses depict depth and volume and how they show the juxtaposition and physical relationship between vari-

FIGURE 3-16 **Rack Focus.**
In the first frame *(a)*, foreground subjects are in focus, and the background subject is out of focus. In the next frame *(b)*, the depth of field is "thrown" to the background, and foreground objects go out of focus. Rack focus enables the director to shift the viewer's attention within the same shot. *(Photo: Jim Morgenthaler.)*

(a)

(b)

(a)

(b)

(c)

FIGURE 3-17 **Using Lenses to Vary Perspective.**
In this series of shots, the subjects remained station-ary. Figure (a) was shot with a medium focal length lens to produce a "normal" perspective as our eyes would see it. In figure (b) a wide-angle lens exagge-rates the distance between the foreground and back-ground subjects. In figure (c) the telephoto lens compresses distance making the subjects appear almost next to each other. Depth of field is also changed, becoming shallower with the telephoto lens. (Photo: Jim Morgenthaler.)

ous foreground and background elements within the shot. This characteristic is called "perspec-tive," and it plays and important role in the audience's perception of the visual information contained in the videospace.

"Normal" perspective shows us the natural depth and juxtaposition of elements as though our eyes were viewing the scene from the cam-era's position.

A wide angle, short focal length lens "forces" perspective, exaggerating the depth dimension of the shot and increasing the size of the fore-ground objects relative to those in the back-ground. Forced perspective also makes subject or camera movement appear faster than it really is. Oncoming subjects appear to grow larger much more quickly, and subjects moving away from the camera recede at increased speed.

A narrow angle, long focal length lens "com-presses" depth perspective, squeezing fore-ground and background elements together visu-ally and apparently reducing the space between the foreground and background in the shot. This compression gives the illusion that subjects mov-ing toward or away from the camera travel much more slowly than they really do. In some extreme cases, the subject can appear to be moving on a treadmill, seemingly covering no distance.

To show you how the difference in perspective affects your perception of a shot, take a look at the three frames in Figure 3-17, which contain the same material in each frame. The first picture was shot with a medium focal length lens, and the depth and perspective appear normal, as though we were standing in place of the camera.

The next frame was taken with a wide-angle

short focal length lens. Notice how the perspective exaggerates the size of the foreground subject relative to the background and how the depth and volume of the space between the subjects appears to be quite large.

Now look at the third frame, which contains the identical information, but which conveys a much different visual impression. The long, telephoto lens has compressed perspective, flattening foreground and background subjects, and suggests little apparent space between them. In addition, the compressed perspective does not exaggerate the size of the foreground subject relative to the subject in the background, but makes both of them appear to be about the same size.

Effects of Perspective on the Videospace

Filmmakers and television directors have long used perspective to enhance the visual meaning of their shots. Most television studios look larger than they really are owing to the director's use of a wide-angle lens to force perspective and exaggerate depth and volume. Auto manufacturers frequently shoot their car commercials and advertisements with a short focal length lens to make the car look longer and more streamlined than it really is. In general, wide-angle lenses which force perspective tend to make space appear larger than in reality and make subjects appear much more powerful and dominant.

The opposite effect is provided by the compressed perspective of the telephoto lens. In this case, the lens flattens out subjects, compressing depth and equalizing the size of foreground and background objects in the picture. This perspective can be quite effective when used with script material which calls for such a visual impression. For example, on a documentary about overcrowding in a college town, the director used a long, telephoto lens to shoot down a row of parked cars. The lens squeezed the cars together and apparently eliminated the space between them, conveying to the viewer the idea of a crowded, congested area which perfectly com-

plemented the program's script. An example of the use of the telephoto lens to slow down subject movement can be seen in the film *The Graduate*. At one point, the main character is shown frantically running toward the camera, but the long lens makes it appear on screen as though the hero is running on a treadmill, huffing and puffing, but seemingly getting no closer for all his effort. (See Figure 3-18.)

Of course, there are some production situations in which we might wish to avoid the particular effects of forced or compressed perspective. In many sports situations, for instance, the long, telephoto lenses which are required to bring distant players close can also unintentionally distort the viewer's spatial perspective. This is why the center field camera which shows the pitcher, batter, catcher, and umpire in a baseball game, makes it appear as though the pitcher is directly in front of the batter when, in reality, they are about 60 feet apart. A similar situation occurs during a football telecast when shooting a play from a camera located at field level. The long lens will pick up the runner, but the flattened depth perspective may give the viewer a distorted impression of the distance between the players.

A wide lens's forced perspective can also occasionally produce unintended distortion, particularly when a subject works too close to the lens and his facial features are distorted into the familiar "banana nose" or "elastic arm" effect. (See Figure 3-19.)

USING LENSES IN PRODUCTION

Whether you will be using a complement of fixed focal length lenses on a turret or a zoom lens, the particular focal length you choose for any shot depends on a number of production considerations as well as the visual perspective you wish to provide for the viewer. We will discuss the use of lenses in production by dividing them into three broad focal length groups: normal angle lenses, wide angle lenses, and narrow angle lenses.

Normal Lenses

FIELD OF VIEW   A normal lens provides approximately the same field of view as our eyes do.

FIGURE 3-18 **Telephoto Depth Compression.** The use of a long focal length lens makes the parking meters appear as though they are so close they are touching each other. *(Photo: Jim Morgenthaler.)*

In 16-mm format, a "normal" lens is approximately 25 to 35 mm.

PERSPECTIVE A normal or medium angle lens produces natural depth perspective with no exaggeration of foreground-to-background subject size, depth, or speed of movement.

DEPTH OF FIELD Normal lenses generally provide a medium depth of field with objects at the extreme foreground and background areas of the picture thrown slightly out of focus, but with a fairly wide area of acceptable focus around the principal subject.

MOVEMENT The movement of subjects toward or away from the camera appears at normal speed. Similarly, camera movement is not exaggerated but appears natural to the viewer. It is rather easy to follow a moving subject on a medium or normal angle lens since the magnification is not very great. Camera movement can be accomplished smoothly on a normal lens although extremely mobile camera moves should not be attempted on this lens since the jitter and shake of a moving camera can appear obvious.

DISTORTION Normal angle lenses produce almost no apparent picture distortion. The very

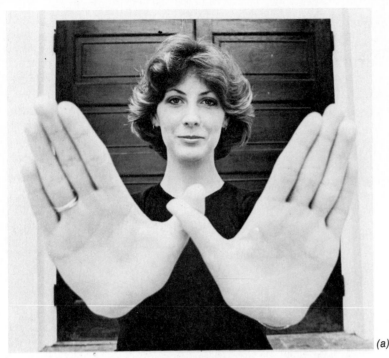

(a)

(b)

FIGURE 3-19 **Wide-Angle Lens Distortion.**
The forced perspective of a wide-angle (short focal length) lens can result in a distorted image. In figure *(a)* the subject's hands appear unnaturally large in relation to her body. In figure *(b)* her body appears elongated, and her head and shoulders appear unnaturally larger than normal proportions. *(Photo: Jim Morgenthaler.)*

slight flattening effect they produce can be quite flattering for many subjects as it tends to de-emphasize some prominent facial features.

Wide-Angle Lenses

FIELD OF VIEW   The wide angle lens provides a very wide horizontal field of view with very little image magnification. Principal subjects appear relatively small in relation to the overall background, and it is somewhat difficult to discern much detail in a very wide shot. In 16 mm format size, most wide angle lenses have focal lengths from about 12 to 25 mm. However, there are some extremely wide angle lenses with focal lengths as low as 6 mm, which are used for special purpose applications.

PERSPECTIVE   Wide-angle lenses force perspective and exaggerate the impression of depth and distance. The forced perspective makes foreground subjects appear much larger than background elements.

DEPTH OF FIELD   Wide-angle lenses produce a very deep depth of field in which virtually all elements within the frame are in focus.

MOVEMENT   The wide-angle lens exaggerates the speed of subject or camera movement. Objects tend to grow in size and to recede very rapidly. Subject movement is very easy to follow on the wide-angle lens, since the wide horizontal view, minimal magnification, and deep depth of field eliminate camera jitter and focus problems. The wide-angle lens is an excellent choice when you must move your camera on the air, particularly for complex movements or for hand-held cameras, since the low magnification and deep depth of field will cover up slight camera shakes and imprecise focusing.

DISTORTION   The forced perspective of the lens can result in a distorted image if the subject works too close to the lens. Wide-angle lenses also tend toward "barrel distortion," in which vertical lines appear to bulge outward in the middle of the picture and to converge at the top and bottom.

LIMITATIONS   The very wide vista on some wide, establishing shots makes picture definition and detail less sharp and may make it difficult for the viewer to see all objects clearly. Also, as the deep depth of field permits little selective focus, foreground, middle, and background objects may compete for the viewer's attention. Finally, the wide-angle lens is highly susceptible to lens "flare" as its wide vista may inadvertently include some studio lights which can produce a distracting, multicolored effect. Flares are usually eliminated by changing the camera's position, by repositioning the lighting instrument, or by attaching a lens hood in front of the lens to block the light.

Narrow-Angle (Telephoto) Lenses

FIELD OF VIEW   A narrow angle lens produces a very narrow field of view with a powerful subject magnification. In 16-mm format, long lenses are generally those with focal lengths running from 100 to about 500 mm. Lenses with focal lengths greater than 500 mm are rarely used, except for special remote applications since their minimum focusing distances are too great for use in the studio.

PERSPECTIVE   The long lens compresses perspective, reduces the perception of space and volume, and makes all subjects on different depth planes within the videospace appear about equal in size.

DEPTH OF FIELD   Long lenses produce a very shallow depth of field with a small area of acceptable focus around the principal subject. The depth of field can be so narrow, in fact, that at close operation distance objects only a few inches from the principal subject may appear out of focus.

MOVEMENT   The extreme magnification and narrow horizontal field of view makes the long lens

FIGURE 3-20 Lens Summary Chart.

Lens	Focal Length	Horizontal Angle	Perspective	Depth of Field	Distortion Characteristics	Production Uses
Wide angle	(16-mm format) 12–25 mm	57°–19°	Forces Perspective. 1. Foreground subject appears larger than background subject. 2. Increases perception of depth; set looks larger than it is. 3. Exaggerates subject/camera movement, making it appear faster than reality.	Very deep. Almost all elements within picture are in acceptable focus.	Barrel distortion. Subjects working too close to lens appear distorted, prominent features are emphasized and can look grotesque and unnatural.	Excellent lens for camera movement, tends to minimize camera shake. Excellent for use with portable cameras which are hand-or shoulder-mounted. Use for wide establishing shots and where cramped quarters require its wide angle. Its short focusing distance makes it a good lens for extreme close-up work on graphics or demonstration material. Use of its forced perspective characteristics make it excellent for moving camera shots.
Normal (medium) angle	(16-mm format) 25–75 mm	20°–9°	Normal Perspective. Relationship between objects appears as it would to our eyes, no exaggeration of depth, subject size, or speed of movement.	Medium. Fairly large area of acceptable focus around principal subject. Objects in extreme foreground and background are out of focus.	No apparent distortion. Slight flattening of facial features tends to flatter most subjects.	Good lens for camera movement and for following subject movement from closer angle than wide lens provides. Usually produces most flattering effects on subject's face and features with minimum of distortion. Good close-up lens for

Lens	Format / Focal length	Angle	Effect on perspective	Depth of field	Disadvantages	Advantages
Narrow angle (telephoto)	(16-mm format) 75–250 mm	9°–3°	Compresses Perspective 1. Squeezes objects together in frame, reduces perception of distance. 2. All subjects appear equal in size. 3. Makes subject movement appear much slower than in reality.	Very shallow. Small area of focus around principal subject. Rest of foreground and background are thrown out of focus.	Severe flattening of features on some subjects may be unflattering. Compression of depth can distort audience's sense of space. Heat wave effect on extreme long lens can produce blurry, annoying effect on scene.	graphics and demonstration material when wide lens cannot get close enough for tight shot. Excellent all-purpose lens with best characteristics of wide and long lenses. Magnification makes it ideal for location shooting, or in-studio for close-ups of subjects at large camera-to-subject distances. Compression of perspective can be used for special effects when warranted. Poor lens for camera movement or portable cameras as it magnifies camera shake and its critical focus makes it difficult to follow action.

difficult to use for most movement situations. Following a rapidly moving subject is very hard since the focus is so critical and the field of view is extremely narrow. The magnification power of the lens also makes on-air camera movement virtually impossible since the magnification of the lens will emphasize the slightest camera shake or jitter. For this reason, long lenses should never be used on a portable camera unless the camera is used on a tripod or on a strong, sturdy camera brace.

DISTORTION The compression of perspective tends to flatten a subject's features and the depth perspective in the shot. The lens may magnify heat waves from the ground when used for extremely long distances on remotes, which can make the image appear wavy and distorted.

LIMITATIONS On-air camera movement is virtually impossible, and following moving subjects must be carefully planned in advance.

Visual Perspective of a Zoom Lens versus a Camera Dolly Shot

There is an important difference between the visual perspective which is achieved with a camera dolly—in which the camera physically moves toward or away from the subject—and a zoom in or out on a scene. Unfortunately, the wide use of the zoom lens and its obvious ease of operation have resulted in some camera operators and show directors automatically substituting a zoom for any camera dolly even though there is a distinct visual difference between the two.

A zoom lens simply *magnifies* (zoom in) or *demagnifies* (zoom out) the image while a camera dolly conveys the more dynamic impression of *actually moving* in or out of the scene. In a zoom, all spatial relationships remain constant as the magnification of the image varies. In the camera dolly, the movement of the camera results in a continually changing set of spatial relationships between objects in the shot as the camera physically draws nearer or pulls farther away from the scene. This visual impression is particularly important when the camera dollies past doorways, arches, furniture, or attempts to convey a subjective view of the action.

The change in spatial relationships and the more active viewer involvement, which the dolly provides as compared with a zoom, are important aesthetic considerations which both the camera operator and the program's director should keep in mind when deciding how to approach such a shot.

Of course, if there is no depth to the subject—such as a photograph or a camera title card—there is no apparent difference between the zoom and the dolly. Since these two-dimensional subjects have no depth perspective to begin with, it is generally preferable to use a zoom lens rather than to dolly into or out of a card or picture, as the zoom will look smoother on the air.

SPECIAL PURPOSE LENSES

Although most studio/field cameras utilize the conventional zoom lens, some production requirements call for special purpose lenses which provide some unique effects impossible to obtain with conventional lenses. These are usually fixed focal length lenses and are conveniently used on portable cameras, where the flexibility of movement combined with the special effect produced can result in some very striking visuals.

FISHEYE LENS The fisheye lens is an ultrawide-angle lens which provides a 180° panorama view. Of course, the lens radically exaggerates the depth perspective, but this distortion is usually the very effect the director wants to achieve. Though the fisheye lens can deliver some very dramatic subjective or surrealistic effects, its distortion is so obvious that it must be used sparingly and with an obvious motivation. (See Figure 3-21.)

DIOPTERS AND LENS SPLITTERS These lenses, adapted from motion picture production, give you the ability to focus clearly on subjects at two different distance planes within the same shot. The lenses work like bifocal eyeglasses in which one-half is a wide-angle lens and the other a telephoto. Using the split lens and carefully positioning the subjects enable you to frame and focus perfectly two widely spaced subjects. (See Figure 3-22.)

STARBURST FILTERS The starburst filter produces a multipointed star effect whenever the camera shoots a highly specular, reflective subject. This can be a visually beautiful effect and is frequently used on musical programs or for some dramatic fantasy scenes. Although called a "filter," the starburst is really a lens since it is specially designed to refract light in multiple directions, creating the starlight effect. (See Figure 3-23.)

FILTERS

Filters are devices which change or modify the nature of light before it enters the camera. Most modern television cameras are equipped with a filter wheel which is positioned between the lens and the pickup tube. The filter wheel holds a number of different filters, which are quickly in-

FIGURE 3-21 **Fisheye Lens.**

(a)

FIGURE 3-22 Split-Field Lens.
The split-field lens enables the camera to overcome normal depth of field limitations. It is important to position the demarcation line between near and distant subjects where it will not show on screen. This makes it difficult to use the split field lens except for shots where a minimum of subject or camera movement is required. *(Courtesy: Tiffen Mfg. Co.)*

FIGURE 3-23 Starburst Filter.
Starburst filters are available with a variety of points and with different spacing between the starburst patterns. In the first frame *(a)* the scene is shot without a filter. In the second frame *(b)* a four-point star filter is used. In the third *(c)*, an eight-point star filter is used. *(Courtesy: Tiffen Mfg. Co.)*

serted into position and removed when they are not needed. On cameras without a filter wheel, you must attach a filter directly to the camera lens.

NEUTRAL DENSITY FILTERS A neutral density filter is a gray filter, which reduces the light level without affecting the color quality of the light. Neutral density filters are used when the camera must shoot under light conditions where the level

of illumination is too high for the camera to handle. The neutral density filter reduces the light intensity by a predetermined factor, giving the shader a greater range of control at the camera's CCU. They are most frequently used on remote telecasts where high light levels from a bright, sunny sky must be reduced to permit accurate camera shading.

Another use of the neutral density filter is to force the camera to operate at a wider f-stop than

(b)

(c)

normal—in effect, decreasing the depth of field. This is helpful in achieving selective focus or rack focus effects, which would be impossible if the camera lens had to be stopped down to reduce the amount of light.

COLOR CORRECTION FILTERS If you do any still or motion picture photography, you know that in order to use "indoor" film outdoors or vice versa, you must use a color correction filter. This is because the film is designed for the color quality of indoor light, which is different than that of outdoor light. Indoor lamps and bulbs produce a "white" light that actually has a reddish tinge. Outdoors, the light from the sun and sky has a bluish tinge. The filter you use on your film camera changes the color quality of the light so that

indoor film can faithfully reproduce colors outdoors.

The television camera works along the same basic principles. It is usually adjusted for indoor lighting instruments which produce light with a slightly reddish tone. Outdoors, on a remote, the camera must either be electronically rebalanced for the bluish light, or a color correction filter must be used to modify the color quality, or "color temperature" of the light.

The filter which changes the color temperature of outdoor light to match the camera's indoor adjustment is an orange, "85" filter. You can get this filter alone or combined with a neutral density filter to both correct for color temperature and reduce the amount of light which enters the camera. These combination filters are called

"85N3," "85N6," and "85N9." The higher the figure following the "N" (for neutral density), the greater the light reduction capability of the filter and the less light that is transmitted into the camera.

DIFFUSION FILTER The diffusion filter softens hard lines in the shot, producing a soft, vaguely blurry image, which is ideal for dreamy, fantasy, or romantic visual effects. Different filters produce varying amounts of diffusion and some are available with a clear center, with the diffusion increasing gradually toward the outer edges of the filter. Slightly diffused filters are sometimes used to eliminate facial blemishes or wrinkles on the performer's face in a close-up.

An inexpensive substitute for the diffusion filter is to use petroleum jelly on a clear glass filter or a neutral density filter. However, the jelly should be applied only to filters which can be attached to the front of the lens. At no time should you apply it, or any other substance, directly to the front of the camera lens. You can remove and clean a clear

FIGURE 3-24 **Diffusion Filter.**
The top photo is without diffusion. The bottom picture shows the scene with the diffusion effect produced by the filter. *(Courtesy: Tiffen Mfg. Co.)*

glass filter, but applying a substance on the camera lens itself can damage the expensive lens permanently. (See Figure 3-24.)

SUMMARY

The function of the camera lens is to collect the reflected light rays from the subject and form the subject's image on the camera's electronic pickup tube. The type of lens that is selected and the way it is used determine the image magnification, the size of the subject, and the field of view that is photographed and establish the viewer's visual perspective of the scene.

A lens is made up of three component parts: (1) the optical elements which form and shape the image, (2) a variable iris, which is opened or closed to control the amount of light which enters the camera, and (3) the mounting system, which attaches the lens to the camera.

The three optical characteristics of lenses which determine the size, view, and perspective of the image are (1) focal length, (2) f-stop, and (3) focus and depth of field.

Focal length is a fundamental lens characteristic which determines the image magnification and field of view. The shorter the focal length, the wider the field of view of the shot and the smaller the size of the subject. The longer the focal length, the narrower the field of view and the larger the size of the subject. Lenses are described in terms of millimeters. Short lenses run from about 12 to 25 mm; normal or medium focal length lenses, from about 25 to 75 mm; and long focal length lenses, from about 75 to 250 mm. Extreme wide and telephoto lenses are also available for special purpose applications.

A zoom lens is a variable focal length lens which can bring a subject closer or move it farther away in the videospace without actually moving the camera. A fixed or primary focal length lens is preset to produce the same degree of image magnification and field of view at all times.

The f-stop is a numerical term used to describe the size of the iris opening of the lens. Changing the f-stop permits more or less light to enter the camera, compensating for variations in light level. The higher the f-stop number, the smaller the aperture opening and the less light which can enter the camera.

A lens is in focus when the light rays from the rear of the lens converge precisely on the face of the camera's pickup tube. Depth of field refers to the area of acceptable focus within the lens's field of view. A large area of acceptable focus is a "deep" depth of field; a narrow area of acceptable focus is a "shallow" depth of field. Depth of field is controlled by three factors: (1) focal length—as focal length increases, depth of field decreases; (2) f-stop—as the f-stop number increases, depth of field increases; and (3) camera-to-subject distance—as camera-to-subject distance increases, the depth of field increases.

Perspective refers to the way the lens portrays depth, dimension, and spatial relationships within the videospace. A "normal" angle lens produces a normal perspective; the videospace appears as if our eyes were in place of the camera. A wide-angle (short focal length) lens forces perspective, enhances the perception of depth, and exaggerates the speed of camera and/or subject movement. A narrow angle (long focal length) lens compresses perspective, reduces the perception of space by flattening depth planes together, and slows down the speed of camera and/or subject movement.

Filters are used to modify or transform light before it enters the camera. A neutral density filter reduces light intensity without affecting its color quality. Color correction filters modify the color temperature of light. Starburst filters refract light from specular objects into star-shaped patterns. Diffusion filters are used to soften hard lines, producing a romantic, dreamlike visual effect.

CHAPTER 4

CAMERA OPERATION

A good camera operator combines a strong visual sense for form and composition, an aesthetic sensitivity for the program's overall concept and approach, and the physical skill and coordination necessary to operate the camera smoothly and with precision. In this chapter we will discuss the basics of camera operation. Keep in mind, though, that effective camera work comes about only through practice and the experience you will gain after working on a variety of different production situations.

CAMERA COMMANDS

As with any specialized field, television production has developed a language and a jargon all its own to communicate directions and commands quickly and accurately. For this reason, it is important that you get into the habit of learning and using the special terminology we will discuss in this chapter and throughout the book. In the case of camera operation, there are a number of special terms to describe its operation.

Camera Movement

The camera can move in two ways: (1) the camera head alone moves atop its stationary pedestal, or (2) the entire camera and pedestal can be moved about.

PAN   The horizontal movement of the camera on a stationary pedestal or tripod.

Command: "Pan right" or "Pan left." Sometimes the director may give you more specific instructions, such as "Pan right with the host as he walks to the table."

A special kind of pan is the *swish* pan or *whip* pan. This is a very rapid move which looks like a blur of light on the television monitor. It is sometimes used for transitions or for other special effects.

A pan across a scene should never wander aimlessly but should have a definite starting point and a definite ending point. If it is at all possible, try to "motivate" the pan. Sometimes you can have a subject look in the direction of the pan and then execute the camera move to reveal to the viewer what the subject sees. If this is neither possible nor desirable, be sure to have definite starting and stopping points fixed in your own mind so the pan movement will be decisive and direct.

TILT   The vertical movement of the camera on a stationary pedestal or tripod.

Command: "Tilt up" or "Tilt down." It is best to use the term "tilt" to refer to all vertical movements and the term "pan" to refer to horizontal moves. This distinction helps eliminate confusion on the part of the camera operator. As with the pan, establish definite start and stop points to help guide your tilt from subject to subject.

PEDESTAL   The movement of the camera up or down as the center telescoping column of the pedestal is raised or lowered.

Command: "Pedestal up" or "Pedestal down." The pedestal control on a camera changes its

point of view just as though you viewed a scene sitting down and then stood up to look around. The pedestal movement can be used to great advantage by the camera operator and the director. Sometimes the director may ask you to vary the pedestal height in order to keep the camera at approximately eye level with the subject.

While it is usually most comfortable for the camera operator to set the pedestal at a convenient operating height, this may not coincide with the best pedestal height for the shot. Good camera operators adjust the height of the pedestal for the best view of the subject, not for their working comfort.

Since you may be asked to change pedestal height frequently during a program, it is best never to lock the pedestal into position completely. If the pedestal has been properly adjusted, the camera's weight should be perfectly balanced, and the pedestal should remain at any given height unless changed by the camera operator. Working with the pedestal lock off allows you to move the pedestal quickly when you have to.

It is important for you to realize that pedestalling up or down is not the same thing as tilting up or down. Tilting the camera simply changes its angle of view from a fixed operating height. Pedestalling actually varies the camera's height and results in a much different visual perspective of the scene which is photographed.

DOLLY   The movement of the camera on its pedestal either toward or away from the subject or scene.

Command: "Dolly in" or Dolly out." The director should usually tell you how fast or slow the speed of the dolly should be.

TRUCK   The lateral movement of the camera on its pedestal.

Command: "Truck right" or "Truck left." The truck can be used to adjust the camera's position to compose and frame a shot better, as well as to follow a moving subject as it crosses the set. If you are following a moving subject, the camera move is sometimes called a "tracking" shot.

Just as there is a difference between a tilt and a pedestal, so too is there a big difference between a truck and a pan. The pan changes the camera's horizontal field of view from atop a stationary pedestal. The truck actually moves the camera, establishing a new shooting angle which results in a much different view of the subject or scene photographed.

ARC A combination of a dolly and a truck, the arc is a semicircular movement of the camera and its pedestal.

Command: "Arc right" or "Arc left." The arc is frequently used to show circular movement or to reveal a view from behind the principal subject.

CRANE A crane is the movement of the camera atop the long arm of a crane.

Command: "Crane up" or "Crane down." Sometimes the command "Boom up" or "Boom down" is used. A horizontal movement of the crane arm is called "tonguing," and the command is "Tongue right" or " "Tongue left."

Of course, it is very common to combine one or

more of these camera movements during normal operation. For example, as you dolly in, the director may ask you to pedestal down. At the same time you will probably have to pan and tilt slightly to keep the subject framed properly. If you expect to undertake a number of highly complex camera movements during a production—as is common in musical, dance, or dramatic programs—you may wish to have the floor manager or a camera assistant work with you to pull cable and help move the pedestal while you operate the camera and the lens controls. Whenever you work with a camera assistant, it is important to establish a carefully coordinated working relationship so you do not work at cross purposes. It is usually best if

FIGURE 4-1 **Camera Movement.**

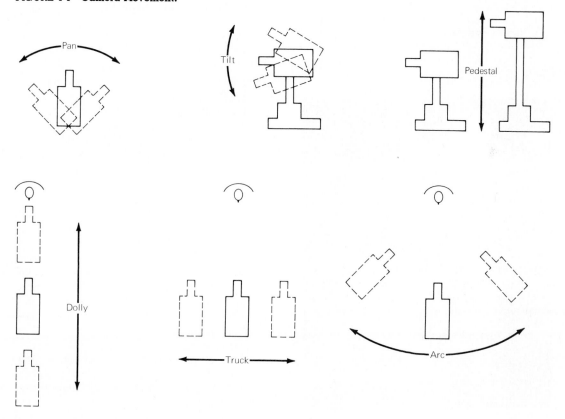

the camera operator takes care of pan, tilt, and lens operations, leaving the camera assistant to steer and help move the pedestal.

Lens Operation

A number of commands refer specifically to the operation of the camera's lens system.

ZOOM The use of the zoom control to continuously vary the camera lens's field of view.
Command: "Zoom in" or "Zoom out." Some directors refer to zooming as "Push in" or "Pull out." For example, "Start with a close-up of the host, and pull out to reveal the guest."

A special effect zoom is the *snap zoom*, which is an extremely rapid zoom in or zoom out used to emphasize a dramatic point or to punctuate an important moment in the program. The effect obviously calls attention to itself and should be used sparingly for maximum impact.

FOCUS Adjusting the camera lens to achieve the sharpest possible picture.
Command: "Focus up."

ZOOM CALIBRATION A zoom lens must always be calibrated or prefocused each time the subject or camera moves to a new position. This should be done automatically as soon as your camera shot goes off the air and before setting up your next shot.

RACK LENSES On a turret-lens camera, each fixed focal length lens must be positioned in front of the pickup tube for the shot.
Command: "Rack lens" or "flip lenses." Usually the director will specifically say: "Rack to your wide lens" or "Rack to a tighter lens."

Portable Camera Movement and Operation

The same commands are used for portable camera operation as for pedestal or tripod-mounted cameras. Of course, the difference is that the camera operator's body performs the dolly or truck instead of a pedestal or movable tripod. The major advantage of a portable camera is the flexibility of movement which a good operator can lend to a production.

If you are going to use a portable camera; have a camera assistant assigned to pull cable, help steady the camera if you must get up, down, or change position; and to help clear the way since your vision may be restricted to the camera's viewfinder. Move your camera on the air on the widest angle lens possible to minimize extraneous camera movement or shaking. Unnecessary camera movement is one of the most distracting problems associated with portable cameras. To minimize this, good camera operators try to make the camera an extension of their bodies. This offers maximum control, cushions your movement against the camera, and reduces shaking and camera jitter.

FIGURE 4-2 **Portable Camera Operation.**
Bracing the camera against your body will help to smooth out movement and avoid excessive jitter or camera shake.
(Courtesy: RCA)

WORKING THE CAMERA

As with all members of the television team, the camera operator's job can be divided into particular responsibilities during each of the four production stages.

Preproduction

Camera operators are rarely included in early preproduction conferences since their presence is not required when early production decisions are being made. Usually, the camera operator will learn about production details on the first day of studio rehearsal. However, on extremely complicated productions or in some unscripted situations—such as sports coverage—the director may wish to have a brief conference with the camera operators prior to rehearsal to explain the overall production approach and to assign each camera its general area of responsibility.

Setup and Rehearsal

Just prior to the rehearsal, the video engineers must register and align the cameras for the best possible picture reproduction. Once the cameras are released for use, here are the things you should do:

1 Put on your headset, and check the intercom system.

2 Unlock the pan, and tilt brakes, and adjust the drag controls so you can operate the camera head smoothly.

3 Unlock the pedestal column brake, and make certain the pedestal is operating smoothly throughout its entire height range.

4 Contact the video engineer, and request your camera to be electronically uncapped. If you also have a cap over the camera lens, remove it, and place it where it will not get lost during the show.

5 Once the camera is uncapped, focus on a medium-wide scene, and hold the camera

stationary while the video engineer shades the picture.

6 Once your camera has been shaded, check the lens controls to be certain the zoom and focus controls work smoothly without sticking or pulling.

7 Check to see if you have enough cable slack to move the camera across the floor without snagging the cable on other equipment.

CAMERA BLOCKING During the first part of the rehearsal period, the director's primary concern will be "camera blocking." This is when the camera shots, angles, and movements are determined and coordinated. Camera blocking is the time to make sure that you can deliver whatever the director asks of you. If you find a shot, an angle, or a movement is causing problems—it may come too soon, the subject may be obscured, or other production equipment is in the way of your camera pedestal—be sure to tell the director so that whatever changes are necessary can be made before the rehearsal time is over.

Consistency is important in how you frame and compose your shots and in the speed of your zooms and dollies. It is no help to the director when your two-shot or your dolly-in keeps changing in screen size or speed each time you rehearse the program. The more consistent you are, the fewer shot corrections the director will have to make during later rehearsals and the actual production.

You will soon find that every director has a personal style in approaching the rehearsal and production and in selecting and setting up camera shots. It is a good idea to find out exactly what your director means for each basic camera shot. For instance, a "medium close-up" for one director might be a "tight close-up" for another. After a short time, however, you should begin to get a feel for the director's ideas and start framing up

FIGURE 4-3
Effective camera operation often requires the operator to work in an unnatural position in order to achieve the right camera angle.
(Courtesy: Visions/PBS)

the shots as he or she visualizes them. Some directors are also extremely precise about the kind of shot and angle they want to see; others are "shot shoppers" who like to give the camera operator a chance to suggest different shots for them. Regardless of how your director approaches the show, it is important never to try to "outdirect the director." You may have a beautiful shot lined up, but only the director, sitting in the control room and watching all of the camera shots, has a complete picture of what is going on. A program can have only one director whose judgment must always take precedence and whose decision is final.

UNSCRIPTED SHOWS You will probably work on many productions which are either partially scripted—such as an interview or demonstration show—or which are completely unscripted—such as news or sports coverage. On such unscripted shows, the director will usually assign each camera operator a particular area to cover. For example, on a homecooking demonstration program, the director may assign Camera 1 to follow the talent, Camera 2 to follow the cooking demonstration in close-up, and Camera 3 to follow talent and the demonstration wide, ready to zoom in or out when necessary. On sports events, the director will usually assign one camera to follow the ball,

another to stay fairly wide on the action, another to "isolate" on a particular player, and so on.

Even shows which are completely unscripted will often follow some sort of basic format. For example, interview or game shows generally follow a regular format which can help you and the director in setting up shots. Of course, an unscripted show means that the director must rely on the camera operator's initiative and judgment to stay alert, follow the action, and quickly set up shots with a minimum of direction. Successful sports directors will tell you that their most important asset is a good camera crew that is able to get the shots and angles which make the director look good.

SCRIPTED SHOWS If the program is completely scripted—such as a drama, musical, instructional, or educational show—the director should have preplanned all of the shots and camera angles before the camera blocking session begins. Many directors like to hand out *shot sheets* to each camera operator. These sheets indicate every camera's shot by number. The camera operator follows along the shot sheet as the director or assistant director readies each upcoming shot by camera and shot number.

If the director does not hand out prepared shot sheets, you should make up your own as the

Viewfinder

Shot
sheet

Focus
control

Zoom
control

FIGURE 4-4
The camera from the camera operator's point of view
shows the viewfinder, shot sheet, and zoom and
focus controls.
(Courtesy: ABC-TV)

director leads you through the camera blocking
session. The only real difficulty with using shot
sheets is that the director is locked into a preset
sequence; if a variation should occur, it may lead
to confusion among the camera operators on the
studio floor. (See Figure 4-5.)

Production

Once you have gone past the dress rehearsal, the
camera operator's most important job is to deliver
the shots and camera moves planned and rehearsed earlier. This is no time to try to discover
new shots and angles. The performers, director,
and other crew members expect to see what has
been rehearsed and have planned their jobs
accordingly. At this point in the production process, no one appreciates surprises. Changing a
previously rehearsed shot or camera sequence
without rehearsing the new one is an invitation to
disaster.

IF SOMETHING GOES WRONG Television is
always subject to Murphy's law, which states with
brilliant conciseness: "If something can go wrong,
it will." No matter how much you have planned
and rehearsed, something can always go awry,
particularly on live broadcasts where there is no
chance for a retake. As the camera operator, you
should try to be alert to possible problems such
as the director's calling for the wrong shot, an
actor's blowing a rehearsed line, or a piece of

FIGURE 4-5 **Camera Shot Sheet.**
On a fully scripted show the director can provide
each camera operator with a detailed list of shots.
The missing shot numbers on the sheet below refer to
camera shots which are being taken by other cameras during the production.

```
           CAMERA 2

              • • •

   2. WIDE SHOT-DEMONSTRATION AREA
              • • •
   4. WIDE SHOT-DEMONSTRATION AREA

              • • •

   7. CU HOST-PULL OUT AS SHE WALKS TO TABLE
              • • •
   9. 2-SHT HOST AND GUEST


              • • •


  14. ECU DEMONSTRATION ON TABLE
              • • •
  16. MCU HOST

              • • •

  19. 2-SHT HOST AND GUEST

              • • •

  23. CLOSING CREDITS ON CRAWL
```

equipment or another performer in the way of your previously planned shot. The first, and most important, thing to remember is, "do not panic." Often, after what may have seemed like the biggest blowup since the H-bomb, you will find that the viewers never even noticed the problem. Remember the good advice of John Litvack, a director of CBS-TV's *Guiding Light*: "No one out there has your script or knows the shots you were supposed to get." If you and the director keep a clear head, you may be able to get out of a tight spot without a viewer's realizing that there was a problem at all.

Postproduction

Once the production has been completed, the camera operator's job is basically over. You are responsible, though, for seeing that the camera head is safely locked, the lens is capped, and the camera and pedestal are stored away properly.

Here is a checklist for postproduction camera operation:

1 Wait until the director gives the all clear before you lock and cap up.

2 Lock the pan and tilt controls, return the pedestal to its lowest height position, and apply the pedestal column lock.

3 If your camera has a lens cap, place it securely over the lens. If you don't have a lens cap, defocus the camera lens completely (so random reflections or lights won't damage the pickup tubes), and request the video operator to electronically cap the camera.

4 Remove your headset, and store it away.

5 Return the camera and pedestal to its storage position. Coil the camera cable in a wide figure-eight.

6 Remove scripts or shot sheets from the camera, and pull up any masking tape you may have used to mark camera positions on the studio floor.

Portable Cameras

The operation of a portable camera is essentially the same as a pedestal-mounted camera. However, there are a few important differences which you should keep in mind when using one.

1 Be sure to check all your cable connections.

2 Check your camera mounting equipment to make certain that the camera head is securely attached and that all straps and buckles are tightened, to give you optimum camera control along with maximum operating comfort.

3 Coordinate your moves with your camera assistant. Be sure your assistant always has sufficient cable slack and will clear a path for you if you must make some major moves during the production.

4 Remember the wider the lens angle, the less obvious camera movement will appear in the picture. If you must use your lens at a long (telephoto) angle, try to brace yourself against a sturdy wall or set piece. If a sturdy area is not available, position your body firmly on the ground.

5 Make on-air moves with the zoom lens in the widest possible position.

6 During camera rehearsal, have your camera assistant mark important positions on the studio floor with masking tape. This will help you to find the exact point for a particular shot and angle with a minimum of readjustment.

7 Even when your camera is off-air, try to stay as ready as possible since you never know when the director might need your shot.

8 Keep in mind the operating limits of your cable length. Never make a major move without first ensuring that you will have sufficient cable slack and that the cable will not get tangled with other equipment.

9 Know the limits of your operating area so you do not walk into another camera's shot or in the way of other equipment.

10 During a break in the production, place your

camera where it will not be damaged or kicked over. Never leave the camera in the middle of the studio floor or the production location. If you must leave the camera for a period of time, be sure to cap the lens to prevent the lens and camera tube from being inadvertently damaged by dirt, spurious reflections, or light.

BUILDING THE VIDEOSPACE: FRAMING AND COMPOSITION

Unlike the theatrical stage where the audience can direct its attention to any part of the stage (as well as to any part of the theater), the television audience must rely totally on the director and the camera operators to organize and compose the sequence of shots which convey the action that is taking place in front of the cameras. Although the director is ultimately responsible for selecting the different shots and angles to cover the program, the camera operator must find and frame the shots. To do this well, you must develop a feel for pictorial composition, which is, to a great extent, learned through the combination of intuition, practice, and experience. Fortunately, a number of excellent guidelines have been developed over many years through the work of painters, sculptors, photographers, and motion picture cinematographers. You can use these guidelines to help you start thinking visually.

One way to think of framing and composition is in terms of building the videospace for the viewer. Since the audience's only measure of what is taking place before the camera is what it sees on the television screen, how you represent this reality in terms of the videospace is of crucial importance. During the course of a production, the director and the camera operators are faced with an almost unlimited number of pictorial choices. Exactly what you decide to include within the television frame, what you decide *not* to include, and how you show this information determines to a large extent how the audience perceives the visual portion of the program.

As an example, take a simple three-person interview. If we shoot all three participants in a wide shot, there is no visual cue for the viewer as to which speaker is the most important. Assuming lighting and other production considerations are equal, the audience will most likely focus on whoever is talking at any particular time. Now, if we isolate one of the participants in a close-up, we automatically exclude from the audience's sight any potentially distracting elements, forcing the viewer to watch the one person who is filling the videospace at that time.

Naturally, many things must be taken into account when deciding exactly how to shoot a subject. However, we can include all of these considerations in two important points: (1) show the viewers what they *need* to see and (2) show the viewers what they *want* to see. Needless to say, you should make every attempt to show both what the viewer needs and wants to see in as interesting and imaginative a way as possible.

Let us take the case of a program demonstrating how to play the guitar. When the host begins to talk about finger positions on the guitar's neck, the audience *needs* to see a close-up of the host's fingers in order to understand and appreciate what is being communicated. If, later on in the show, the host plays a tune to show how a practice exercise ought to sound, the audience *wants* to see a good close-up of the guitarist's fingers in order to enhance their enjoyment and appreciation of the show.

While these two "rules" may seem overly general, if you keep them in mind during a production, you won't go wrong. There are times, however, when you may want to violate the rules for special effect. At the start of a murder mystery, for instance, the viewers might want to see the face of the killer, but they are willing to suspend—for a time—seeing the villain in order to enjoy the whodunit. If you frustrate the viewers too often or in the wrong places, however, you run the risk of confusing or annoying them to the point where they no longer understand or enjoy the show.

As we determine what the audience needs and wants to see, we are building the videospace. It is convenient to look at this building process from the camera operator's point of view as a combination of two different, but related, functions: (1) *framing* the shot and (2) *composing* the shot. Although these terms are often used interchangeably, for now it would be a good idea to differentiate between them.

Framing refers to the inclusion (and exclusion) of various pictorial elements within the videospace and how these elements are shown to the viewer. For example, in the interview program we just mentioned, the framing decisions include who among the participants we show in a particular shot, the kind of shot we use (a wide shot, close-up, etc.), and the particular camera angle (normal, low angle, high angle) that we select.

Composition refers to the organization of these pictorial elements in such a way as to present a picture to the audience which is unified, shows the relationship between the elements within the frame, and which provides an aesthetically pleasing visual image. Of course, the two concepts of framing and composition are very closely related to each other. While we will discuss them independently, for now, once you get to the point where you are selecting shots and angles and composing the shot, you must automatically combine the two as you build the videospace.

FRAMING THE SHOT

As you stand behind the camera and frame a shot, there are a number of technical decisions as well as aesthetic judgments to make. The technical matters include such obvious points as which subjects to include and which to exclude, how to present the subject with maximum visual clarity, and how to frame the scene to include the most important elements within the essential area of the picture frame. The aesthetic decisions include the particular field of view that the shot will encompass, the camera angle you will use, and how to frame and follow subject movement. Together these factors play an important role in building the videospace.

Visual Clarity

A primary consideration when framing any shot is its *visual clarity*. Obviously, a shot which does not show the important action clearly to the viewer cannot be an effective shot. Undoubtedly, at one time or other, you have watched a television program in utter frustration because the shots the director selected or the camera operator framed simply did not deliver the visual information you wanted and needed to see to enjoy the show. For example, a profile shot is usually weaker and conveys less information than a full-face shot. This is particularly important when the subject is speaking directly to the viewing audience, as in a commercial or an instructional program. In this case, the better shot would be a straight-on. In almost every production situation, regardless of content, there are usually some shots and compositions which are more effective than others. Needless to say, it is impossible to give you any rules or guidelines since each situation is different. Your intuition and experience must determine the framing which will provide the best communication to the viewing audience by making optimum use of the videospace.

Essential Area

The picture you see in the camera's viewfinder is not the same picture the television audience ultimately sees. This is because, as the signal from the camera passes through the television system, the outer portions of the picture are "cropped" or lost, reducing the size of the picture and eliminating the subject material along the outer borders. The central portion of the camera's picture which is transmitted to the viewer is called the *essential area*, and every camera shot should be framed to include the most important visual information within this space.

A good rule of thumb to use in framing a shot is to assume the outer 10 to 15 percent of the camera viewfinder's shot will be lost to the viewer. Some camera operators like to draw a square on their viewfinders with a grease pencil to remind them to keep all important shot elements—the performers, the sponsor's product, graphic titles, and pictures—within the essential area. Of course, the program director must ultimately determine the framing and composition, but it is a lot

Image on
studio-camera
viewfinder

Image on
control room
monitor

"Safe area"
for average
T.V. receiver

FIGURE 4-6 Essential Area.
The camera viewfinder overscans the image, showing more information than
what the viewer will ultimately see. The camera operator must compensate for
this cropping effect by composing shots in the viewfinder so as to keep the most
important visual information within the essential area.

easier on everyone if you automatically frame each shot with the essential area in mind. (See Figure 4-6.)

Field of View

One of the most important choices facing the director and the camera operator is exactly how large to show the principal subject and his or her surroundings: the *field of view* of the shot. You might think of the field of view as the balance between the principal subject and his surrounding background area. A very wide field of view will take in a great deal of background but makes the size of the subject rather small in the shot. On the other hand, a tight close-up gives dominance to the subject who fills the screen, excluding any distracting background elements from the shot.

In the early days of motion pictures, many directors followed a basic shot progression sequence. The convention was to begin each scene with a wide, establishing shot to help orient the viewer to the spatial relationships among the elements in the scene. Once the audience was sufficiently comfortable, the shots would become progressively tighter, isolating the principal subjects and eliminating more and more background. Although a strict adherence to this formula is no longer necessary since audiences can follow the action without such an obvious orientation, the various shots within the progression make a convenient way to break down and organize the component parts of the videospace.

EXTREME LONG SHOT (ELS) A very wide field of view in which the camera takes in the entire playing area. The principal subject or subjects are small in relation to the background and tend to compete with the surroundings for the viewer's attention.

LONG SHOT (LS) A slightly closer field of view than the extreme long shot, but the subject remains dominated by the much larger background area.

MEDIUM SHOT (MS) The subject becomes

FIGURE 4-7   **Field of View.**
The sequence of shots illustrates the progression
from an "extreme long shot" to "extreme close-up."

much larger and more dominent. The background is still important but now shares the videospace with the subject.

CLOSE-UP (CU)   The subject becomes the primary focus of interest within the shot. Only a small portion of the background is visible.

EXTREME CLOSE-UP (ECU)   The subject virtually fills the screen and is clearly the central focus of the shot. Some directors call an extreme close-up on a subject's face a "slice shot" since the shot is so close it literally shows only a portion or a slice of the subject's face.

Needless to say, these descriptions of the camera's field of view are relative and can vary depending upon the other shots in the sequence, the individual director, or the program's subject matter. For example, in an intimate dramatic show, the director may tighten all the shots so that a normal medium shot is the widest shot in the show. The notion of what constitutes a "long," "medium," or "close-up" shot can also vary. One director's close-up may be another's medium shot. A good camera operator learns to adapt quickly to each director's personal style in framing the field of view. Finally, a shot can be long, medium, or close in relation to the size of the subject. A close-up of Mount Everest would probably be considered an establishing long shot for most other subjects. Similarly, an establishing long shot of an ant farm would be, under most other circumstances, an extreme close-up.

CUTOFF LINES   Another way to designate image size is to refer to subject *cutoff lines*. These are natural lines which occur at various points of the body and produce aesthetically pleasing shots. The most common cutoff shots are the *full shot* (FS), the *knee shot* (KS), the *waist shot* (WS), and the *bust shot* (BS). (See Figure 4-8.)

MULTIPLE SUBJECTS   When you must frame more than one person in the shot, most directors call the shot by the number of subjects to be included. For example, there is a *two-shot* and a *three-shot*. Anything wider than a three-shot is generally called a "wide" or "long" shot. Another grouping that is commonly used is the *over-the-*

shoulder shot (OS). This is an effective shot because it establishes a relationship between characters and, at the same time, enhances the depth of the shot. (See Figure 4-9.)

FIGURE 4-8   **Cutoff Lines.**
A convenient way to organize framing and to refer to shots is by using natural cutoff lines. This can also help to eliminate confusion because a "medium shot" may mean different things to different people, but a "waist shot" is a more specific shot description.

Head shot/Close-up

Bust shot

Waist shot

Knee shot

Full shot

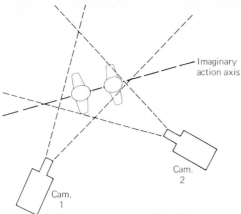

FIGURE 4-9 Reverse Angle Shots.
Frequently two cameras will shoot a reverse angle of a scene to show the viewer various points of view. In this case, all cameras must remain on the same side of the subject's "action axis" to avoid inadvertently reversing screen direction in the videospace when cutting between two cameras.

Camera Angles

The angle from which the camera photographs a subject or a scene is another important factor in building the videospace. Not only can a variety of camera angles provide the viewer with the most advantageous or interesting viewpoint, but certain camera angles produce a unique perspective which can affect the audience's perception of what is happening on the television screen.

NORMAL ANGLE The normal angle positions the camera at approximately the subject's eye level, shooting the scene as we would normally view the world. Of course, the "normal" angle depends to a large extent on the subject being photographed. If we were shooting a group of small children at play, our "normal" adult angle would be too high. In this case, the camera should be lowered to shoot at the children's normal angle.

As you operate your camera, stay aware of the normal angle. Remember that this refers to the subject's position, not to the camera height which will be most convenient and comfortable for you. If a subject is seated, the camera must be pedestalled down to eye level. If the subject gets up, the camera should pedestal up to maintain a normal angle. This is why most interview shows build their sets and chairs on risers to raise the seated height of the participants so that the camera operators can shoot at a normal angle without stooping or bending during the production.

HIGH CAMERA ANGLE A high camera angle positions the camera above eye level, with the camera shooting down on the subject. A high camera angle is very useful when trying to show the audience an overview of the set or playing area. In addition, a high camera angle tends to make the subject appear to be smaller in size and stature. Looking down at a subject from a high camera angle suggests a feeling of loneliness, lack of power, and loss of dominance.

LOW CAMERA ANGLE A low camera angle positions the camera below eye level so the camera shoots up toward the subject. Shooting up from a low angle tends to increase the audience's perception of the subject's size and suggests a feeling of power, dominance, and dynamism. This is why so many political candidates prefer to have their commercials shot from a slightly lower camera angle to give the audience the impression that they are physically taller and psychologically powerful and dominant. The same psychological effect can be used in product advertising. For example, many new-car commercials are shot with the camera at a low angle position, making the car seem larger, sleeker, and more powerful to the viewer.

CANTED ANGLE A canted angle is produced by tilting the camera on its horizontal plane. The picture that is produced is dynamic, exciting, and unstable. It suggests a feeling of fantasy, sus-

pense, or unreality to the viewer. A canted angle is very easy to achieve with a portable camera by simply asking the camera operator to tilt the camera to the desired angle. On studio pedestal cameras, which are not easily tilted horizontally, you can achieve the same effect by using a special prism device fitted over the lens. Turning the prism will tilt the image at any desired angle producing the canted effect.

The canted angle is clearly dynamic and exciting and shows us the world in a way we don't ordinarily see it. But for this very reason, the shot calls attention to itself and should be used sparingly to be most effective. (See Figure 4-11.)

SUBJECTIVE CAMERA ANGLE A subjective camera angle puts the camera in the place of a character and shows us the scene from the character's point of view. When used effectively in a dramatic production, the angle can make a great impact on the viewer. For example, let us consider a scene in which a woman is running away from an attacker. Shooting with a hand-held portable camera which travels along the same route as the woman, will involve the viewers in the action, making them feel a part of the play. The same technique can be used in instructional programs or in documentaries. Shooting a racing car speeding around a track from an objective angle may be exciting, but it cannot compare with placing a camera inside the driver's cockpit, showing the audience what the driver actually sees as the car races around the course. (See Figure 4-12.)

Framing Movement

The camera operator must frequently deal with a moving subject. Since both the subject and the camera itself can move, framing movement is an important dimension in building the videospace.

It is generally true that movement toward or away from the camera is more forceful, dynamic, and interesting than movement parallel to the camera. This is because movement toward or

FIGURE 4-10 **Camera Angles.**
Notice how our perception of the subject is changed
by varying the camera's shooting angle.

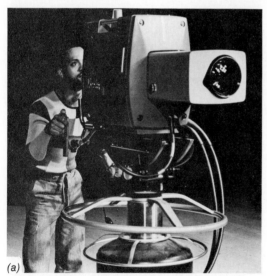

(a)

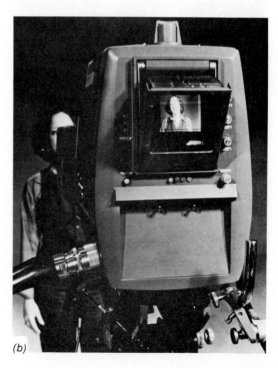

(b)

FIGURE 4-11 **Canted Angle.**
A canted angle suggests instability, excitement, or tension.

FIGURE 4-12 **Subjective/Objective Camera Angles.**
In the first frame (a), we see the camera operator from
an objective angle. In the second frame (b), we see a
subjective shot of the same scene in which we view
the scene from the camera operator's point of view.
Subjective shots are useful when you want to bring
the viewer into the action.

away from the camera tends to increase or decrease the subject's size in the videospace. This, in turn, enhances the depth perspective of the shot, producing a visually interesting and exciting picture. Parallel or lateral movement does not change the subject's size and, consequently, is not as dynamic or exciting.

When you follow a moving subject, particularly one moving across the screen parallel to the camera, try to lead the subject by giving him or her more space in front than behind. This will prevent the illusion of a moving subject's being forced against the leading edge of the picture frame.

Framing and following movement are not easy,

and they get even harder when you are shooting a tight shot. Unless you can carefully rehearse the move, you may have to shoot slightly wider to keep the subject properly framed and to reduce camera shake during the move. (See Figure 4-13.)

COMPOSITION

Composition is the organization of the visual elements within the videospace. A well-composed shot (1) is unified, (2) establishes the spatial and psychological relationships between the various components, (3) directs the audience's attention to the most important elements within the scene, and (4) produces a picture which is aesthetically pleasing and interesting to watch.

FIGURE 4-13 **Framing Movement.**
When panning with a moving subject, try to lead the subject by providing more space in the direction of the movement.

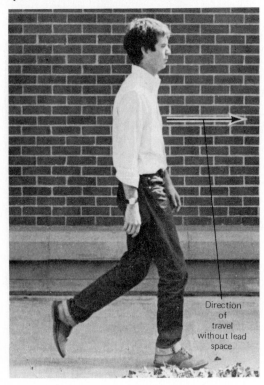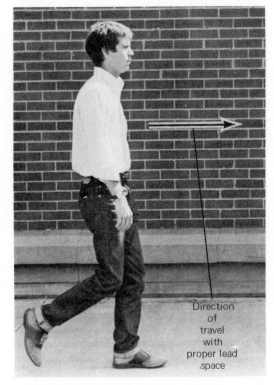

FIGURE 4-14 Composing in Depth.
The first frame is shot from a perpendicular angle and provides no depth to the scene. The second frame was shot from an angle to induce depth and produce a more interesting and effective shot.

Composing in Depth

A television picture is two-dimensional, providing information about height and width. Whatever depth appears in the shot must be achieved through the careful use of camera angles; performer blocking; such production variables as set design, lighting, and makeup; and picture composition.

When you compose a picture, try to enhance the depth of the videospace as much as possible. You might think of the scene as consisting of a number of different depth planes: the foreground, the middle ground, and the background. If a shot contains information from more than one depth plane, its depth perception is enhanced and the shot looks more interesting and alive to the viewer.

One way to do this is to shoot subjects from an angle rather than straight-on. Just as moving subjects appear more dynamic when seen moving toward or away from the camera, so too is the depth increased when shot from a slight angle. Another way to enhance depth is to make good use of the background area. A neutral background, which has little or no depth itself, will not convey to the viewer the feeling of space and distance between the principal subject and the background. Using the foreground plane by in-cluding foreground interest in the shot is another excellent way to increase depth. (See Figure 4-14.)

Organizing the Videospace into Groupings

Whenever we view a scene in our everyday lives, we unconsciously attempt to organize the various elements into some sort of cohesive, unified whole. This makes it easier for us to comprehend and to follow the action as it takes place. The great masters of painting, sculpture, and photography have capitalized on this trait to help focus and direct our attention to the most important elements within their works.

Although the number of possible groupings is virtually limitless, we can organize them into three major categories: (1) balanced and unbalanced groupings, (2) triangular groupings, and (3) foreground-background groupings.

BALANCED AND UNBALANCED GROUPINGS
A balanced grouping organizes the picture into a symmetrical arrangement in which all the major pictorial elements are equal in visual attraction or dominance. This does not necessarily mean that the subject is positioned directly in the center of the frame although this is one way to achieve

(a)

(b)

FIGURE 4-15 **Balanced Composition.**
In frame (a), the subject is balanced within the shot, and the impression is one of
equilibrium and stability. The second frame (b) is unbalanced, the left side of the
frame having greater visual weight.

balance. Rather, it means that when the various elements of the shot are combined in the videospace, the picture conveys a feeling of balance or symmetry to the viewer.

A balanced composition suggests stability, the lack of movement or dynamism, and the feeling of equilibrium.

An unbalanced shot or grouping achieves the opposite impression by organizing the videospace into an asymmetrical or unbalanced picture. The subject might be positioned to the side of the frame without a counterbalancing element to equalize the visual "weight" of the subject. The picture appears to be off-balance, and the asymmetrical arrangement draws the viewer's attention.

An unbalanced composition conveys the impression of instability and the potential for dynamic movement and often focuses the audience's attention on that particular element or portion of the shot which has the greatest visual weight. (See Figure 4-15.)

TRIANGULAR GROUPINGS The triangular grouping, as its name suggests, organizes elements within the frame into a triangular fashion, the visual dominence usually occurring at one of the apex points of the triangle. Although a triangle has three sides, this does not mean that the grouping can be used only with three subjects. Fewer or more than three subjects can be grouped into a pyramid design. The number of subjects which are organized is unimportant; it is the overall shape and grouping within the videospace that matters. The triangular grouping imparts a feeling of strength and stability within the videospace.

The triangular approach is a very effective grouping for a number of reasons. First, it enables you to include more subjects within a tighter shot than would be possible with a less compact grouping. Secondly, dominance can be easily achieved through effective positioning of the subjects and using the natural apex points to direct the viewer's attention. Finally, the triangular grouping can position subjects on different depth plans and enhance the depth dimension of the shot. (See Figure 4-16.)

FOREGROUND-BACKGROUND GROUPINGS
A foreground-background grouping arranges

FIGURE 4-16 Triangular Groupings.
Although three subjects form the most common type of triangular grouping, fewer or more than three subjects can be organized into a triangular grouping. *(Courtesy: Yorkshire Television Ltd.)*

subjects on two or more depth planes within the videospace. Needless to say, this grouping is very effective in enhancing depth perspective since the composition naturally uses foreground

FIGURE 4-17 Foreground-Background Composition.

and background planes. It is also an effective way of conveying visual dominance. Although the foreground element is usually the stronger in the shot, because of its enlarged size and prominence within the frame, this dominance can be modified through the use of selective focus and the effective use of depth of field. The composition also produces a very interesting and dynamic shot, which is one of its advantages.

There are two common variations on the foreground-background composition: (1) an over-the-shoulder shot, and (2) using foreground treatment.

OVER-THE-SHOULDER SHOT The over-the-shoulder shot frames a scene from over a subject's shoulder. This is usually used when two or more characters are speaking since it helps to establish the relationship between them. It is also frequently used on interview shows for the same reason. The over-the-shoulder shot is effective because it induces depth in the shot, establishes a relationship between subjects on different depth planes, and offers a great deal of versatility in framing different shots from the same basic camera and subject positions.

FOREGROUND TREATMENT Positioning an object or subject in the foreground of the shot to help frame background elements is an effective, interesting, and visually strong compositional technique. Sometimes the foreground element is an important thematic device—for example, the whiskey bottle which tempts the reformed alcoholic seated in the background. Used a different way, the foreground element might simply be a set piece, prop, doorway, or window which helps to induce depth and provide a visually interesting shot. (See Figure 4-17.)

Compositions to Avoid

Any shot which does not show the subject clearly or which unintentionally confuses the viewer is poor composition. Some of the most common pitfalls to avoid are the following:

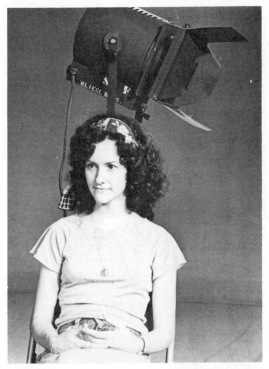

FIGURE 4-18 Poor Juxtaposition of Elements Within a Shot.
The poor composition makes it appear as though the lighting instrument is growing out of the subject's head. Moving the camera right or left or repositioning the subject would eliminate the problem.

POOR JUXTAPOSITION OF SUBJECTS The flower pot growing out of a person's head or a set piece extending from a subject's ear are unintentional compositions that you should avoid. Repositioning either the camera, the subject, or the background element should correct the problem. (See Figure 4-18.)

LACK OF HEADROOM Remember to consider the essential area, particularly when framing close-ups. A head shot, which is tightly framed in the camera's viewfinder, will usually mean insufficient headroom on the home receiver. While all four sides of the frame are cropped by the TV system, insufficient headroom is the most common problem.

LOOK SPACE When shooting a subject at an angle or in profile, give the subject some "look-

space" to compensate for the psychological impression that the look conveys. Composing the shot to give a bit more room in the direction of the look helps balance the shot and conveys a better visual impression to the viewer. (See Figure 4-19.)

DIFFERENCES IN SUBJECT HEIGHT Unless you are shooting an extremely wide shot, be careful to compensate for obvious differences in subject sizes. Either the taller subject should come down to the lower level—as in the host of a children's show coming down to the child's height—or a riser should be used to raise the shorter subject higher in the frame.

UNSUITABLE CAMERA ANGLE Be certain that the camera angle you are using to photograph a subject complements the visual impression you want the viewer to receive. For example, inadvertently shooting down on seated subjects may give the audience the erroneous impression that they lack dominance, authority, and power.

TOO MUCH SCREEN SPACE If you frame a shot with too much space in the center of the screen, you may draw the viewer's attention away from the principal subject over to the background area. This can be a problem in two-person interviews where each participant frames the edges of the screen, directing the audience's attention to the center of the videospace. Either move the subjects closer, shoot from an angle, or modify the set, lighting, or both to avoid having the central area compete for the viewer's attention. (See Figure 4-20)

DISTRACTING MOVEMENT All things being equal, a moving subject will invariably draw the audience's attention away from a static subject. Veteran performers know this only too well and sometimes use little movements to steal a scene from another actor. When you compose a shot, keep this factor in mind so the audience's focus is not inadvertently drawn away from what should be the dominant portion of the videospace.

FIGURE 4-19 Look Space.
Whenever a subject looks offscreen, frame him or her with more room in the direction of the look. In the first frame, the subject looks as though she were being pushed by the right side of the television frame. In the second shot, the added "look space" produces a more effective composition.

FIGURE 4-20 **Too Much Screen Space.**
The subjects are positioned along the edges of the screen, and the most important central screen space area is empty. The solution is either to move the subjects closer together or to change the camera's shooting angle to reduce the empty space in the middle.

PROFILE SHOTS In general, a full-face shot is far more effective and much more flattering than a profile shot of the subject. Whenever possible, avoid stark profiles or "ear shots" which do not show the viewer much of the subject's face. Naturally, there may be special circumstances which call for a profile shot, but in most cases cross-shooting or arcing the camera around slightly will provide you with a much more effective full-face shot instead of the weaker profile.

SUMMARY

Camera operation commands are designed to convey information quickly and accurately between the camera operator and the program's director. Commands for camera movement on a stationary pedestal are: pan, tilt, and pedestal. Commands which refer to the movement of both the camera and its base are: dolly, truck, and arc. A zoom produces the effect of motion although the camera itself does not actually move.

Of the four production phases, the camera operator is most concerned with two: rehearsal and production. In rehearsal, the director will set up shots and angles during camera blocking. On scripted shows the camera operator will usually receive a shot sheet which indicates each camera shot by number. If the show is unscripted, the camera operator will be told the basic shots which are expected, but the director must rely heavily on the operator's initiative and judgment to follow the action.

The camera operator's most important job is to build the videospace through framing and composition. Framing is the inclusion of various elements within the videospace and how they are shown to the viewer. One of the most important framing considerations is the essential area—the part of the camera's picture which will not be cropped or cut off during transmission. The field of view determines the subject size and varies from the extreme long shot (ELS) through the long shot (LS), medium shot (MS), medium close-up (MCU), and close-up (CU) to the extreme close-up (ECU). The camera angle is another important factor in framing the shot. A normal angle shows the scene as we would usually see it. A high angle positions the camera above the action, shooting down on the subject. A high angle tends to make the subject appear small, isolated, and psychologically weak. A low angle positions the camera below the action, shooting up at the subject. A low angle shot tends to make a subject appear large, strong, and psychologically more dominant. A canted angle tilts the camera on the horizon and produces a disorienting, dynamic and tension-filled shot.

Composition is the second major element in building the videospace. Composition refers to how the various pictorial elements are arranged within the videospace. A well-composed shot is unified, shows the relationship between the elements in the picture, concentrates the viewer's attention on the most important aspects of the shot, and produces a picture which is aesthetically pleasing and interesting to watch.

Elements within the videospace can be organized into any one of three major groupings: (1) balanced-unbalanced groupings, (2) triangular groupings, and (3) foreground-background groupings. Since television is a two-dimensional medium, shots which are composed to enhance the viewer's perception of depth are generally more interesting and more dynamic.

CHAPTER 5

TELEVISION LIGHTING

Equipment and Operation

Lighting is a fundamental part of any television production. Without proper illumination the television system cannot operate, and the camera will not reproduce an image clearly and accurately. Lighting is also a creative element in a television production because the picture itself is made up entirely of light. How a subject is illuminated contributes in large part to how the image looks in the videospace.

Since lighting is a mixture of both science and art, it is sometimes viewed as an almost magical area of production, its secrets shared among only a few practitioners. Though it is true that effective lighting requires creativity and imagination, it also demands the same careful and systematic attention to planning and execution as do all other aspects of a production. In this chapter we will discuss some of the important properties of light you will need to know in order to work with television lighting and lighting equipment and its operation. In the next chapter we will see how to use the equipment to light a subject or a scene.

OBJECTIVES OF TELEVISION LIGHTING

There are six basic objectives in television lighting:

To fulfill the technical requirements of the system. The lighting must provide a sufficient level of illumination to permit the television camera to operate and to pick up and reproduce faithfully the photographed image.

To provide a three-dimensional perspective. Television is a two-dimensional medium which reproduces an image with height and width. Depth must be provided through the use of camera angles, performer blocking, set design, and the careful use of lighting. Our perception of depth can be enhanced by the proper use of light to emphasize texture, shape, and form.

To direct attention to important elements in the scene. The use of light and shadow can help the program director by guiding the viewer's attention toward the most important elements in a shot or scene.

To establish the mood of the scene. Lighting can provide the viewer with a clue to the overall emotional mood of a scene. Dark, shadowy lighting will convey a feeling of mystery, tension, or drama. Brightly lit scenes will impart a feeling of happiness, gaiety, or fantasy.

To fix the time of the action. Lighting can set the time of day in which the scene takes place—daytime, evening, nighttime, and so on.

To contribute to the overall aesthetic composition of the shot. The use of lighting to accent various components within the picture will help the program director's pictorial composition. Lighting which is designed in coordination with the set designer and program director provides an aesthetically pleasing image for the viewer.

Of course not all six objectives are called for in every production situation. The design of the lighting must ultimately serve the program director's concept of the show. Lighting which is inconsistent with the program's overall objectives may be a visually artistic triumph but a practical failure.

THE NATURE OF LIGHT

Light is *electromagnetic radiation* which is transmitted from a wide variety of natural and artificial sources. The light which our eyes are able to see, visible light, is only a small fraction of the entire electromagnetic spectrum.

Our eyes perceive this visible portion of light as white light although it is really made up of a number of different colors. We are all familiar with the operation of a prism which separates white light into a series of colors called "spectral hues." These hues are actually different wavelengths of light which vibrate at various frequencies. We perceive them as colors.

Properties of Color Light

In an earlier chapter we discussed the properties of color light as they related to the operation of the color television camera. You should recall that the three basic attributes of color light are *hue*, *saturation*, and *luminance*, or "brightness." Hue describes the color of the light itself. Saturation describes the intensity of a particular color, or hue. Saturation depends primarily on the amount of white light which is mixed in with a pure hue; a 100 percent saturation represents the undiluted color which has little or no white light added. Luminance is the overall brightness value of the color. The brightness of a particular color depends upon how much light the color reflects. If we view colors of varying degrees of brightness on a monochrome camera, those colors with more brightness will appear to be white or light gray and those with less brightness will look dark gray or black.

To a large extent, hue, saturation, and brightness are subjective evaluations which are difficult to determine accurately. One of the biggest problems in measuring them objectively is that they are quite interdependent with each other and surrounding objects. A relatively light subject placed against an even brighter background will appear to be darker than the same object placed against a dark background.

Nevertheless, there is some convenience in accurately describing the properties of a particular light quantitatively, and a system has been devised to permit us to do this. It is called the *Munsell system* and was designed by the American painter, Albert Munsell. The system utilizes twenty basic hues, which revolve around a brightness axis arranged in nine steps from pure white to pure black. Each hue has a graduated saturation scale so that it is possible to determine a color's exact hue, brightness, and saturation from its position on the Munsell scale. (See Color Plate 7.)

Although the Munsell system is handy for designing the colors for costumes, sets, and graphics for color television, it would not be all that important if we were only going to reproduce our picture in color. While it is true that most programming is produced in color, a significant number of viewers continue to watch in black and white. In this case, the Munsell system can play an important role by showing us how various colors will appear to a viewer who is watching in monochrome.

All hues which lie on the same horizontal plane of the Munsell scale will reproduce in black and white with the same degree of achromatic brightness. Viewers watching in monochrome will see all of these different hues as the same shade of gray. This is a vital consideration when planning the use of colors for costumes, sets, and graphics since we must provide a certain amount of contrast between the foreground and background elements of our picture. Similarly, the use of colored light on a cyclorama or set piece must be planned with its brightness value in mind. If we select one hue for a performer's costume and another for the background and both appear on the same horizontal Munsell plane, viewers watching in color will easily perceive the difference but viewers watching in black and white will see both costumes and background as the same shade of gray. The necessary contrast between foreground and background elements will be lost.

You must also consider brightness values when you intend to use a color change as a thematic cue for the audience. For example, a close-up of a light changing from red to green will be obvious to the color viewer but meaningless to the viewer watching in black and white. In this case, you must be sure to change not only the hue but also to modify the brightness levels between the two color sufficiently so that all viewers will understand that a change has occurred.

Color Temperature

The "white" light emitted by various light sources is, in fact, made up of different mixtures of spec-

tral hues. *Color temperature*—expressed in "degrees Kelvin" (K)—is a convenient way to describe the particular mix of spectral hues which a given light source emits. Think of an early morning sunrise. The sky is reddish orange, and the color of everything is affected by the warm, golden tone of the light. By noontime, as blue light tends to overpower the reddish hues, the sky appears a bluish white and this, in turn, affects the colors of everything illuminated by sky light. Toward late afternoon, as the sun begins to set, reddish hues again predominate as the sky assumes a rosy glow and objects illuminated by sky light reflect a much warmer color tone.

Whenever you see this change in the color of the sky and the resulting color changes of objects illuminated by the sky, you are seeing nature's demonstration of color temperature.

A useful analogy to explain the concept of color temperature is to imagine an iron poker which is put into a furnace. The tip of the poker, which was black when cool, begins to glow a dark red as it heats in the furnace. As the poker grows hotter, the color of the tip changes from red to a bright yellow. If we leave the poker in the furnace long enough, the tip will continue to change from yellow to blue and then to white-blue.

If we were to measure the Kelvin temperature of the light from the poker with a Kelvin meter, we would see a steady increase in color temperature as the tip changed from red to yellow and to blue. A reddish-yellow color would read about 2800° K. The bright yellow would read about 3200° K, and the blue about 5500° K.

Although our analogy required the tip of the poker to be heated in order to change color, in reality the color temperature of light and an object's physical temperature need not be related at all. A firefly emits a glow which measures about 5000° K on a meter, but obviously the actual temperature of the firefly is not that hot. A blue sky which reads about 10,000° K has an actual temperature which is well below freezing. The Kelvin scale indicates the color quality of the light, not its actual temperature. As the color quality of light changes from yellow to blue, the color temperature rises.

Color temperature is important because the color temperature of a light source will affect the color of a subject which is illuminated by that light. You may have had the experience of trying to match two articles of clothing under fluorescent lights in a department store only to find that once you arrived home, the colors no longer matched as well as they had in the store. That is because the color temperature of flourescent light is about 4800° K and the light is blue in color. Tungsten light bulbs emit a much warmer, yellowish color, and the difference between the two light sources reflected from the clothing changed your perception of the clothing's color. The same idea is used in a variety of makeup mirrors which reflect a number of different color temperatures. The color quality of light in a candle-lit restaurant is quite different from the color temperature of a flourescent-lit office.

When television was exclusively monochrome, strict adherence to color temperature was not very important since the camera was insensitive to slight variations in the color temperature of different light sources. But once color cameras were developed, the color temperature of each light source became critical to the camera's operation. If every light instrument were to emit light with different degrees of color temperature, skin tones would be constantly changing depending upon the light source. Consequently, color television studios require that each instrument emit light with a precise color temperature so that the color camera can be balanced for a light of a specific color quality. Theoretically, the color camera can be balanced for any source. However, 3200° K is standard when using tungsten sources, while "daylight" sources are balanced in the range of 5000°—7000° K.

Hard and Soft Light

The quality of light produced by natural or artificial sources can be classified as either "hard" or "soft."

HARD LIGHT   Hard light is intense and directional and creates strong shadows. The light which is produced by a bright sun in a clear, cloudless sky and which results in strong, distinct shadows is a prime example of hard light. Most artificial light sources designed to produce hard light use a smooth, shiny, reflective surface to produce a "specular" reflection which results in a hard quality light.

SOFT LIGHT   Soft light is a uniform light which spreads in all directions and creates few, if any, shadows. The light we see on a cloudy, overcast day where the illumination is relatively uniform without creating any distinct shadow is an example of soft light. Soft light is produced by reflecting light from a rough or matte surface, which results in a "diffused" reflection. Owing to its diffused, shadowless quality, soft light is used to fill in shadows, to blend in illumination, and to provide a smooth, even lighting effect on a subject. (See Figure 5-1.)

CONTRAST RANGE

No television system yet devised is as sensitive as the human eye or can handle brightness variations as the eye can. At best, television cameras can reproduce a brightness or "contrast range" of about 30:1. This means the lightest area of a picture should be no more than thirty times brighter than the darkest area in order for the television system to reproduce the scene accurately. The contrast range is expressed as a brightness ratio and is a very important factor to consider in television lighting.

Strictly speaking, the contrast range is a relative concept because it refers to the lightest and darkest areas within a single camera shot. However, the camera's ability to handle excessively bright or high reflectance subjects or very dark or low reflectance subjects is also limited. Normally, the camera works best when it photographs a scene with no more than 60 percent white reflectance and no less than 3 percent black reflectance. These two brightness extremes are called *television white* and *television black*, respectively because this is how the camera reproduces these shades in the videospace. Using the two extremes as the ends of the brightness scale, we

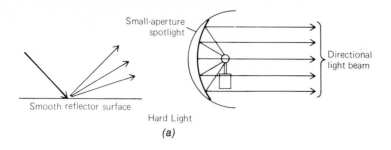

Small-aperture spotlight

Smooth reflector surface

Directional light beam

Hard Light

(a)

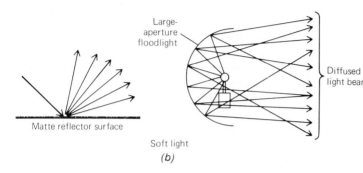

Large-aperture floodlight

Matte reflector surface

Diffused light beam

Soft light

(b)

FIGURE 5-1 **Hard and Soft Light.**
Hard light *(a)* is intense and directional and creates strong shadows. It is usually produced by a small-aperture light source utilizing a specular or highly reflective surface. Soft light *(b)* is a more diffused light, which spreads in all directions and produces few, if any, shadows. Soft light is produced with large-aperture light sources utilizing a matte surface reflector.

can progress from TV black through shades of gray to TV white. This is called the *gray scale*, which has been standardized using either seven or ten distinct gray scale steps. (See Figure 5-2.)

Although it is best for all lighting, costumes, and other production elements to adhere to the optimum picture contrast range, starting with the whitest elements at 60 percent reflectance and progressing downward to the darkest at 3 percent reflectance, there are times when this is impossible. In this case, the video engineer must compensate for the increased contrast range by "stretching" either the white or black ends of the picture. If the black end of the picture is stretched, the system will reproduce more detail in the darker ends of the picture at the expense of detail in the lighter end. On the other hand, if the white end is stretched, the engineer provides more detail in the lightest elements and "crushes" the black, losing contrast and detail in the darker portions of the picture.

The most important consideration when determining the contrast range of a picture is to ensure that the major subject of interest is included within the contrast range. This is usually the performer's face and, since skin tone reflects approximately 35 to 40 percent reflectance, or #3 on the gray scale under usual circumstances, skin tones will

normally fall well within the normal contrast limits. Using the face as a basic starting point, we can light the background brightness level about one and one-half or two times darker than the face for conventional scenes or as much as three or four times darker for more dramatic effects.

Even seemingly insignificant props or background pieces can inadvertently upset the contrast ratio. For example, you may have carefully lit a set within the proper contrast range only to find that a performer is using a script written on pure white paper. Since the white paper reflects well over the 60 percent reflectance limit for normal operations, the video operator must shade down the white end of the picture and this will result in compressing shadow detail. In this instance, the white paper should be replaced with beige or blue paper which reflects much less light and will keep the picture within the proper contrast range.

MEASURING LIGHT INTENSITY

Light intensity is measured with a *light meter* which is calibrated in footcandles (fc). The light measurement technique most commonly used in television is the *incident* light reading. An incident reading is taken with an incident light meter positioned at the subject's location with the meter facing toward the camera. The meter is designed to measure the amount of light, from a single light source or from a number of sources, which illuminates the subject. The incident technique has the advantage of being relatively easy to use, provides an average reading of the illumination on a scene, and permits the lighting director to measure the illumination on the set before the performers actually appear on the floor. (See Figure 5-3.)

The average reading obtained with an incident light reading is usually sufficient for most televi-

sion production situations since the main purpose in taking a light reading is to establish a basic operating level of illumination for the television camera. A more precise reading of the illumination on a subject or area can be taken with a second measurement technique, a *reflected* reading. This is taken with a reflected light meter held facing the subject from the camera's point of view. The reflected meter reads the level of illumination reflected from the subject and into the camera's lens. Reflected readings usually read the average brightness of the entire scene although some special meters called *spot meters* are designed to measure the reflected illumination from a very specific area, usually less than a 3° horizontal angle. Spot meters are rarely used in television except for special applications since the detailed information they provide is often

FIGURE 5-2 **Television Gray Scales.**
By using "television white" (about 60 percent reflectance) and "television black" (about 3 percent reflectance), it is possible to build a stepped brightness scale. A minimum of two or three gray scale steps is usually necessary to provide sufficient contrast on a color or monochrome receiver. Average skin tone is about 35 percent reflectance, or #3 on a ten-step gray scale. Step #5 on the ten-step scale—18 percent reflectance—is a standard photographic measure of brightness, which is frequently used to take average light readings.

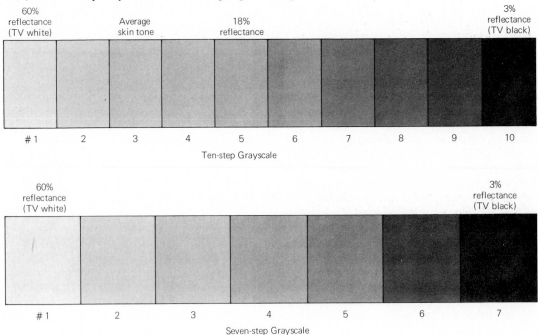

FIGURE 5-3 **Incident Light Meter.**
This meter measures light intensity in footcandles.
The white globe is designed to measure the average
light intensity on a scene or subject with the meter
facing the camera.

than if you light at 400 fc. A good rule of thumb is
to work with a footcandle level which is as low as
possible while still providing the video engineer
with sufficient illumination to operate the cameras
at a proper signal-to-noise ratio for a technically
good picture.

LIGHTING AND VIDEOSPACE

Many people are surprised to walk onto a brightly
lit set only to discover, once they view the picture
on the television monitor, that it appears as
though nighttime. This is because our eyes and
the television camera do not perceive the effects
of lighting in exactly the same way. For this reason
it is vitally important that the lighting director
always make a final check of the lighting on
camera since this is the audience's only measure
of reality.

While we must often make accommodations for
the limitations of the television camera, such as
using television white and television black instead
of pure white and black, the use of videospace
offers some distinct advantages when used properly by the lighting director. For example, lighting
can help portray to the viewer a reality which
does not actually exist. The effective use of videospace permits a 2,000-watt instrument to produce
the illusion of sunlight streaming through a window or enables us to light a scene which appears
to be nighttime and yet provides sufficient illumination for the technical operation of the cameras.
You must always remember that the way the
lighting appears in the studio is irrelevant. What is
crucial is how the lighting is translated by the
camera onto the videospace which is the viewer's
only reference point. It must look "right" on screen
to be judged effective.

LIGHTING EQUIPMENT

Television studio lighting systems include three
components: (1) lighting instruments or "lumi-

unnecessary for most television lighting requirements. (See Figure 5-4.)

The light meter is used in television during the
hanging and focusing of the instruments to establish a basic operating level of illumination. Determining what this level will be depends upon the
camera you will be using and the level of illumination which the video engineers determine as
necessary. Plumbicon, saticon, and similar modern tubes can operate quite comfortably between
75 and 250 fc. Even more sensitive cameras are
capable of reproducing acceptable images with
light levels as low as 5 or 10 fc. This is usually not
recommended, however, since the low light level
does not provide the video operator with a wide
enough range of control and the picture may
exhibit a deterioration in quality.

All things being equal, a lower light level is
preferred since smaller instruments can be used
and the lower level affords the lighting director
greater control in balancing the illumination of the
various instruments. It is a lot easier to accent or
highlight an area when you are lighting at 125 fc

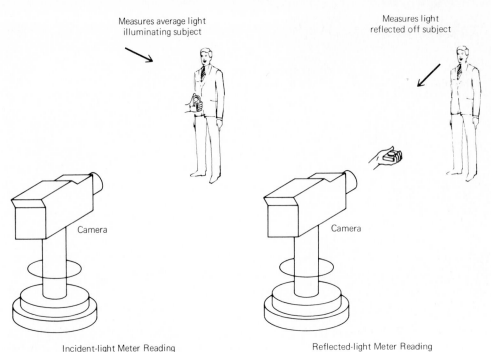

Measures average light illuminating subject

Measures light reflected off subject

Camera

Camera

Incident-light Meter Reading

Reflected-light Meter Reading

FIGURE 5-4 Light Meter Readings.
Incident readings measure the average illumination on a subject and are taken with the meter facing toward the camera. Reflected readings measure light, reflect from a subject into the camera, and are taken by pointing the meter toward the subject.

naires," (2) hanging or mounting devices, and (3) lighting control equipment.

Lighting Instruments

Lighting instruments are classified according to the type of light they produce. *Spotlights* produce a hard, directional beam of light, and *floodlights* produce a soft, diffused beam of light. All lighting instruments, whether spotlights or floodlights, utilize a lamp which serves as the actual source of illumination.

LAMPS For many years, the basic light source had been the incandescent tungsten lamp. Although the tungsten lamp provides a sufficient output of light when first installed in a lighting instrument, as the lamp ages there is a marked decrease in light output and a change in the color temperature of the lamp.

The development of the *tungsten-halogen lamp* (originally called the "quartz-iodine" lamp) solved both of these problems. The tungsten-halogen lamp maintains a constant color temperature by recycling the tungsten particles which are burned off the filament during the lamp's normal operation. This prevents the particles from settling on the glass enclosure, blackening the glass envelope, and decreasing the light input.

Tungsten-halogen lamps offer another advantage in that the size of the lamp itself is much smaller than older tungsten lamps. This has resulted in the design of tungsten-halogen lighting instruments which are smaller than their tungsten counterparts, yet provide an equal or greater amount of light output than larger tungsten instruments. For these reasons, tungsten-halogen lamps have become the standard of the lighting industry and are found in both color and monochrome television studios. (See Figure 5-5.)

All professional lamps are designed to operate at a color temperature of 3200° K. Although theoretically almost any color temperature could have been selected as the standard, 3200° K was chosen as an effective compromise between the level of light output and the life span which could be expected from an average lamp. As the color temperature of a lamp increases, the light output increases but the life span of the lamp is considerably shortened. For a sufficient light output with a relatively long lamp life, 3200° K has proved satisfactory.

An important point to remember when using tungsten-halogen lamps is never to touch them with your bare hand even when they are cool. This is because your hand and fingers leave an oily deposit on the glass envelope which interferes with the recycling operation and will lead to rapid deterioration and premature lamp burn-out. When installing or replacing a tungsten-halogen lamp, always remember to use a cloth or a plastic holder to prevent your bare hand and fingers from touching any part of the lamp's glass enclosure.

Lamps are available in a variety of light intensities which are rated in *watts*. Each lighting instrument has a specific lamp wattage capacity which must never be exceeded. For example, a spotlight rated at 1,000 watts (1 kw) can never use a lamp with a wattage greater than 1 kw.

SPOTLIGHTS A spotlight is an instrument designed to produce a relatively narrow beam of hard, directional light. The intense light produces strong shadows on areas not directly illuminated and permits the lighting director to control light and shadow. The types of spotlights most frequently used in television are (1) fresnels, (2) ellipsoidals, and (3) lensless or "openface" spots.

FRESNEL SPOTLIGHTS The most commonly used spotlight in television lighting is the fresnel

spotlight. These instruments are equipped with a specially designed fresnel lens which is made relatively thin and lightweight to resist overheating and cracking and yet provides a very even and directional beam of light. (Figure 5-6.)

Inside the fresnel spotlight, the lamp and reflector ride on a special gear track which permits you to vary the position of the lamp and reflector relative to the fresnel lens. When the lamp and reflector are positioned up front, close to the lens, the rays of light diverge, widening the width of the light beam and producing a less intense illumination. This is called "flooding out" the beam. When you move the reflector and lamp further away from the lens, toward the rear of the housing, the light rays converge, narrowing the width of the light beam and producing a more intense light. This is called "spotting down" (or "pinning") the beam. This focusing operation is one of the primary advantages of the fresnel spotlight. (See Figure 5-7.)

Fresnel spotlights are available in a wide variety of sizes and intensities from minispots of 100 to 150 watts through 5,000- and 10,000-watt giant spots. The most commonly used fresnels in most small and medium-sized studios are the 750-watt, 1,000-watt ("ace"), and 2,000-watt ("deuce") sizes. Natually, the higher the wattage, the stronger the beam's intensity and the greater the available light throw distance.

The fresnel spot is the "workhorse" of television spotlights due to its enormous flexibility. Because

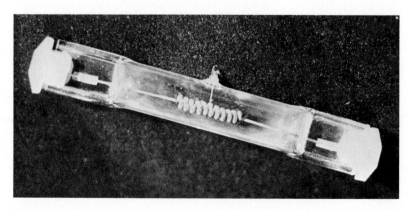

FIGURE 5-5 **Tungsten-Halogen Lamp.**

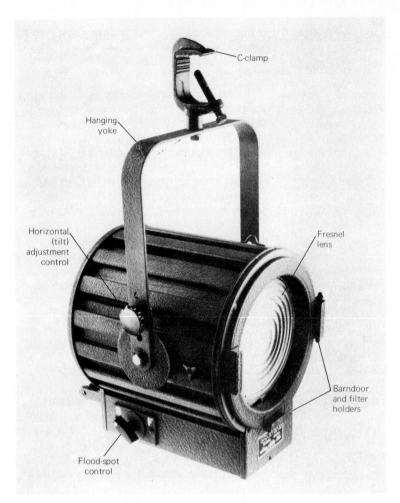

C-clamp

Hanging
yoke

Horizontal
(tilt)
adjustment
control

Fresnel
lens

Barndoor
and filter
holders

Flood-spot
control

FIGURE 5-6 **Fresnel Spotlight.**
(Courtesy: Kliegl Bros.)

of its ability to focus the light beam through a range of beam sizes and intensities and because the beam produced is relatively even and free of variations in intensity or "hot spots," the fresnel can be used in a wide variety of lighting situations. It can be used to illuminate performers, sets, and background areas as well as for special purpose applications.

ELLIPSOIDAL SPOTLIGHTS The ellipsoidal spotlight (sometimes called a "leko") projects a high intensity beam of hard, directional light with a rapid falloff in intensity at the edges of the beam. Ellipsoidals are equipped with internal shutters which enable you to "trim" the beam's

shape and size. Although you can focus an ellipsoidal spot by moving the lens toward or away from the fixed lamp and reflector, its focusing range is not as flexible as the fresnel's, and this is one reason why the ellipsoidal spot is rarely used to light performers. Another problem in lighting with the ellipsoidal is the fact that its very hard light beam tends to produce harsh and unflattering results on the subject. (See Figure 5-8.)

A unique feature of the ellipsoidal spot is the ability of some units to accept patterns for projection. These patterns are called "cucalorus," or "cookies," and are available in various patterns and designs.

Except for its use as a pattern projector, the ellipsoidal is not used in most common television lighting situations. It is primarily a special purpose unit which is utilized when its distinctive light characteristics are necessary for a particular lighting application. The most commonly used ellipsoidals are the 750-, 1,000-, and 1,500-watt sizes.

LENSLESS SPOTLIGHTS A number of spotlights are available which do not utilize a lens to focus the light rays. Although these lensless, or "open face" spots are capable of limited beam-focusing, their flexibility and range of control cannot compare with a fresnel spotlight. The primary advantage of the lensless spot is its lightweight portability and efficient light output, two features which are particularly important when lighting remote events outside the studio.

Since size and weight are generally not crucial in-studio considerations, where maximum focusing flexibility and control are of paramount importance, open face spots are rarely used as primary studio lighting instruments. However, their small size and highly efficient light output do make them ideal auxiliary instruments which can be used during studio productions in those situations where conventional instruments cannot be easily positioned. For example, you might need to illuminate an inaccessible portion of a set where conventional fresnel instruments cannot be hung or mounted. The lightweight lensless spot can be mounted behind a corner of the set piece out of

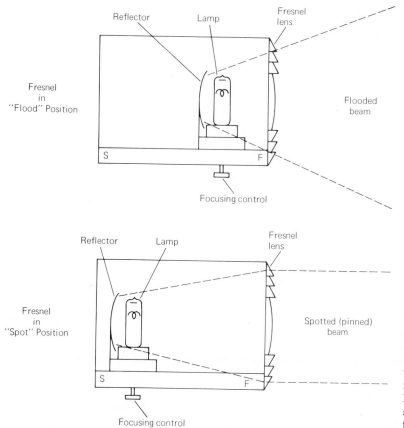

FIGURE 5-7 **Flood-Spot Focusing of Fresnel Spotlight.** Moving the lamp and reflector toward the lens spreads out the beam. Moving the lamp and reflector away from the lens concentrates the light rays into a more intense and directional beam. Spotting the beam is sometimes called "pinning the beam."

FIGURE 5-8 **Ellipsoidal Spotlight.**
(Courtesy: Kleigl Bros.)

FIGURE 5-9 **Lensless Spotlight.**
The lensless spotlight is sometimes called and "open face" spot.

camera range where it can supply the necessary extra illumination. (See Figure 5-9.)

SPECIAL PURPOSE SPOTLIGHTS A number of spotlights are used for special purpose applications and are rarely found in most small or medium-sized television studios unless there is a regular need for the type of lighting effect they produce.

HMI INSTRUMENTS An HMI lamp is a gas-filled bulb which operates in a manner similar to fluorescent lamps. The unit requires a special ballast or starter for proper operation. The advantage of the relatively new line of HMI instruments is greatly increased light output using smaller, lighter weight, and lower powered units as compared with conventional tungsten-halogen instruments.

An HMI unit produces about five times the light output of a comparable tungsten-halogen unit. For example, a 200-watt HMI unit produces approximately the same amount of illumination as a 1,000-watt tungsten-halogen instrument but with greatly lower power requirements, and produces 50 percent less operating heat. The high efficiency of the units delivers more light with smaller and lighter instruments.

HMI lamps are color balanced for daylight quality light (about 5600° K), which makes them ideal for use as booster lights for outdoor remotes and for lighting interior remote locations where only a few instruments can be utilized. The low heat output is another advantage since HMI units will not increase room temperature as rapidly as tungsten-halogen lamps do. Of course, color cameras must be rebalanced for daylight operation if they are used indoors with HMI units, but this can be done rather quickly. Finally, HMI units

FIGURE 5-10
The small and lightweight open face spot is frequently used for illumination on productions which are shot outside the studio on location.
(Courtesy: Berkey-Colortran)

from 200 watts to 1,200 watts (which produce illumination equivalent to 1,000-watt to 6,000-watt tungsten-halogen instruments) can be powered off conventional 120-volt/60-cycle circuits. (See Figure 5-11.)

FOLLOW-SPOTS The follow-spot is a large, high-intensity spotlight which has been borrowed from the theatrical stage. In the theater, the follow-spot is a quick way of directing the audience's attention toward a particular performer or area on stage. Follow-spots are equipped with controls which enable you to "iris out" or "iris in," enlarging or diminishing the size of the beam. The follow-spot is mounted on a floor stand, and the operator can pan and tilt the instrument to follow the action on the studio floor.

PAR LIGHTS The PAR ("Parabolic Aluminized Reflector") light is an integrated unit which includes a lamp, lens, and reflector. Different PAR units are available with beam spreads ranging from "wide" to "medium" to "spot." In addition, PAR units are available balanced either for 3200° K indoor tungsten or for daylight use.

Although the multiple PAR lamps are contained

FIGURE 5-11 **HMI Spotlight.**
(Courtesy: LTM Corp.)

123

on a single mounting, each unit can be positioned individually to distribute the light evenly over a large area. The highly efficient PAR units offer high light intensity with a long-range throw capability. They are very useful for lighting large, indoor areas on remote situations, especially when there is a large distance between the instrument position and the subject. They are also frequently used on daylight remote productions to provide fill light to lighten dark shadow areas and reduce an excessively wide contrast range on a subject. (See Figure 5-12.)

FLOODLIGHTS   Floodlights emit a beam of soft, highly diffused light which produces few, if any, shadows. One of the common characteristics of all floodlights is their use of a "large aperture" to provide a diffused light over a widely spread area. The three most common varieties of floodlights used in television are (1) scoops, (2) soft lights, and (3) broads.

SCOOPS   Just as the fresnel is the "workhorse" spotlight, the scoop is the most commonly used floodlight. Aptly named for its shape, the scoop consists of a lamp which is mounted inside a semicircular housing and surrounded by a matte-surfaced reflecting material. The scoop emits a very soft light which is highly diffuse and which cannot be directed to any one specific area as is possible with a spotlight. The soft, even illumination provided by the scoop makes it an excellent instrument to use for fill lighting and to even out areas of highlights and shadows, blending all the illumination on a set smoothly. (See Figure 5-14.)

The most commonly used scoop in television is about 18 inches in diameter and equipped with either a 1,000- or a 1,500-watt lamp.

SOFT LIGHTS   The soft light is a very large aperture instrument which produces the most diffused light of any floodlight. Although the soft light is used extensively in still and motion picture photography, it has only recently been adopted for use in television as a floodlight. Soft lights are ideal for producing virtually "shadowless" lighting, and they can be used to fill in shadows which

FIGURE 5-12   **PAR Light.**
(Courtesy: Berkey-Colortran)

FIGURE 5-13
Banks of PAR lights are useful for filling in shadow areas when producing outdoors on location. The instruments pictured here are balanced for daylight to match the color temperature of the natural illumination.
(Courtesy: Berkey-Colortran)

FIGURE 5-14 **Scoop Floodlight.**

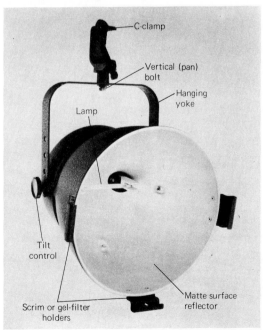

C-clamp

Vertical (pan) bolt

Hanging yoke

Lamp

Tilt control

Scrim or gel-filter holders

Matte surface reflector

FIGURE 5-15 **Softlight.**
(Courtesy: Kliegl Bros.)

Kliegl Bros.

are produced by directional instruments such as fresnel spots. The biggest disadvantage of the soft light is its large size and the fact that the large aperture of the soft light does not permit the lighting director to isolate its beam toward one specific area. Nevertheless, the soft light is becoming an indispensable source of soft, diffused illumination and is very valuable for special applications when diffused lighting is required.

Soft lights are available in sizes ranging from 750 watts to 5,000 watt (See Figure 5-15.)

BROADS The broad, sometimes called a "pan light" owing to its shape, is a rectangular floodlight. Its high light efficiency and convenient size have made it popular for both in-studio and remote lighting applications. The broad does not produce a light which is as soft and diffused as that from a scoop or soft light, but its beam can be controlled by trimming with metal flap "barndoors." Most broads used in television are equipped with 1,000-, 2,000-, and 5,000-watt lamps. (See Figure 5-17.)

SPECIAL PURPOSE FLOODLIGHTS There are two additional types of floodlights which are used for special purpose lighting applications.

STRIP LIGHTS Strip lights are long rows of broads which are used to illuminate a large area evenly. They are most commonly used to light background pieces, curtains, or cycloramas. Strip lights are available in a variety of combinations and sizes and can be either hung from above or positioned on the studio floor behind a set piece and angled upward toward a curtain or cyclorama. Strip lights are easily fitted with color media to produce colored lighting effects. (See Figure 5-18.)

CYCLORAMA LIGHTS Almost all television studios are equipped with a cyclorama—a large, continuous piece of muslin or scrim material surrounding the background edge of the studio. A number of manufacturers have developed a special cyclorama (or cyc) instrument which employs a highly efficient reflector to spread a very even

FIGURE 5-16 **Softlights Mounted from Studio Lighting Grid.**
(Courtesy: ABC-TV)

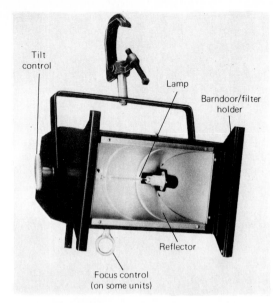

Tilt control

Lamp

Barndoor/filter holder

Reflector

Focus control (on some units)

FIGURE 5-17 Broad.
(Courtesy: Berkey-Colortran)

beam of illumination across a wide area of the cyc. The cyc projectors are hung from the studio battens on the ceiling and, when properly focused, provide an even wash of illumination across the entire cyclorama. Cyc lights are usually fitted with color material to light the cyc in a variety of hues, which is very effective in color production. (See Figure 5-19.)

Hanging and Mounting Systems

All television studios are equipped with a series of battens or grids which are suspended from the

ceiling, usually in a crisscross pattern. The lighting instruments are mounted and hung from these battens. The most common type of batten is the *pipe grid*, which is a series of metal pipes hung from the studio ceiling. In some studios, the battens are on a counterweight or motor-driven system which permits them to be easily lowered to the floor for mounting instruments and then raised to any desired operating height. This is a great advantage since various types of programs may require different batten heights for optimum lighting. The major disadvantage of the movable batten is its increased cost over fixed position grids. (See Figure 5-20.)

The mounting device which is used with the pipe batten is the *C-clamp*. The clamp itself is attached to the yoke of the lighting instrument and once the C-clamp screw is fastened tightly around the pipe, the instrument is secured firmly in position. Once the C-clamp has been tightened in place, the instrument can be panned or tilted to permit focusing the light beam. (See Figure 5-21.)

A variation of the C-clamp/pipe-batten system is the *lighting track*. Instead of a metal pipe, a track with two flanges is used to hold the instrument in place on the batten. A special mounting device replaces the C-clamp and is designed to ride within the track. A thumbscrew on the track clamp is used to tighten the instrument securely into position. The lighting track system has not been widely used because of its increased cost

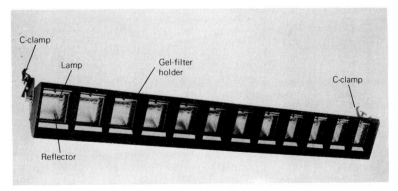

C-clamp

Lamp

Gel-filter holder

C-clamp

Reflector

FIGURE 5-18 Strip Lights.

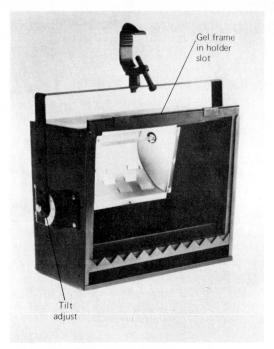

FIGURE 5-19 **Cyclorama Projector.**
The cyc projector is usually fitted with a colored gel
filter to "paint" a cyc for color productions.

FIGURE 5-20 **Lighting Grid.**

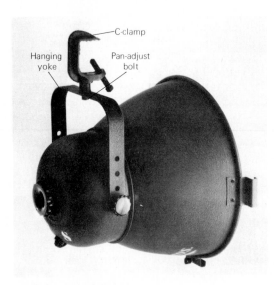

FIGURE 5-21 **C-Clamp.**

over the conventional pipe batten and because
hanging and removing instruments are more
cumbersome than with the C-clamp.

PLUGGING STRIPS Running parallel along
each pipe batten or track is a duct or "power rail"
containing cable which supplies electrical power
for each instrument. Hanging from the plugging
strips, at equal intervals, are short cables with
connector plugs called *pigtails*. The pigtails are
the wires into which each instrument is plugged in
order to connect it into the electrical power system. Above each pigtail is a number corresponding to a patch cord on the dimmer patch panel.
This permits you to connect each instrument to a
pigtail and then to patch the specific instrument
into a particular dimmer circuit. (See Figure 5-22.)
    Depending upon the system employed in your
studio, the ends of the pigtails are equipped with
either three-pin connector plugs or with twist-lock
connectors. The male counterpart of either plug is
attached to the connector cable of every lighting
instrument. Twist-lock plugs require a half-turn to

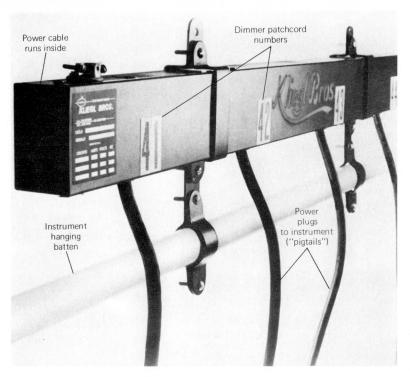

Power cable
runs inside

Dimmer patchcord
numbers

KLIEGL BROS.

Instrument
hanging
batten

Power
plugs
to instrument
("pigtails")

FIGURE 5-22 **Plugging Strip.** The electrical power for lighting instruments runs through cables and terminates in individually numbered "pigtails." The numbers along the strip refer to individual dimmer patch cords and permit each instrument to be connected to any dimmer in the lighting system. *(Courtesy: Kliegl Bros.)*

FIGURE 5-23 **Lighting Connector Plugs.** Instruments utilize either the three-pin plug *(a)* or the twist-lock connector *(b)*. The twist-lock connector requires a half-twist to secure or to unplug a connection.

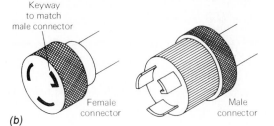

(a)

Keyway
to match
male connector

Female
connector

Male
connector

(b)

complete the connection, and you must be sure to line up the keyway properly before attempting to push them together. (See Figure 5-23.)

AUXILIARY HANGING DEVICES Although the majority of instruments are hung directly from the studio battens, you may sometimes need to hang certain instruments at lower heights than a fixed grid will permit. In this case, an adjustable hanging pole is attached to the batten, and the instrument is hung at the bottom end of the pole. The length of the aluminum pole is variable and allows you to position the instrument at the necessary operating height. (See Figure 5-24.)

A commonly used alternative to the telescoping pole is the *pantograph*, and accordionlike mounting device which permits an instrument or a ganged pair of instruments to be raised or lowered to a particular operating height. For a pantograph to be used effectively, the fixed grid must be at least 15 to 20 feet above the studio floor. Although the pantograph does permit you to raise or lower the height of the instrument relatively easily, it is heavy, cumbersome, and difficult to move from batten to batten. In addition, pantographs have a tendency to slip, rotate, or swing, which can alter the position of a previously focused instrument. (See Figure 5-25.)

Sometimes a lighting director must position an

instrument in a place where neither battens nor pantographs will permit. In this case, a *gaffer-grip* clamp can be used to fasten a lightweight instrument to sturdy set pieces or studio flats. The clamp is equipped with heavy, spring-loaded jaws which firmly grasp any solid surface. The ideal instrument to use with a gaffer-grip clamp is a lightweight lensless spotlight which can be held securely in place by the clamp.

There are a variety of additional hanging devices which can be employed for special lighting

FIGURE 5-24 **Telescoping Hanging Light Pole.**
The aluminum light pole permits instruments to be hung lower than the normal grid height allows. The pole has the advantage of remaining stationary in positon; it will not sway or vary in height unlike the pantograph.
(Courtesy: Mole Richardson Co.)

FIGURE 5-25 **Pantograph.**
Although most studios are replacing pantographs with telescoping poles and other alternative hanging devices, a number of studios still use them owing to their convenience.

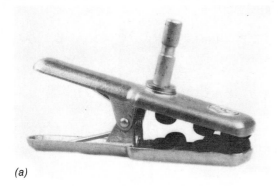

(a)

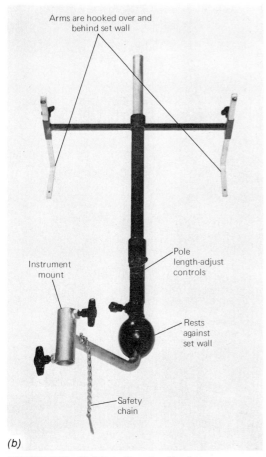

Arms are hooked over and behind set wall

Instrument mount

Pole length-adjust controls

Rests against set wall

Safety chain

(b)

FIGURE 5-26 Lighting Hanging Devices.
The gaffer-grip *(a)* permits a lightweight instrument, such as an openface spot, to be positioned anywhere necessary. The trombone *(b)* permits an instrument to be positioned over a studio set wall, enabling the set wall to support the weight of the instrument.
(Courtesy: Mole-Richardson Co.)

applications. Among these are *trombones, set wall brackets,* and *wall sleds.* (See Figure 5-26.)

Floorstands are telescoping metal poles which are mounted on casters for easy movement. Since so much equipment is constantly moving across the studio floor during a television production, it is wise to try to hang as many instruments as possible on the studio grid where the instruments and cables will not interfere with the operation of cameras, booms, and other production equipment. However, floorstands are used in those instances where the proper positioning of the light requires it to be mounted on the studio floor. Floorstands are commonly used when lighting tabletop objects or to illuminate windows or backings from behind the set where the floorstand will not interfere with most studio operations. (Figure 5-27.)

Lighting Control Equipment

One of the most important elements in effective lighting is the ability to control the light from every instrument. When we talk about lighting control, we refer to three distinct, but related, operations: (1) controlling the distribution of the light, (2) controlling the intensity of the light, and (3) controlling the power distribution of electricity to the various lighting instruments.

CONTROLLING LIGHT DISTRIBUTION There are four ways to control the distribution of light: (1) the positioning of the instrument, (2) barndoors and shutters, (3) beam control on the fresnel spotlight, and (4) lamp-to-subject distance.

INSTRUMENT POSITIONING The most obvious method of controlling light distribution is the placement of the instrument relative to the subject. Positioning an instrument directly overhead at a 90° perpendicular angle produces the smallest area of light distribution. As the vertical angle decreases, the light distribution and the size of the beam increase. (See Figure 5-28.)

BARNDOORS AND SHUTTERS Barndoors are metal flaps which are attached to the front of fresnels, lensless spots, and to some broads. Barndoors come in two basic varieties, two-door and four-door. Four-door units are more flexible as they permit you to control spill light on all four sides of an instrument. All barndoors permit you to rotate the entire barndoor 360° for accurate lighting control. Barndoors are one of the most convenient and commonly used methods of distribution control. Sometimes a slight adjustment of the barndoor is sufficient to provide the proper placement of shadow and highlight on a performer or background area, reduce unwanted shadows, including boom shadow, and eliminate lens flare in the camera. (See Figure 5-29.)

Shutters are built into ellipsoidal spotlights and permit you to trim the light beam into a variety of shapes and sizes. Manipulating the four shutters will produce different circular and rectangular light beams.

There are a number of additional devices which can be used to control the light distribution of an instrument. A *snoot* is a long, cylindrical tube

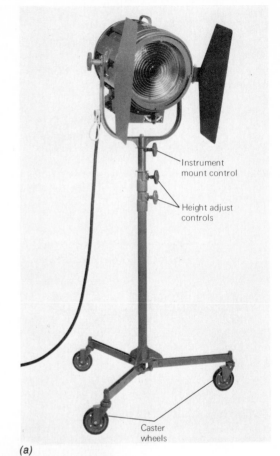

Instrument mount control

Height adjust controls

Caster wheels

(a)

(b)

FIGURE 5-27 **Floorstand.** The studio floorstand *(a)* is a heavy-duty stand which will accommodate almost all standard lighting instruments. For remote operation, collapsible floorstands *(b)* are usually more convenient and offer greater flexibility when used with lightweight openface instruments. *(Courtesy: Mole-Richardson Co. and Berkey-Colortran.)*

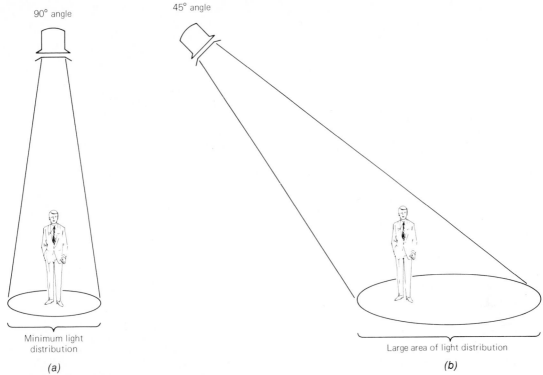

90° angle

45° angle

Minimum light
distribution

Large area of light distribution

(a)

(b)

FIGURE 5-28 **Light Distribution and Lamp Position.**
As the lamp-to-subject angle decreases, light distribution increases.

FIGURE 5-29 **Barndoors.**

which is mounted over a spotlight to control the light. A *flag* is a black, scrimlike piece of material which is usually mounted on an instrument or mounting device with a flexible arm which allows you to position the flag to best control the instrument's light distribution. These last two devices are commonly used in film production, but they can be used effectively in television for special applications.

BEAM CONTROL ON THE FRESNEL SPOTLIGHT Flooding out the beam on a fresnel spotlight will enlarge the light beam's area of distribution. Spotting down the beam will reduce the size of the light beam and decrease light distribution.

LAMP-TO-SUBJECT DISTANCE As the lamp-to-subject distance increases, the distribution of the light will increase. The reverse, of course, is also true: as the lamp-to-subject distance decreases,

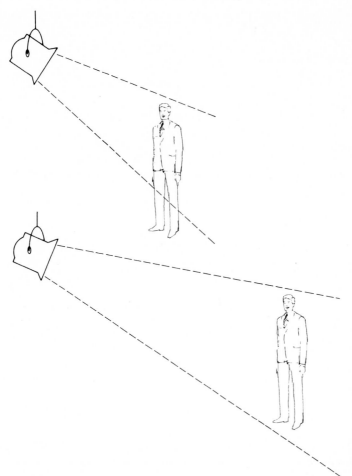

FIGURE 5-30 **Lamp-to-Subject Distance and Light Distribution Control.**
The closer the subject is to the lighting instrument, the more concentrated the light. The farther the subject is from the light source, the greater the beam spread and the wider the light distribution.

the size of the light beam is reduced. (See Figure 5-30.)

CONTROLLING LIGHT INTENSITY   There are five ways of controlling the intensity of light. They are (1) lamp wattage, (2) lamp-to-subject distance, (3) scrims and screens, (4) beam control on the fresnel spotlight, and (5) dimmers.

LAMP WATTAGE   The most obvious method of controlling light intensity is to select an instrument which will produce approximately the correct level of intensity when operating at full power. Naturally, all other factors being equal, a 500-watt fresnel will produce much less light than a 2,000-watt fresnel.

LAMP-TO-SUBJECT DISTANCE   The distance between the subject and the instrument is another means of controlling the light intensity. The greater the subject-to-lamp distance, the less intense the light. An important point to remember about lamp-to-subject distance is the *inverse square law*. Simply stated, as the lamp-to-subject distance increases, light intensity tends to fall off much more rapidly in proportion to the distance. For example, doubling the lamp-to-subject distance decreases the light intensity to about 25 percent of the original brightness level. Practically speaking, we must remember that after a certain lamp-to-subject distance has been reached, the falloff of light will diminish much less rapidly and other means of controlling light intensity must be employed.

SCRIMS   A scrim is a translucent material, made

of gauze or a glass fiber, which is used to diffuse and soften the light. Scrims also decrease light intensity without affecting the color temperature. Scrims are commonly used on scoops and other floodlights to increase the soft-light quality and to allow their illumination to blend in smoothly with other instruments.

SCREENS Screens are made of wire mesh and reduce light intensity without diffusing the light or altering the color temperature. For this reason, screens are commonly used on spotlights to reduce their intensity while not affecting the hard, directional light beam. Both screens and scrims can be cut to any desired size before mounting in an instrument. This permits you to diffuse or reduce the intensity of a portion of an instrument's light on one area while still maintaining the original brightness and light quality on another area. For example, cutting a scrim in half is an effective way to decrease a light's intensity on a performer who is positioned close to the spot while maintaining the spotlight's intensity for those areas farther away. (See Figure 5-31.)

BEAM CONTROL ON THE FRESNEL SPOTLIGHT
The beam control on the fresnel not only controls the size or distribution of the light but also affects the intensity. When the beam is spotted down, the "hot spot" produced by the instrument is about twice the intensity of a flooded out beam from the same instrument. Most lighting directors like to work with a fresnel which is completely or almost completely flooded out since this will produce a relatively even spread of light across the entire throw with a much more flattering effect on the subject. A spotted down beam has a very obvious hot spot with a rapid light falloff which can produce uneven lighting on a subject or area.

When focusing a fresnel, it is a good idea to spot the beam down completely, using the hot spot to center the beam on the subject. Once the instrument has been firmly locked into place, the beam can be flooded out as far as necessary to achieve the proper light distribution and intensity level, and any additional spill light can be trimmed by using the barndoors.

DIMMERS The most accurate and flexible means of controlling light intensity is to use dimmer circuits. There are three common types of

dimmers, (1) resistance dimmers, (2) autotransformer dimmers, and (3) electronic dimmers. Although they operate in slightly different ways, their purposes and functions are exactly the same. The dimmer controls the amount of electrical power which reaches each lighting instrument. The operator can use the dimmer control to increase or decrease gradually the amount of voltage and, in so doing, vary the intensity of the light. One hundred percent power produces an instrument's full intensity; zero percent power completely blocks all power and cuts off the light. Dimmers are indispensable in balancing the intensity of each instrument to achieve the desired lighting effect.

Resistance dimmers are rarely found in operation today as they are large, unwieldy to operate, and generate a substantial amount of heat and noise during operation. They have been replaced by the *autotransformer dimmer*, which is still rather large in size when compared with newer, electronic dimmers, but offers much easier operation than resistance dimmers and produce far less objectionable noise and heat. (See Figure 5-32.)

The most advanced dimmers available today

FIGURE 5-31 **Screen.**
A screen is useful for decreasing light intensity without affecting color temperature quality or overly diffusing the light beam.

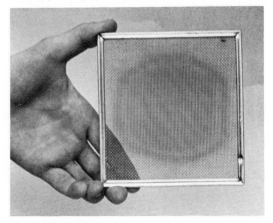

are electronic dimmers, sometimes called "silicon controlled rectifier," or SCR dimmers. These dimmers employ solid state circuitry and a small "pilot" voltage to control the electrical power. Since only a low pilot voltage is needed through the dimmer control board, the actual dimming circuits can be installed far from the studio, backstage or in the basement, where they will not get in the way or create noise and electrical interference problems. The control panel itself can be situated in either the studio or in the control room. Some systems use a portable dimmer board which can be plugged into a connector in the studio during setup and focusing opera-

FIGURE 5-32 **Autotransformer Dimmer.**
(Courtesy: Kleigl Bros.)

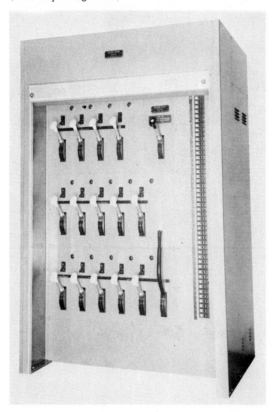

tions and then moved to the control room and plugged into another connector during the actual program production. SCR dimmers offer added flexibility in their ability to be interfaced with a computer memory system which stores information about each dimmer setting. These settings are then coded as "presets" and can be immediately recalled when the operator simply presses a button. (See Figure 5-33.)

DIMMER OPERATION Dimmers operate just like the faders on an audio console. When they are in the "down" position, they stop all current flow and the instrument connected to the dimmer circuit is "off." As the dimmer is moved upward, the electrical current begins to flow to the instrument and the lamp starts to glow. When the dimmer is in the topmost position, there is no resistance across the circuit and the lamp functions at full intensity. Dimmers are marked in increments called "points" from 0 to 10, with "0" being "off" and "10" being full power, or full light intensity.

A single dimmer control will operate one instrument or a number of instruments. The total number of instruments a dimmer can safely handle is called the *dimmer capacity* and is commonly rated in terms of watts rather than in actual number of instruments. Thus, a 12,000-watt dimmer is capable of controlling a maximum of 12,000-watts. This could be a combination of twelve 1,000-watt instruments, or six 2,000-watt instruments or three 2,000-watt fresnels and four 1,500-watt scoops, and so on. It is important for you to know the available dimmer capacity before planning your lighting so that you do not overload the dimmer and cause the safety circuit to shut off power.

Depending upon the size and sophistication of the lighting system, some dimmer boards are equipped with a number of dimmer *presets*. The preset enables you to set the exact operating level on a number of different dimmer circuits and then simultaneously fade up or fade out all the dimmers with one control. The advantage of the preset, aside from its obvious operating convenience, is that it permits you to fade up a number of dimmers, each set at a variety of points, in proportion with one another. For example, you may have balanced the lighting on a set by using instruments connected into three dimmers. The

FIGURE 5-33 SCR Dimmer Board.

When the dimmer controls are in the bottom position, the current is off, and no electricity reaches the instruments which are connected to the dimmer. When the dimmer control is in the uppermost position, full current flows to the instruments and they glow at full intensity. The numerical scale is measured in "points" from 0 to 10. The submaster switches permit the dimmer board operator to assign individual dimmers to various submasters so that multiple dimmers can be operated with only one control.

first dimmer is set at "10," the second dimmer is set at "5," and the third dimmer is set at "8." Fading these three dimmers up on a preset handle will bring up each dimmer in direct proportion with each other so that when the master preset control is up to half, the first dimmer will effectively read "5," the second dimmer will read "2½," and the third dimmer will read "4."

DIMMING AND COLOR TEMPERATURE Dimming any lamp, even a tungsten-halogen lamp, will affect the color temperature of the light. Although we must be careful not to dim lights too far when lighting for color or we will destroy the color balance of the lighting, it has been found that an instrument can be dimmed to one-half its normal brightness before the color shift becomes objectionable. If you find that you must operate an instrument dimmed beyond this point, you should probably choose an alternate method of reducing the light intensity such as adding a scrim, increasing the light-to-subject distance, or selecting a less powerful instrument.

CONTROLLING POWER DISTRIBUTION TO INSTRUMENTS You will recall that along each studio batten runs a cable conduit with pigtail connectors for each instrument. Every pigtail is numbered, and a connector cable leads directly to the patch board system, which is usually located backstage of the television studio. The "patch board" resembles a large telephone switchboard with a number of patch cords and an assortment of dimmer circuit connector outlets.

Each patch cord has a number engraved on the plug corresponding to the pigtail number on the studio grid. Thus, if we have 100 pigtails in the studio, we will find 100 patch cords at the patch panel. The patch cord for each pigtail can be plugged into any dimmer circuit on the patch board. For example, if we wish to connect an instrument which is plugged into pigtail #35, and assign this to dimmer #1, we would find patch cord #35 and plug it into one of the connector outlets for dimmer #1. This completes the electri-

cal circuit and permits the dimmer control #1 to regulate the intensity of every instrument which is patched into its circuit. (See Figure 5-34.)

Many patch panels also have a number of circuits which are not connected to any dimmer controls. These *nondim circuits* operate with a simple on-off switch and are used with those instruments which will not require any dimming during a production.

SLIDER CROSS-CONNECT SYSTEM Another system of assigning instruments to dimmer circuits actually eliminates all patch cords by employing a crosspoint matrix with slider switches. Each dimmer circuit is represented by a horizontal bus with a series of numbered contacts along the bus on which the slider switch travels. Every pigtail outlet in the studio grid has a corresponding contact along each horizontal dimmer row. The operator simply slides the switch vertically across the horizontal buses, permitting the connection of any pigtail outlet to any dimmer circuit in the system. You can assign as many instruments to a dimmer as you choose, provided you do not exceed the dimmer's capacity. The slider cross-connect system becomes impractical to use for a large number of dimmer circuits and outlets because it is difficult to scan the matrix quickly to see how many loads are patched into a particular dimmer.

COMPUTERIZED PRESET SYSTEMS The ultimate in dimmer control can be found in systems which utilize a computer to store information about dimmer settings and lighting cues and even to control the execution of a number of multiple cues on the lighting director's command. Some of these systems utilize a light pen and a CRT, which the operator uses to instruct the computer in virtually every lighting function. The intensity for each dimmer is controlled with the light pen and can be automatically activated by the computer on command. Once the lighting director is satisfied with the dimmer settings for all instruments, the computer is instructed to store all dimmer settings in its memory. These settings are instantly retrievable by a simple light-pen command from the control board operator. (See Figure 5-35.)

The ability of the computer's memory to store

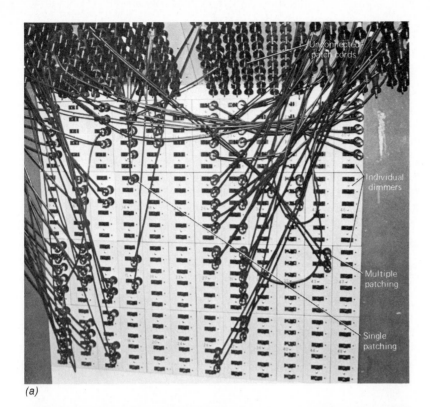

(a)

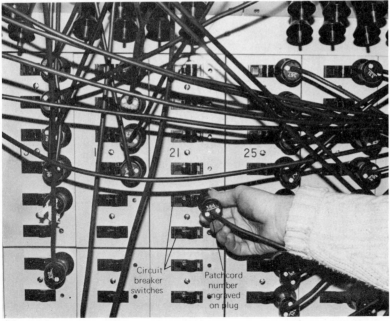

(b)

FIGURE 5-34 Lighting Patch Board.
Every pigtail connector in the studio terminates at its own numbered patch cord. Plugging the appropriate patch cord into the proper dimmer completes the electrical circuit and enables the dimmer board operator to control the light intensity of every instrument.

CRT/light-pen unit
to control dimmer settings

FIGURE 5-35 **Computer Assisted Lighting System.**
The operator controls the dimmer intensities by using a light pen and a special CRT. Once the various dimmer intensities have been established and the lighting is properly balanced, different dimmers can be assigned to various preset cues for flawless execution of a complex series of lighting changes.

and retrieve a massive amount of precise information permits lighting directors to incorporate many complex lighting cues with the assurance that they will all come up exactly as planned. The computerized systems not only save time during setup and balancing, but also free lighting directors to express their imagination and creativity to the fullest.

SUMMARY

Lighting is a fundamental part of any television production. Although the most important objective in lighting is to provide sufficient illumination for the operation of the television cameras, it is also used to enhance three-dimentional perspective, direct attention to important elements in the picture, establish the mood and time of day, and contribute to the overall aesthetic composition of the shot.

The quality of light can be either hard or soft. Hard light is intense, directional, and creates strong shadows. Soft light is diffuse, spreads uniformly in all directions, and produces light shadows. Spotlights supply hard light and are used as the primary source of illumination. Soft light is supplied by large aperture floodlights, and their diffused illumination is ideal for filling in shadows created by hard light and for blending in illumination on the set.

Light intensity is measured with a light meter. The most commonly used technique is the incident reading, which is taken from the subject's position and reads the average light that falls on the subject. A more specific reading of light intensity can be obtained with a reflected meter, which is aimed at the subject from the camera's position. Light readings are used to establish the proper light level for camera operation and to help determine the limits of the contrast range, which is about 30:1.

The fresnel spotlight is the workhorse television lighting instrument. Its focusing ability and the evenness of its light beam make it ideal for illuminating performers, sets, and background areas. Other special purpose spotlights are the ellipsoidal, lensless spots, follow spots, and PAR lights.

Floodlights such as the scoop, pan, and soft light are used to provide diffuse, shadowless illumination. Special purpose instruments such as strip lights and cyc projectors are used to illuminate large areas of sets or draperies. These instruments are often fitted with color material to provide colored lighting effects.

Instruments are usually hung and mounted on overhead battens, which are suspended from the studio ceiling. For special purpose applications, instruments can be hung from the battens on telescoping rods or pantographs, mounted on floorstands, or attached to set pieces with gaffer-grip clamps, trombones, or set wall plates.

Effective lighting requires the ability to control accurately the distribution and intensity of the light. Light distribution is controlled through (1) the positioning of the instrument, (2) the use of barndoors and shutters, (3) the beam control on the fresnel spotlight, and (4) varying the lamp-to-subject distance. Light intensity can be controlled through (1) selection of lamp wattage, (2) varying the lamp-to-subject distance, (3) the use of scrims and screens, (4) beam control on the fresnel spotlight, and (5) dimmers.

The distribution of electrical power to each instrument is controlled through the patch panel, which enables each instrument to be assigned to a particular dimmer circuit. Some sophisticated distribution and dimmer systems incorporate a computer which memorizes instrument-dimmer circuit assignments and controls the operation of the dimmer board during the production.

CHAPTER 6

TELEVISION LIGHTING
Methods and Techniques

The television image is created entirely with light. How we choose to illuminate a subject or a scene will determine how the image will ultimately appear in the videospace. When we light a scene, we are actually painting with light to create shape and texture, highlight and shadow, accent and detail. As with most other visual arts, a number of lighting guidelines and conventions have evolved over the years. In this chapter we offer them as a departure point for you to use in beginning to work with television lighting. They are not the only solutions or approaches to a particular lighting situation, and all successful lighting directors have broken the "rules" at one time or another to achieve a particular effect.

In the final analysis, all lighting must be evaluated by how it appears in the videospace, on the television screen. Only if the lighting complements the production's objectives, adds texture and depth to a scene, produces an aesthetically pleasing image, and provides enough illumination for the television system's operation can it be judged effective lighting.

One of the best ways to learn about television lighting is to observe your everyday surroundings carefully. Most television lighting tries to replicate natural lighting within the television studio. For the next few days, try to be especially conscious of all lighting, both natural and artificial, which is present in a variety of everyday situations. For example, how does the light in your living room at night differ from the light in the living room during the day? Where does the principal source of light come from? Where do shadows fall? Is the lighting in a classroom or office different from lighting at home? What is the light like outdoors in the early morning? At noon? In the evening? At night?

The point is that lighting and its effects on a subject change depending upon the location, the time of day, and the principal source of illumination. We cannot begin to light for television convincingly until we develop an eye and a feel for light as it exists naturally all around us.

Television lighting is, for the most part, an in-studio operation. However, as more and more productions move outside the studio, the ability to light for remote situations is becoming an increasingly important part of the lighting director's job. In this chapter we will cover the fundamentals of lighting for studio productions. In chapters 12 and 21 we will discuss lighting for on-location remotes. Although the challenges may vary somewhat, the fundamentals of effective television lighting remain essentially the same whether the production is shot inside the studio or at a remote location.

THREE-POINT LIGHTING

Television is a two-dimensional medium which provides us with a picture containing information about height and width, but with little depth perspective. Lighting is a valuable tool through which we can provide the viewer with additional information about the shape, volume and texture of an object. Much of this depth information is conveyed by light and shadow, which reveal a subject's form and dimensions.

We use both light and shadow on a television subject to reveal as much depth perspective as possible. Many beginning lighting directors are overly concerned with the appearance of shadows and try to remove them either by repositioning the instruments or by washing out the shadows with additional light. Yet it is the very presence of a shadow which helps to define space and to reveal the dimensions and form of the subject. Flat, diffused, and shadowless lighting produces a very uninteresting image which lacks any depth, modeling, or "punch."

You have probably seen indoor home movies which were illuminated by an instrument attached directly to the movie camera. Although the scene is lit sufficiently to produce technically acceptable pictures, the subjects appear flat and uninteresting and the lighting does little to flatter their appearance. That is because the light source is positioned directly in front of the subject and can provide few shadows or highlights to help the viewer define space and texture. (See Figure 6-1.)

The same home-movie effect would appear in television if we simply lit a scene with a bank of floodlights. We would achieve the technical level necessary for camera operation, but the pictures would appear lifeless and would convey little depth perspective. Early television cameras required high levels of illumination and could tolerate only a small contrast range between light and dark areas. This technical limitation forced lighting directors to light with few, if any, shadows. As soon as the cameras became more sensitive to light and were able to accommodate a wider

(a) (b)

FIGURE 6-1
In (a), the subject is illuminated with a lighting instrument positioned head-on. Compare this with picture (b), in which lighting highlights and shadow are used to bring out the depth dimension in the subject and to produce a much more flattering and interesting image.

contrast range, lighting directors quickly took advantage of this increased sensitivity and began lighting their subjects more realistically and with much more convincing and effective results.

We take advantage of shadows and highlights by considering the function, position, quality, and intensity of the light which will illuminate a subject. The three-point lighting techniques does this by using instruments in three basic lighting positions. They are commonly referred to by their primary functions: key light, back light, and fill light.

Key Light

FUNCTION The key light is the principal source of illumination for a scene. As such, it provides the primary modeling effect, determines the basic camera operating level (f-stop), and acts as the reference point against which all other instruments are positioned and their light intensities balanced relative to the light level of the key.

POSITION The key light is positioned in front of the subject, slightly offset to the side and at an elevated angle. (See Figure 6-2.)

It is difficult to provide specific horizontal and vertical angles for the key light since the positioning of the key will vary depending upon the particular lighting effect the lighting director wishes to achieve. Generally, when we decrease the horizontal angle of the key light and position it more frontally, we tend to flatten out a subject's features and provide less modeling. Increasing the horizontal angle dramatically increases modeling and will produce an aging effect on the subject by emphasizing features and skin texture. Changing the key light's vertical angle will also

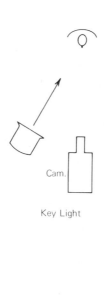

Cam.

Key Light

FIGURE 6-2 **Key Light.**
The key light serves as the principal light source on the subject. Varying the horizontal and vertical angles of the key light determines the modeling of the subject.

produce different shadows and highlights. As we increase the vertical angle, shadows begin to appear under the eyes, nose, chin, and lips of the subject. The entire modeling effect becomes quite pronounced and can be rather unflattering. The additional shadows tend to make the eyes recede into their sockets and produce an overall aging effect on the subject. If we lower the vertical angle, the effect is reversed, and ultimately shadows will begin to appear above lips, nose, eye sockets, and the like. Since we are most accustomed to light sources emanating from above a subject, this shadow reversal can produce some very weird effects, much like Halloween trick-or-treaters who hold flashlights under their chins. (See Figure 6-3.)

A good starting point for placing the key light is approximately 30° to 35° to the side of the camera and at a 30° to 35° elevated angle. But remember

that this is only a starting point; the exact horizontal and vertical angles can be determined only after you have decided on the lighting effect you wish to achieve. Some production situations may require you to reduce the modeling and flatten out texture—for example, many women performers look better with the key light positioned more frontally. Other situations, such as a dramatic program, might require very pronounced modeling with severe shadows to produce a lighting effect which will complement the script. (See Figure 6-4.)

A key light can be either *motivated* or *unmotivated*. A motivated key must be positioned in such a way as to suggest that its illumination on a set or subject is coming from the direction or source which is determined by the setting or the script. For example, a character in a drama who walks over to a window in daylight should be lit

keys for all subjects on approximately the same side.

LIGHT LEVEL   The key light will determine the relative intensities of all other instruments, which are balanced in proportion to the key's light level. Since the key will determine the basic camera operating level (f-stop), you should take a light reading to ensure that the light level produced is within the video operator's range of control. Too low a level will not provide sufficient technical illumination for camera operation. Too high a level on the key, relative to the other light sources used in a scene, may produce "blooming" or other undesirable effects in the camera's picture.

INSTRUMENTS   Fresnel spotlights are almost always used for key lights. This is because the hard, directional light produced by the fresnel provides the necessary modeling effect on a subject by creating shadow areas. In addition, the flexibility of the flood-spot control on the fresnel and the evenness of the light beam provide the lighting director with greater control than with any other lighting instrument. If necessary, inserting a scrim or screen will slightly soften and reduce the intensity of a fresnel's beam. The size and wattage of the fresnel chosen for the key light depend primarily on the throw distance. Most medium-sized studios customarily use 1-kw and 2-kw fresnels as their basic key light instruments.

Back Light

FUNCTION   The back light is used to cast a rim of light around a subject's head and shoulders to separate the foreground subject from the background and to increase the apparent definition by providing contrast.

POSITION   Ideally, the backlight should be positioned directly behind the subject and at an angle that will prevent it from directing light on the front of the subject. The backlight should not illuminate the top of a subject head since this will create an unnatural and unattractive lighting effect. (See Figure 6-5.)

LIGHT LEVEL   The light level of the back light,

with a key which is positioned from the same general direction as the window. Similarly, a subject who is sitting near a lamp or light fixture at night should have his key light hung from the side where the set lamp is placed.

Unmotivated keys are used when there is no need to consider the exact direction of the light or if a program is obviously taking place within a television studio. News programs, interview shows, and quiz programs are examples of production situations in which a motivated key is unnecessary. In this case, the key light could be positioned on either side of the subject although it is good practice to remain consistent in placing

FIGURE 6-3   **Key Light Positioned Beneath the Subject.**
The shadow reversal obtained with the spotlight positioned under the subject produces an eerie and unnatural lighting effect.

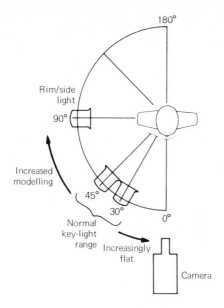

Horizontal Placement

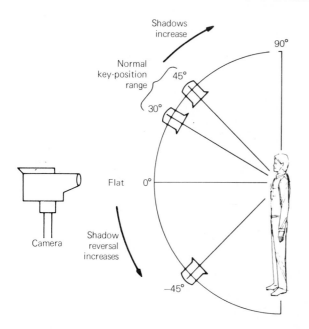

Vertical Placement

FIGURE 6-4 **Key Light Positioning.**
The horizontal and vertical placement of the key light determines the modeling of
the subject. The greater the horizontal and/or vertical angles, the more shadows
and the greater the modeling effect.

as with all lights, must be set relative to the intensity of the key. There are a number of different opinions as to what the correct light level of the backlight should be. While it is generally agreed that a rim of light across a subject's shoulders and hair will tend to enhance foreground-background separation and provide more depth to the picture, raising the intensity of the back light to too high a level can create a very unnatural lighting effect which calls undue attention to itself. Heavy back light was used quite frequently in black and white television since there was no color information to help establish depth perspective. It was not infrequent to set the level of the back light as high as one and one-half times the key light's intensity. As color cameras supply color information to help establish depth perspective, this has reduced the necessity for high back-light-to-key-light ratios. Today, some lighting directors are of the opinion that, except in rare instances—such as a romantic setting or a nighttime scene—back light should be kept to a bare minimum, if used at all.

A good rule of thumb is to start with the back light level set at about half the key light's intensity and then adjust the level up or down depending upon the effect produced on a television monitor. Remember, too, that different subject and background combinations will require different back-to-key ratios. For example, a blonde foreground subject placed against a dark background will usually require less back light than a brunette against the same dark set. The best approach is to plan your back lighting according to each production situation and to balance its intensity relative to the key light, the subject, and the background, once you can see how all the factors interact on camera.

Sometimes a carefully positioned key light can also serve as back light for another camera's subject. While this approach leads to economy in

Back Light

FIGURE 6-5 **Back Light.** The back light is used to cast a rim of light around the subject's head and shoulders; it adds to the depth dimension and helps to separate the subject from the background.

hanging and focusing instruments, it also decreases the lighting director's control and flexibility.

INSTRUMENTS Since back lighting requires a hard, directional light source, the best instrument choice is usually a fresnel spotlight. To avoid any hotspots which could ruin the effect of a back light, it is best to flood out the fresnel completely and then to trim off any spill light by using barndoors. The fresnel's hard light can be softened slightly, if necessary, by inserting a scrim.

Fill Light

FUNCTION Fill light is used to lighten or to completely eliminate the shadows produced by hard light sources. Of course, completely washing out shadows with a fill light would destroy the modeling effect which was created by the key light.

POSITION The fill light is positioned on the side

of the camera opposite the key light. For best effect, the quality of the fill light should be diffused, soft light with no apparent direction. This will permit you to blend in the soft light smoothly with the key light's illumination. (See Figure 6-6.)

LIGHT LEVEL The intensity at which the fill light is set depends on the lighting effect you are trying to create. However, the fill light should never be more intense than the key light, or it would have the effect of replacing the key as the primary source of illumination. Some program situations require a very brightly lit scene in which shadows are transparent or almost invisible. Other productions may require very heavy, dark shadows which create a somber, moody effect.

When we light a scene with heavy fill light and attempt to lighten or eliminate many of the shadow areas, we are lighting for *high key*. A scene in

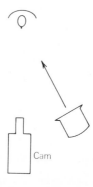

Fill Light

FIGURE 6-6 **Fill Light.** The fill light is positioned on the side opposite the key and is used to fill in shadows created by the key light.

which there is little, if any, fill light and many dark shadow areas is said to be lit for *low key*. It should be noted that the terms "high" and "low" key are used differently in television than they are in motion pictures. Since a motion picture is shown in a completely controlled viewing environment, it is possible to utilize low-intensity key lights for low-key scenes. This is not possible in television since the proper operation of the television system requires the fullest use of the white to black level. In television, low key refers to the application of a very high contrast ratio. In other words, far more key-light illumination than fill light. (See Figure 6-7.)

FIGURE 6-7 **Low-Key Lighting.** Low-key lighting uses shadow and highlight to create a distinctive modeling effect on the subject. *(Courtesy: PBS-Visions)*

High-key lighting in television is usually associated with most nondramatic programs such as news, interviews, or panel programs, or on musical variety shows where the lighting director tries to create a bright and lively look for the picture. (See Figure 6-8.)

Low-key lighting sharply increases the key-to-fill contrast ratio and produces a very moody, dramatic look. For a very dramatic effect, the LD (lighting director) might choose not to employ any fill light at all, letting the fill side of the subject go completely dark.

INSTRUMENTS Unlike key and back lighting, which usually utilize only fresnel spots, fill lighting affords the lighting director a wider choice of instruments. For those situations where you must carefully control for spill light, as in most dramatic programs, the best choice for fill lighting in this situation would be a fresnel spot which is completely flooded out with a scrim or screen attached to diffuse and soften the light. Barndoors are employed to trim off any excess fill light and to control the light's distribution. This approach is particularly effective when you are lighting for maximum control of picture elements.

High-key lighting, which is characterized by the absence of dark shadow areas and requires a bright and evenly lit set, can often be achieved by using floodlights since their diffused, soft light will blend in smoothly with existing instruments. Of course, it is almost impossible to control spill light from a scoop or soft light, and this is a major consideration in selecting instruments for fill lighting.

Background Light

Once the key, back, and fill lights have been positioned and balanced, the basic three-point lighting plan is complete. The principal subject is illuminated although, for most situations, the background area must still be lit.

FUNCTION Background lights are used to illuminate, accent, and model the scenery and background.

POSITION The placement of background lights is determined by the particular set you will be using during the production. It is always a good

(a)

(b)

(c)

FIGURE 6-8 **Key-to-Fill Ratio.** The amount of fill light relative to key light produces variations in scene contrast. In the first frame *(a)*, the ratio of key-light illumination to fill light is fairly close. In the second frame *(b)*, the fill light's intensity is reduced relative to the key-light illumination, and shadows become more distinct. In the third frame *(c)*, the fill light is completely eliminated. The subject's contrast ratio expands as shadows go completely dark.

idea to try to isolate the illumination between the foreground subject and the background so that the lighting director can exercise maximum control in balancing intensities on the different areas of the set.

In general, background light should develop as much texture and dimension in the background as possible. Not only will this help to provide depth to the camera shot, but it also makes the setting appear more realistic and interesting to the viewer. Naturally, not all backgrounds have texture with which the lighting director can work. A painted flat or a cyclorama has little or no depth, and, in this case, there is no point trying to induce depth on a background which really does not have any.

Most set backgrounds look best if you accent certain areas, letting others fall into darker shades or even to black. Again, the overall lighting plan will guide your use of background lighting. (See Figure 6-9.)

If you are using pleated draperies in the background, spotlights positioned at a sharp angle to the curtain will tend to lend highlights and texturing to the pleats. Positioning the instrument at a more direct, frontal angle will de-emphasize the pleats, and the lighting will flatten out the curtain's appearance. Occasionally, lighting directors will employ spotlights at very severe angles to the curtain to produce a "slash" of light across the drapery.

Plain wall flats are another commonly found background in many television productions. In this instance, the lighting director must decide whether to light the entire flat evenly or to let portions of the lighting fade into gray or black.

LIGHT LEVEL The light level of the background lights should be adjusted only after the foreground lights are positioned and balanced. Remember that the intensity of the background light will also affect the viewer's perception of the foreground subject's brightness. A dark background will make a brightly lit subject stand out much more prominently than a lighter background.

The light level of the background lights is usually governed by the intention of the scene and the necessity of keeping the background area secondary in visual importance to the action in the foreground of the shot. As a general rule,

FIGURE 6-9 **Background Light.** The background should be illuminated to reveal whatever depth and texture exist. Notice how the ceiling is lit with highlight and shadow to texture the recessed paneling. In addition, the background illumination slowly grows darker as your eyes move from the floor level to the ceiling. *(Courtesy: PBS)*

the background brightness level should be about two-thirds the brightness of the principal subject in the foreground.

Background lighting can play a major role in setting the time of day and establishing the overall mood of a scene. For this reason alone it is not satisfactory for the lighting director to concentrate on lighting the foreground, with the background illuminated later almost as an afterthought. A well-designed and executed lighting plan for the background is equally as important as well-lit foreground areas.

INSTRUMENTS   Fresnel spotlights are excellent instruments for lighting background areas. Not only does the hard, directional light cast shadows and define texture and form, but the barndoors permit accurate control of the light beam.

Ellipsoidal spots can be used in special applications to cast a projection against a cyclorama, curtain, or even a background flat. Using a cucalorus with an ellipsoidal spot is a very inexpensive and yet effective way to break up the monotony of a cyclorama or curtain, and, in a dramatic situation, an ellipsoidal fitted with a specially designed pattern can throw an image of a window, jail cell bars, and the like onto another portion of the background set.

Floodlights and strip lights are useful instruments to use when you wish to illuminate a background area evenly. Fitting them with color media will produce a colored light effect which, in special instances, can be quite effective. You can even change the color of a background, such as a cyclorama, by controlling the balance and intensity of a number of differently colored instruments.

Special Purpose Light Sources

The key, back, fill, and background light sources are the most frequently used in television lighting. However, there are a number of additional light sources which can be employed for special applications.

SIDE LIGHT   The side light is placed directly to the side of a subject and augments the fill light by providing accents and highlights. Side lighting also enhances the modeling effect on a subject, but unless you are trying to create a very special

effect, be careful to adjust the intensity of the side light so it will not call undue viewer attention to the light source or produce unflattering shadows on the subject. Side light is commonly used on dance productions, where the most important element is the outline and form of the body and not the dancer's face.

KICKER LIGHT   A kicker is a special position for the backlight. The instrument is positioned at the rear and to the side of the subject. Its purpose is to add additional accents to the head and hair, and the effect it produces is usually quite pronounced. Kickers are often used for glamour effects where the halo produced augments the overall mood of the scene. (See Figure 6-10.)

FIGURE 6-10   **Kicker Light.**
The kicker is a variation on the backlight. It is used to accent and highlight the hair and is frequently associated with a glamour effect.

EYE LIGHT The continuous action production technique which is used in television doesn't permit the application of a single instrument which will add sparkle to the eyes during close-ups. In television lighting, to obtain the eyelight glint, the keylight must double as the eyelight. Since the eye light will be effective only during a close-up, it is important to mount the key light relative to the close-up camera's position and to make sure it is hung at a vertical and horizontal angle where its beam will reflect directly into the subject's eyes, producing the desired sparkle and filling in shadows around the eye socket.

Modifying the Three-Point Lighting Approach

Our discussion of the three-point lighting approach has considered the placement of each instrument relative to fixed positions for cameras and subjects. A good question you might ask is, "What happens to the three-point lighting plan during a program in which both cameras and subjects must constantly move to a variety of positions. How, then, do I position the key, back, and fill lights?"

The point is well taken and illustrates an important distinction between conventional television lighting and lighting for still photography or for motion pictures. The two latter situations provide the lighting director with the luxury of lighting a subject for one particular camera angle at a time. Television does not allow this when a program is shot with multiple cameras, which are continuously moving to new shot positions throughout a production. Unless you are producing your program on videotape, with a great deal of postproduction editing where you can stop and relight for different camera angles, one lighting setup must be adequate for all shots and for all angles—from a wide, establishing shot, to a tight close-up. As with so much of television production, compromise is the key word.

Television lighting directors generally light each subject with the close-up shot primarily in mind since the majority of any program's shot will usually be as close as a waist shot, if not tighter. The wide shots are left to take care of themselves with slight touch-ups here and there to keep the overall lighting effect consistent.

The most effective approach is to light each major camera and subject position individually with its own key, back, and fill light. This job is made considerably easier if the director will provide you with a detailed script and floorplan indicating camera positions and actor blocking. From this point it is relatively easy to determine where the key light and subsequent instruments must be positioned for each camera angle. While this approach often requires using a greater number of instruments than alternative lighting methods, the lighting director gains a considerable amount of control over the lighting, particularly the ability to balance and focus individual instruments without affecting other camera or subject positions.

The alternative approach is to light a large playing area rather than specific positions. Using this method, the lighting director refers to the director's floorplan and delineates a number of major playing areas, lighting each one with its own key, back, and fill lights. Depending on the size of the area, you may need to use a number of instruments for each function. For example, the area may require two or three fresnel spotlights, all of which serve as the key light. While this technique usually requires fewer instruments than the camera and subject approach, it does have the disadvantage of not providing as much control for the lighting director. The most important factor in the area approach is to be sure that a multiple number of instruments, which are all functioning as, say, the key light, produce a smooth and even spread of illumination across the entire area as though it were illuminated by a single instrument.

Although the area approach does have its disadvantages, some production situations lend themselves quite readily to this method. For example, a musical-variety program or a demonstration show which requires relatively even lighting over wide areas of the studio floor might be approached using the area method. On the other hand, dramatic productions which demand more

FIGURE 6-11 **Lighting Summary Chart.**

Function	Instrument	Position Relative to Primary Camera	Effects and Special Applications
Key	Fresnel	Approx. 30° to 40° vertical angle and to side of camera	Principal source of illumination. Basic reference for balancing intensity and position of other instruments. Positioning more frontally reduces modeling effect, steeper angle both vertically and horizontally increases modeling effect and produces more texture in subject.
Back	Fresnel	Either directly behind subject or at slight angle, vertical angle should be between 30° and 45°	Produces a rim of light around subject's head and shoulders to separate foreground subject from background and to enhance perception of depth. Heavy backlighting is used to create nighttime scenes or for special effects. Avoid too much intensity or backlight may produce artificial effects.
Fill	Fresnel Scoop Broad Soft Light	Approx. 30° to 40° vertical angle on opposite side of key light	Used to fill in shadows which are created by key light on subject and to fill in shadow or dark areas on sets, backgrounds, and overall playing areas. Fill light intensity is set relative to the key light. A low key-to-fill ratio produces few shadows; a high key-to-fill ratio produces many shadows leading to more texture and modeling. When you are using fresnel instruments, beam should be flooded out and scrim or screen used to soften light.
Background (BG)	Fresnel Broad Scoop Ellipsoidal	Position depends upon desired effect. Frontal position flattens out background. Steeper vertical and horizontal angle increases modeling and texture on background.	Used to light backgrounds on sets, curtains, cycloramas, intensity is always balanced relative to foreground lights on subject. Lack of any background lighting creates a cameo effect. Background lights are usually the last to be focused and balanced since they depend heavily on foreground illumination. BG lights should usually attempt to bring out whatever modeling and texture exists in the background.
Side light and kicker	Fresnel	Approx. 90° angle on either side of subject.	Used to accent highlights on hair and shoulders and to outline the form of the body on such programs as dance or gymnastics. Can be used effectively to enhance nighttime lighting effects. Effect can be somewhat artificial if not properly positioned and intensity is not correctly balanced.
Cyclorama lights	Scoops Strip Lights Cyc Projection Lights	Can be hung from battens near cyc and/or placed on floor facing upward behind ground row or cyc row.	To illuminate a cyclorama. Usually, the instruments are fitted with color media to color the cyc. A number of different color effects are possible by gelling different instruments and varying intensities from the lighting control console.
Pattern projections	Ellipsoidal (fitted with cucalorus or other pattern)	Perpendicular frontal angle will project pattern without distortion. As the angle of instrument becomes more severe, projected pattern is distorted.	To project a pattern against a cyclorama set or other backing.

precise lighting would best be lit by using the individual camera and subject technique. Sometimes a lighting director might wish to combine the two techniques, lighting high-key areas with the general area method and illuminating low-key or strongly motivated scenes by lighting for camera and subject positions.

SPECIAL LIGHTING EFFECTS

Special lighting effects are used to augment conventional lighting techniques to provide a unique reality to the television viewer. As with any other special effect, they rely heavily on the effective use of videospace to suggest a reality which may not actually exist within the studio.

Lighting for Daylight

INDOOR SCENE   A crucial consideration when lighting an interior set for daylight is to motivate the major source of illumination. Be certain that the main key lights are positioned in such a way as to throw light from the direction of doorways or windows. Certainly, the brightest area in the background should be a window or open doorway, and a strong fresnel spotlight can be used to throw a convincing shaft of light through the window or door to suggest sunlight. If you observe a normal room in daylight, you will see that not all the walls are brightly lit. Usually, walls surrounding windows are rather hot, and, of course, the stream of light through a window will illuminate those portions of the room on the opposite side. However, many other wall areas still exhibit shadows and dark spots, and this should be used by the lighting director in the studio to help break up the monotony and flatness of a uniformly bright background set.

OUTDOOR SCENES   To suggest exterior daylight inside the television studio requires the use of hard, directional lighting and the judicious

placement of shadows. Of course, you might argue that the sun produces only a single shadow while a group of instruments in the studio will result in multiple shadows which may not produce a completely realistic effect. While this is a valid argument, a single source of light will not always produce the proper modeling and depth for a subject. The appearance of multiple shadows, if they are seen by the camera at all, will have to be written off under the heading of "poetic license." When planning lighting, remember to consider the time of day of the scene as suggested by the script and the director. High noon in the desert requires a much different approach from a late afternoon scene in a city park.

Nighttime Lighting Effects

Lighting for night, whether interior or exterior scenes, requires a low-key approach to produce a high contrast ratio between light and dark areas. The lighting director must use the key light to produce enough illumination for an acceptable picture while permitting dark areas to go completely black. It is a mistaken notion to approach nighttime effects by simply dimming down all instruments. Not only will this create a noisy and unacceptable image, but the scene will simply appear washed out and will not convey the feeling and mood of night convincingly.

Night effects are achieved by heavy back and rim lighting, with dark backgrounds punctuated by a few rays of light to relieve the monotony and to provide additional interest and depth perspective. While motivating the light is often an important consideration, sometimes we must cheat slightly to provide sufficient illumination to light the actor's faces properly for close-ups. In reality, the light from the moon streaming through a bedroom window may not be as bright as the key light we are forced to use in the studio. Audiences tend to overlook this fact, however, and it is important to light actors' faces sufficiently so the viewer can see their features clearly. Careful control of spill light is essential in creating an effective nighttime effect, and, for this reason, floodlights should be avoided since their light cannot be easily controlled and the diffused, flat lighting they produce will invariably wash out some of the shadows which contribute to the overall mood of the scene. An

(a)

(b)

(c)

FIGURE 6-12
In the first frame *(a)*, the living room set is illuminated with floodlights which produce a very flat and uninteresting image. The second frame *(b)*, shows the same set lit for daytime interior. Notice how the use of shadow and highlight add depth to the scene and give it more "punch" and visual interest. The third frame *(c)*, shows the set lit for nighttime with heavy shadows and only selected areas fully illuminated.

effective technique in lighting for night is to light each actor and camera area carefully and then let the performers walk through dark shadows with little or no fill light. Exterior night scenes are approached in much the same way. Heavy back and rim lighting will enhance the nighttime effect and deep shadows will convey the mood and time of day to the audience.

Sometimes the addition of a blue-colored gel to the instruments will provide a convincing "nighttime" quality to the light although the color should be carefully selected and applied sparingly.

Limbo and Cameo Lighting

Limbo lighting is a technique in which the foreground subject is seen against an undefined, infinite background. One of the most commonly used variations on the limbo effect is called "cameo" lighting. This is when a foreground subject is seen isolated against a totally black background. Obtaining a cameo effect requires very careful control of light distribution to keep as much spill light as possible from falling on the background area and inadvertently illuminating it. Placing the foreground subject as far forward as possible from the background will help the effect by isolating much of the ambient light from bouncing off the floor and lighting the background or drapery. A black velour curtain can be used in the background to achieve absolute black.

Spotlights are the only instruments which should be used in cameo lighting since the distribution of their light can be most carefully controlled. On the other hand, if you are lighting for white limbo, a bank of floodlights will provide even background illumination which will enhance the effect of infinite space.

A commonly used variation on the cameo effect is to begin a scene with the foreground subject lit against a black background. On the proper cue, the background lights are slowly faded up, revealing actors and set pieces, which are located behind the foreground performer. Of course, the reverse of this is also possible by fading out the

background to leave the foreground performer lit against an infinite background. (See Figure 6-13.)

Projecting Patterns

Projecting patterns of light onto sets, cycloramas, draperies, and other background elements is an effective and inexpensive method of using lighting to provide visual variety.

We have already discussed the ability of an ellipsoidal spotlight to accept a cucalorus pattern. These patterns are usually fashioned out of metal but some more expensive ones are made out of Pyrex or a synthetic called Vycor glass, which resists cracking. Patterns come in a wide assortment, including stars, sunbursts, abstract designs, and clouds. You can even make your own special patterns by using .005-inch aluminum foil or the bottom of pie tins and cutting out a pattern with a knife or razor blade. (See Figure 6-14.)

The size and shape of the patterns can be varied by changing the horizontal and vertical position of the ellipsoidal spot relative to the background area. A steep horizontal/vertical angle will distort the pattern's shape while a directly perpendicular angle will produce an evenly shaped design. Experimenting with a number of different patterns and instrument positions will enable you to arrive at the best combination for your production needs.

Lighting with Color

Lighting with color offers an opportunity to use color and light in a variety of creative and imaginative ways. Colored light is produced by attaching *color media*, commonly called "gels," to an instrument. A few years ago, gelatin was the material which was used in the manufacture of all color media (hence the term, "gels"), but gelatin deteriorates quickly under hot lights and over long periods of use. More recently, manufacturers have developed newer plastic materials which are much more effective than gelatin in resisting heat and color deterioration.

A number of manufacturers produce complete lines of over fifty colors in the three materials: *gelatin, acetate,* and *polyester.* Of the three,

(a)　　　　　　　　　　　　　　　　(b)

FIGURE 6-13　Limbo and Cameo Lighting.
Limbo lighting *(a)* shows the subject against an undefined, seemingly infinite background. Cameo lighting *(b)* shows the subject against a completely dark background.

polyester is by far the sturdiest and most durable. Although polyester material costs more than either acetate or gelatin, it was specially developed for use in television, where high heat levels and long periods of use tend to deteriorate most other materials quickly. Polyester media are sold under a variety of trade names, including Berkey-Colortran's "Geletran" and Roscoe's "Roscolux".

A swatchbook, containing small samples of the entire line of color material, is used to select a particular color for each lighting situation. When selecting a color, do not merely hold the swatch of material up to the light since this will not provide an accurate representation of how the colored light from an instrument will appear onstage. Rather, shine a light through a small piece of the material onto a white cardboard or, better yet, on a piece of the cloth or set material which will actually be used in the production. (See Figure 6-15.)

Color media are available from the manufacturer in either single sheets or in large rolls. Either is easily cut to fit inside any sized gel frame holder, which is then attached to the front of the lighting instrument and held in place by three protruding brackets. Gel frame holders get very hot after only short periods of use, so be sure to use asbestos gloves when handling frame holders to avoid burning your hands.

All color media absorb a certain amount of light during normal operation, and some deeply saturated blue and green colors can reduce light output by as much as 95 percent. You should keep this in mind when deciding on the size and wattage of the instruments which you intend to fit with gels.

LIGHTING PERFORMERS WITH COLOR Under normal circumstances, color media are not used on those instruments which will illuminate performers since their illumination will upset the color balance of the light and produce a very unnatural

FIGURE 6-14 Pattern Projected on Cyclorama.
A variety of patterns is commercially available (see opposite page), or you can make your own for special purpose applications.

229 CLOUD #6   230 CLOUD #7   256 CLOUD #8   257 CLOUD #9
250 CRESCENT MOON   231 REALISTIC STARS   245 EVENING STAR   232 LARGE STARS
244 SUPER STARS   234 STARBURST   243 STAR BREAKUP   222 SMALL BREAKUP
223 LARGE BREAKUP   254 ZIG ZAG   235 RADIAL LINES   233 SPIRAL
258 DAISY PATTERN   253 ROSE   259 APPLAUSE   260 NO SMOKING

Pattern Projection.
Patterns can be projected onto a background wall, curtain, or cyclorama. These are only a few of the many which are commercially available. *(Courtesy: The Great American Market)*

picture. Of course, there are always exceptions to the rule, and some musical, fantasy, or even dramatic situations, may call for just such an effect. In this case, it is important to supply a motivation for coloring the light to avoid confusing the viewer.

LIGHTING A CYCLORAMA FOR COLOR The use of colored light with a cyclorama can produce some very effective and varied lighting for color television. To light the cyc evenly, most studios utilize a series of strip lights hung above the cyc on battens and another strip positioned below the cyc on the floor, hidden behind a ground row. An alternative is to utilize a cyc projector which is designed to throw an even wash of colored light across wide areas. Regardless of which instruments you choose, each luminaire must be fitted

with a gel. You may wish to alternate colors and assign instruments with the same color to similar dimmer circuits. This enables you to "paint" the cyc with different hues by adjusting the balance of the variously colored lights. Highly saturated gels are best for cyc lights, and you can modify the color which is produced, either by adding white light, which will desaturate the color, or by adding complementary colors on different instruments.

Sometimes, lighting directors prefer not to light the cyc evenly but, instead, to produce an effect where the center of the cyc is highly saturated with color which then gradually recedes into gray or black.

One of the most difficult problems in lighting a cyc is to keep the foreground illumination from spilling onto the cyc and washing out the color.

This is a particularly difficult problem in smaller studios, where little space is available to separate the foreground subjects from the cyclorama. In this case you may have to modify the position of some foreground instruments and light with a steeper angle than usual in order to control as much spill as possible from hitting the cyc. A good alternative for small studios is to use a lower reflectance cyc—one which is made from material about 60 percent gray rather than from material which is completely bleached out.

Changing the color of a cyc during a show is a very effective way of increasing the visual variety of the set while adding little in additional cost or time. For example, a musical program might use an amber cyc for uptempo numbers and then change to blue for slow, moody songs. The variations possible are limited only by the number of instruments you have available. On some occasions, adding a projected pattern to a colored cyc will provide even more visual variety. A blue cyc with a cloud projection can be used for an outdoor scene, and a variation on this is to use a star pattern with a dark blue cyc to suggest nighttime.

Some lighting directors have used tiny "pea" lights with a cyc for a blinking star effect. The

string of lights is hung between the scrim and the muslin backing and the thin power cord disappears on camera once the cyc is properly illuminated.

Lighting for Chroma Key

The use of electronic matting, often referred to as *chroma key,* has become increasingly popular because it provides the program director with a large measure of creative flexibility. Conventional chroma key operates by using a particular hue as the key color. Whenever the scanning system detects this color in the shot, it electronically inserts another video source in place of the key color. A more sophisticated chroma key system, called "luminance chroma key," operates by combining brightness information and color in producing the "keying" effect. Luminance chroma keyers are particularly effective as they will permit you to key in transparent objects such as

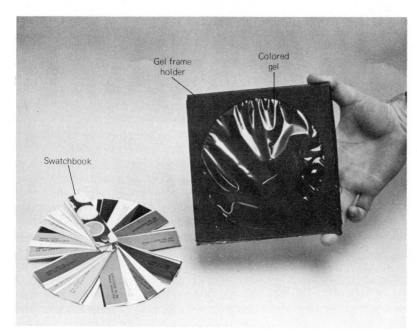

Gel frame holder

Colored gel

Swatchbook

FIGURE 6-15 **Color Media and Gel Frame Holder.**
Color media are available in a wide assortment of colors. The lighting director uses the swatchbook to select the exact shade for the particular production situation.

glass or the shadow of the object being keyed. This key operates by reading both the brightness of the background area as well as the hue and requires very precise lighting to operate properly. The use of chroma key is discussed in Chapter 11. However, because lighting is a fundamental part of the chroma key process, we will consider the proper lighting approach for keying in this section.

The industry has designated two colors, green and blue, as the primary key colors. Although other hues may be used, it has been found that shades of blue and green are farthest from the colors of skin tone, and these colors tend to key best over a wide range of production situations. Since the chroma key matting operation relies on the appearance of the key color to trigger the insert circuitry, the key area must be evenly lit so shadows and spurious highlights will not interfere with the matting operation, producing objectionable "crawling" or "tearing" around the edges of the insert. Contrary to popular belief, chroma key does not require excessively high brightness levels on the background area to operate properly. In fact, if the background is lit normally with the proper foreground-to-background brightness ratio, the key should work well. The important consideration is not so much a high intensity, but a relatively even spread of light across the background chroma key area.

Since luminance chroma keyers read both the color and the brightness values of the background to trigger the insert circuitry, the threshold for the key operation is very narrow; a difference of approximately 10 percent brightness alone in the background is sufficient to trigger the insert. Therefore, it is vitally important when using a luminance chroma keyer to keep light intensity levels even across the entire background area. This can be achieved by using multiple sources which blend well together or by using a single source with a very flat field to illuminate the background.

The foreground subject of the key should be as far away from the background as possible for two reasons. First of all, greater separation decreases the likelihood of shadows falling on the background and creating tearing or bleeding in the key. Secondly, increased separation decreases the possibility of colored light bouncing off the background and contaminating the foreground subject, resulting in crawling around the edges of the key. Increased separation and the proper balance of higher foreground intensity to background illumination will keep contamination and its effects to a minimum.

In the event that you cannot increase separation between the foreground and background, a number of lighting directors suggest using a yellow gel on the back lights to minimize bounce-back contamination. The yellow light acts as a filter and absorbs blue light, decreasing the bounce light's effect. If you are using a green background, a magenta filter will produce the same results. While this is a theoretically convenient solution, it should be used only as a last resort since the filtered light can sometimes create more problems than it solves.

As producers and directors become more familiar with chroma key, they are using it in a wide variety of production situations. For example, chroma key is often used to insert an entire figure or group of people into another picture. While this can be quite effective from a production standpoint, it creates tremendous problems for the lighting director. The entire set area where the performers are positioned must be completely covered with a chroma key material, including all the floor area which will appear on camera. The lighting director must not only light the actors, but must also provide even illumination across the entire set so that the keying operation can function properly. Sometimes this requires the lighting director to modify radically or even to ignore some of the basic lighting conventions. For example, on one network special, the lighting director was required to light a number of actors across a large area, who would then be keyed into a second camera source. The effect was to produce "Heaven," with angels ethereally floating above the set. Two hundred and twenty scoops were hung at equal intervals to illuminate the "key" set evenly. No fresnels were used at all since their hard light might have produced hot spots which would have ruined the matte effect.

This is simply one of the many examples of how the mechanics of a production must always supersede aesthetic considerations. In this particular situation, depth perspective was secondary to an effective key, and the set was lit to achieve the program's primary objective.

Although such extreme cases are rarely encountered by most lighting directors, proper lighting is one of the most crucial factors in achieving a good key effect—regardless of how simple or complex the program. Often, only time and experimentation will provide the solution to effective lighting for chroma key. If your production will be using chroma key, be prepared to spend some time in the studio working with lights, the set, camera angles, and performers and then checking the results on a camera and monitor, where you can see the interaction of all the production elements.

Firelight

The flickering effect of a fire on a subject can be simulated by attaching narrow strips of cloth to a long stick of wood. When a lamp is placed at an

appropriate level (usually close to the ground and facing upward), the stick is gently waved in front of the light, producing a flickering effect on the subject. A yellow or orange gel placed in front of the light will provide the warm hue associated with firelight. Depending upon the intensity and position of the instrument and the speed of the stick moving in front of the lamp, you can simulate a gentle fireplace or a raging forest fire. Of course, the lighting effect alone will rarely be convincing. You must augment the lighting with appropriate sound to achieve the most realistic effect. (See Figure 6-16.)

Moving Lights

Moving lights, such as from a car's headlamps, can be simulated by using a fresnel spotlight mounted on a floorstand. A stagehand is instructed to pan the light across the set and this,

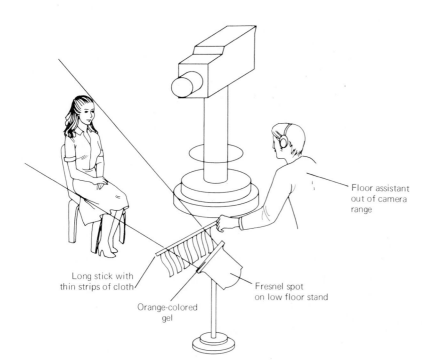

Long stick with thin strips of cloth

Orange-colored gel

Fresnel spot on low floor stand

Floor assistant out of camera range

FIGURE 6-16 **Firelight Effect.**
Flickering light can be simulated by waving a long stick with thin strips of cloth attached in front of a spotlight which is focused on the subject. Adding an orange gel to the light will enhance the fire effect.

combined with appropriate sound effects, creates the effect.

Reflected Water

A large pan, in which you have placed some small pocket mirrors covered with water, will cast a reflected ripple effect of water from a pool or lake. You must carefully position a spotlight beam to shine into the pan and, at the same time, gently agitate the water to produce the effect. Be careful when lighting to avoid spilling the performer's light into the water, for it may wash out the reflected light and ruin the illusion.

Lightning

The effect of a burst of lightning can be achieved by utilizing a number of low wattage, high-intensity lamps, such as photofloods. The lights are assigned to their own dimmer and are quickly switched on and off, producing the short bursts of light. Lightning is most effective when it briefly overexposes the scene. Again, the use of sound effects coupled with the lighting will enhance and reinforce the overall illusion.

Practical Lights

Practical lights are self-illuminating scenery devices, such as table lamps and wall fixtures, which must actually function during a production. Under most normal studio lighting conditions, the brightness of ordinary household lamps and light fixtures will wash out on camera. For this reason, you may have to reinforce the effect of a practical light by focusing "special" instruments to illuminate a set or area which is supposed to be lit by the practical lamp. It is important for the actor and the dimmer board operator to rehearse the lighting cue so the studio lights will switch on and illuminate the set at the same time the performer turns the switch on-camera. The easiest way to coordinate this is to have the dimmer board operator take a visual cue from a video monitor during the production.

PLANNING THE LIGHTING

Planning and executing the lighting for a program are the responsibilities of the lighting director. All networks and most large stations and production studios employ a staff of individuals whose exclusive job is lighting. At smaller stations and studios, however, the task of lighting may be delegated to someone who must combine lighting duties with other production jobs. Whether your studio has someone whose sole responsibility is lighting or who combines lighting with other production functions, the basic steps in planning and executing the lighting for a television production are the same.

Preproduction Conference

As with other areas of television production, the first step in planning the lighting is the preproduction conference. This is when the lighting director has the opportunity to discuss the program with the director, technical director, audio engineer, and set designer to plan concepts and objectives.

During the meeting the LD will learn how the producer and program director intend to approach the program. Highly complex shows usually require more planning than a simpler interview or panel show. The director and the scenic designer should have already planned the floorplan, and a copy should be available to the LD. The program director will use the script and floorplan to indicate major actor and camera positions, and the LD should make notes on both his script and floorplan. The audio engineer will discuss the best method of picking up sound—if hand, desk, or lavalier mikes will be used, there are few audio complications for the lighting director to be concerned with. However, if a boom will be used, the LD must know its approximate position and operating range so he can anticipate boom shadows.

THE LIGHTING PLOT   After discussing the program objectives with the key members of the production team, the LD must use the floorplan, the script, and information about the performer and camera positions to design the lighting plot.

Planning the lighting, even for the simplest productions, requires attention to a large number of details. The LD must select the type and size of lighting instruments, determine their function and position in the studio, supervise their hanging and focusing, balance the overall intensity of the lighting, and provide whatever special lighting effects are necessary. The only logical and efficient way to approach this task is to plan the lighting with pencil and paper long before you ever enter the studio. Working with a floorplan and lighting plot not only gives you a better overall view of the entire set and equipment positions than you can get from standing in the middle of the studio floor, but preplanning with paper and pencil also permits you to experiment with a number of different lighting approaches without wasting expensive crew and studio time. It is a lot easier to erase a mistake on the lighting plot than to reposition completely instruments on the lighting grid in the studio.

The lighting plot indicates the position of every instrument which will be used in the production. Some LDs like to use tracing paper superimposed over the floorplan. This permits you to make changes on the lighting plot without obliterating the original floorplan. Of course, you can draw directly on the floorplan, but mistakes and revisions, which are inevitable, may result in a messy and unreadable light plot.

Studio floorplan sheets usually include the lighting grid which is superimposed directly on the floorplan and not only gives the set designer and program director a good idea of space and size, but also permits the LD to position instruments on the battens which directly correspond to those in the studio. A plastic template is a convenient way of drawing in the size and type of lighting instruments. If you don't have a template available to use as a stencil, you can easily devise your own system for indicating fresnels, scoops, strip lights, ellipsoidals, and the like on the lighting plot. The important thing is to remain consistent so that everyone concerned with the plot—the LD, the lighting crew, and so forth—will understand what each symbol means. (See Figure 6-17.)

The first step in plotting the lights is to determine which lighting approach you will use: the specific performer-and-camera method or the major-area technique. We have already suggest-

ed that the type of program will usually determine the best approach. If you must also consider motivating the light, be sure to keep in mind the major source or sources of illumination which will be on the set such as windows, doors, lamps, and so on. While you plot the position of the instruments, remember that, although you are working with a two-dimensional floorplan, the lights will be operating on a three-dimensional set. It is relatively easy to determine the proper horizontal placement for each instrument, but much more difficult to visualize the effect of the instrument's vertical angle on the subject. Instruments which are hung on battens about 12 to 13 feet high should be approximately 10 to 14 feet from a subject standing on the studio floor to achieve approximately a 35° vertical angle. Of course, this is only a rough guideline and will vary considerably depending upon the height of the battens in your studio and the height of the performer. It is a good idea to survey the studio you will be using prior to planning your lighting plot so that you can determine the necessary combination of horizontal and vertical angles in advance.

Working with the director's script and with the studio floorplan, the LD plots the position of the instruments for each important performer and camera angle or for all major playing areas, depending upon the approach chosen. It is best to start by plotting out subject key and back lights first, before planning the background lights. The use of fill light is often difficult to plot accurately in advance since fills are balanced relative to the key, back, and background lights. Nevertheless, it is wise to try and approximate the number of fill lights you think you will need and to include them on the floorplan. This will make the job of hanging and focusing the instruments much easier for the lighting crew.

As you plot out each instrument, you should write a small number next to it, starting with #1 and continuing until every instrument is numbered. On a separate piece of paper, write down the size and function of each instrument. For example, if Instrument #1 were a 2,000-watt

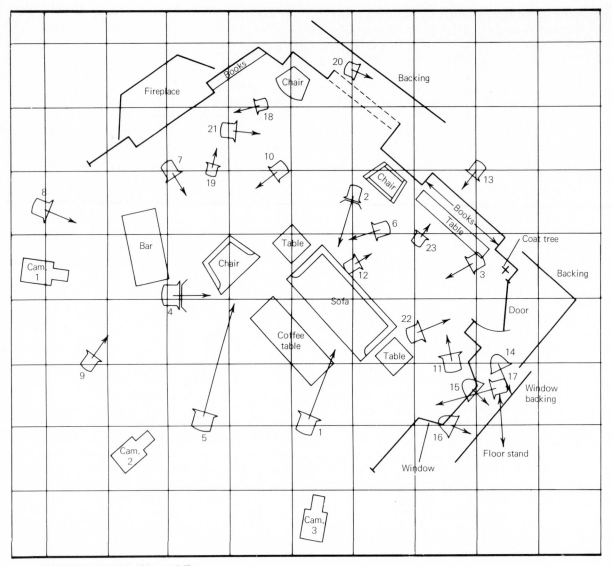

FIGURE 6-17 Light Plot and Key.
The light plot is used by the lighting director to plan the lighting design. The plot
indicates the type of instrument and its position for every light source used in a
production. The key, shown on page 167, uses the instrument number from the
light plot to identify each instrument's size, function, and dimmer assignment.

fresnel used to key-light an actor on a sofa, you
would write: "#1: 2kw, John's sofa key." By the
time you have completed plotting all the instruments, you will have a complete listing of every
light, its size, and its function. This information will
prove invaluable in locating individual lights dur-

ing studio setup and rehearsal and will help you
find troublesome instruments quickly by simply
looking up the instrument's function on the chart
and referring to the appropriate number on the
lighting plot.

Devising the lighting plot is a complex and

LIGHT PLOT KEY

INSTRUMENT NUMBER	INSTRUMENT	FUNCTION	DIMMER
1	2K	SOFA KEY	1
2	1K	SOFA BACK (left)	2
3	1K	SOFA BACK (right)	2
4	2K w/SCREEN	SOFA FILL	3
5	2K	CHAIR KEY	1
6	1K	CHAIR FILL	4
7	1K	CHAIR BACK	5
8	2K	BAR AREA KEY	1
9	1K	BAR AREA FILL	6
10	1K	BAR AREA BACK	7
11	2K	DOOR ENT. KEY	8
12	1K	BOOK SHELF BG	9
13	1K	REAR SOFA AREA BK	7
14	1500 w SCOOP	WINDOW BACKING	10
15			10
16			10
17	2K - ON FLR STND	WINDOW LIGHT	11
18	750w	FIREPLACE BG	12
19	750w	SLASH ON F.P./BOOKS	12
20	750w	HALLWAY	13
21	1K	HALLWAY ENT. AREA	13
22	1K	DOORWAY ENT. AREA	14
23	750w	TABLE/BOOK SHLF BG	15

tedious procedure, but care and attention to detail during this phase of the planning will pay off in enormous savings in time, energy, and money once you enter the studio to hang and focus the instruments.

DIMMER AND CUE SHEETS Once the lighting plot is completed, the LD must assign instruments to dimmer circuits. You should already know how many dimmers are available and what their wattage capacity is. With these factors in mind, the LD begins to assign instruments to dimmers, carefully noting which instruments, by number, are assigned to each circuit. This information is recorded on a dimmer sheet. If you are working with a limited number of dimmers, it is best to assign a number of key lights to one dimmer and save other individual dimmers for fill and background lights, which usually must be balanced relative to the key light's intensity. Of course, the lighting director must also keep special lighting cues in mind, so that groups of instruments can be faded up or down without affecting the rest of the lighting.

Once the LD has completed the dimmer sheet, the last step is to plan the lighting cue sheet. This will indicate the various lighting cues which must be made during the production and should specify which dimmers are involved and what the "point" setting or intensity of each is. Of course, the exact point setting cannot be determined until you have arrived in the studio and you are able to balance the lighting on the set. (See Figure 6-18.)

Setup and Rehearsal

The program producer will usually notify the LD when the studio will be available for setup—the time when instruments can be hung and focused. The size of the job and the program's budget will determine the amount of time and the size of the crew available to the LD.

Just as the program director must plan the best way to use rehearsal and production time, so too must the LD work out a schedule for rigging and focusing the lights. Compounding the personnel and time limitations further, while the lighting crew is hanging lights, another crew will usually be working on the set too. To assure as smooth an operation as possible, the LD should confer with the art director or set designer in advance to work out the best method of approaching the many tasks which must be completed in a relatively short amount of time.

HANGING AND FOCUSING THE LIGHTS

Using the lighting plot as a guide, the LD instructs the crew where to position each instrument. Sometimes instruments will already be hanging from the grids and need only be repositioned. Other times, the crew will have to start completely from scratch. There are a variety of ways to approach hanging and focusing. Some LDs like to hang all the instruments in roughly the correct position and then go back and focus each instrument individually. Others prefer to hang and then focus each instrument before proceeding to the next, although this is a much slower procedure. In either case, key lights should be aimed and focused before you begin working on the fill and background lights.

While the instruments are being hung, a crew member must be positioned at the patch panel to connect the instruments to the proper dimmer circuits. If you are familiar with the studio in which you will be lighting, you may wish to assign outlets to dimmer circuits even before you hang the lights. Once this is done, the entire patch board can be patched and ready before any of the instruments are actually in the air.

FOCUSING AND BALANCING THE LIGHTS

Once the instruments are hung and patched and once the necessary color media or scrims are inserted into the instruments, the set is ready for final light focusing and balance. The LD must use a light meter to set the intensities of the key lights for the standard camera operating level. If a camera and studio monitor are available, the LD can check the lighting on the monitor from time to time to see how it actually appears onscreen.

The key light is almost always focused first since it is used as a reference point for all other instruments. The best way to focus a fresnel is to spot it down completely, open the barndoors, and point the hot spot directly at the center of the subject. Once the hot spot is in the middle of the subject or area to be lit, the instrument is firmly tightened in place. Now the beam can be flooded out as far as necessary, and the barndoors can be used to trim the excess spill light. During focusing, never look directly into the light, for the intense beam can not only damage your eyes, but it will affect your vision and impair your judgment for some time afterward. If a crew member passes his or her hand in front of the instrument, you can observe the area it covers without looking at the light itself.

Once the key light has been properly focused, you can proceed to focus the additional lights, including the back, fill, and background instruments. During the focusing process, you will invariably discover some problems which will require revisions in instruments and positioning. Consulting your lighting plot and instrument chart should make these touch-ups relatively easy and painless.

Once all the instruments have been positioned and focused, it is time to balance the illumination on the set. The most effective method of balancing lights is to start with all the instruments off and add them one by one, watching to see what each instrument contributes to the overall lighting. If a studio camera is available, the LD may wish to watch the balancing on a monitor, instructing the dimmer board operator to vary the point levels of different dimmers to achieve the proper lighting effect. At this point, the lighting should come together, and the picture should begin to show the subject's depth and form while also creating the appropriate mood, setting, and time of day. If the program director is available, he or she should be consulted to approve the final lighting balance.

At this time, the dimmer board operator or the LD must record the point settings for each dimmer so that the lighting balance can be re-created by simply matching dimmer settings to the information on the dimmer sheet.

CAMERA REHEARSAL Although the lighting may now look perfect on a stationary camera, it is only during the camera rehearsal period that the LD will be able to see how the various camera shots, performers, and lighting interact. This is

also when the boom operator and audio engineer will be working with their equipment, and any additional problems can be discovered and resolved.

No matter how carefully the LD has designed, focused, and balanced the lighting, there are invariably trouble spots which crop up as the camera rehearsal proceeds. A performer may look too hot in one position, or the background may look wrong in another shot. A boom shadow may appear across an actor's face, or a lighting cue may not look as good as it should.

While the director works with the production crew and performers, the LD and the lighting crew will be touching up the lighting to solve whatever problems appear. The actor's hot spot is resolved by inserting a screen into the key light, slightly reducing its intensity. A change in dimmer settings on some background lights helps to bring out more detail in the set. The boom shadow is removed by using barndoors to trim off some excess spill light, and the lighting cue is rehearsed to correct errors in timing.

Production

By the end of the camera rehearsals, the lighting should be as close to perfect as possible. Individual problems will have been discovered and solved, and light cues timed and rehearsed. During the production, some program directors will have the assistant director cue the light changes. Others will ask the lighting director to cue all light changes based upon the script and actor movement. In those studios where an electronic dimmer board can be set up in the control room, the lighting commands can be received directly from the director or AD. Otherwise, the LD and the control board operator are connected to the control room via the production intercommunication system.

Once the production is completed, the LD and crew must strike the lighting, Screens, scrims, and patterns should be removed and stored safely where they will not be damaged or mis-

DIMMER/PATCH SHEET

DIMMER #	SETTING	PATCH NUMBERS	FUNCTION	PRE SETS
1	[10]	368, 305, 329	SOFA KEY/CH KEY/BAR KEY	
2	[8]	349, 339	SOFA BK (L)/(R)	
3	[8]	327	SOFA FILL	
4	[9½]	364	CHAIR FILL	
5	[10]	315	CHAIR BACK	
6	[8]	308	BAR FILL	
7	[10]	313	BAR BK/REAR SOFA	
8	[8]	374	DOOR KEY	
9	[6½]	345	BOOK BG	
10	[10]	354, 355, 356	WINDOW BACKING	
11	[8½]	412	WINDOW LIGHT	
12	[9]	317, 325	FIREPLACE BG/BOOK BG	
13	[10]	311, 351	HALLWAY, DOORWAY AREA	
14	[7]	326	DOOR ENT.	
15	[9½]	335, 338	TABLE/BOOK BG	

(a)

FIGURE 6-18 **Dimmer and Cue Sheets.**
The dimmer sheet (a) indicates which instruments are to be patched into a
particular dimmer. The information is necessary for keeping track of instruments
during light balancing and for production lighting cues. The cue sheet (b)
indicates each lighting cue and the dimmer settings involved. The dimmer board
operator follows the information on the cue sheet during the production as he or
she controls the lighting changes.

CUE #	DIMMERS	DESCRIPTION
1	1-15 @ BALANCED SETTING - NORMAL AS ON DIM PATCH SHEET	DAY
2	1^{10}/2^{5}/5^{10}/7^{10}/8^{5}/9^{5}/11^{4}/12^{5}/13^{8}/14^{7}/15^{9}	NIGHT/FULL
3	5^{10}/7^{10}/8^{8}/9^{6½}/11^{4}/14^{7}/15^{9}	NIGHT - JOHN ALONE
4	1-15 NORMAL as on DIM PATCH SHEET	DAY

(b)

placed. Color media must be evaluated, and those which are too badly burned or discolored to be reused should be destroyed and the reusable gels filed away.

COMMON LIGHTING SITUATIONS

There are a number of production situations which are frequently encountered on all levels of television production and which require fairly standard lighting approaches. This section deals with a number of such situations and offers some suggested ways to approach the lighting. You should remember, however, that these are only examples to use as a starting point in developing your lighting plans. No two production situations are ever exactly the same, and there is never one, best way of lighting any program.

News Programs

Most news sets utilize a desk for the newscaster and some device for projecting graphics behind the commentator. The most commonly used techniques for utilizing graphics with news are (1) chroma key and (2) rear screen projection. Since each works on an entirely different principle, the lighting approaches for each will vary.

First, we will light the foreground subject—the newscaster—and then consider the background. Since the newscaster will invariably be lit for high key and without any regard for motivating the key light, we can place our key on either side of the camera. Let us say we decide to position the key on the left side of the camera. We will position a 2-kw fresnel approximately 30° to the left and at a 35° vertical angle.

The backlight will also be a fresnel, and we have selected a 1,000-watt instrument with a barndoor. Each of these instruments has been assigned to a separate dimmer circuit.

We have selected a 1-kw fresnel with a barndoor and scrim for the fill light, since the fresnel and its barndoors will permit accurate control of spill light, and the scrim will diffuse the light and reduce its intensity. Distribution control is particularly important because both chroma key and rear screen projection require a very carefully lit background for the best operating results. The fill light is also assigned to a separate dimmer, so its intensity can be balanced relative to the key light.

Determining the proper background lighting depends primarily upon the visual graphics device we will employ in the program. Since chroma key requires an evenly lit background in order for the system to read the "key" color, we have used a group of scoops to provide illumination for the chroma key background.

Rear screen projection, on the other hand, works by using a strong overhead projector positioned behind a rear screen. If you will be using this graphic technique, you must be sure that no spill light falls onto the background screen where it will wash out the projected image. To light the background for rear screen projection, we have

selected a number of 750-watt fresnels with barndoors. The lights are positioned and focused to illuminate the background set while keeping all spill off the rear projection.

Figure 6-19a shows the newscaster and set lit for chroma key, and Figure 6-19b shows the same set lit for rear screen projection.

Interview Programs

In an interview program, the primary focus is naturally on the participants in the discussion. The best approach for this production situation is usually to light each participant with his or her own key, back, and fill lights. However, if the number of participants on a panel becomes too large to make this approach practical, you can substitute area lighting for individual camera-and-subject-position method. In this case, a number of individuals would be lit with the same key, back, and fill lights. The biggest problem in using this technique is to avoid having the person seated closest to the key light being excessively brighter than the panelist who is positioned further away.

For the three-person interview illustrated in Figure 6-20, we have lit each with individual three-point lighting. We used three 2-kw fresnels for keys, three 1-kw fresnels for back lights, and floodlights for fill. We selected scoops in this instance because the diffused light will blend in well with the key light, and there is no need to control spill light over the set area.

An important consideration in lighting the interview program is to determine the direction in which the panelists will be facing during the show. Although the chairs may be positioned outward, during conversation participants have a tendency to face the person to whom they are speaking. In this case, we have slighly modified the position of the instruments to take this factor into account.

The background is a cyclorama, which we have lit blue. We used blue color media, which were fitting over the cyc strips at both the top of the cyc and along the ground row, for even illumination. We are also using an ellipsoidal spotlight with a cucalorus to throw a pattern onto the cyc.

Lighting for Charts and Demonstration Programs

When a program will utilize either a demonstration or the use of graphic materials, such as a chart,

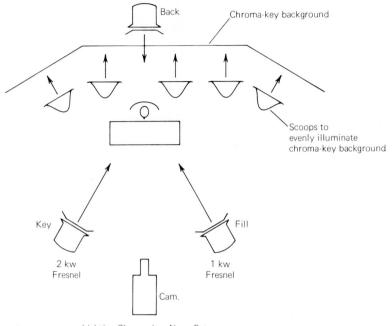

1 kw Fresnel

Back

Chroma-key background

Scoops to
evenly illuminate
chroma-key background

Key

Fill

2 kw
Fresnel

1 kw
Fresnel

Cam.

Lighting Chroma-key News Set
(a)

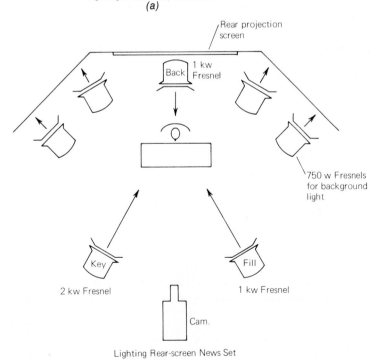

Rear projection
screen

Back 1 kw
Fresnel

750 w Fresnels
for background
light

Key

Fill

2 kw Fresnel

1 kw Fresnel

Cam.

Lighting Rear-screen News Set
(b)

FIGURE 6-19 Lighting Design for Typical News Set.
The basic lighting approach is identical for both chroma-key *(a)* and rear screen *(b)* projection. Since the chroma key operation requires a fairly evenly lit background surface, we used scoops to illuminate the background area. In the rear screen set, spill light would wash out the projected image so a number of fresnel spots are used to illuminate the background, and barndoors are used to keep all illumination off the rear screen area.

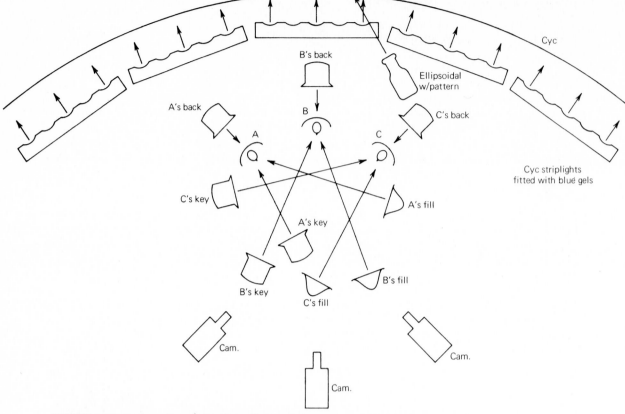

FIGURE 6-20 Lighting Design for Three-Subject Interview.
Each subject is lit with a separate key, back, and fill light. Scoops were used for
fill since their diffuse light will lighten shadow areas and because lighting
distribution control is not critical. The background cyc is lit with a series of strip
lights, which are fitted with blue gels to paint the cyc with color. A pattern is
projected on the cyc with an ellipsoidal spot to break up the flat background and
to add visual variety.

blackboard, or weather map, the lighting director must not only light for depth and texture, but must also ensure maximum visibility and detail of the demonstration or graphic. Demonstration programs invariably use close-ups, and the LD must consider this factor when planning the lighting.

Often best results are obtained by lighting the demonstrator and other performers separately from the demonstration area or the graphic board. Performers can be lit with conventional three-point lighting, and the demonstration area (table-top, stove or range, floor, and so forth) can be lit with diffused light from scoops or broads. The

diffused light will eliminate many distracting shadows, which might interfere with the audience's view of the demonstration, and will also help to minimize specular reflections from such items as cooking untensils.

Lighting for a blackboard or chart requires positioning the key light at an angle so that the shadow which is produced by the performer's body and arm will fall away from the board. Figure 6-21 illustrates how to position a key light to throw the shadow behind the subject and the camera's area of interest, keeping the board and graphic free from distractions.

Lighting for the Boom Microphone

Lighting for the boom microphone is one of the most difficult and challenging assignments facing a lighting director. Just when a set appears to be lit perfectly for the cameras, performers, and setting, a boom shadow appears in the middle of the camera shot! Unfortunately, there is no instant cure for boom shadow. The placement of the boom, the position of the lighting instruments, the movement of actors and camera, and the dimensions of the set all interact to create the problem.

In trying to deal with boom shadow, it is important to realize that it is practically impossible to eliminate the shadow completely. Simply turning off the instrument which is creating the problem is rarely satisfactory since it can result in poor lighting. On the other hand, flooding the shadow with enormous quanities of soft light in an attempt to wash out the shadow may result in overlighting the set and possibly ruining the lighting balance you worked so carefully to achieve. The best solution is actually twofold. First of all, you must determine with the program director and audio engineer where the boom will be positioned and

how it will cover the action. This information can help you find the best positions for your lighting instruments to keep shadows to a minimum. Secondly, since boom shadows cannot be eliminated, you must throw the shadow out of the camera shot so it will not be visible to the audience. Placing the key light opposite the boom mike will tend to throw the shadow away from the camera's area of interest, but this solution is far from foolproof. Experienced lighting directors will tell you that the mechanics of the production always supersede aesthetic considerations. In this case, you might have to cheat the position of your key light—even if it is motivated—to eliminate boom shadows from the camera range.

Sometimes it is possible to throw the boom shadow onto a dark area of the set, where it will be obscured by the darkened background. Other times, the boom will have to be repositioned

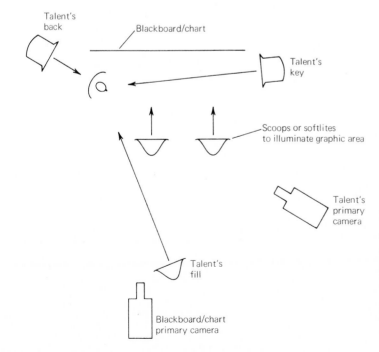

FIGURE 6-21 **Lighting Design for Subject and Blackboard or Chart.** The key light is positioned so that it will throw its shadow behind the subject and away from the graphic. The scoops are used to illuminate the graphic area since their light produces very even illumination without strong shadows which might distract the viewer.

where it can still pick up acceptable audio, but will create fewer lighting problems. The best general advice in dealing with the frustrating problem of boom shadows is to have a good working relationship between the program director, the audio engineer, and the lighting director. Boom shadow problems are not insurmountable, but they require a certain amount of flexibility and compromise within the production team.

Lighting for Camera Graphics

The most important consideration in lighting camera graphics or title cards (sometimes called "flip cards") is to avoid flares or undesirable reflections which will ruin the shot. Many flip cards are produced on highly reflective paper, which can cause troublesome specular reflections. The best approach in lighting a camera graphic is to use a single fresnel instrument, from a relatively far distance, which is steeply angled toward the

graphic. Then, if any reflection does result, the reflected angle will be so steep as to be outside the camera's range.

Once the program director has indicated where he or she intends to position the graphic card stand, you must decide how to hang or mount the instrument. Occasionally, a director will position a flip stand in a corner of the studio where the camera can easily pick up the graphic without too much additional camera movement. If there are no overhead battens available at this location, you may have to mount the instrument on a floorstand.

A good rule of thumb is to place the fresnel about 10 feet from the graphic at a very steep angle, about 75° to the graphic's surface. Let the light skim the graphic and illuminate the entire card evenly. (See Figure 6-22.) Sometimes a scrim inserted in a fresnel is helpful in reducing the intensity of the beam and diffusing the hard light.

Be sure to check the brightness level on the camera card to be certain that the intensity approximates the brightness level on the studio set. This will permit the camera to be quickly focused on the flip card without the video engineer's

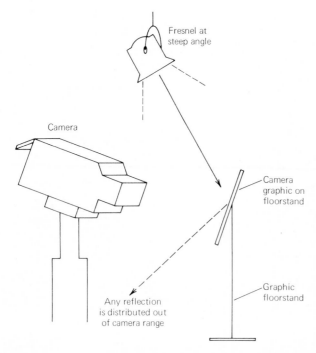

Fresnel at steep angle

Camera

Camera graphic on floorstand

Any reflection is distributed out of camera range

Graphic floorstand

FIGURE 6-22 **Lighting a Camera Card Graphic.**
Positioning a fresnel spotlight at a very steep angle will throw any reflection or glare outside camera range. Scoop lights can also be used to illuminate camera graphics, but the lack of distribution control can cause problems in small studios where there is insufficient separation between the graphic and the studio set. Scoops can also produce glare on highly reflective graphics, and their uncontrollable light distribution can create reflections directly into the camera lens.

having to adjust exposure and it will allow the director to cut to the camera much more quickly if necessary.

Lighting for Remote Productions

The lighting techniques which we have discussed apply equally to productions which are produced at remote locations as well as to those produced inside the studio. However, remote production presents a number of special lighting challenges, especially because few locations come with the kind of fully equipped lighting system which you would find inside a studio. Usually, the remote unit must bring its own lighting equipment and may have to make special arrangements to provide the necessary electrical power.

The lighting requirements for a remote vary considerably depending upon the type of production. ENG (electronic news gathering) pro-

ductions usually use very simple lighting to supplement available light in order to provide the necessary illumination for the operation of the television camera. At the other extreme, sophisticated Electronic Field Production or multiple-camera remotes may involve a great deal of precision lighting. In all cases, it is essentially a matter of applying the equipment and techniques which we have discussed in these two lighting chapters to the particular requirements of the remote production. In Chapter 12, which covers ENG production, and in Chapter 21, which is about remote production, we will discuss in detail how to apply the skills you've learned in these chapters to remote operations.

SUMMARY

Standard television lighting utilizes the three-point approach, which uses key, back, and fill light to illuminate a subject. The key light is the principal source of illumination and serves as the reference point against which all other lights are positioned and balanced. Key lights are usually positioned at about a 30° to 35° angle to the side and above the subject. Fresnel instruments are almost always used for key lights. Back lights are positioned behind the subject and are used to increase depth perspective by creating a slight rim of light around the head and shoulders. Fill light is used to lighten shadow areas and to blend in illumination on subject and set areas.

Although the three basic positions are most commonly used in television lighting, some additional sources such as side light, kicker, and eye light are used for special purpose applications.

Since television production uses a multiple camera approach, the three-point lighting plan is often modified, with the most attention given to the close-up, which is the predominant shot in almost all television programs.

Special effect lighting can be used to enhance the lighting on a set and to provide a special illusion for the audience. Among the most common special effect lighting situations are the use of color media to produce colored light, and the use of lighting for electronic chroma key.

Preplanning the lighting is an important factor in successful lighting. During the preproduction conference, the LD will receive a copy of the script and the director's floorplan indicating basic camera and performer positions. The LD confers with the director and the audio engineer to determine if any additional production elements will involve special lighting considerations. Then the LD prepares a lighting plot—which indicates the size, function, and position of each instrument—to be used as a guide during hanging and focusing.

Once in the studio, the LD must supervise the crew in positioning the lights and then in balancing each instrument to create the desired lighting effect, which should ultimately show depth and texture, create the appropriate mood and time of day, and, of course, provide sufficient illumination for the operation of the camera.

CHAPTER 7

TELEVISION AUDIO

Equipment and Operation

It has become a well-worn cliché that audio has always been a poor step-child to the video portion of a television program. Like the weather, poor-quality television sound was something that many complained about, but until recently there wasn't much anyone could do about it.

For the most part during television's early years, just the ability to reproduce a picture was considered so remarkable that both viewers and production people were not very concerned about the quality of television sound. Manufacturers produced television sets with cheap amplifiers and tinny speakers, but nobody cared much because their attention was focused on the picture anyhow. As time went on and sound quality could be improved, stations and production personnel figured that there was not much point in taking the time and expense necessary since the audience could not hear the difference, given the poor audio quality of their sets. Since the quality of program sound remained so marginal, there was little incentive for set manufacturers to produce better quality equipment and so the vicious cycle continued.

It is only recently that a number of developments have occurred to break the cycle. First, a generation has grown up with high-fidelity stereo equipment, and they are now demanding the same quality in television sound that they are used to hearing from records, from tapes, and in movie theaters. Second, the rapid growth of home video has produced equipment such as video discs, video cassettes, and wide-screen television units which can offer viewers excellent sound reproduction. Finally, many cable systems are now offering their subscribers programming with high-quality stereo sound, and the trend toward top-quality audio is increasing rapidly.

Even in conventional programming which will not be reproduced in stereo, there is a recognition that no matter how well produced a show's video might be, if the audio production is weak, the entire show will suffer. Audio is no longer considered to be the horse that trails behind the cart but is now an integral part of any production. As such,

audio must be planned and produced with the same care and attention to detail that are given to the visual elements of a show. In this chapter we will discuss the equipment used in audio production and its operation. In the next chapter, we will see how to use this equipment in the production of television sound.

THE NATURE OF SOUND

We cannot discuss the operation of audio production equipment without spending some time on the characteristics of sound itself. A sound is produced by a vibrating body. As a body rapidly vibrates back and forth, it creates pressure variations in the surrounding air, which is pushed outward in waves. These sound waves radiate from the source as waves ripple on a pond when a rock is tossed into the water.

When we talk about sound from a production standpoint, we are primarily interested in the way our ears perceive sound. That is because there are often important differences between the physical properties of sound and how we actually *hear* it. While a detailed discussion of these various factors is not possible here, there are two important sound characteristics which you should be familiar with in order to work with television audio: sound frequency and sound intensity.

Sound Frequency

The sound waves produced by a vibrating body are cyclical and can be graphically illustrated by the "sine wave" in Figure 7-1. These waves vary in both frequency (the number of cycles per second) and in intensity (the strength or "amplitude" of the wave).

The faster a body vibrates in the air, the greater the number of cycles produced and the higher the sound's frequency. We perceive sound frequency as *pitch*. A high-pitched sound, from a piccolo or flute, for instance, produces a great many sound waves per second. A low-pitched instrument, such as a bass or a tuba, produces far fewer vibrations per second, and we hear a much lower sound. We measure a sound's freqency in *hertz* (Hz), interpreted as "the number of cycles per second produced by a sound source."

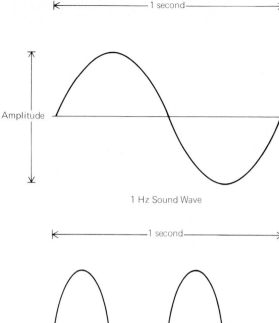

1 Hz Sound Wave

2 Hz Sound Wave

Amplitude

1 second

1 second

FIGURE 7-1 Sound Waves.
A sound wave can be diagramed as a "sine wave." Each cycle is measured from the start of one wave to the beginning of the next. The number of complete waves or "cycles" per second is expressed in *hertz* (Hz) and determines the frequency of the sound. The height of the wave is its *amplitude* which we perceive as loudness.

People are capable of hearing frequencies between 16 and 16,000 Hz, depending on age and health. As we grow older, our sensitivity toward sounds at the upper and lower extremes of the frequency range diminishes, and we lose our ability to perceive them. In addition, normal hearing is most sensitive to sounds between 500 and 4,000 Hz, the range of sound frequencies most important for understanding speech.

Professional audio equipment can reproduce sound frequencies from about 16 to 20,000 Hz, which is the full audio spectrum necessary to pick up and reproduce sound and music accurately.

Sound Intensity

The intensity of a sound depends on the amount of energy used to produce the pressure variations in the air. The more energy used, the greater the pressure variations, and the more intense the sound. We perceive sound intensity as loudness.

We sense variations in sound intensity by comparing one sound's volume to another. Our sense of loudness is more a relative measure than an absolute scale, so we are really working with a ratio of the intensity of one sound to that of another. We measure this intensity ratio in *decibels* (dB).

Simply stated, a decibel is a comparison between the loudness of two sounds. It is a logarithmic scale which enables us to express very large physical values with a rather small and convenient scale of numbers. For example, the decibel scale is designed so that a doubling of intensity is expressed as a change of 3 dB. In other words, whenever we double or halve the intensity of a sound, we vary it by 3 dB. Therefore, a change from 3 to 6 dB is a doubling of the sound, as is a change from 6 to 9 dB or from 15 to 18 dB. The reason you should be familiar with the decibel scale is its frequent use in audio production to refer to sound level and to describe various equipment characteristics. For example, the VU (volume unit) meter used to indicate the strength of an audio signal visually is calibrated in decibels.

TYPES OF MICROPHONES

All microphones contain two basic components: a diaphragm and a generating element. The *diaphragm* is a flexible device which is sensitive to the air pressure variations of a sound wave. The diaphragm vibrates according to the pressure variations of the sound. Attached to the diaphragm is the *generating element*, which converts the diaphragm's vibrations into electrical energy. The current that is produced is routed to the audio control console, where it is amplified and regulated by the audio engineer.

The construction of the microphone and the type of generating element used are important factors which determine a microphone's operating characteristics. There are three types of microphones which are used in television production: (1) dynamic microphones, (2) ribbon microphones, and (3) condenser microphones.

Dynamic Microphones

In a dynamic microphone, the diaphragm is attached to a coil of wire located close to a permanent magnet. As the diaphragm vibrates according to the sound waves which reach it, the "voice coil" moves back and forth inside the magnetic field. This creates an electrical current, which is directly proportional to the movement of the dia-

phragm. The louder the sound, the greater the diaphragm's movement and the greater the electrical energy produced. Since dynamic microphones utilize an internal coil, they are sometimes called "moving coil" microphones.

Dynamic mikes are capable of producing excellent sound fidelity, and their rugged construction makes them durable and relatively insensitive to the harsh handling production mikes are subject to in daily operation. (See Figure 7-2.)

Ribbon Microphones

The construction of a ribbon microphone, sometimes called a "velocity" microphone, is in many ways similar to the dynamic or moving coil microphone. Again, a permanent magnet is used to provide a magnetic field, but, instead of a voice coil, the ribbon mike uses a thin metal strip or ribbon, which serves as both the diaphragm, to receive the sound waves and as the generating element, to produce the electrical current.

Ribbon mikes tend to be more fragile than most dynamic mikes. This limits their use outdoors or whenever the microphone is subject to much movement or handling. They produce a very warm, rich, and mellow sound, which is often desirable for miking announcers, singers, and musical instruments.

Condenser Microphones

Condenser microphones use a condenser, more accurately called a "capacitor," as the generating element. A capacitor is a device consisting of two plates which can hold an electrical charge when current is supplied. The charge depends on the physical distance between the two plates and varies as the space between them grows larger.

The diaphragm of a condenser microphone is actually one of the capacitor plates, which moves relative to the other stationary plate. When air pressure variations from the sound source move the diaphragm, the space between the two capacitor plates varies and this produces an electrical output voltage.

Condenser microphones offer excellent audio response characteristics, but they require a power supply to both charge the capacitor and to

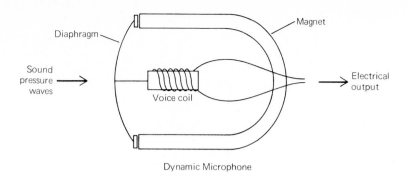

Dynamic Microphone

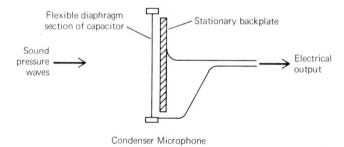

Ribbon Microphone

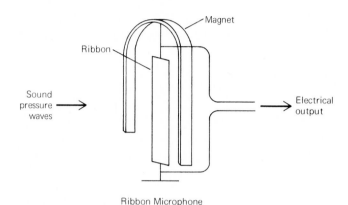

Condenser Microphone

FIGURE 7-2 **Microphone Generating Elements.**

amplify the tiny output current. The power supply is large and bulky, making the condenser microphone impractical for most television audio applications.

ELECTRET CONDENSER Electret condenser

microphones are capable of the same excellent performance characteristics of the normal condenser mikes, but do not require large and bulky power supplies. An "electret" is a special capacitor element which is charged once during manufacture and is designed to hold the charge indefinitely. This means that it does not require external power to charge the element for use. Although the electret condenser still requires a power supply, solid state technology has reduced the size of the electronics and necessary power to the point where a tiny battery and associated electronics are built into the microphone itself or housed in the cable connector jack. (See Figure 7-3.)

Electret condenser mikes can be made very small and unobtrusive and yet deliver excellent audio fidelity. The internal battery supply will last over 1,000 hours, but it is a good idea to have spare batteries always on hand since the microphone cannot function without battery power.

MICROPHONE PICKUP PATTERNS

Audio pickup refers to how the microphone responds to sounds coming from different directions. While no microphone can be designed to accept sounds from one direction and totally reject sounds emanating from another, the pickup pattern of a microphone indicates its area of maximum audio sensitivity. There are two types of pickup patterns: (a) omnidirectional, and (2) unidirectional.

Omnidirectional Patterns

A microphone with an omnidirectional (sometimes called "nondirectional") pickup pattern is designed to receive sound from all sides equally without discriminating or favoring sound from any particular angle.

A *polar diagram* is a convenient way to illustrate a microphone's pickup pattern. The diagram is interpreted by assuming the microphone is in the center of the graph facing the 0° axis. The two 90° axes are at the sides of the microphone, and the 180° axis is at the rear of the mike. As you can see from Figure 7-4, an omnidirectional microphone has a 360° sensitivity range. Sounds emanating from any direction will be picked up equally well using a microphone with this type of pickup pattern.

Unidirectional Patterns

Unidirectional microphones are designed to be most sensitive to sound emanating from one direction and to suppress or reject sounds coming from the rear and sides of the mike. Unidirectional pickup patterns are available in three general categories: (1) bidirectional, (2) cardioid, and (3) supercardioid.

BIDIRECTIONAL PICKUP As its name suggests, the bidirectional mike has two "live" sides of equal size. Bidirectional pickups are valuable

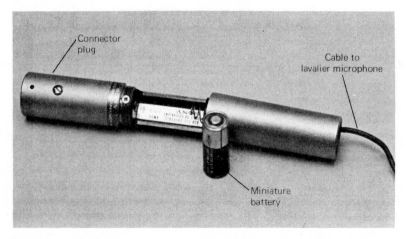

FIGURE 7-3 **Electret Condenser Power Supply.**
Electret condenser microphones require a battery-supplied electrical power for operation. Most lavalier mikes enclose the battery inside the microphone connector plug assembly. For larger hand-held mikes, the batteries are usually located inside the microphone case itself. Whenever you insert a battery inside its holder, make certain you have positioned the battery correctly with the positive (+) end facing in the proper direction, or microphone damage can result.

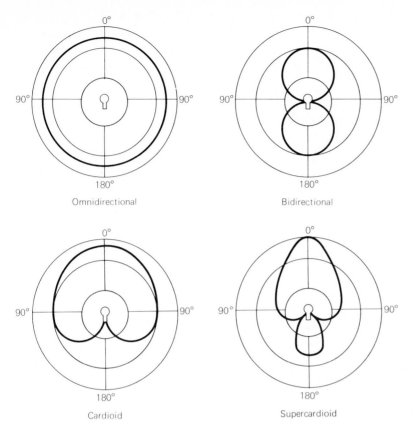

Omnidirectional

Bidirectional

Cardioid

Supercardioid

FIGURE 7-4 Polar Diagrams.

when two performers wish to work a mike simultaneously and must see each other. These mikes were often used in radio drama, so that the actors could watch each other as they worked the mike. The problem with its use in television is that the side opposite the performer naturally faces the cameras and crew, and the mike has a tendency to pick up extraneous production noise in the studio. For this reason, bidirectional mike pickups are rarely, if ever, used in television production.

CARDIOID PICKUP The pattern of a cardioid microphone, as its name suggests, is shaped like a heart. Although the "live" area pattern varies, most cardioid mikes are highly sensitive within a 120° area facing the microphone. The cardioid pattern is very commonly used since it offers some directional isolation yet has a wide audio pickup range on the live side of the microphone. Another advantage of the cardioid pickup is that

the performer or subject can work farther from the microphone and still produce better quality sound than is possible using an omnidirectional microphone.

SUPERCARDIOID PICKUP These highly directional microphones are designed to pick up sound within a very limited range. Since their polar diagram resembles an extreme cardioid pickup, they are called supercardioids. One misconception about supercardioid microphones (sometimes referred to as "shotguns" because they use a long tube to gather the desired sound and to suppress sounds from the side and rear of the microphone) is that they are specially constructed. Actually, the microphone itself is quite similar to cardioid or omnidirectional mikes; it is the design of the tube combined with the design of the cardioid microphone which produces the highly directional pickup pattern.

Because of their pickup characteristics, highly directional microphones can be used at greater distances from the subject than other mikes and still provide good sound quality. They are frequently used mounted on a boom or hand-held for special pickup applications.

The main problem with a highly directional mike is its very directionality. Unless the microphone is pointed directly at the sound source, poor audio quality can result. But at the same time, do not think a shotgun microphone has a completely narrow pickup pattern which will reject all spurious noise outside a very narrow-angled cone. While the microphone's pickup sensitivity is greater at larger distances, it is also highly susceptible to ambient noise, reverberation, and echo, which can ruin the overall sound quality. Even with a supercardioid microphone, the closer you can position the microphone to the sound source, the better the sound quality will be.

The use of a *parabolic reflector* permits an ordinary cardioid mike to assume highly directional pickup characteristics. The parabolic reflector is a metal or fiberglass dish about 3 to 4 feet in diameter. It is equipped with a microphone mounted in the center of the dish. The microphone does not face outward, but instead is pointed toward the center of the reflector and picks up the concentrated sounds as they are reflected off the parabolic dish. Many remote productions use parabolic reflectors to provide an additional dimension to the program audio. For example, parabolic reflectors are often used on football remotes to pick up the quarterback's signals. A technician standing on the sidelines points the reflector toward the quarterback while the signals are called out. (The audio engineer in the control room must be careful to turn on the parabolic mike only when necessary since it can pick up a considerable amount of unintended audio.) On most games, the audio engineer will use the reflector mike to pick up the quarterback's signals and the sounds of the two lines colliding after the ball is snapped. The mike is then immediately cut off to avoid the possibility of

the player's on-field "remarks" from inadvertently being transmitted. As with any highly directional mike, much of the sound quality depends upon how accurately the technician who is operating the mike points it toward the sound source. (See Figure 7-5.)

The 3- to 4-foot diameter of the parabolic reflector makes its size impractical for in-studio use. Since it is rare that the audio engineer will have to pick up sounds from as great a distance inside the studio (where conditions can be controlled), the use of parabolic reflectors is usually limited to outside the studio unless very special circumstances require its use indoors.

Using Microphone Pickup Patterns in Production

There is no such thing as a "better" microphone pickup pattern. Each is designed to be applied to a particular audio situation. The idea is to let the pickup pattern work for you by maximizing the microphone's sensitivity to sound that you want to cover and minimizing extraneous sound which would interfere with the audio quality.

For example, if you were recording a musical group with a number of microphones, you would probably choose cardioid pickups since each microphone should be positioned to pick up only a part of the total sound. An omnidirectional microphone would pick up audio from all areas and make mixing and balancing the overall sound from multiple microphones difficult. Cardioid microphones are also helpful when you wish to minimize reverberation, extraneous noise, and other ambient sounds. An omnidirectional microphone is perfect to use when extraneous noise is not a problem and when one microphone must cover a wide area.

FREQUENCY RESPONSE

The frequency response of a microphone is an important characteristic which determines how sensitive the microphone is to the various audio frequencies which it must pick up and reproduce. Frequency response is usually provided as a range (for example, 45–18,000 Hz) which indicates the lower and upper frequency extremes that the mike is capable of reproducing accurately.

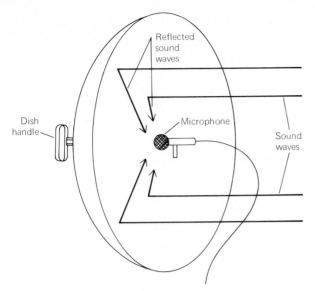

FIGURE 7-5 Parabolic Reflector.
The bowl-shaped reflector gathers sound waves and concentrates them at the live side of the microphone.

A microphone with a "flat" response would theoretically be equally sensitive to all audio frequencies within its response range without unduly emphasizing or suppressing certain frequencies. A mike with a flat response is ideal for recording musical instruments where the audio engineer wants to reproduce all of the frequencies of the instruments with maximum accuracy.

Some microphones are designed with special frequency response characteristics which enhance the mike's usefulness in certain situations. For example, many lavalier microphones are designed with increased sensitivity to the frequencies which make up the human voice and to suppress low frequencies to eliminate unwanted noise and to decrease the "barrel" effect of being located close to the performer's chest cavity. Some mikes permit the audio engineer to adjust the microphone's frequency response characteristics to best match the production's audio requirements. A common feature on many mikes is a "low frequency roll-off" control, which reduces the microphone's bass response and is helpful in eliminating wind noise or the artificially increased bass effect when a performer works very close to the mike. The point is that a microphone's frequency response can be a useful tool which the audio engineer should take into account when matching a particular microphone to a specific audio production situation.

IMPEDANCE

The fourth important microphone characteristic is its *impedance*, which refers to the amount of electrical resistance in a circuit and is expressed in *ohms* (often abbreviated with the symbol Ω). Low impedance means there is relatively little signal resistance and, therefore, a relatively large current flow. High impedance means a greater resistance and a smaller current flow. Professional microphones, such as those used in television production, are always of low impedance (referred to as "low Z"). This is because low impedance microphones permit the use of very long cables without any loss of high frequency signals. Since most professional production applications require mikes to be used with cables longer than 15 or 20 feet, only low impedance microphones can be used without degrading the sound quality.

Low Z microphones run anywhere from 30 to 250 ohms; high impedance mikes run from 10,000 ohms and above. A microphone's impedance must always match the input connector impedance on a tape recorder or audio mixer. This is never a problem in professional studios since all microphone inputs are designed for low Z operation. However, some remote tape recorders may utilize a high Z input connector, which means you must either use a high impedance microphone or a transformer designed to match the low Z mike output to the high imped-ance input on the recorder or mixer.

ON-CAMERA MICROPHONES

A convenient way to distinguish between different microphones is to describe them by how they are used in production. On-camera microphones are those which are seen in the camera shot and are used when their presence on screen will not distract the viewer or interfere with the picture. Off-camera microphones are those which must be used when the microphone must remain out of the camera's view because inclusion in a shot would destroy an illusion or interfere with the action.

There are five types of on-camera micro-phones: (1) lavalier, (2) hand, (3) desk, (4) stand, and (5) headset.

Lavalier Microphones

The lavalier is probably the most commonly used microphone in television production. The lavalier, as its name suggests, can be clipped to the lapel of a coat or jacket; pinned to a tie, dress, or blouse; or hung around the neck like a pendant. All lavalier microphones are designed with an omnidirectional pickup, but since they are posi-tioned relatively close to the speaker's mouth, extraneous noise is rarely a problem. (See Figure 7-6.)

Unobtrusive, "tie-tack" lavalier microphones are electret condensers which can be made so small because the battery and miniature electron-ics are contained in the cable connector jack out of camera range. Because of the small size and appearance of the lavalier, some audio engineers use two in a *dual-redundancy* system. This simply means that the performer wears two mikes on the same clip. Only one microphone is actually on at any time, but should the mike fail during the production the audio engineer can immediately switch to the back-up mike without audio interrup-tion. (See Figure 7-7.)

(a)

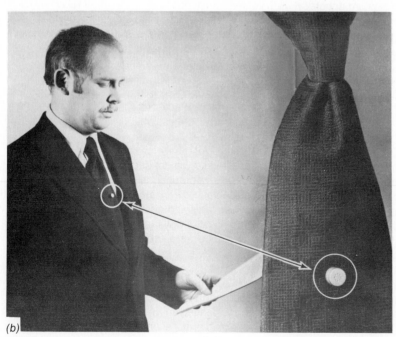
(b)

FIGURE 7-6 **Lavalier Micro-phones.**
Lavalier microphones are small and unobtrusive. Some are worn on a cord around the neck *(a)*, and others are attached to a tie clip or pin *(b)*.
(Courtesy: Electro-Voice and Shure Bros, Inc.)

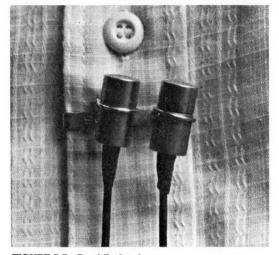

FIGURE 7-7 Dual Redundancy.
Two lavalier microphones are worn on the same clip.
Only one is actually "live," but, in the event of
microphone failure, the audio engineer can quickly
switch to the back-up mike without interrupting the
audio pickup.

USING THE LAVALIER IN PRODUCTION The
lavalier provides excellent audio quality when
used for general speech pickup and eliminates
many audio-associated production problems.
Booms require additional personnel and a great
deal of care in lighting and staging. Hand mikes
require the talent to hold the mike at all times.
Stand or desk mikes restrict talent movements. A
lavalier microphone, however, requires no additional production personnel, eliminates lighting
problems, and provides hands-free operation and
freedom of movement.

There are a number of disadvantages to using
the lavalier, however. First, its small size makes it
easily damaged or misplaced unless care is
taken to safeguard it. Because of its omnidirectional pattern, it will pick up extraneous sound
unless positioned correctly. It is a good idea for
the audio engineer or a crew member to actually
place the lavalier on each performer—especially
on nonprofessional talent—before production begins to ensure proper handling and the correct
placement of the microphone. If the performer will
face in one direction during the show, position the
microphone on the side where talent will normally

look during conversation. Take care in positioning
a lavalier on any performer wearing a necklace or
pin which might jostle the microphone and create
annoying sounds during the show. Some directors ask the audio engineer to hide the lavalier
under a tie or blouse. While this is possible, the
audio engineer should make a test before production to ensure that the hidden lavalier is not
rubbing against the talent's clothing or undergar-
ments, and the sound pickup is clear and unmuf-
fled.

With all of its obvious advantages, there are still
some production situations where a lavalier is not
the best microphone choice. A show with a number of on-camera participants is one example
where too many individual lavaliers would be
necessary. An audio engineer who is assigned to
a children's show where the host must interview
twenty children would be better off choosing a
hand microphone or even a boom rather than
attempting to provide each child with a separate
lavalier. Lavalier microphones are also not recommended for production situations where the best
quality audio is necessary. While the reproduction is excellent for general speech, a lavalier is
not the ideal microphone to use for covering a
singer or musical instruments.

Desk Microphones

A desk microphone is appropriate when talent will
be seated behind a table or desk or does not
have to move and the microphone can appear in
the shot. News programs, panel discussions, and
news coverage of speakers from a podium or
stand are common production situations for a
desk mike.

Dynamic, electret condenser, and even ribbon
mikes with a cardioid pattern are all used for desk
microphones. Dynamic mikes are very popular
because of their durability, especially when the
desk mike doubles as a hand microphone for field
use. (See Figure 7-8.)

USING DESK MICROPHONES IN PRODUCTION
As with a lavalier, the audio engineer should

position the desk mike in the direction talent will normally face during the show. Sometimes one desk mike, properly positioned, can cover two people. Since the desk mike is positioned on the desk or table top, it is highly sensitive to noise around the table area. Remind talent to avoid tapping their hands or feet, and make certain that no physical actions are planned on the same surface where the desk mike is located.

Hand Microphones

The hand mike is commonly used by emcees and singers in the studio and by reporters and interviewers in the field. The advantages of the hand mike are no special lighting or staging requirements, complete talent control over mike position, and good pickup characteristics. The disadvantages are that the performer must hold the microphone and is partially responsible for how well or how poorly the sound is covered on mike. (See Figure 7-9.)

Hand microphones are available with either omnidirectional or cardioid pickup patterns. Although most hand-held mikes take advantage of the rugged durability of dynamic construction, electret condenser hand-held microphones are also available.

Stand Microphones

A stand microphone is simply a desk or hand-held mike which is mounted on a mike stand. Stand microphones are frequently used by singers, for miking musical instruments, and for announcer microphones both on-camera and off. The stands come in many different shapes and sizes, from the usual straight mike stand to flexible neck stands and stationary boom-arm stands.

FIGURE 7-8 **Desk Microphone.**

FIGURE 7-9 Hand Microphone.
(Courtesy: WSB-TV)

The latter two are frequently used for miking musical instruments or to position the microphone where a singer-musician can work the mike and simultaneously play an instrument. (See Figure 7-10.)

The major disadvantage with a stand microphone is that the talent's movement is severely limited to an area around the mike stand. The advantages are hands-free operation, no staging or lighting problems, and the ability to position the microphone exactly where it is needed and to keep it at a constant distance from the subject.

Headset Microphones

There are certain production situations in which it is important to keep a performer's hands free while at the same time positioning the microphone very close to talent's mouth. This is especially true for remote situations such as sporting events where the crowd noise is so loud that it can drown out the announcer's voice unless the mike is very close. The headset microphone is a dual headset with a high-quality mike attached. Talent can hear the program audio in one ear, the director's commands in the other ear, and keep his or her hands free to handle various papers during the production. With the microphone positioned so

FIGURE 7-10 Stand Microphone.
Stand microphones used to mike guests on the *Mike Douglas Show*.
(Courtesy: Mike Douglas Show-Group W Productions)

FIGURE 7-11 Headset Microphone.
An omnidirectional, high-quality microphone, which is positioned directly in front of the announcer's mouth. The "biaural" headset enables the announcer to hear program audio in one earpiece and receive the director's instructions and cues in the other earpiece.
(Courtesy: Telex Communications Inc.)

close to the talent's mouth, crowd noise is kept to a minimum. In addition, the padding in the headset permits talent to hear the production commands which might otherwise be drowned out. (See Figure 7-11.)

OFF-CAMERA MICROPHONES

Off-camera microphones are used whenever the microphone cannot appear in the camera shot. There are two types of off-camera microphone operation: (1) boom mountings and (2) hidden microphones.

Boom

The term "boom" refers to a family of microphone mounting devices which are used to position the microphone just out of camera range, but where it

can still pick up the sound. Although almost any directional microphone can be used as a boom mike, specially designed dynamic and electret condenser, cardioid, and supercardioid microphones are manufactured especially for use on booms. These microphones produce excellent audio response while their highly directional pickup characteristics enable the mike to work at a greater distance from the subject and to reject extraneous noise. (See Figure 7-12.)

There are three types of boom mounting devices: (1) the hand-held fishpole, (2) the medium-sized tripod, or "giraffe" boom, and (3) the large, perambulator boom.

HAND-HELD FISHPOLE The simplest kind of boom is the hand or fishpole boom. This is a long aluminum tube with a microphone mounted on one end. The telescoping tube can be extended or retracted depending upon how much reach is necessary to cover the designated playing area.

FIGURE 7-12 Boom Microphone.
The boom microphone is a highly directional microphone which can be positioned out of camera range and still pick up program audio.

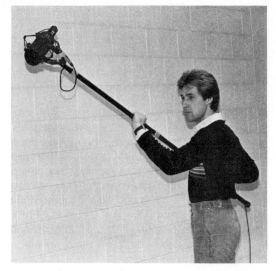

FIGURE 7-13 **Fishpole Boom.**

Fishpoles are frequently used on remote location, because of their portability, and in the studio when a larger boom cannot be used for lack of space. The problems associated with using the fishpole boom are these: (1) the boom becomes heavy for the operator to hold after awhile, and (2) the boom operator does not have maximum control over the microphone placement. Once the fishpole length has been set, the operator can extend or retract the boom only by moving his or her body toward or away from the set. (See Figure 7-13.)

TRIPOD ("GIRAFFE") BOOM The tripod or giraffe boom is basically a fishpole attached to a pedestal and mounted on a tripod with wheels so the unit can be moved easily around the studio. In addition, most medium-sized booms are equipped with a rotating device so that the mike can swivel without moving the boom itself. The major problems associated with a giraffe boom are these: (1) once the length of the boom has been set, it cannot be extended or retracted—instead the entire tripod must be rolled in or out; (2) the boom's reach is rather short and has a limited operating range when compared with a large, perambulator boom. In a small studio, this can be an advantage since space is at a premium and a large boom might cause more problems that it

may solve in terms of space and camera movement. (See Figure 7-14.)

PERAMBULATOR BOOM The large boom is a sophisticated and expensive device which utilizes a series of belts and pullies to control the movement of the boom and the microphone. The two most commonly used booms in television are the Mole-Richardson boom and the Fisher boom. The Mole-Richardson has a disadvantage in that the boom arm can be swiveled only about 180° before the operator is forced off the platform. The Fisher boom has a circular platform which permits operating the boom in a 360° circle. The Fisher boom is also equipped with a seat which can make long production periods easier for the boom operator. (See Figure 7-15.)

BOOM OPERATION The entire platform and boom are mounted on wheels. A steering device at the rear enables a crew member to position the boom and to move it about the studio during production. The boom arm can be panned horizontally left or right or tilted up or down. The retraction and extension controls permit fast and silent extension or retraction of the boom and microphone either out toward talent or in toward the boom operator. It is important for the audio engineer and the director to know the maximum extension range of the boom (it varies from model to model) in order to plan performer blocking and audio coverage. The microphone can be rotated almost 360° without moving the boom arm itself. This enables the boom operator to cover a wide area without having to pan the entire boom arm back and forth physically.

All large booms use a counterweight to balance the weight of the microphone and to compensate for the boom arm's extension or retraction. The proper balance can vary depending on the type and weight of the microphone and mount you are using. Always check the boom's balance before a production. If you do not, the boom operator must physically bear the weight of the boom rather than sharing it with the counterweight system. This can

FIGURE 7-14 **Tripod ("Giraffe") Boom.**

tire the operator quickly and lead to inferior audio pickup on the show.

Probably the most difficult boom operation to master is judging accurately the distance between microphone and performer. The microphone must be positioned above and *in front* of the talent, or you will not achieve the best audio pickup. A good way to locate this position is to have talent raise his or her hand upward in a 45° angle. The boom mike should just touch the performer's outstretched fingertips. A good rule of thumb for the boom operator to follow is to keep the microphone as close as possible to the performer without dipping the mike into the camera shot. This means that in close-ups the microphone can be lowered and positioned relatively close to the talent. This will provide not only good sound pickup but also an accurate "sound perspective." In other words, we see the performer in close on the shot, and his or her voice sounds close to us. On the wide shot, the boom must necessarily go up to avoid being included in the wider shot. At the same time, the sound presence will change, and as we see the actor from a greater distance, we will also perceive the sound as coming from farther away. (See Figure 7-16.)

Many studios have added a small television monitor to the boom platform. This permits the boom operator to see the "air" shot and to position the boom accordingly. In addition, the boom operator usually wears a private line (PL) headset, which keeps him in contact with either the audio engineer, the director, or both. Some studios have dual headsets so that the boom operator can hear the director or audio engineer in one ear while listening to the program audio in the other ear.

Because the boom operator must position the microphone correctly for proper sound reproduction, the boom operator, audio engineer, and director must carefully coordinate boom moves, shot changes, and talent blocking before production. Of course, in an unrehearsed program, the boom operator must be doubly alert to any sudden moves talent might make. In this case it is

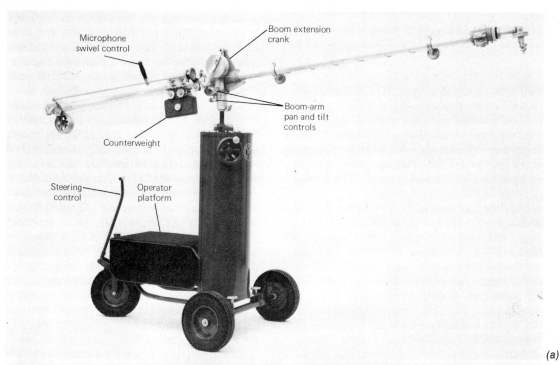

(a)

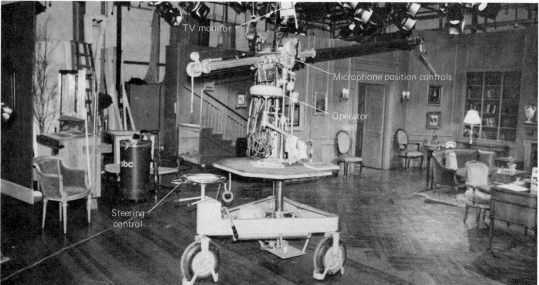

(b)

FIGURE 7-15 Perambulator Boom.
The Mole-Richardson boom *(a)* only permits a 180° arm swivel, so it must be positioned properly or moved if the microphone must be positioned outside the 180° range. The Fisher boom *(b)* offers a much wider arm swivel and has a seat which enables the operator to work in greater comfort.
(Courtesy: Mole-Richardson Co. and ABC-TV)

important to impress upon the performer the necessity of "telegraphing" any moves before actually making them. Talent should say something like, "Let's get up and go over here," to alert the director, audio engineer, and boom operator of the unplanned move. There is nothing more embarrassing than to have a performer rise unexpectedly and hit his or her head on the boom microphone; getting the boom into the camera shot is only slightly less embarrassing.

As with most production equipment, the boom offers its share of advantages and disadvantages. The advantages of the boom are these: (1) the microphone is out of the picture, (2) talent need not worry about holding a microphone or having

FIGURE 7-16 **Positioning the Boom Microphone.**
The boom microphone must be positioned in front of the talent's mouth in order to pick up the best sound. A convenient method to determine the optimum boom position is to ask talent to raise an arm at a 45° angle. The microphone should just touch the fingertips. From this point the microphone can be raised or lowered slightly, depending on how wide the director's camera shots are.

to contend with the trailing mike cables, (3) talent has more mobility within the pickup range of the boom, and (4) one boom can cover a large number of performers although not all at the same time. The disadvantages of the boom are as follows: (1) the size of the boom itself—in many small studios, the size of the boom may severely curtail camera movement since the boom platform takes up so much space, (2) The boom requires additional crew members, which can be costly, (3) The boom necessitates careful lighting to reduce the possibility of boom shadow. While boom shadow can never be completely eliminated, it is possible to light a set so as to throw the boom shadow out of camera range. This requires more planning and setup time to adjust the lights properly for boom shadow. (4) The boom cannot cover performers who are widely dispersed. Unless you can use more than one boom (a very expensive proposition), it is best to attempt to cover this audio problem in a different way.

The choice of whether to use a boom mike and the specific type of boom mounting depends upon the nature of the program, the size of the program's budget, and the equipment available. Because the use of the boom requires coordination with other units of the production team, it is an audio decision which should be made relatively early in the preproduction stage.

Hidden Microphones

There are times when it is either impossible or impractical to use a boom, and yet the microphone must be kept out of the camera shot. In this case an alternative would be to hide the microphone either on the set, or in some cases, on the performer. The hidden set mike can be positioned almost anywhere as long as it can cover the necessary audio during the production. Some common hiding places are inside special props, behind flowers or books, or on mike stands just outside camera range.

There are a number of advantages to using a hidden mike. Unlike booms, you need no additional crew members, and there is no problem in lighting for boom shadow. The biggest problem with the hidden mike is that you must position your actors very close to the microphone and make certain that their movement is limited to the area in which the microphone can acceptably pick up the sound. For this reason it is difficult to use only hidden mikes for an entire dramatic program since the positioning of the hidden mike severely restricts actor blocking. However, the hidden mike can be a valuable addition to a boom microphone if used in a particular set where the boom cannot reach and where actor movement is not essential.

Here is an example: *The Doctors* is a daytime series, which is produced by NBC in a medium-sized studio in New York. The studio can accommodate about five or six separate sets, which are positioned very close together. These sets are covered by two booms. Tucked away in a corner of the studio is a permanent set, which is the nurses' station at the hospital. Because the studio is so small, it is almost impossible for a boom to cover the actors who must work at the nurses' station. A hidden mike is used to provide audio coverage for this set. Since the set is small and actor movement is limited, a desk mike, placed on the nurses "desk" and hidden from camera view by a plant, can adequately cover the actors' voices. The actors are blocked around the microphone, and this leaves the two booms free to cover the larger sets which can accommodate their size more easily.

There are a number of considerations to take into account when planning to use hidden microphones:

1   Position the microphone in a location where it will pick up the audio with acceptable quality.
2   Do not place the microphone inside a prop or piece of furniture which might be jostled by talent during the program. This would result in the microphone rubbing against the furniture producing extraneous noise which would mar the audio.
3   Dress the mike cord out of camera sight, and secure it with gaffer tape so that the performers will not trip over it during the production.
4   During the program, the audio engineer should keep the fader pot on the hidden mike closed until it is actually going to be used. This will eliminate extraneous noise from entering the sound track and prevents the "echo"

effect which occurs when indirect sound reaches a second mike an instant after the primary microphone picks up the direct sound.

HIDDEN HAND MIKES   Sometimes the simplest solution to a problem is often the best one. If you have a situation in which you need to cover a performer's audio, but a boom microphone is impractical and hiding the mike presents other problems, an alternative is to have a floor assistant simply hold a hand mike near the talent and out of camera range. As long as the floor assistant holds the mike steady and avoids mike noise, the sound pickup will be acceptable. The director can help the situation by bringing his cameras in close to permit the hand mike to be held as near the talent as is possible. Although a highly directional shotgun mike is ideal for this purpose, a cardioid hand mike can serve as an inexpensive and generally adequate substitute. Be sure that the floor assistant holds the mike steady, avoids creating mike noise, and keeps the mike still until certain that the audio engineer has closed the mike's fader on the audio console, before placing the mike down.

HIDDEN LAVALIER MICROPHONES   Lavalier mikes are now made so small and unobtrusive that it is not difficult to hide one on a performer's body under his or her clothing. Of course, you will have to make certain that the microphone cable is out of the camera shot and that the performer will not have to move around much while trailing the mike cable.

The cable problem is eliminated if you use a wireless lavalier microphone which doesn't require any trailing cable. In this case, the mike's transmitter unit must also be hidden on the performer's body, but these are small enough to hide easily under a jacket or behind the performer's back and out of camera range.

HANGING MICROPHONES   Hanging a microphone from the lighting batten should be the last alternative considered by the audio engineer. Except in rare situations, a hanging mike will often produce more problems than it solves. With the mike overhead, there is always the possibility of picking up extraneous production noise. Since the mike's position cannot be varied during the show, unless talent is located in the exact position, the audio may be off-mike. The microphone's height also remains constant throughout the show. If the mike is hung too low, it may appear in a wide camera shot; if too high, poor sound quality will result. Finally, the lighting director may have to contend with shadows created by the hanging mike.

The only time a hanging microphone is useful is when you must cover a large group of performers who remain stationary throughout their performance. For example, a musical group, a choir, or an orchestra can sometimes be miked successfully with one or more hanging microphones.

CABLES AND CONNECTORS

The output of a microphone must travel through a cable in order to reach the audio control console. This requires the use of special audio cable, connector plugs, and connector boxes, which are located throughout the studio.

Cable

Audio cable consists of two wires surrounded by a protective insulating sheath. On small lavalier microphones, a thin wire is permanently connected to the mike and terminates in a connector plug. On most other hand, desk, stand, and boom microphones, there is no permanently attached cable. Instead there is a receptacle which accepts one end of an audio extension cable. The other end is designed to be connected to an input box to feed the signal into the audio control system.

Audio cable has a natural coil or curve which you should follow when winding it up. It is never a good idea to wrap the cable around your elbow or to tie it together by its ends. This can damage the internal wires and create audio problems. Instead, follow the natural coil and use a plastic twist-holder or even a pipe cleaner to secure the coiled cable together.

When you are running audio cable, be sure you have provided a sufficient length for talents' movements or for the crew member to operate the microphone or boom. Audio cable should be dressed out of camera sight and away from the path of cameras, other equipment, and technical crew members. Gaffer tape is useful in securing the cable underneath tabletops, along the edges of the set, and even to the studio floor to prevent someone from tripping over it.

Connectors

Professional microphones and mike cable all use a standard connector plug or "jack" known as a "Cannon XLR" or simply a "cannon plug." This is a three-pronged plug with a male and female version. As a general rule, most audio outputs utilize a male plug and most inputs utilize a female receptacle. Cannon connectors lock together when they are mated, which is a safety device to prevent the connection from accidentally coming loose during a production. To disconnect the male plug from the female outlet, you must first push the little tab button before attempting to pull out the plug. (See Figure 7-17.)

CONNECTOR BOXES   In most studios, audio connector boxes are located at regular intervals around the wall of the studio. The boxes look like electrical outlets, but they contain female cannon plug receptacles. Each jack receptacle is labeled with a number or letter which corresponds to its input on the audio control console inside the control room. This permits the audio engineer to route each microphone signal into the proper channel on the audio control console in the control room. (See Figure 7-18.)

WIRELESS MICROPHONES

Wireless microphones operate without any connection cables between the microphone and the

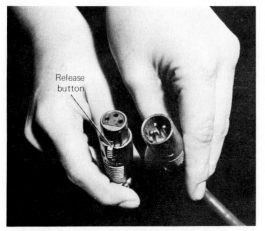

FIGURE 7-17 **Microphone Connector Plugs.**
To unplug a Cannon connector, you must depress
the tiny release button before attempting to pull the
connectors apart.

control room. The performer and the audio engineer gain all the advantages of a lavalier or hand
microphone without the disadvantage of trailing
mike cables, which can limit a performer's mobility.

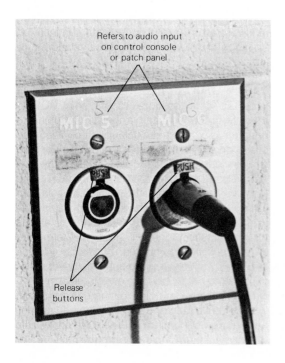

The wireless mike is a conventional lavalier or
hand-held microphone connected to a small
battery-powered radio transmitter. A small antenna from the transmitter unit relays the signal via an
FM radio frequency to a receiver connected to the
audio control console. From this point, the output
of the receiver is fed to the audio control console
and regulated by the audio engineer as would be
the case with any conventional cable microphone.

Many hand-held wireless microphones include
the transmitter and battery inside the mike case
itself. For lavalier mikes, the transmitter and battery are contained inside a pocket-sized unit,
which can be easily hidden under a performer's
clothing. (See Figure 7-19.)

Earlier models of the wireless mikes were notoriously unreliable. Sudden audio drop-outs and
interference with program audio by spurious police or taxi calls were not uncommon. Newer RF
(radio-frequency) mikes utilize a "diversity receiving system" which is designed to eliminate these
problems. Wireless mikes with the diversity system can be depended upon to provide reliable
operation. The transmission range from the microphone to the receiver is anywhere from 50 to
1,000 feet depending on operating conditions. As
a rule of thumb, try to position the receiver as
close to the microphone as possible. This will
ensure even better and more reliable quality.

FIGURE 7-18 **Audio Connector Box**
Audio connector boxes are positioned around the
walls of the studio so that microphones can be
connected wherever they are located. The female
connector in the wall requires a male plug for a
connection. The release buttons above each wall jack
must be depressed as you pull the male plug to
disconnect the microphone from the box. The numbers above each connector jack correspond to the
appropriate input on the audio control console or to
the patch panel jacks in the control room. When
plugging or unplugging audio connectors, always
hold the metal connector, and never exert force on
the cable itself.

FIGURE 7-19 Wireless Microphones.
The pocket-sized transmitter on the left can be used with most lavalier or conventional hand-held microphones. The hand mike on the right is a self-contained unit, which includes a battery power supply and a miniature transmitter. The output of the receiver station is connected to a conventional microphone input jack and regulated by the audio engineer as with any other normal microphone.

Depending on the microphone and battery system, the transmitter can operate from four to sixteen hours before a new battery is needed.

The advantages of the wireless mike are obvious: freedom from cables both inside the studio and outdoors on remote location. The biggest disadvantage is the cost, particularly the fact that for each separate microphone you need, you must provide a separate transmitter and receiver, which can become very costly.

AUDIO CONTROL CONSOLE

The audio control console is the central coordination point for all audio signals in a television program. It is at the audio console where all sound sources including microphones, tapes, records, sound tracks from film and videotape, and remote audio feeds are regulated, mixed, and sent out to the station's transmitter or to a videotape recorder.

Audio consoles vary in their size, configuration and sophistication, but there are four features common to all audio control consoles: (1) individual volume controls for all sound sources, called faders or potentiometers; (2) a volume unit (VU) meter which visually displays the signal strength of the sound as it passes through the audio console; (3) master level controls to regulate the

FIGURE 7-20 Audio Control Console.
A rotary fader console with two output channels. It can be used to record stereophonic sound or, more commonly for television production, to send two different monaural signals from a single control board when necessary.

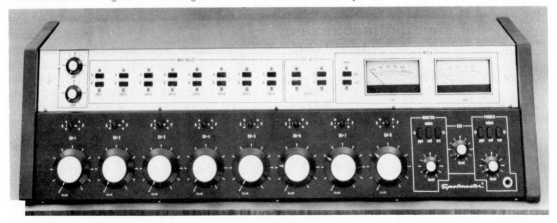

sound level of the entire output of the audio console; and (4) a monitoring system which is used both to cue upcoming sources and to listen to the mixed program sound via loudspeakers or headsets in the control room. (See Figure 7-20.)

Faders

Every individual sound source must be controllable by the audio engineer for a proper audio level. Consequently, each sound source which enters the audio board must be regulated by its own, separate control. These controls are called *faders* or *potentiometers* ("pots" for short), and they are used to vary the volume level of all sound sources. Faders can either be linear slide faders or round, circular knobs. The slide faders are pushed up to increase the sound and down to reduce or "attenuate" the volume. Potentiometer knobs are turned clockwise to increase the volume and counterclockwise to attenuate the sound. Every fader regulates its individual sound source independently of the other microphone or audio inputs. (See Figures 7-21 and 7-22.)

LOW-LEVEL AND HIGH-LEVEL INPUTS The terms "low-level" and "high-level" refer to the

signal voltage which enters the audio console. A microphone produces a low-level signal of less than a thousandth of a volt (a millivolt). The output of audio tape or cartridge machines, soundtracks from film or videotape, or remote audio feeds from outside the studio measure 1 volt or more, which is a high-level signal. Often, low-level signals are called "microphone level," and high-level signals are called "line level."

Most audio control consoles have faders and associated circuitry which are designed specifically for either low-level or high-level audio inputs. Because of the large difference in signal strength, you cannot input a high-level signal into a low-level channel or vice versa. On some modular audio consoles, it is possible to switch a channel electronically to accept either a microphone or a line input, depending on which is needed for a particular production situation.

MICROPHONE PADDING There are times, especially when recording live music, when the output signal from a microphone is too high and

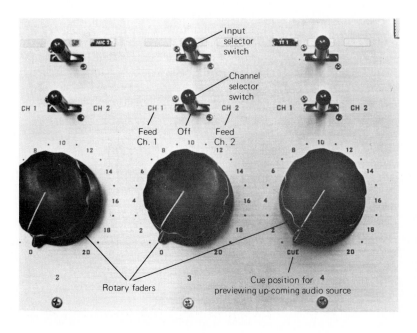

FIGURE 7-21 **Rotary Faders.**
Turning the fader clockwise increases the sound level. The switches above each fader are for assigning the fader output to one of two separate channels. The uppermost switches are input selectors, which increase the board's operating flexibility by permitting one fader to serve multiple functions.

FIGURE 7-22 **Slide Faders.**
Pushing the fader upwards increases the sound level. When the fader is at the bottommost position, the channel is off.

may overload the audio preamplifier in the control console. To prevent this, the microphone output can be *padded* by adding electrical resistance to the line to decrease the output voltage and provide greater control with the audio fader. Otherwise, the slightest fader opening would result in a large and uncontrollable burst of audio which is difficult to mix with other sources and can produce a distorted sound.

VU Meter

All audio consoles have at least one VU (volume unit) meter, which visually displays the intensity of the sound level leaving the audio console. VU meters are calibrated in two scales: "volume units" in decibels and "percentage of modulation." Figure 7-23 shows a typical VU meter with both scales. The numbers on the bottom scale—which run from −20 to +3—are volume units measured in decibels. The design of the scale on a VU meter takes into account the logarithmic progression of sound intensity and human perception of loudness. The numbers above the scale—which run from 0 to 100—refer to percentage of modulation. When the needle reads "100" or "0 dB" we are sending 100 percent signal strength through the audio board and into the sound system. When the needle reads over 100 percent, we are "overmodulating" the signal, and we risk transmitting too strong a level, which can

overload the circuits and result in distorted audio.

Before we consider the aesthetics of sound mixing, we must be certain that our sound level is technically acceptable. A sound level which is too low or "in the mud" will result in technically poor recording. Similarly, a level which is too high or "in the red" will produce a distorted sound since we are sending out a larger electrical signal than the system can handle. Keeping the sound at an acceptable level is called "riding gain," or "riding levels." In general, the VU meter should register somewhere between 80 and 100 percent modulation (or −2 to 0 dB) with occasional peaks into the red (+1 to +3 dB). It is important to remember that the VU meter will display only the intensity of the console board's output. It does *not* display the proper audio balance between various sound

FIGURE 7-23 **VU Meter.**

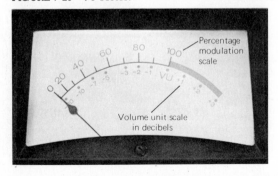

sources. It is entirely possible to send out a *technically* acceptable sound level which is totally wrong *aesthetically*. For example, if an announcer is mixed with a musical track so that the announcer's voice is so low he can hardly be heard over the music, the sound level might be technically perfect according to the VU meter, but totally incorrect according to the aesthetic requirements of the program. The audio engineer should use the VU meter as a guide for technical levels, but the meter can never substitute for an audio engineer's sensitivity in achieving the proper balance among the various sound sources.

Master and Submaster Controls

All audio control consoles have at least one *master control fader*. This fader controls the final mixed output of the entire audio console. The master gain control is usually preset before the production by the audio engineer and is rarely touched during a production since it controls the entire output level of the board and not individual sound sources. It can be used as a convenience, though, when you must completely fade in or out a number of sound sources simultaneously. For example, if you are ending a program with the simultaneous fade of a number of sound inputs, you can simply fade down the master fader and all the sources will fade out at the same time and in the same relative proportion to each other.

Many consoles provide for subgrouping of inputs. These controls are called group masters or *submasters*. With these controls you can assign a number of separate sound sources to one submaster fader and control the entire group with a single fader rather than with many. For example, say you must use four separate microphones to cover an orchestra. It can be very difficult to ride all of the faders accurately and simultaneously and also to mix in a singer's mike. If you work with submasters, you can simply preset all four audio mikes for the proper orchestral audio balance and assign them to one submaster. Then you need ride only two controls; the submaster (that is, the orchestra level) and the singer's mike.

Monitoring System

The audio console's monitoring system is designed to allow the engineer to listen to the

program audio which is being mixed through the board while cuing or previewing the sound from upcoming audio sources. These two systems are controlled with the program monitor and the cue and audition sections of the audio control console.

PROGRAM MONITOR The program monitor control works like the volume control on a radio or television set. It permits the audio engineer to adjust the volume of the sound coming from the control room and studio loudspeakers independently of the sound level which is being transmitted from the audio console.

The monitor level should be preset before a production at a volume which provides for comfortable yet clear listening. Some consoles permit you to adjust the volume in the studio, audio control area, and production control area separately so that you can provide the proper monitoring loudness where needed. Once the monitor level is set, leave it alone because it provides your benchmark for evaluating the overall sound mix.

Unless you are careful, the monitor volume may give you a distorted impression of the actual sound level leaving the board. For example, it is possible to have a microphone fader set so low that the VU level is in the mud, but by boosting the setting on the monitor control, you can achieve the aural illusion of a technically acceptable audio level. The VU meter and the monitoring control system are partners which must be used simultaneously. First, make certain the meter is peaking correctly; then listen to make sure the audio mix between sources sounds right.

Most studios have an automatic program monitor cutoff circuit which is engaged whenever a microphone is opened inside the studio. This kills the monitor speaker as long as the mike is live and prevents sound emanating inside the studio from looping back via the speaker and entering the microphone again creating an annoying feedback squeal. That is the loud whistle you sometimes hear when a microphone is held too close to a public address speaker in an auditorium.

AUDIO FOLDBACK   The monitor cutoff circuit is a great convenience, but it can also be a hindrance when talent in the studio must hear portions of the audio while working with a live microphone. For example, a singer may be accompanied by a prerecorded music track. With the cutoff circuit in operation, there would be no way for the singer to hear the music while singing into the microphone. The solution to this is audio foldback, which enables the engineer to feed the studio monitor selected portions of the audio while deleting the output of the open studio microphones. By folding back, or returning only certain parts of the audio, the talent can sing to the prerecorded track and not produce the feedback loop.

CUE OR AUDITION CONTROLS   All broadcast audio controls are equipped with a separate cue or audition system of amplifiers and speakers. This permits the audio engineer to monitor a source before actually mixing it into the program sound. On most boards certain faders (usually those controlling turntable, tape, cartridge, or remote inputs) have a click-stop switch at the "off" position of the fader or pot which engages the fader into the "audition mode." This permits the engineer to hear the output on a tiny speaker or over a headset to permit accurate cuing or to check that the source is ready for air.

Modular Control Consoles

As television audio requirements have become increasingly complex, many production facilities have installed modular audio consoles which are similar in design and operation to the kind used in recording studios. Each module on the console is a horizontal strip containing a variety of audio controls which enable the audio engineer not only to control the sound level, but also to manipulate the quality of any audio source independently from other inputs. For example, echo can be added to a singer's microphone without affecting any other mikes. The value of the modular console

is the enormous flexibility it offers the audio engineer to control and shape each of the various inputs which, when combined and mixed together, make up the program's audio. (See Figure 7-24.)

Although specific features vary from manufacturer to manufacturer, each modular strip usually contains the following controls: (1) slide fader, (2) equalizers, (3) echo/reverberation controls, (4) bus assignment switch, and (5) pan-pot control.

SLIDE FADER   Each module receives an individual audio input, which can be a microphone, the audio track from a videorecorder, a phonograph turntable, and so on. The slide fader is the device used to control the level of the incoming sound source. Each of the other controls along the module are used to alter the quality of the sound which its corresponding fader is controlling.

EQUALIZER   An equalizer is a device which alters the frequency response of a sound signal. Using the equalizer controls, the audio engineer can modify very specific portions of the audio signal while leaving other portions intact to produce a precise change in the original frequency response and color the texture and sound quality of the original audio signal.

Equalizers permit the audio engineer to "cut" or "boost" bass, midrange, and treble response of a sound source independently of one another. This capability can be used to modify the original sound during production to correct for acoustical problems, to enhance an aural effect, or to provide for special sound effects. Equalizers are also commonly used during postproduction mixing to produce the exact quality of sound desired. Equalizers will be discussed in detail in the next chapter.

ECHO/REVERBERATION CONTROLS   The echo/reverberation controls on the module enable the audio engineer to add as much or as little of the desired echo effect to the original sound source. Again, the degree and quality of the echo can be precisely determined by the audio engineer and will affect only the sound source which is being controlled by the specific module. Thus, echo can be added to certain microphones in varying degree without adding echo to other sound sources during a recording.

FIGURE 7-24 Modular Control Audio Console.
The modules include a slide fader, equalizer, echo/reverberation controls, and
pan-pot controls for use when recording in stereophonic sound.
(Courtesy: ABC-TV)

BUS ASSIGNMENT CONTROLS The output of
each individual module is routed to a particular
submaster or master output control using the bus
assignment controls. This permits the audio engineer to have tremendous flexibility in assigning
the output of any number of modules to a particular output source. For example, all of the microphones which are used to cover a musical band
can be assigned to one submaster control while
the microphones which pick up the singers are
assigned to another submaster. Similarly, when
we wish to produce a stereophonic mix-down,
certain modules are assigned to the left channel
and others are assigned to the right channel.

PAN POTS Pan pots are used to position the
audio "image" in aural space. This is possible, of
course, only when a production is being recorded
for stereophonic sound. With the use of stereo
becoming increasingly more common, however,
the pan pot is becoming an indispensable audio
tool. With the pan pot in its center position, the

sound source appears to come from the middle,
directly inbetween the left and the right speakers.
Rotating the pan pot to either the left or the right
will move the "image" of the sound a corresponding amount to position the sound in dimensional
space. The further the pan pot is turned in either
direction, the more pronounced the directional
effect until the sound emanates only from either
the left- or the right-hand side. Of course, it is
possible to employ the pan pot during a recording
literally to move a sound from one side to another.
We will discuss ways in which the pan pot is used
in audio production in the next chapter.

Portable Mixers

With remote production becoming increasingly
commonplace, a variety of portable audio mixers
have been designed specifically for use in the
field. The modern generation of portable audio
mixers offers extremely lighweight portability with
a powerful audio production capability enabling

the audio engineer to control the input of a number of microphones and to feed one or more outputs to videotape recorders, videocassette recorders, and audio tape recorders in the field.

Most portable mixers can be run either off conventional ac current or with a battery pack to provide complete mobility at the remote location. Monitoring is done through headsets which the audio engineer wears and is usually switchable to permit monitoring of the signal at various stages of the mix. For example, it is possible to monitor the sound being picked up by microphones or to switch to monitor the output signal of the mixer unit as it feeds an audio or videotape recorder. (See Figure 7-25.)

PATCH PANEL

Audio control consoles are purposely designed to provide for maximum flexibility of operation. Consequently, the various faders or pots can be assigned to a number of different functions. Since it is impossible to prewire a sound console to fit every potential production requirement, a patch panel is used to route various sound sources to designated fader positions on the audio console.

The *patch field*, or *patch panel*, is like a telephone switchboard. There are a large number of connector holes called "jacks," which can be interconnected by means of a special connector cable or *patch cord*. Every microphone input in the studio is wired directly into the patch panel. At the patch panel the audio engineer can connect a particular mike output to a particular preamplifier input and fader on the audio board. Sometimes certain mike outputs are automatically wired into

FIGURE 7-25 **Portable Audio Mixer.**
Front view *(a)* shows VU meter and rotary faders for controlling four separate inputs. The rear view *(b)* shows mike connector jacks for each input. The mixer can be run either off ac power or on battery power if ac is not available. *(Courtesy: Shure Bros.)*

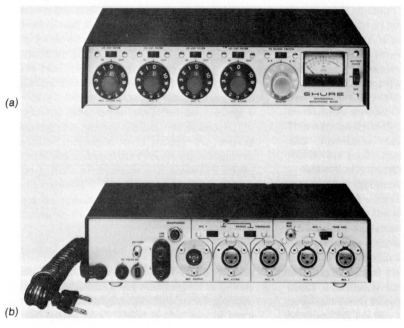

(a)

(b)

FIGURE 7-26 Patch Panel.
The patch panel is used to control the routing of audio signals in and out of the audio control console and through auxiliary audio equipment.

the control console since these are most commonly used. This is called a *normal wiring*. By inserting the patch cord into the appropriate position on the patch panel, you disconnect the normal wiring and can connect the mike output to any other microphone fader you wish. (See Figure 7-26.)

By properly patching each microphone into the control console, the audio engineer can position the microphones for maximum ease of operation. For example, if you are the audio engineer for a talk show with five participants, you should patch

each mike into the control console so the position of the mikes on the audio console corresponds to the way the participants are seated on-camera. In other words, the person at the far left would be patched into fader #1, the next person patched into fader #2, and so on. This will make riding gain during the production much less confusing.

Because every possible output from a studio and input to the console must appear on the patch panel, patch fields can appear very intimidating. They are not really as complicated as they seem if you keep in mind the fact that the patch field is usually organized by sound source and function. All mike outputs are located in one area, all tape and record turntable outputs in another, and all remote lines are in another area. By properly utilizing the patch panel, the audio engineer has far more flexibility and control over sound sources than would be otherwise possible.

RECORDS AND TURNTABLES

There are few television productions which do not utilize material from phonograph records at one time or another. In addition to commercially available music and dramatic records, a number of companies produce sound effects and musical effects on records specifically for use in television production. These services provide an almost unlimited variety of music and sound effects at a very reasonable price.

The operation of the record *turntable* is essentially the same as for a home player. Most professional turntables are built to meet very exacting technical specifications and are designed to operate for long periods of time. Turntables for professional use are typically designed for two-speed play: 33⅓ and 45 revolutions per minute (rpm). A major difference between home turntables and professional turntables is the fact that professional models are never equipped with a record changer. The operator must play each record manually. Professional turntables are also equipped with a clutch which enables the operator to engage the motor and start the turntable revolving at the proper speed immediately.

Cuing Records

A record must be *cued* so that the sound will be available the instant the director calls for it during the show. The most common technique for cuing phonograph records in television is the *slip cue*. The procedure for slip cuing is as follows:

1   Place the turntable clutch or speed selector in the "neutral" position.

2   Place the turntable stylus just before the appropriate cut or band on the disc.

3   Turn the turntable fader on the audio console to the "cue" position. This will engage the cue amplifier system and permit you to monitor the record without sending its signal to the program channel.

4   Slowly turn the turntable clockwise by hand until you hear the first bit of desired sound or *modulation*. This is the cue point.

5   Once you have found the cue point, back up the record counterclockwise until you just

pass the modulation. The stylus should now be in a silent groove immediately before the start of the record cut.

6   Just before the director needs the record, gently hold the outer lip of the record and start the turntable by engaging the clutch. Professional turntables are designed with a felt surface which allows the turntable to spin while the record is held in position by the audio engineer. Hold the record lightly enough so you do not disturb the turntable's motion, yet firmly enough to prevent the record from revolving with the turntable.

7   At the director's command, release the record and simultaneously bring up the turntable's fader on the control console. The sound will begin immediately. The reason you fade up the control pot is to avoid the possibility of a record beginning when not yet up to speed. This is called "wow in" and should be avoided. Even if the record wows in or has skipped a groove in the cuing process, fading in the pot can disguise the error. With a little care and practice, you should be consistently able to slip cue records exactly when needed, without any wowing or miscues. (See Figure 7-27.)

There is another method available for cuing records. You cue the record as you do with a slip cue, but instead of holding the disc while the turntable spins, back the record counterclockwise about half a turn. Then, on the director's command, engage the turntable and simultaneously turn up the fader on the audio board. The sound will come in without a wow since many professional turntables are designed to reach a steady speed within a half-turn. The problem with this cuing technique is that a brief period of time elapses before the sound begins. While this is acceptable in some radio productions, most television audio is coordinated with the video, and the audio engineer is expected to provide audio precisely on the director's command. For this reason, slip cuing is the preferred method of record cuing.

Records are a common source of music and sound effects. A good studio will have a record library containing both commercially available records and recordings specially produced for

FIGURE 7-27 **Slip Cuing a Record.**

broadcast purposes. The library should have a cross-indexed file indicating title, artist, and length of each cut on the record. This will make music selection easier and less time-consuming.

AUDIOTAPE RECORDING

Audiotape recording enables the engineer to record electronic audio information, store it, and then replay the sound later. Audio recording is also an important creative tool which permits sound to be manipulated and edited to vary its sound or sequence or to eliminate mistakes or imperfections.

Audio recorders are available in a variety of sizes and formats, but they all work the same way. Audiotape is made of a plastic base coated with a metallic oxide. The audio signal which enters the tape recorder activates the *recording head*, which functions like an electromagnet. The recording head produces a magnetic field which aligns the oxide on the tape into a specific pattern corresponding to the original audio information. During playback, another head decodes this magnetically arranged pattern on the tape and produces an electronic current which duplicates the original signal. The output of the *playback head* is amplified and fed to the audio console, where it is regulated by the audio engineer.

There are two types of audio tape formats which are used in television: (1) reel-to-reel machines and (2) cartridge machines.

Reel-to-Reel

Reel-to-reel, or "open reel" machines utilize magnetic tape, which is supplied on spools. Conventional audio tape is ¼-inch wide and comes in a variety of reel sizes and tape lengths. The tape must be threaded onto the machine, where it runs past the heads at a speed of either 7½ or 15 inches per second (ips). The faster tape speed delivers a much higher quality recording but also uses more tape. In general, 7½ is the standard for most audio recording requirements; 15 ips is used only when the situation demands the highest fidelity possible. (See Figure 7-28.)

TAPE TRACKS The tape recorder organizes the recording information in a specific area of the tape called a "track." Most monaural, reel-to-reel professional machines are *full-track* recorders, which means that the entire ¼-inch width of the tape is used for recording and playback. A professional stereophonic tape recorder would use two tracks; one for the left channel and another for the right channel. You may be familiar with home stereo recorders which divide the tape into four tracks: two tracks for tape running in one direction and two tracks for the tape running in the opposite direction when you turn it over.

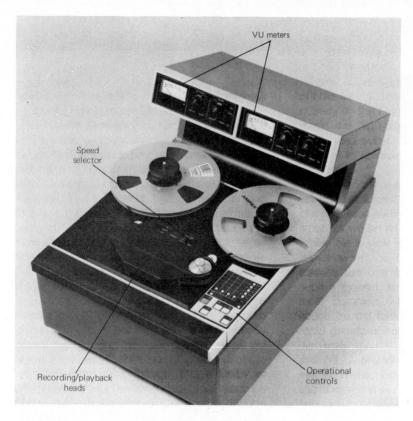

VU meters

Speed
selector

Recording/playback
heads

Operational
controls

FIGURE 7-28 Reel-to-Reel Audiotape Recorder.

Because a full-track machine utilizes the full width of the tape and in only one direction, you cannot take a tape which was recorded on a two-direction, two-track or four-track machine and play it back on a full-track machine. If you do, the full-width playback head will pick up the track or tracks going in the right direction as well as those which were recorded in the opposite direction. Figure 7-29 shows the arrangement of audio record and playback heads and various tape track configurations.

It is always best to try to record your program material on the same type of tape machine you will use for playback. If this is impossible and you must use a multiple track machine, erase your tape with a "bulk eraser" before recording. Then record on either one or two tracks in only one direction. Remember, too, that most professional machines will not play back any tape recorded at a speed slower than 7½ ips.

BULK ERASING A major advantage of audio-

tape is its capacity to be erased and reused. Audiotape is erased by using a strong magnetic field to completely rearrange the tiny magnetic charge on the tape. This is called *bulk erasing*. Although it is possible simply to rerecord over a reel-to-reel tape, it is better to bulk erase the tape first. This will prevent any possibility of the prior material on the tape being incompletely erased by the tape recorder, resulting in an echo or "ghost" effect.

To bulk erase a tape, you must pass it through a *tape degausser*. Rotate the reel or cartridge on the degausser and, while you keep the degausser on, slowly remove the tape to a distance of at least 2 feet. You can then turn off the degausser, and one side of the tape will be erased. Having it 2 feet away from the magnetic field before turning off the degausser will prevent a low frequency hum from appearing on the erased tape. To be absolutely certain that the tape is completely erased, turn the reel or cartridge over and repeat the procedure. You should remove your wrist-

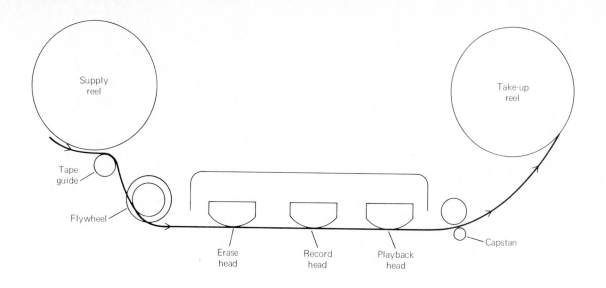

Supply reel

Take-up reel

Tape guide

Flywheel

Erase head

Record head

Playback head

Capstan

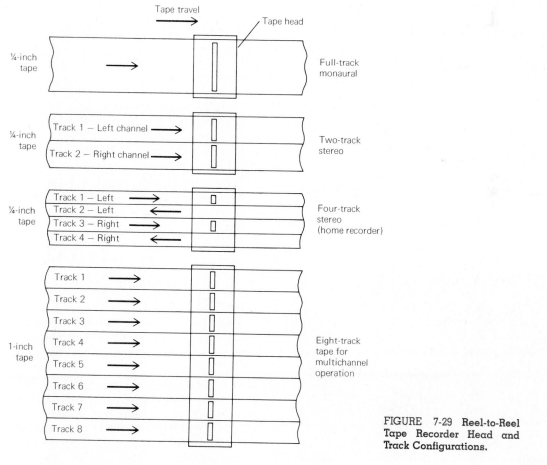

Tape travel

Tape head

¼-inch tape

Full-track monaural

¼-inch tape

Track 1 — Left channel
Track 2 — Right channel

Two-track stereo

¼-inch tape

Track 1 — Left
Track 2 — Left
Track 3 — Right
Track 4 — Right

Four-track stereo (home recorder)

1-inch tape

Track 1
Track 2
Track 3
Track 4
Track 5
Track 6
Track 7
Track 8

Eight-track tape for multichannel operation

FIGURE 7-29 **Reel-to-Reel Tape Recorder Head and Track Configurations.**

watch before degaussing tape since the strong magnetic field can interfere with the operation of the metal parts in a conventional wristwatch.

It is always a good idea to bulk erase a tape reel or cartridge prior to a recording session to eliminate any previously recorded material which could ruin a new recording. (See Figure 7-30.)

CUING REEL-TO-REEL TAPE Reel-to-reel tape can be tightly cued so that the program material will begin on the director's command. As with cuing a record, the audio engineer must utilize the cue system on the audio console to prevent the audition system from being sent over the air. To cue a reel-to-reel tape:

1 Place the tape fader on the audio console into the "cue" position.

2 Play the tape until you first hear sound. This is the cue point. Stop the machine. You can rock the tape back and forth manually as long as you are careful not to stretch or break it.

3 Once you have found the cue point, manually back up the tape about 1 inch. The tape is now cued.

4 If you need a particularly tight cue, you can spin the flywheel, which is located to the left of

FIGURE 7-30 **Bulk Eraser.**

the tape heads, counterclockwise with your thumb just before starting the machine. If you spin the flywheel immediately before pressing the start button, you will achieve the necessary momentum to start the tape at the correct speed and avoid the possibility of a wow in.

Tape can be visually cued by using the white or colored leader which is commercially available. By splicing several feet of the leader directly before the cue point, you can cue the tape visually without having to use the cue amplifier system. If your tape reel has a number of different cuts, the leader can also visually indicate the number of each cut. Use a felt-tipped pen to write the name and number of each cut directly on the leader.

TAPE EDITING One of the advantages of audio tape is the ability to edit the program material. Editing permits the audio engineer to splice in leader tape, to remove mistakes or "fluffs" without retaping the entire sequence, to add or delete program elements, and to shorten or lengthen a particular taped cut.

1 Find the correct "out" point. This is the point at which you wish to make an edit. To find this point, gently rock the tape back and forth by hand, listening to the output until you hear where the particular sound ends.

2 Lightly mark this point, which is directly over the playback head, with a colored grease pencil on the base side of the tape. (See Figure 7-31.)

3 Gently remove the tape from the machine, and place it in an editing block with the oxide side facing down. The editing block has a ridged slot, which will hold the tape in place.

4 Position the marked point over the diagonal slot, and cut the tape with a razor blade. Be sure you slice the tape; do not "chop" or "guillotine" it.

5 Rethread the tape from the supply reel, and manually move the tape over the playback head until you find the correct "in" point. This is the point where you wish to rejoin the tape pieces. Mark this point, which is directly over the playback head, with a grease pencil.

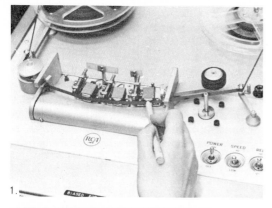

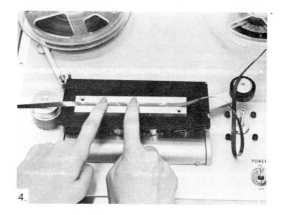

FIGURE 7-31 Audiotape Editing.
The steps are these: (1) Mark edit point with grease pencil directly over the playback head. (2) Place tape into editing block. (3) Position edit point over diagonal slot, and cut cleanly with razor blade. (4) Slide tape ends together, and join splice with a piece of editing tape. (5) Gently remove tape from editing block.

6 Remove the tape, and place it in the editing block. Cut again at the grease pencil mark.

7 Place the "out" side into the edit block, and slide the two pieces of tape together. They should just touch but should *not* overlap.

8 With a piece of 7/32-inch splicing tape, join the two ends and press down firmly to get a secure splice. You should never use ordinary cellophane tape since the adhesive will ooze out and leave a deposit on the tape heads.

9 Gently remove the audio tape from the editing block. Use both hands to avoid tearing or peeling the tape.

10 Rethread the tape, and listen to your edit.

With a little practice, you will find that you can cut between words, pauses, even music. A helpful hint: if you think you will need to do a great deal of very close editing, it is a good idea to record the material at 15 ips instead of the conventional 7½ ips. The increased speed will doubly space out the magnetically encoded sound on the tape and give you more "edit room."

MULTITRACK RECORDERS Multitrack recorders were first introduced for use in the commercial recording industry. Professional multitrack recorders permit the audio engineer to lay down individual parallel tracks which may be controlled independently of each other. In professional music recordings, the artists can first record the rhythm on one track, the lead instruments on another track, and the vocal on a third track. The multitrack recorder permits the audio engineer and talent to listen to some tracks while recording another. The advantage of multitrack recording is that one track can be recorded while not disturbing the recordings on the other tracks. After the initial recording session is completed, the audio engineer can then balance and "mix-down" the tracks into either a monaural or stereo version. This process permits both performer and audio engineer great flexibility. Instead of having to worry about the final audio mix at the time of initial performance, the engineer can concentrate on recording each track and after the session is complete, mix-down the many separate tracks into the final version. As many as twenty-four separate audio tracks are used for some recording sessions. (See Figure 7-32.)

In order to accommodate such a large number of tracks on one tape, multitrack machines usually use wider tape than the conventional ¼-inch tape used on monaural machines. Tape widths from ½ to 2 inches are used, depending upon the number of tracks desired.

Although conventional television audio is monaural, the use of multitrack recorders has become commonplace on very complex productions. Television audio engineers use a number of individual tracks for recording the sound portion of an elaborate program. They accomplish this by synchronizing the audio tape recorder with the videotape recorder. The two machines are electronically interlocked and travel at exactly the same speed. Instead of the audio console's sending its output directly to the television videotape recorder, a number of separate or *discrete* tracks are fed to the multitrack recorder, where they are

FIGURE 7-32 **Multitrack Audiotape Recorder.**
A twenty-four–track audiotape recorder, which uses 2-inch-wide audiotape. The units mounted on top of the bridge are used to synchronize the operation of the tape recorder with a videotape machine so that the sound will match the picture. Multitracking offers enormous production flexibility since every track can be manipulated independently without affecting any other track. *(Courtesy: Ampex Corp.)*

recorded on individual audio tracks. The audio tape recorder is synchronized or *slaved* to the videotape machine. When the videotaped picture is played back, the audio tape recorder also plays back in perfect sync with the video. In a mix-down session after the initial production has been recorded, the audio engineer combines all the audio tracks into either a single monaural track, which can be laid onto the videotape for later broadcast, or into a two-track stereophonic tape, which can be used for FM simulcasting of the television sound track. (For a more complete discussion of this synchronizing process, see Chapters 9 and 10.)

The multitracking provides the audio engineer with flexibility. During the actual taping, a number of audio sources such as performer dialogue, orchestra, sound effects, and audience applause can each be recorded on a separate track without the immediate need for final balance and mixing. After the production, music or sound can be added or altered without holding up the entire production cast and crew. Admittedly, this is an expensive process but as less costly hardware is developed, the multi-track technique is becoming a realistic option for many moderate-sized production facilities.

Cartridge Machines

Although it is possible to cue reel-to-reel tape tightly for television production, audio cartridge machines are designed to make this task even easier and more accurate. Audio cartridge machines utilize ¼-inch magnetic tape similar to that used in reel-to-reel machines. However, this tape is loaded into a continous loop and inserted into a plastic cartridge. When the "cart" is recorded, a cue pulse is recorded on a separate cue track just before the start of the program material. The audio engineer simply inserts the recorded cart into a machine and presses the "play" button on the director's cue. Provided the appropriate fader is open, the cart will begin immediately without wowing in. The cart is designed to play through the program material and to keep playing until it reaches the cue pulse, at which time it stops and waits for another "play" command. (See Figure 7-33.)

Carts are available in a variety of play lengths

FIGURE 7-33 **Audio Cartridge Machine.** *(Courtesy: Broadcast Electronics Inc.)*

from ten seconds to five and one-half minutes. The cart you select for recording should be only slightly longer than the program material it contains. This is because the tape must play through until it reaches the cue pulse before it will stop and recue itself. If you need to record the sound track to a one-minute commercial, do not use a five-minute cartridge. Although some cart ma-

FIGURE 7-34 **Audio Cartridge.**

chines do have a "fast forward" operation, it will still take time for the cart to recycle completely. It is also a dangerous practice to remove a cart which has not yet recued since another engineer may insert the cart into a machine and expect it to have been correctly recued. Whenever you have time, preview all carts before air time to ensure they are correctly cued and to preset a cart's audio input level on the control console. (See Figure 7-34.)

RECORDING CART MATERIAL Recording material on cartridge is relatively simple. You can record material from any sound source, but reel-to-reel tape is the preferred method because you can edit the program material first before transferring it to the cart for air use.

The operation for cart recording is:

1   Select a cartridge which is slightly longer than the length of the program material.

2   Bulk erase the cartridge. This is especially important because cart machines do not automatically erase previously recorded material as open reel machines do.

3   Place the cart into the machine, and press "record." The built-in VU meter will now register any audio input.

4   Cue your reel-to-reel tape with the program material. Set the proper input level on the cart machine according to its VU meter.

5   Mark a reference line on a reel spoke with a grease pencil. Note the "clock" position of the mark, and then rotate the reels backward—or clockwise—exactly one or two full turns.

6   With a finger on the cart "record" button, use your other hand to start the reel-to-reel playback, and keep your eyes on the reference spoke of the reel.

7   Push the cart record button when the reference line has completed three quarters of the circle. The cue will be tight but not upcut.

8   When recording is completed, close the input source fader and recycle the cart. Check the

cart by playing back to ensure proper cue start and audio level.

Sometimes you may wish to record a number of very short music selections or sound effects on cart. You can record each on a separate cart, but this can become burdensome if you have ten or fifteen different carts to be used on a show. There is an alternative. Select a cart with a relatively long play time, say two and a half minutes. Let us assume you wish to record a series of door bells for use in the program. Follow the procedure for recording, but at the end of each cut, stop the cart machine instead of permitting it to recycle. Next, press the record button again. This will ready the cart for another recording. Again, follow the recording procedure, and after each cut is finished, press "stop" and activate the record button. What you are doing is placing a number of stop-cue pulses on the cart after each cut. Thus, when you play back the cart, it will stop immediately after each cut and will ready itself for the next cut. The only problem with this method is that you must completely recycle the cart if you need to go back to a cut you have just played. This sort of multiple-cue cartridge recording method is best suited for identical cuts, such as a series of telephone rings, door bells, voice-over announcements, or program themes.

Because carts are so easy to cue, many television audio engineers prefer to prerecord all music and sound effects from records and reel-to-reel tape onto carts for use during the production. There are a number of good reasons for this practice: (1) As the sound quality of carts is equal to that of records or reel-to-reel tape, a properly recorded cart will exhibit no noticeable audio deterioration; (2) There is always the danger of playing the wrong cut on a record, particularly sound effect records which can have as many as twenty-five or thirty separate cuts on one side (ever try to count eighteen cuts in?); (3) Records or tapes can be inadvertently played at the wrong speed, but carts cannot; (4) Records or tapes can wow in, and discs can skip a groove; carts will always start on cue and will never wow in if properly inserted into the cart machine; (5) Each sound effect or music selection can be on a separate cart. If these are labeled and stored in order of play, there is little chance for confusion and the audio engineer can easily repeat a sound selection or delete a cut during rehearsal or production.

CASSETTES Audio cassettes have become very popular for home recording owing to their small size and automatic threading. Unlike a cartridge, which employs an endless loop of tape, a cassette consists of tape about ⅛-inch wide on two reels enclosed inside a plastic case. Although the technical level of some cassette recorders approaches broadcast quality, their use in production is limited by the inability to cue the tape accurately. It is advisable to copy or "dub" any cassette program material either to reel-to-tape or to cartridge for all production work.

DIGITAL AUDIO RECORDERS

When we use a microphone to pick up a sound, we are transforming acoustical sound vibrations into electrical energy. In effect, what we are creating is a series of electronic "waveforms" which represent the original sound's frequency and intensity in electronic terms. The conversion of acoustical vibrations into a corresponding electronic signal is called an *analog* process because we are creating an electrical "analogy" of the original sound.

The problem with an analog system is that whenever we must work with the electronic waveform itself—to process it, record it on tape, or reproduce it—we are inviting inevitable distortion and the buildup of unwanted noise, which results in a deterioration of the sound quality. These problems can be eliminated through the use of digital technology. In a digital system, the original electronic waveforms produced by a microphone are converted, or *digitized*, into a computerlike numerical code which represents the original sound. These computerlike numbers are then stored on the recording tape. Because only numbers are recorded on the tape and not the complex waveform itself, as we do with analog recorders, noise and distortion are virtually eliminated. Further, we can process and manipulate the digital signal and produce multiple generation copies of the original recording without any loss of quality.

During playback, the digital audio recorder transforms the numerical code back into analog waveforms to reproduce the sound with a quality as rich and transparent as the original sound source. The experience of listening to a digital recording is astonishing, and it is virtually impossible to tell the difference between a well-recorded digital tape and the original sound source.

One of the major differences in working with digital audio is the fact that we cannot edit digital tape with a razor blade and editing block as we do with conventional audio recorders. That is because the signal on the tape is encoded into the digital code. Instead, digital audiotape must be edited electronically, much as we electronically edit videotape.

Despite the tremendous advantage which digital audio recording offers, the technology does present some drawbacks. First, the equipment is exceptionally expensive to purchase and com-

FIGURE 7-35 Digital Audio Recorder.
A twenty-four–track audio recorder which uses digital technology to record and reproduce extremely high-quality sound.
(Courtesy: Sony Corp.)

plex to maintain. The result is that only a handful of production facilities have digital recording capability at the present time. Also, the technology is so new that there is no industrywide standardization. Various manufacturers have produced digital audio recorders in a variety of configurations, so that a tape which is recorded on one manufacturer's equipment must be played back on the same type of recorder. These drawbacks are only temporary, however, and reflect the fact that digital technology is still in its earliest stages of development. Indeed the advantages that digital recording offers audio production are so enormous that it is only a matter of time before the equipment becomes widely accessible, with the result that digital recording will be the only way to record and reproduce high-quality television sound. (See Figure 7-35.)

FIGURE 7-36 **Microphone Summary Chart.**
(Courtesy: Electro-Voice; HM Electronics, Inc.; Neumann/Gotham Radio; Sennheiser Electronic Corp.; Shure Bros.; Telex Communications Inc.; Vega Inc.)

Manufacturer/Model	Type	Pickup Pattern	Production Applications
Lavalier Microphones			
Sony ECM-50 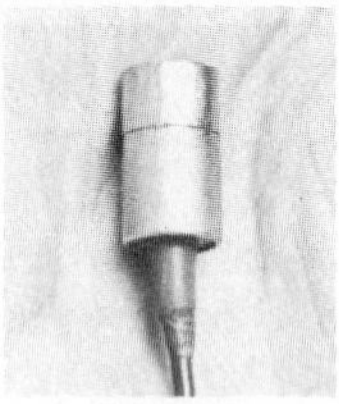	Electret condenser	Omnidirectional	Excellent sound quality and small, unobtrusive appearance make it good choice for all lavalier applications.
Shure SM 11 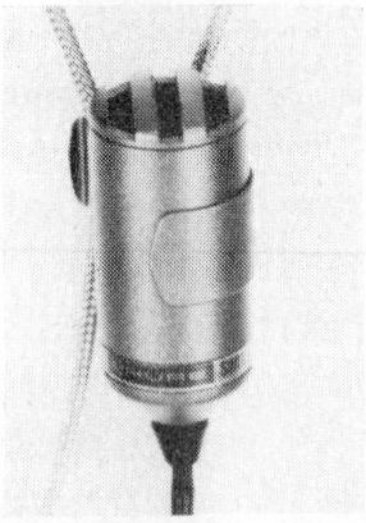	Dynamic	Omnidirectional	A small, rugged microphone which provides excellent sound quality. Can be used with lavalier cord, clip-on tie clasp, or tie-tack assembly.
Electro-Voice CO 85 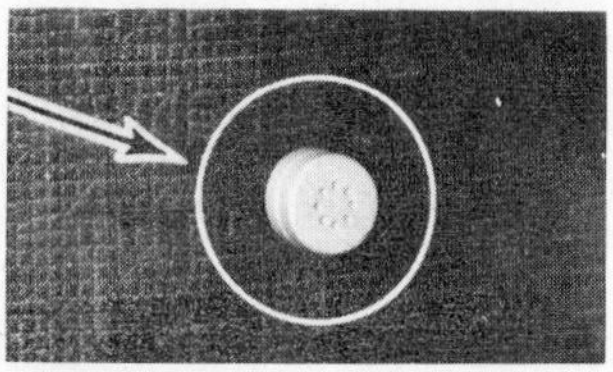	Electret condenser	Omnidirectional	Extremely small and unobtrusive tie-tack which is virtually invisible when worn. Excellent lavalier-type microphone for use in hiding mike out of camera range.

FIGURE 7-36 **Microphone Summary Chart (continued).**

Manufacturer/Model	Type	Pickup Pattern	Production Applications
RCA BK-6B 	Dynamic	Omnidirectional	Rugged, lavalier-cord microphone. Larger in size compared with many other lavalier mikes.

Hand/Desk/Stand Microphones

Manufacturer/Model	Type	Pickup Pattern	Production Applications
Electro-Voice RE 55 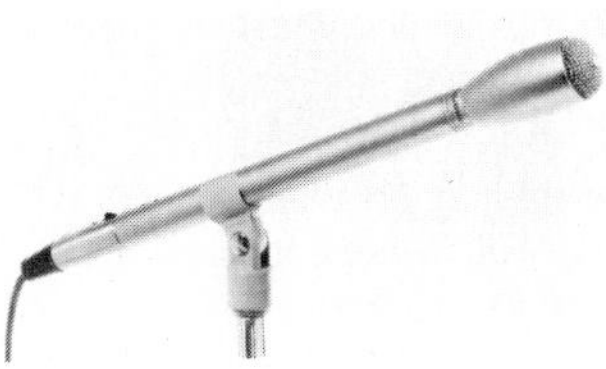	Dynamic	Omnidirectional	Exceptionally smooth response for critical recording applications. Used on stand for musical instrument pickup or hand-held for performers. Frequency response is flat from 40-20 kHz.
Electro-Voice CS-15 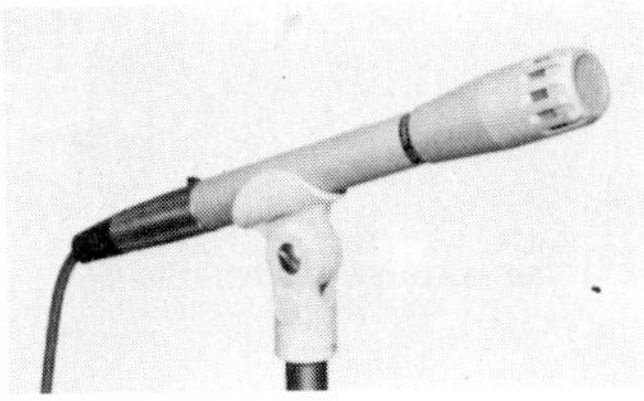	Electret condenser	Cardioid	Wide frequency range for music or voice applications.
Shure SM 53/SM 54 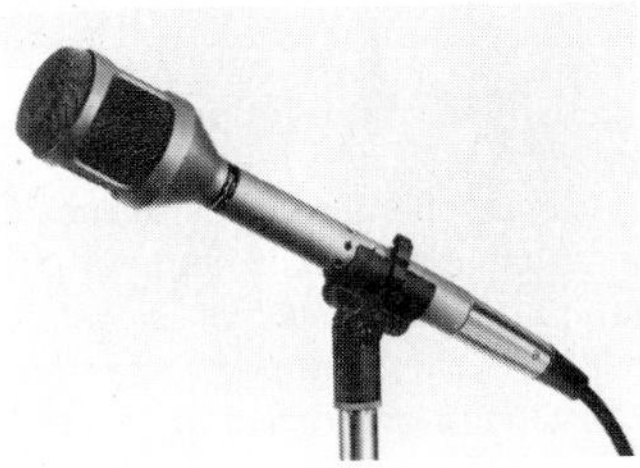	Dynamic	Cardioid	Rugged all-purpose microphone. Useful for instrumental and vocal pickup. SM 54 is identical to SM 53 but includes a built-in wind-blast filter to suppress breath and wind noise. Excellent for close miking of instruments, vocalists, and for outdoor remote use. Can also be used mounted on fishpole boom.
Electro-Voice RE 15/RE 16 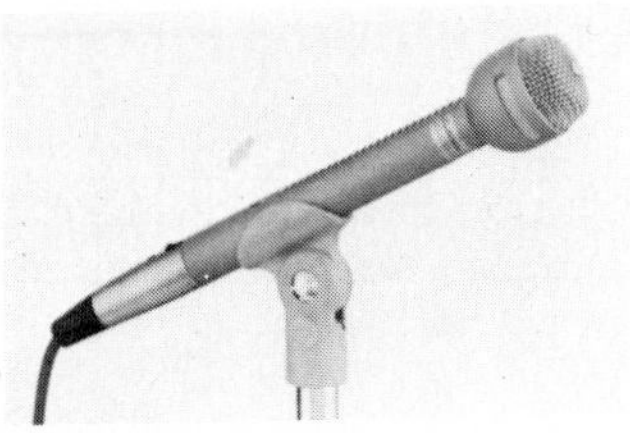	Dynamic	Super cardioid	Rugged, all-purpose microphones with a smooth frequency response. RE 16 is identical, but includes built-in windscreen and pop-filter.

FIGURE 7-36 **Microphone Summary Chart (continued).**

Manufacturer/Model	Type	Pickup Pattern	Production Applications
Electro-Voice 635 A 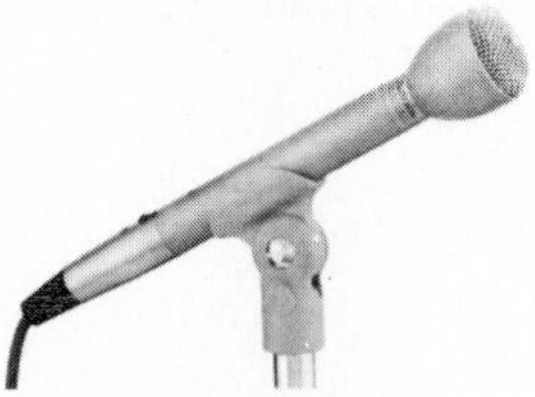	Dynamic	Omnidirectional	A rugged microphone which has been consistently used in a variety of studio and remote applications over the years. Good for all-purpose voice work.
Shure SM 7 	Dynamic	Cardioid	Excellent, wide frequency range with smooth response. Internal switches permit four different microphone responses: flat, bass frequency reduction, mid-frequency boost, and combination bass roll-off and mid-range boost. Used on stand for instrumental and vocal applications.
Electro-Voice RE 20 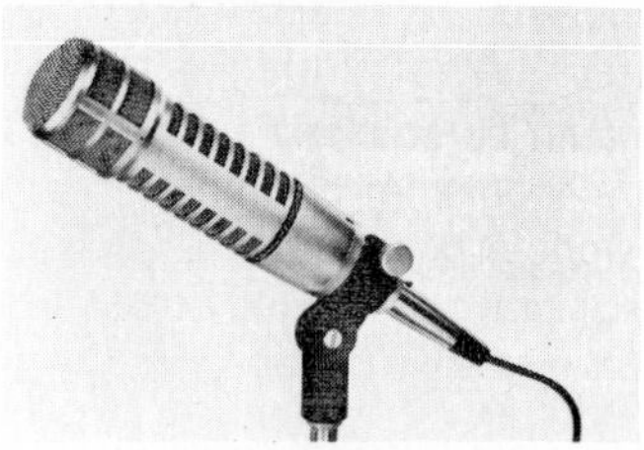	Dynamic	Cardioid	Excellent frequency response for a variety of instrumental and vocal recording applications.
Sennheiser MD 421 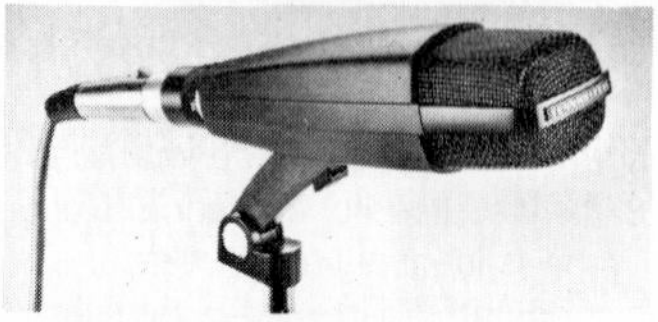	Dynamic	Cardioid	General, all-purpose studio microphone.
RCA 77-DX 	Ribbon	Cardioid/Bidirectional	Classic microphone which is still used for off-camera announcers and for musical application. A built-in switch enables microphone pickup pattern to be varied from directional-cardioid to a bidirectional pickup.

FIGURE 7-36 **Microphone Summary Chart (continued).**

Manufacturer/Model	Type	Pickup Pattern	Production Applications
Neumann U 87 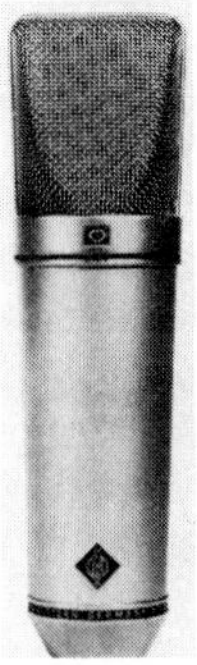	Condenser	Cardioid	Highly sensitive condenser microphone which is often used for critical studio recording applications. Large size and need for power supply limits its use outside the studio and for some on-camera applications.
Shure SM 58 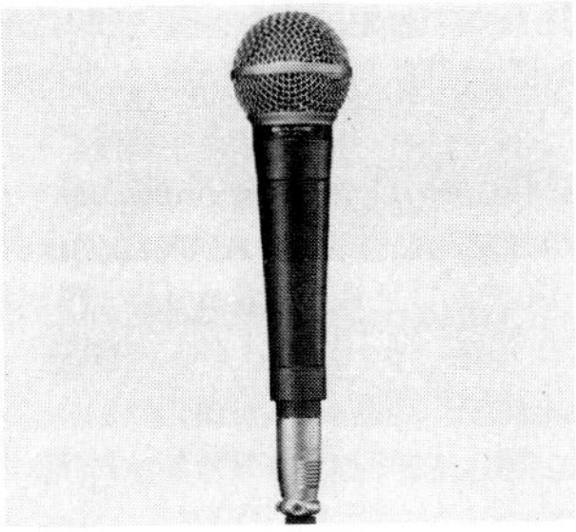	Dynamic	Cardioid	All-purpose microphone among the most widely used for musical, vocal, and speech applications. Moderate proximity effect and built-in windscreen make it an ideal microphone to use for singers where close miking is desired. Also excellent all-purpose microphone for remote field production.

Boom and Shotgun Microphones

Manufacturer/Model	Type	Pickup Pattern	Production Applications
Electro-Voice DL42 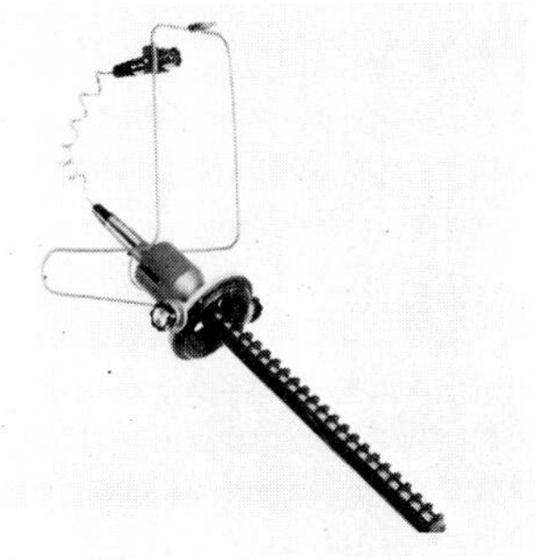	Dynamic	Super cardioid	Use with all boom mountings. Can also be fitted with handle for hand-held applications. Windscreen is advisable especially when used outdoors.
Sennheiser MKH 815T 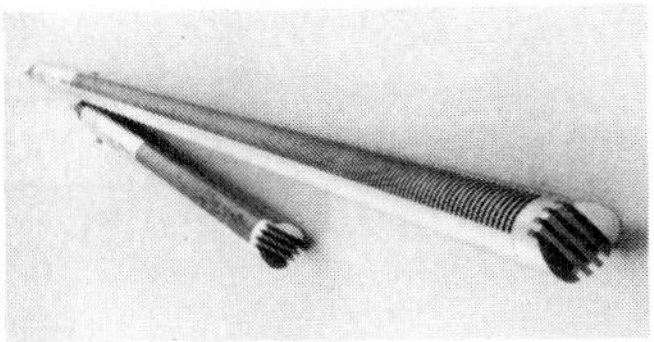	Condenser	Super cardioid	Highly directional microphone which is ideal for boom mounting and for hand-held use when microphone must remain out of camera range.

FIGURE 7-36 **Microphone Summary Chart (continued).**

Manufacturer/Model	Type	Pickup Pattern	Production Applications
Headset Microphones			
Shure SM12 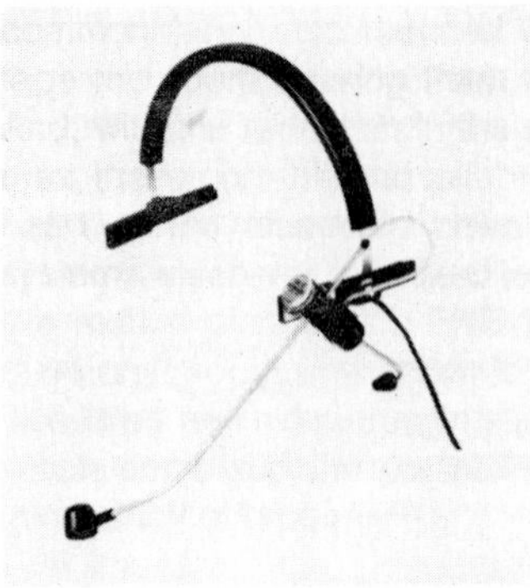	Dynamic	Cardioid	Lightweight microphone and headset combination which is ideal for remote sports and news events where talent's hands must remain free and mouth-to-microphone distance must remain as consistent as possible.
Telex CS-90 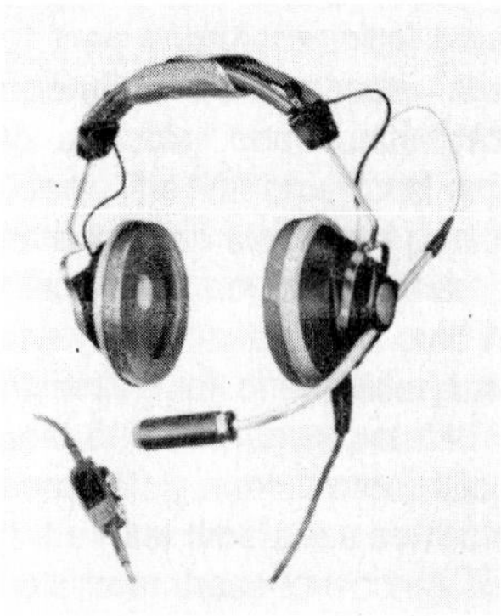	Dynamic	Omnidirectional	Dual headset permits monitoring program audio in one channel and director's production cues in the other. Foam earcushions provide a good sound seal in noisy remote locations.
Wireless Microphone Systems			
Vega Diversity 63 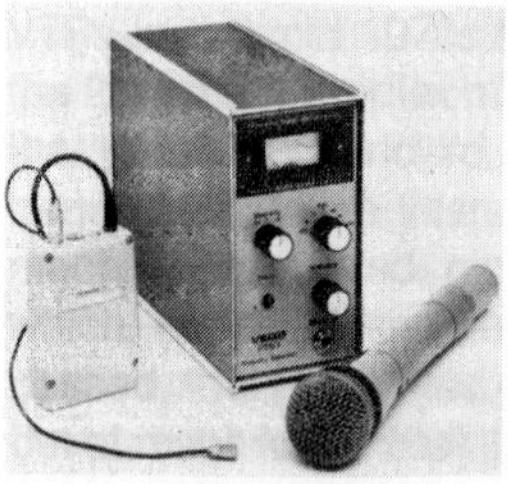	Dynamic	Cardioid	Hand-held microphone includes miniature transmitter, battery, and antenna. Transmitter unit accepts any low-impedance microphone and is most commonly used with a lavalier. Diversity system eliminates audio drop-out or interference problems.
HME System 22 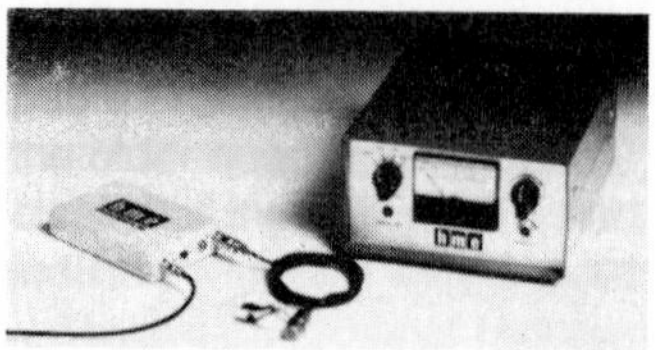	Electret condenser	Omnidirectional	Wireless system utilizing Sony ECM-50 microphone.

SUMMARY

Equipment used in the production of television audio can be divided into three categories: (1) microphones and mounting equipment, (2) audio control consoles, and (3) recording and reproduction devices.

A microphone converts acoustic energy into electrical energy. The three types of microphones used in television production are dynamic (moving-coil), ribbon (velocity), and condenser microphones. Mikes are often described by their pickup characteristics. The two basic pickup patterns are *omnidirectional* (the microphone is sensitive to sounds from all directions), and *directional* (the mike is most sensitive to sound emanating from a particular direction).

Microphones are used either on-camera (where the television audience can see the mike) or off-camera (where the mike is concealed from view to preserve the illusion the director and performers are trying to create). The most common on-camera mikes are the lavalier, the hand mike, the desk mike and the stand mike. Off-camera mikes are usually mounted on a boom which permits the microphone to be suspended over the performers and out of camera range. Other uses for off-camera mikes are to hide them in props or furniture, under performer's clothing, or held by a floor assistant out of camera range.

The audio console is the central control point for television sound. The console's functions are to amplify every audio input and to permit the audio engineer to mix and balance all audio inputs for proper transmission or recording. Audio consoles all have the following features: (1) *faders* to control the sound level of all audio sources, (2) *VU meters* to indicate visually the intensity of the sound, (3) *master and submaster* controls to regulate the level of mixed audio, and (4) *monitoring systems* to permit the audio engineer to listen to the program audio and to cue or audition upcoming sources before adding them into the program mix. Modular control consoles enable the audio engineer to control and aurally shape each sound source independently.

Recording and reproduction devices consist of records, reel-to-reel tape recorders, and audio cartridge machines. Phonograph records are often utilized in television production. In addition to commercially available recordings, a number of companies provide music and sound effects which can be used in production. Records must be cued before use so that the program material can be introduced instantly upon the director's command.

Audiotape permits the audio engineer to record music or sound and to play it back during the production. The two types of audiotape machines used in television are reel-to-reel and cartridge. Reel-to-reel tapes can be edited to remove mistakes, add or delete program material, or splice in leader tape.

Cartridge machines permit program material to be placed on audio carts and offer the advantage of automatic cuing.

Digital audio recorders transform analog signals into a digital code which enables the sound to be recorded, processed, and reproduced without any noticeable degradation in the quality of the original sound.

CHAPTER 8

TELEVISION AUDIO
Production Techniques

Producing good television audio is a difficult job. The problems of sound recording and reproduction which face the television audio engineer are similar to those faced by recording or radio engineers. However, these challenges are compounded by the fact that the television audio engineer is expected to produce high-quality audio while minimally interfering with the program's visual elements. The recording engineer can select a microphone and position it without regard to the mike's appearance or whether its position obscures the performer from an audience's view. Yet these and other similar considerations are of paramount importance to the TV audio engineer in addition to the basic requirement of producing good sound.

To make matters worse, the recording or radio engineer has the advantage of a studio which was acoustically designed for sound production. Even the best designed television studio is an acoustical nightmare with high ceilings, hard floors, and personnel and equipment constantly moving about during the production. These facts are not cited to excuse audio engineers, but to represent the reality they face. To produce consistently good audio under these difficult conditions requires skill, imagination, and talent. It also requires the audio engineer to understand how to use the audio equipment which is available for each production situation.

In this chapter, we will discuss audio production techniques, special audio effects, and audio operation in the four production stages.

CREATING THE AUDIOSPACE

You might think of the *audiospace* as the aural equivalent of the program's videospace. Just as the videospace serves as the audience's reference of visual reality, what is heard over the speakers is the audience's only measure of aural reality. This means that the audio engineer must evaluate the program audio for the way it will ultimately sound to the audience and not as it sounds inside the studio or even in the control room, where expensive and high-quality monitoring systems may present a distorted impression of how the audience actually hears the sound.

Audiospace can also be manipulated for a variety of special effects. As you will see later in the chapter, there are many electronic and mechanical means of modifying or shaping sound in subtle or dramatic ways. In general, the audio should complement the video. Ideally, the two should interact so the combination of sound and picture presents a message which is more effective than either would convey individually. Sometimes this means that the audio should directly relate to the video. Other times, the audio might be more effective as a counterpoint to the visual image. An example of the first case would be high-energy, uptempo music to accompany an edited montage of former professional athletes as we see them performing their greatest accomplishments. Taking the same example with an audio counterpoint, we might play a slow, moody, and sentimental ballad which would produce an interesting conflict between images of athletes in their prime and nostalgia for a time that has passed.

Naturally, these are creative decisions which must be made by both the director and the audio engineer. The point is that audio is both a creative element in a production as well as a technical operation. A program that is planned only for video, with little if any consideration for the audio, is like a meal prepared with only half the necessary ingredients. Sound can play a very important role in any production, but only if you approach building the audiospace with the same care and planning as you would the videospace.

Naturally, everything about audio goes into building the audiospace. But there are two important concepts which go a long way in determining how the audience perceives the sound portion of the show: sound presence and sound perspective.

Sound Presence

As you walk toward a person who is speaking, you naturally expect to hear the sound of the voice change. The closer you get, the louder the voice sounds. This is the intensity of the sound. But something else about the sound quality changes as you approach the speaker. The quality of his or her voice becomes fuller and richer and appears to be closer. This characteristic is called *sound presence*. Sound presence is a function of many related audio elements, including the type of microphone you use, the room acoustics, the volume of the sound source, and the ratio of direct to indirect sound waves which are picked up by the mike. (See Figure 8-1.)

Do not confuse sound presence with *sound intensity*. The closer a performer works to a microphone, the louder the sound intensity. Of course, the audio engineer can compensate for any fluctuation in microphone-to-subject distance by adjusting the mike fader on the audio control

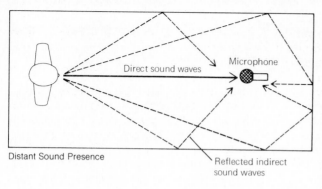

Distant Sound Presence

Direct sound waves

Microphone

Reflected indirect sound waves

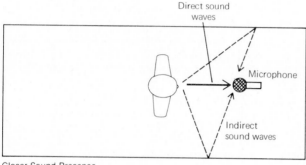

Direct sound waves

Microphone

Indirect sound waves

Closer Sound Presence

FIGURE 8-1 Sound Presence.
Sound presence increases as the subject-to-mike distance decreases because the microphone receives more direct sound waves and fewer reflected waves. The pickup characteristics of the microphone you are using will also affect sound presence. A highly directional microphone, which rejects reflected waves from the sides and from behind the microphone, will produce greater sound presence than an omnidirectional mike. Sound presence is also affected by what sound engineers call the "proximity effect." This is an increase in low frequency tones as a performer works close to the microphone. Many performers intentionally work the mike close to produce the proximity effect which gives their voices a fuller, richer sound. Omnidirectional microphones are impervious to the proximity effect. Most cardioid and bidirectional microphones exhibit the proximity effect in varying degrees, depending upon the microphone model, its construction, and its frequency response.

console. If you wish to keep the sound intensity fairly constant throughout the program, you must ride level as the mike-to-subject distance varies. But raising or lowing the microphone gain will not, by itself, change the sound presence. This is because sound presence also depends upon the ratio of direct to indirect sound waves which reach the microphone. The closer a performer works the mike, the higher the ratio of direct sound waves to indirect sound waves. As the subject works farther from the microphone, more and more indirect sound waves are picked up. The closer the mike is to the subject, the greater the sound presence, which we perceive as sounding full, rich, close, and intimate. As the mike-to-subject distance increases, the less sound presence, and the sound quality is perceived as thin, distant, and lacking intimacy.

A room which is acoustically "dead" absorbs indirect sound waves, preventing them from bouncing off walls and ceilings back into the microphone. An acoustically "live" room generally has many smooth wall, floor, and ceiling areas where sound bounces off, indirectly reaching the microphone and affecting the sound presence.

The microphone is also an important factor in sound presence. An omnidirectional microphone will pick up sound from all over and is highly susceptible to indirect sound waves which bounce off various surfaces. A directional microphone is more sensitive to sounds from one direction and, by suppressing other sounds, cuts down the indirect sound which reaches the microphone.

Sound presence plays an important part in how we perceive the audio of a show. When we want to attract a viewer's attention to a particular part of the videospace, we can use a number of lighting, lens, and camera-angle tricks to make a particular subject more dominant on screen. Sound presence offers a similar approach for the audio. A performer who is recorded with more sound presence sounds closer and more dominant than someone recorded with less sound presence.

Tony Schwartz, an expert in sound production, offers an example of how sound production can affect an audience's attention and perception of television subjects.[1] Schwartz suggests that one reason the television audience perceives Johnny Carson as being the dominant performer on the *Tonight Show* is because he uses a ribbon desk mike which is positioned very close to his mouth. The rest of the panel is covered by a boom microphone suspended overhead. Because Carson is physically closer to a microphone which produces a very rich, warm, and mellow audio response, his sound presence is fuller and more detailed. He is perceived by the audience as being "closer" and more important than the other performers, who are miked from much farther away.

Sound Perspective

Sound perspective refers to the relationship between the sound and the picture. Together, the sound presence and the image in the videospace determine the sound perspective. If you do not establish the proper relationship between the sound presence and the picture, the credibility of the program may suffer. When we see a tight close-up of a performer, we expect that the sound presence will be closer, fuller, and more intimate than the sound presence associated with a long shot.

Sometimes sound perspective is created naturally by the television production process. When the director shoots an actor in a tight close-up, the boom operator can position the microphone very close to the speaker since the mike will not be visible in the shot. The sound will be perceived by the audience as coming from a relatively close distance and will complement the image. You might say that the sound presence of the audio is the same as the "visual presence" of the close-up shot. When the director cuts to a wide shot of the actor, the boom operator must naturally "boom

[1]Tony Schwartz, *The Responsive Chord*, Anchor Books, New York, 1974, p.37.

up" to make certain that the mike will not appear in the shot. This will increase the mike-to-subject distance and will change the sound presence. The audience will perceive the sound as emanating from a longer distance, and the sound perspective will match the shot. In both cases, sound perspective has been maintained through boom positioning.

Dramatic programs require a considerable amount of coordination between the program director and the audio crew since the sound perspective must change during the show. If you are producing a program where all the performers should be perceived as sounding equidistant from the viewer (a round-table discussion, for instance), it is important to create a sound presence for each performer which will convey this aural impression to the audience.

SOUND MIXING

Sound mixing is an art which combines an audio engineer's creativity with the technical skills and abilities necessary to produce a sound track that complements the visual elements of a production. In the preceding chapter, we discussed the use of the audio control console to regulate the level of incoming sound sources in order to produce technically acceptable sound quality. This, of course, is only the barest minimum required for a television production. The audio engineer must also mix together the various sound sources so that the relative intensity of each is in direct proportion to their importance in the overall audio message. For example, a singer and a back-up band must be mixed properly to produce the correct audio balance between the two sound sources. But mixing doesn't refer exclusively to music. Virtually every television production has a variety of sound elements which must be mixed and balanced to produce an aesthetically pleasing and effective sound track. Take a newscast, for example. In this case, the audio engineer may have to mix the sound levels from reporters in the studio, sound tracks from various videotape and film sources, a live feed from a remote news location, and music and sound effects which occur throughout the production.

Sound Levels

Before beginning any television production, the audio engineer must set the basic sound levels for each audio input. This requires listening to each sound source individually and adjusting its corresponding pot or fader for an approximate operating level. We can divide sound sources into two large categories: (1) *microphones*, which pick up live sound, and (2) *premixed* inputs, which include the previously mixed sound from records, audiotapes, and the tracks of video recorders and film chains. Premixed sources also include the feed from another audio console, such as the output from a remote production like a live news unit at the scene of an event.

MICROPHONE LEVELS Here is the procedure for setting microphone levels:

1 Close all studio microphone faders.

2 Open one mike, and have the talent who will use the microphone speak a few lines of dialogue or read from the script. Professional actors and announcers will usually provide a stable sound level throughout the entire production. Nonprofessionals have a tendency to project differently during the sound level check than they will during the actual production. In this case, have the performer read a few lines from the program script or talk to the host of the program in order to give you a more accurate representation of how he or she will speak once on the air. Under no circumstances should you permit talent to give you a simple "Testing, one-two-three" for a level check since this sort of sentence fragment is unrepresentative of how the performer will actually speak during the program.

3 While the talent speaks, adjust the fader until the VU meter indicates the proper level. The meter should read between 80 and 100 per-

FIGURE 8-2 **Audio Control Area.**
A well-designed audio control area places all necessary audio equipment within easy reach of the audio engineer. A glass-enclosed booth is ideal because it permits the audio engineer to mix the sound and to preview upcoming audio sources without interfering with other control room operations.
(Courtesy: WXIA-TV)

cent modulation with very occasional peaks into the red. As you check for sound level, listen carefully for sound presence and adjust the position of the subject or the mike if necessary. If one microphone must cover a number of different performers, be sure to check each performer's level individually. You should also check any ensemble levels which might be necessary if one microphone must pick up multiple sound simultaneously.

4 Close the fader on the audio console, and repeat the procedure with all other microphones, one at a time.

This level check will provide you with only an approximate setting for each microphone. During the rehearsal you will have to readjust the levels for the proper balance once you hear how all the sound sources interact with each other.

Setting levels on a boom microphone is a little trickier since both the boom and the talent will be moving during the course of the production. In this case, obtain a basic level with the boom in a fairly normal position. You should also check levels for those times in the script where the performer must speak very loudly or must whisper softly. During rehearsals it is especially important to listen for sound presence and to correct both fader levels and boom positioning.

PREMIXED SOUND LEVELS Since prerecorded material should have been properly recorded when it was originally produced, establishing levels for these sources is usually easy. The best way is simply to preview each sound source to establish the proper fader position on the audio control board. Premixed feeds from remote locations are already balanced, but, again, you must preview the feed to establish the proper fader position before the production.

Although premixed sound sources generally remain constant in their level since they have already been mixed, you will often use them in combination with other sound sources, so you will have to vary the level of both the premixed source and the source from the studio to achieve the correct overall sound mix during production.

MARKING FADERS On most audio boards there is room for the audio engineer to attach a

strip of plain masking tape across the row of faders. This allows the engineer to identify each control's function by name. As you run through microphone level checks, be sure the name or identification on the masking tape strip corresponds to the proper microphone. This makes it easy to identify each microphone's fader quickly and accurately. Once the production is over, pull off the masking tape to leave a clean board for the next production.

If you know you will have to combine a live microphone with another sound source—a musical theme and an announcer, for example—it is a good idea to rehearse this mix and to set preliminary levels. You can mark the fader setting for each control with a grease pencil or with a tiny piece of masking tape so you can quickly return both faders to the predetermined positions.

Equalization

The use of equalizers offers the audio engineer an added dimension in sound control. Equalization, or EQ, enables you to shape and texture sound by boosting or cutting input signals at specified frequencies through the use of special electronic filters. At one time equalization was used primarily in the music recording industry, but the growing popularity of modular multichannel control consoles with built-in equalizers for each channel has made the use of EQ a fairly common practice in television production as well.

Modular consoles with built-in equalizers for each channel enable the engineer to apply different equalization to each sound source as necessary. This makes the use of equalization especially flexible since it is applied specifically to individual sound sources, leaving others unaffected. Control consoles without built-in equalizers can use an outboard *graphic equalizer*, which is a separate unit that is patched into the audio control system. Although the graphic equalizer works in the same way as a built-in equalizer, it can only equalize one channel at a time, making it somewhat less flexible than modular equalizers which can equalize individual sound sources independent of each other.

Equalization, whether applied through a modular unit or a graphic equalizer, can be used on any sound source, including microphones, film and tape tracks, prerecorded audiotapes, records, and so on. (See Figure 8-3.)

USING EQUALIZERS IN PRODUCTION Equalization is used for two basic reasons: (1) to correct for existing acoustical problems and (2) as a creative means of sound modification.

EQ FOR ACOUSTICAL PROBLEMS Despite the audio engineer's best attempts, there are times when inferior room acoustics or poor recording situations on location result in sound quality which is less than ideal. For example, on location shooting there is sometimes no way to eliminate an annoying buzz which is produced when audio lines are run too close to electrical power cable. In this case, equalization can be used to cut low frequencies in the original sound signal to eliminate the hum without unduly affecting the other frequencies of the sound signal. Similarly, a very boomy room can sometimes produce a sound quality which is just "not right." Adjusting the EQ of the sound source by boosting some frequencies and cutting others may help to produce a more pleasing sound quality.

Equalization is also useful in matching separate sound recordings which were produced at different locations or under varying conditions. For example, a voice-over narration which is recorded inside the studio may have to match a report-

FIGURE 8-3 **Equalizers.**
The modular equalizer *(a)* is a part of the multichannel audio console's fader module. Each channel's input can be equalized to boost or decrease specific selected frequencies. The graphic equalizer *(b)* is a separate unit which enables the audio engineer to manipulate selected frequencies of a single output. The graphic equalizer is so named because the position of the various control faders is a graphic representation of the processed audio signal.

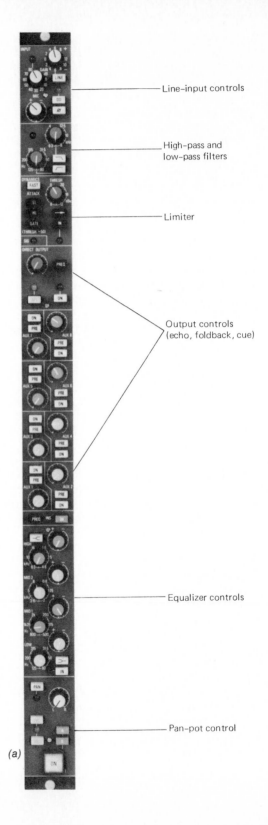

Line-input controls

High-pass and low-pass filters

Limiter

Output controls (echo, foldback, cue)

Equalizer controls

Pan-pot control

(a)

(b)

er's on-camera dialogue which was recorded on location in the field. Equalizing one or both of the sound tracks can smooth the transition from studio voice-over to the on-location sound track.

Although equalization can do some wonderful things in correcting poor quality audio, it should never be viewed as an all-purpose cure for audio problems. There is no substitute for a properly recorded sound track, and, despite the seemingly magical tricks we can play with sound using EQ, there is no way to produce a high-quality sound track from an inferior original through equalization. If equalization is overused, it can produce a thin, unnatural sound which lacks richness and presence. You are far better off changing microphones or experimenting with the subject to mike positioning than attempting to improve mediocre sound quality solely through the use of equalization.

EQUALIZATION AS CREATIVE SOUND MODIFICATION The creative application of EQ as a means of modifying original sound to produce a particular sound quality has become a standard procedure for all music recording sessions and the technique is easily applied to television audio production too. Boosting or cutting certain frequencies can improve the sound of an instrument or vocalist, help to shade or color sound quality and texture, or make a particular instrument or performer stand out more prominently without actually increasing the microphone sound level within the overall audio mix.

As an example, we can add sound presence to thin or transparent sounding musical instruments such as the acoustic guitar by boosting frequencies in the 100–300 Hz region. We can make a voice-over narrator stand out more prominently

from a track which might combine music and live "natural" sound by boosting the lower-midrange frequencies from about 800 to 2,500 Hz.

It is difficult to provide any hard-and-fast rules concerning the creative use of equalization since the ultimate sound quality depends not only on the particular sound source you are working with but on the interaction of all the sound sources which are mixed together to produce the overall sound track.

Most audio engineers begin by broadly applying some equalization to an individual sound source. Then they refine the EQ by gradually adding more or less at various frequencies until they achieve the particular effect they are looking for. But it is crucial to check the equalized sound source along with all other sources which will be added into the mix to make certain that the combined effect achieves the exact quality sound desired.

Although creative use of equalization offers tremendous flexibility to the audio engineer, as is true of any good thing, too much can cause problems. Not every sound should be equalized, and not every sound source will benefit from equalization. The use of EQ is only one of many techniques which the audio engineer should use in producing television sound. The proper selection of individual microphones and the positioning of the subject and the microphone can often produce the exact quality of sound desired without the need for additional sound modification. As a general rule of thumb, use equalization sparingly, and, when in doubt, use less rather than more.

EQUALIZATION IN POSTPRODUCTION The use of equalization during postproduction multitrack mixing offers the audio engineer increased control and creative flexibility in producing the optimum final sound track in which all sound sources are combined to produce the most pleasing and effective final mix. During a postproduction mixdown, individual sound tracks can be further modified both to improve the quality of the original audio and to integrate it better into the final program sound track. The advantage of multitrack postproduction mixing is that various sound problems can be corrected and eliminated and the overall track shaped and balanced through equalization and the application of other special audio processing techniques without the entire studio production crew's waiting until the audio is perfected.

Mixing Live Music

Live music presents one of the most difficult challenges to the audio engineer. The final sound balance depends not only on the mix, but also on the selection, number, and positioning of the microphones you use. The number of microphones is determined by the size of the musical group and by the audio engineer's approach. Close miking involves using multiple microphones with each covering a very specific number of instruments. A more general approach uses fewer microphones but offers much less audio control. The pickup pattern and the audio frequency response characteristics of a microphone are also very important. Select a microphone with a frequency response which complements the instruments to be covered. The mike position also plays a key role; closely miked instruments provide greater presence, sound as though they are physically closer, and stand out more prominently than instruments which are miked from a greater distance.

Figure 8-4 illustrates some common musical situations and suggested methods for positioning the microphones. The objective in balancing music is to blend all the individual microphones together to produce a good-sounding mix. This is usually a matter of trial and error since studio acoustics, the position of the band, and the "sound" of the music will vary for each production. Remember the VU meter will reflect only the amplitude level of the audio. It is up to the audio engineer to listen to the music carefully and to adjust each microphone until the desired balance is achieved.

If you are using an audio console with a submaster system, you can balance the individual orchestra mikes and leave them in their preset positions. Then assign all the orchestra microphones to one or two submaster faders. You might want to divide the band, assigning the rhythm section to one submaster and everything else to one or two additional submasters. This allows you to preset individual mike levels and ride gain on the entire orchestra with a minimum number of fader controls.

When mixing a vocalist with an orchestra, listen carefully for the proper balance between singer and music. If necessary, the audio engineer can alter the sound quality of either instruments or singer by adding reverberation and equalization.

There are a number of excellent books which deal with music mixing and recording in far more detail than we can cover here. The bibliography in the back of the book indicates a number of suggested sources for more detailed information.

PRERECORDING PROGRAM MATERIAL

There are situations when the television director or the audio engineer may decide to prerecord certain audio segments for use during the actual production. The three most common types of

FIGURE 8-4 **Microphone Placement for Music Pickup.**
(a) Drum Set. (1) Dynamic mike for kick drum. (2) Condenser or dynamic mike mounted on boom for toms and cymbals. (3) Dynamic mike mounted on boom to pick up snare and hi-hat. Position microphones to achieve the balance and overall sound quality desired. *(b) Piano.* Condenser, ribbon, or dynamic mike mounted on boom and positioned over second or third hole in soundboard. *(c) Electric Amplifier for Guitar, Bass, Piano.* Dynamic mike mounted on stand in front of amplifier speaker. Aim at the center for bright sound; off center for a fuller,

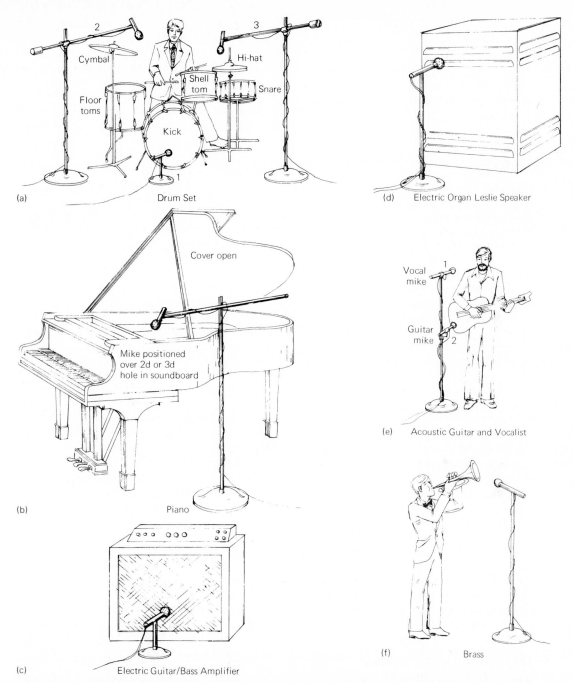

(a) Drum Set

(d) Electric Organ Leslie Speaker

(b) Piano

(e) Acoustic Guitar and Vocalist

(c) Electric Guitar/Bass Amplifier

(f) Brass

more mellow sound. *(d) Electric Organ Leslie Speaker.* Mount dynamic or condenser mike on stand or boom. Placing mike at top, bottom, or in the middle of the sound louvers will vary the sound quality. *(e) Acoustic Guitar and Vocalist.* (1) Dynamic or condenser mike for vocalist. Use microphone with windscreen to prevent popping and to permit vocalist to work close to mike. Proximity effect of microphone should also be considered when selecting a particular mike for vocalist. (2) Dynamic or condenser mike mounted on stand to pick up guitar. *(f) Brass.* Dynamic or condenser mike mounted on stand, or boom, and positioned slightly off center of instrument's bell to reduce high frequencies, wind noise, and excessive brilliance.

prerecorded program material are (1) the voice-over, (2) prerecorded musical tracks, and (3) lip sync.

Voice-Over

Voice-over refers to playing a previously recorded voice track over the visual portion of the program. A common example of the voice-over is the use of narration over a segment of silent film. The narration is recorded prior to the program, sometimes with the talent watching a playback of the film, and permits greater control over the synchronization of audio track and video.

On some dramatic shows, the director might utilize a voice-over segment to give the viewing audience the impression that an actor is thinking. This effect is accomplished by prerecording the audio segment and inserting it into the program over a close-up of the actor deep in thought. The actor hears the voice-over track through the studio monitor speakers and can act and react accordingly.

Sound perspective should always be considered when recording a voice-over. Because the voice-over is often used in combination with live audio, it is a good idea to record the voice-over in the same studio which will be used for the actual production. This will keep the sound quality of the voice-over similar to the quality of the live audio. Of course, if audio matching is not important, because the sound will be altered through filtering or reverberation, then the voice-over can be recorded in any convenient studio. Voice-overs can be played back from reel-to-reel tape recorders or dubbed to an audio cartridge for ease in cuing during the production.

Prerecorded Music Tracks

Sometimes it is impractical or impossible to provide a singer with a live musical accompaniment. There are many possible reasons. The limited size of the studio may prevent setting up a full orchestra or band, or the program's budget cannot afford the expense of a full orchestra. Other times a singer may wish to re-create his or her unique

"sound" and wants to use the musical track from the original commercial recording.

Whatever the reason, it is not uncommon for an audio engineer to have to mix a live performer with a prerecorded sound track. This is accomplished by playing the musical track through the audio console and simultaneously feeding the output of the prerecorded track into the studio monitor speaker system so it can be heard by the singer. The performer sings into a live, directional microphone, and the mike input is then mixed with the prerecorded track at the audio board. In order for talent to hear the music while a studio microphone is live, you must either bypass the monitor cutoff system or else employ audio foldback. The foldback feeds only the musical track into the studio to eliminate the possibility of creating a feedback loop.

Since many singers must hear the music at high volume levels in order to sing on key, it is important to rehearse the monitor's volume level setting prior to the production to find the best compromise between the singer's volume requirements and the point where possible audio feedback begins. In addition, some singers insist on hearing themselves as well as the track over the monitor system. In this case the monitor speakers must be carefully positioned with the directional mike's dead side facing the speakers to avoid producing feedback. Wall-mounted speakers are usually much too far away from the singer to be helpful in this case, and they can increase the chances of creating a feedback loop. Floor speakers which can be precisely positioned relative to the microphone, singer, and cameras are a much better choice for use with live microphones.

Aside from the obvious advantages in terms of studio space and budgetary savings, prerecording the music track relieves the audio engineer from the task of balancing each instrumental mike and then mixing the entire orchestra with the singer. Since the orchestra is already premixed, the engineer need only balance the singer with the single orchestral track.

Lip Sync

Lip sync, short for lip synchronization, is accomplished by having a singer pantomime to a previously recorded composite vocal and musical sound track. There are times when it is impossible for a vocalist to sing live at all during a television production. The singer may be involved in an elaborate dance or production number, and the boom might not be able to cover the area adequately. Also, it is difficult for a performer to dance and sing at the same time or to sing while recovering from an energetic dance routine. The program director might ask the singer to prerecord the entire song in advance and then to fake singing while the mixed track is played back on the studio monitor speakers. A remote production, where good-quality music reproduction is very difficult to achieve, is another example where lip sync may be required.

Lip sync requires very careful coordination between director, talent, and audio engineer to diguise any possible synchronization errors and to maintain a smooth transitional sound quality between live program audio and prerecorded lip sync material.

MUSIC IN TELEVISION

Music plays a very important role in television production in a number of ways. Of course, music is often the focal point of a show itself, but music is commonly used in subsidiary but nevertheless important ways as (1) theme music to open and close a show and to provide a transition between program elements and (2) as background music to establish the atmosphere and to complement the action.

Theme Music

A program's opening theme is usually the first audio element the audience hears. The music should be selected to catch the audience's attention, stimulate its curiosity, and set the mood or tone for the show. A program's theme often becomes its trademark, so it is worth whatever time and effort it takes to find the particular selection which works best.

You can find theme music from a number of sources. The commercially available production records are a quick and easy way to find specific instrumental tunes which convey a particular mood because they are specially composed and recorded for production purposes. If an index is used, it is easy to locate possible selections under such categories as "action," "documentary," "industrial," "religious," or "weird, eerie, and

space." Unfortunately, these selections tend to sound trite and canned, and you may be better off spending more time with contemporary or classical selections which are harder to find but usually sound better.

As a general rule, avoid overly familiar tunes—both popular as well as classical—which have been used time and time again. The opening theme from the movie *Star Wars* was a brilliant use of music to establish a mood and to set up the audience for the story to come. But it has become so overworked since then that its use sounds clichéd. It is best if your show's theme is identifiable only with your program and does not carry along dozens of other impressions for the viewer. Of course, there are situations where you may want this familiarity to work to your advantage, but these are generally special circumstances.

Sometimes a piece of a larger work can be excerpted and modified for use through tape editing. A short section can be extended in play length by recording it a few times and editing the pieces together. Similarly a long selection can be edited down to a shorter segment.

Background Music

Background music is designed to enhance a production subtly without upstaging the primary program content. Often we are unaware that music is present on a program unless we listen especially for it. Yet is has been shown repeatedly that, properly used, background music can contribute a great deal to setting the overall tone or mood of a scene or program segment. Since background music is designed more to establish a mood than to be an identifiable piece of music, the easily accessible selections from production records are usually very useful for this purpose. Looking under the various categories in the catalogue should lead you to some appropriate selections of varying lengths. (See Figure 8-5.)

Sometimes music is used as a thematic element itself by associating a certain piece or even a few notes with a particular action. The idea works something like Pavlov's dog experiments. By playing the music a few times and associating

Catalogue No.	Title	Composer	Timing
5249 A2.	**FOLLIES NUMBER** In the manner of a chorus line, with a kick routine before the footlights, but lighter in texture without bawdiness.	George Chase BMI	1:27
5249 B1.	**METEOR'S TRIP** Fast motion into the unknown. Speed in outer space. Modernistic laboratory of bubbling retorts and flashing lights.	Roger Roger BMI	2:01
2.	**DREAMING AWAKE** Awakening into strange surroundings. Lost memory.		:41
5250 A	**ROLLING MILL** Suggesting activities in an industrial plant or logging operations in the big woods. Melodic treatment.	George Chase BMI	2:32
5250 B1.	**MARTIAN'S PATROL** Weird march, with a definite drum beat, suggesting approach of strange or unearthly creatures.	Roger Roger BMI	1:53
2.	**STRATOSPHERIC DREAM** Strange and bewildering sights. An unknown fairyland of fantastic beauty. Sights never seen before.		1:56
5251 A1.	**SPELL OF THE UNKNOWN** Awesome approach to new worlds. A new planet looms into view.	Roger Roger BMI	1:27
LP59A 2.	**TOWARD DISCOVERY** Cautious approach to unknown places. Fear of danger.		1:39
5251 B1.	**ESCAPE IN THE NIGHT** Excited action in weird circumstances. Flight from terror.	Roger Roger BMI	1:07
2.	**AGGRESSION** Attack of strange unknown forces. Interplanetary strife.		:38
3.	**NEW THREAT** Menacing discovery. Approaching danger. Ominous portent.		1:02
5252 A	**FANFARES** (a) Grotesque (b) Down Pyramid (c) Dramatic (d) Brassy Curtain (e) Court Scene (f) Court Scene Proclamation (g) Shorter Proclamation (h) Light Fanfare Bridge (i) Menace	Roger Roger BMI	 :10 :10 :09 :09 :12 :11 :05 :11 :09
5252 B1.	**CEREMONIAL ENTRANCE** A big grandiose fanfare opening—ending unresolved.	Roger Roger BMI	:52
LP59B 2.	**ATOMIC MONSTER** Heavy, weird situation. Storm activity. Menace on a large scale.		2:10
5253 A1.	**TRIUMPHAL FULFILLMENT** A romantic theme presented in heroic style. Suggesting the successful conclusion of a hazardous undertaking.	Roger Roger BMI	2:05
2.	**AFTERMATH** Desolate landscape. Victim of some weird catastrophe, with a melancholy theme interposed.		1:30

FIGURE 8-5 Production Music Catalogue.
A page from the Thomas J. Valentino production music catalogue. The various selections are described in terms of mood and tempo; the running times are provided to facilitate music selection.
(Courtesy: Thomas J. Valentino Co.)

it in the viewer's mind with a particular mood or action, you need only play the theme later on to trigger the same reaction in the audience. Many suspense and mystery films are able to keep the audience guessing through the clever use of music.

MUSICAL APPROACH There are a number of factors you must consider when selecting background music. These include the musical style, the atmosphere or mood, the instrumentation, and the tempo. *Style* refers loosely to the genre of music: baroque, classical-romantic, modern jazz, hard rock, and so on. *Instrumentation* refers to the number and type of instruments used. Not every production situation lends itself to a full orchestral treatment. Perhaps a simple guitar, flute, or piano would be more appropriate. *Tempo* is the speed of the beat or rhythm; whether the music is fast or slow. *Mood*, of course, is the overall atmosphere which the piece conveys.

Music selection for any production is never an easy task. It takes time, patience, and perserverance. However, music is one of the most effective ways to build the audiospace, and the right music selection used at the right time can be the little touch which enhances a production and takes it above the ordinary.

SPECIAL AUDIO EFFECTS

Audio effects can be used to modify or enhance the sound track and provide the sound engineer with great creative and technical flexibility. Special audio effects can be divided into two categories: (1) sound effects, which includes both live and recorded effects; and (2) special electronic effects.

Sound Effects

Sound effects, like music, can often be among the subtle but effective elements which make an ordinary production better than average. Sound effects contribute to creating the overall mood or atmosphere for a program and help convey a special illusion to the audience. Often, sound effects are combined with special visual effects to manipulate both the videospace and the audiospace simultaneously. For example, a number of special lighting effects can provide the visual illusion of a fire, but the sound of crackling wood completes the overall effect.

Sound effects must be planned and coordinated carefully, however, because nothing is worse than producing a sound effect at the wrong time in a show. A classic case of mistiming occurred during one of the early live television dramas in which an actor playing a detective was supposed to shoot another actor on-camera. The sound of the gun was on a record which the audio engineer was to play from the control booth. For one reason or another, the record was miscued and, when the actor fired his gun, no sound accompanied

the action. The actor tried again and again, but there was no gunshot sound. Finally, in desperation, the actor pointed the gun at his adversary for a final time, cocked the hammer, and, as he fired the gun, yelled out "Bang!" A split second later the audio engineer finally got the record cued properly and sent out the sound effect of the gunshot!

MECHANICAL SOUND EFFECTS During the heyday of radio and in the early days of television, many sound effects were created live through mechanical means. A technician was positioned in a corner of the studio surrounded by odd-looking equipment which, in skilled hands, produced an array of sound effects. Unfortunately, mechanical sound effects have a number of significant disadvantages: they take up valuable studio space, they need an isolated area so the sounds are not picked up by the performer's mike, and they require at least one additional sound technician.

The availability of almost every conceivable sound on record or tape, combined with the use of an audio cartridge which provides accurate cuing and timing has virtually eliminated the use of most mechanical effects. However, we have listed a number of mechanical effects which are used frequently and which are relatively easy to produce:

Fire can be simulated by crumpling cellophane close to a live mike. Depending on the speed and intensity of the crackling and how close you work the mike, you can produce sounds ranging from a cozy fireplace to a raging forest fire.

Footsteps can be produced by walking over a bare, uncovered "riser." This is a set piece made of a plywood sheet attached to a frame of 2-×4-inch wooden studs. Depending on the type of shoes you wear, how you walk, and how you mike the sound, you can get a variety of effects. Covering the top of the plywood with gravel or sand increases the number of possible effects.

Telephones, bells, and *buzzers* are produced by rigging up a small box which includes a number of electrical bell and buzzer devices.

These can be either battery- or ac-powered and will provide a variety of bell-type sounds on cue.

RECORDED SOUND EFFECTS   With an unlimited number of effects available on records and tapes, most of the sound effects used in television production are played from the control room by the audio engineer. At many production studios, recorded sound effects are listed on index cards, which are cross-referenced to specific records or tapes. This permits the audio engineer, director, or production assistant to locate quickly whatever sounds are needed. You will often find a number of different versions of the same sound so you can select the particular one which best fits your needs.

Sound effects records have as many as thirty different cuts on each side. To eliminate any possible confusion during production, most audio engineers prefer to dub the effects from a record to open reel tape or to audio cartridge. Dubbing first to tape permits you to edit the sound effect.

A very useful property of prerecorded sound effects is the ability to produce variations on the original sound by playing the record (or tape) faster or slower than normal speed. Never forget the concept of audiospace; the only consideration is how the sound will be perceived by the audience. Your viewers will never know that the roar of a building explosion, which sounds so convincing, was really produced by playing a recording of a cannon blast at half speed.

Another technique is to modify the original prerecorded sound effect through equalization or filtering. Both mechanical and recorded sound effects can and should be manipulated until they create the specific aural "reality" which you wish to convey to the audience.

Special Electronic Processing

There are a number of ways in which we can electronically process the audio signal to modify or enhance the original sound. Earlier we discussed equalization, which is one way of elec-

tronically altering or shaping the audio quality of a sound input. In the following section we will discuss a number of other methods of specially processing the audio signal.

FILTERS   A filter is an electronic circuit designed to pass only selected frequencies while eliminating all others. Filters are frequency-adjustable, so you can vary the high and low frequency cutoff range, and switchable, so that the audio engineer can insert or remove the filter effect whenever necessary. The most common use of a frequency filter is to produce the illusion of someone speaking over a telephone, through a radio or television speaker, or over a public address system. The effect is achieved by cutting off all low frequencies around 250 to 300 Hz and all high frequencies above 3,000 Hz. These are only approximate guidelines, however, The filter effect should always be set by the audio engineer while listening to the sound and varying both the high and low cutoff controls until the correct effect is produced. Any audio source can be run through a filter, although microphones are the most common sources. (See Figure 8-6.)

ARTIFICIAL REVERBERATION   The terms "reverberation" and "echo" are often used interchangeably although, strictly speaking, they refer to two very distinct sound characteristics. *Echo* is the repetition of a sound. *Reverberation* is the persistence of a sound until it totally "decays" or fades away. When we talk about adding echo, or "reverb," to a recording, we are really referring to the addition of artificial reverberation to add a controllable amount of reverberation to the sound. This may be done to compensate for a very "dry" and "reverbless" recording environment or to modify the sound quality for creative purposes.

Artificial reverberation is produced using either electronic or electromechanical units. Controls on the reverb unit enable the audio engineer to vary the amount and the duration of the reverb effect. On modular multichannel audio consoles, built-in reverb circuits enable the engineer to assign the output of any audio channel through the reverb unit and to control the type and amount of reverb effect individually for each sound source. The range of reverb effects runs from a subtle coloration, as though the sound were recorded in a

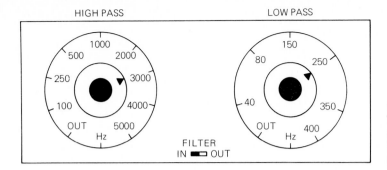

HIGH PASS   LOW PASS

FILTER
IN ◼◻ OUT

FIGURE 8-6   Frequency Filter.
The frequency filter enables the audio engineer to cut off high and low frequencies to produce a variety of special aural effects.

warm, reverberant room, to a canyonlike echo effect.

TAPE REVERB If a special reverberation unit is not available, there is an inexpensive way to provide reverb using reel-to-reel tape machines. Since all professional recorders have separate playback and record heads which are spaced out along the tape path, there is a brief time interval from the time a sound is recorded until it reaches the playback head. By operating the tape machine in "record" and feeding the output of the play head back to the record head, you can produce an echo effect.

To produce tape reverb, prepare to record as you would normally. However, take the output of the tape machine and patch it into the audio board on a separate fader. This will provide a delayed feedback loop from the recorder,

through the console, and back again to the recorder. Once you start recording, adjusting the amount of gain on the tape input to the audio console will vary the echo effect. Too much gain will result in producing a feedback loop which will distort the sound. After a little experimentation you will find the exact position on the tape recorder fader which will give you the desired reverb effect. (See Figure 8-7.)

One problem with this method is that you must record continuously to produce reverb, and this means that you must have enough audiotape for as long a time as you will need the effect. One way of solving this problem is to use a "tape loop." This is a strip of audiotape, about 4 feet long, which is spliced together into a continuous loop. The tape is threaded through the heads normally, but an empty reel of tape is used to maintain tension as the loop hangs over the

FIGURE 8-7   Tape Reverb.
Adding reverberation with an audio tape recorder is accomplished by taking advantage of the brief time delay between the newly recorded sound and its playback since there is a space between the record head and the playback head. Varying the amount of tape feedback by controlling the sound level of the tape recorder's input into the audio console produces a variety of reverberation effects.

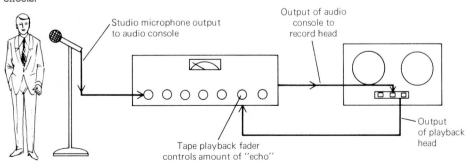

Studio microphone output to audio console

Output of audio console to record head

Output of playback head

Tape playback fader controls amount of "echo"

machine. (See Figure 8-8.) The loop will record continuously as long as the tape machine is running.

The tape machine reverb technique has another problem associated with it. Since the space between the record and playback heads is permanently set, you cannot achieve a variety of echo effects. You can play the tape at a different speed to vary the interval between the original sound and the repeat, but the much greater flexibility available with special reverb units is impossible when using the tape reverb method.

Reverberation is frequently used to modify a program's sound for a psychological or emotional effect. We associate reverb with unreal settings such as fantasy or dream sequences. Reverb is also used to distinguish between a character's prerecorded "thoughts" and normal dialogue. Of course, reverb is also employed to produce a logical sound perspective. If the director shows a performer inside a large, cavernous hall, the sound should reverberate in order to produce the proper sound perspective and to improve the overall illusion.

Reverberation is commonly used in music recording to enhance the sound of vocalists and instruments. Both singers and instruments sound richer, fuller, and more dominant with the right amount of reverberation added. Unfortunately, there is no guideline or recipe to follow in using reverb. The amount and type of reverberation is a matter of the audio engineer's taste and judgment. Reverb should be used sparingly, or you risk ruining the sound quality since there is no way to remove or decrease reverb once added. This is one instance where multitrack recording techniques offer a significant advantage. All of the individual tracks can be recorded "dry," or without reverb. The reverberation is added later in the mix-down so the audio engineer can experiment until she finds the right amount for each channel and not worry about ruining a recorded track by adding too much or too little.

COMPRESSION Virtually every sound source entering the audio console has a fairly wide dynamic range, meaning that there are occasional peaks and valleys in the overall level of the input signal. You can actually see the dynamic

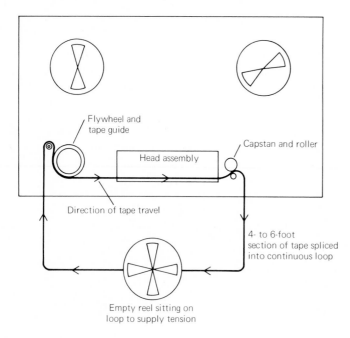

Flywheel and tape guide

Capstan and roller

Head assembly

Direction of tape travel

4- to 6-foot section of tape spliced into continuous loop

Empty reel sitting on loop to supply tension

FIGURE 8-8 **Endless Loop for Tape Reverb.**

range by watching the fluctuation of a VU meter needle as you record an announcer speaking into a microphone. The audio engineer establishes a proper level so that the needle will not peak too high into the red nor remain too low at the bottom end of the scale, but under normal circumstances the needle will still vary considerably, depending upon the loudest or softest parts of the announcer's voice as he or she speaks into the microphone.

If we apply electronic *compression* to the announcer's voice, what we are doing is essentially limiting the dynamic range by pushing the highest peaks downward and pushing the lowest parts upward in level. You can see the effects of compression by watching the VU meter. A compressed signal will show much narrower fluctuations as we force the peaks and valleys of the signal toward a midpoint. Theoretically, if we apply a great deal of compression, we will get to the point where the VU meter hardly fluctuates at all but remains relatively still at the point on the VU meter where the audio engineer has set the microphone's level.

An electronic compressor is a processing device which enables the audio engineer to vary automatically the amount of dynamic restriction of an audio signal; that is, determining how wide or narrowly the VU meter will be permitted to fluctuate, which represents the degree of constriction or "compression" of the signal's dynamic range. This is very useful in two common situations: (1) when we are faced with a sound source which varies considerably in level and requires constant attention to regulate the signal and (2) when we wish to make a particular sound source more prominent by raising its level in the overall mix without risking overloading the signal during occasional peaks or bursts of high intensity which would produce a distorted sound.

In the first case, imagine a vocalist who tends to move toward and away from the microphone continually during a performance. The audio engineer must constantly ride gain on the performer's mike in order to smooth out these variations and produce a quality sound track. This is possible, of course, but the audio engineer is usually busy with so many faders at one time that it is difficult to focus his or her attention completely on the singer's mike. In this case, by feeding the

singer's mike through a compressor, it is possible to smooth out these variations automatically and produce a better-sounding track without constantly riding the singer's microphone fader.

In the second case, we can use compression to take a sound which is relatively low in level and boost its gain without risking distortion. For example, say we are producing an industrial training tape and we wish to mix in the natural sounds of a factory with narration and music. The factory sounds are generally low in level, but there are occasional peaks which must be smoothed out or we risk distorting the signal. By compressing the track, we can raise its gain resulting in a more prominent sound and yet eliminate any sudden intensity peaks which might result in a distorted signal.

Compressors are "outboard" devices which means they must be patched into the audio control system depending upon which signal you wish to process. Any signal—microphone, record, sound track from film or videotape, and so on—can be compressed. Controls on the compressor enable you to preset the amount of compression, in other words, how severely you wish to constrict the natural dynamic range of the sound source and when the compression starts and ends. Usually, the audio engineer will employ only a slight amount of compression, which smooths out the overall signal imperceptibly to the listener. However, severe compression can result in some interesting special sound effects which are useful in special situations. As with any sound processing, however, use only the minimal amount of compression since any constriction of a sound source's dynamic range can adversely affect the quality of the final sound track.

NOISE REDUCTION SYSTEMS One of the undesirable, but inevitable, consequences of manipulating an audio signal is the production of unwanted "noise" which appears as hiss and distortion. With the increase in postproduction editing and various sound-processing and manipulation techniques, increasing amounts of noise are generated and can result in a final track which has an objectionable amount of noise.

Noise reduction systems such as the *Dolby A* and the *dbx* systems are designed to suppress much of this noise and produce a final track which is cleaner and much closer in fidelity to the original sound. Although the various systems work in slightly different ways, the underlying theory is the same for all. The original signal is encoded, and those parts of the signal which are most susceptible to noise are processed to separate the desired sound from the unwanted noise. During decoding, the signal is once again restored to its original sound, but much of the extraneous noise is reduced, or eliminated. These systems can be placed anywhere along the audio processing chain, but they are most frequently used at the output of the audio console before the signal is fed into either the audiotape or videotape recorder. Remember, though, that a Dolby or other noise reduction encoded tape must be played through a similar unit in order to decode the signal and faithfully reproduce the original sound.

STEREOPHONIC SOUND   One of the most exciting and significant audio developments in recent years has been the ever increasing use of stereophonic sound for television. Stereo not only provides the illusion of depth to the audiospace, but since it is played through a high fidelity sound system, it offers unsurpassed quality to the entire audio track. However, because it adds a great deal of complexity in terms of both technical operation and production procedures, its use must be carefully planned to be effective.

TRANSMITTING STEREO SOUND   The most common method of transmitting stereophonic sound is through *simulcasting*. This is the simultaneous broadcast of the television video signal with a synchronized stereo sound track broadcast over an FM station in the same local market as the television station. Viewers who wish to enjoy the program in stereo simply turn off their television audio and listen to the sound track over their FM receiver. Those viewers who do not wish to listen in stereo can, of course, simply watch the show and listen to the monaural television sound which is transmitted conventionally with the video.

Many cable systems are now offering stereophonic sound through the use of a vacant frequency on the FM spectrum. The cable operator feeds the high-fidelity stereo signal along the cable, and subscribers tune into the FM stereo channel to receive the stereo sound along with the video.

The obvious solution to the complexities involved in simulcasting would be the development of a television audio signal which is "multiplexed" much as we do now with FM stereo broadcasts. While this is now a regular feature of Japanese television service, in the United States television stations are only beginning to experiment with such a system, so simulcasting will remain the most common method of transmitting stereo sound for the foreseeable future. Of course, the rapid growth of home video cassette and video disc units which can reproduce stereo is another method of providing high-quality stereo sound into the home, and there is little question that the use of stereo in audio production for television will continue to increase.

PRODUCING A STEREO TRACK   In order to produce a stereo sound track, you must use a control console which has at least two outputs, one for the right channel and one for the left channel. Typically, a multichannel audio console is used combined with a multitrack audiotape recorder. During postproduction, the various tracks are mixed down to provide a monaural track and a two-channel stereo track. Of course, on a live broadcast the stereo and combined mono tracks are simply transmitted directly from the audio console without being recorded on tape.

Stereo adds depth and perspective to the audiospace, and it is the positioning of the various sound elements in aural space which is one of the most important creative aspects of stereo audio production. Earlier, we mentioned the pan-pot control which enables the audio engineer to locate the position of any sound source between the left and right speakers. Using the pan-pot controls permits the audio engineer to "build" the audiospace through the positioning of the various sound sources which together make up the total sound track For example, if we were mixing a musical group, we might position the drum kit and the bass dead center. The lead guitar might be positioned on the left side; and the rhythm guitar, on the right side. Keyboards and percussion could be positioned on either side and moved "forward" or "backward" in the audiospace depending upon their prominence during the performance.

The spatial perspective which stereo offers is sometimes a temptation to create artificial positioning, which is called "Ping-Ponging." As the name suggests, this is when we move various elements across the audiospace or have one character's dialogue emanate from the left speaker and the other character emanate from the right speaker. When used to excess, the effect can sound like some of the stereo demonstration records which were produced during the early days of high fidelity when bongos would be heard first on the left speaker and then on the right. Most audio engineers who are experienced in working with stereo for television advise against such obvious techniques. They recommend that stereo be used extensively for music and sound effects but that dialogue be positioned directly in the center of the audiospace. Of course, each situation demands an individual approach, but, in general, stereo should be used naturally and not in such a way that it calls undue attention to the panorama effect.

It is crucial for the audio engineer who is mixing in stereo to monitor the mix both in monaural as well as in stereo. Remember that a substantial portion of the audience will still receive the sound track monaurally, and there are times when an excellent stereo mix produces a very poor mono track. The secret is to produce an effective stereo track but not at the expense of the mono version.

AUDIO IN PRODUCTION

Audio is one of the key elements in any show, and the audio engineer is a primary member of the production team. The audio engineer's involvement in a production spans all four production stages from preproduction planning through to postproduction operation.

Preproduction Planning

As in all phases of television production, the proper care and attention given to preproduction planning will pay off in a smoother technical and creative operation during production. Before the actual production, the audio engineer must meet with the program director to determine the audio requirements for the production:

1 *How complex will the audio portion be?* Interview programs, instructional presentations, and talk shows are usually of low audio complexity and require less crew and setup time. Musical or dramatic programs are usually highly complex and require a larger audio crew, more setup time, and considerable rehearsal.

2 *What kinds of microphones and mounting equipment are necessary?* Can the mikes be seen on-camera, or must they be kept out of camera range? Knowing this will help determine the kinds of microphones you will use and the type of microphone mounting.

3 *Are there any special audio pickup problems?* For example, how much talent movement can be expected? Will talent have to share the mike with other performers? Will you need additional mikes to cover other studio areas? If you expect a great deal of talent movement, a boom or wireless lavalier microphone might be the best choice.

4 *Are there any music requirements?* If you will be covering live music, what is the composition of the band or orchestra, and where does the director intend to place them; on-camera or off-camera? Do you need to use prerecorded sound tracks? If so, will special monitoring requirements, such as audio foldback, be necessary?

5 *Are there any special audio effects?* Will you need to use sound effects? If so, will they be live or prerecorded? If they will be live, where will you place the sound effects technician

and how will you mike the live sounds? Will you need any special electronic effects such as frequency filters, equalizers, or echo? Which mikes must be used with the special effects? How are the effects to be used during the production? For example, will the effects have to be synchronized with on-camera talent moves, or will they be used in the background?

6 *Miscellaneous audio requirements?* Will the program require an opening musical or sound effects track and a closing track for credits? Who will select theme music? How long must the theme music run to cover the open and close of the production?

Will you have to patch in audio from the film chain, videotape recorder, or remote feeds? How will these audio sources be used during the production?

Will you require any special communication system, such as a headset PL to the floor crew, sound effects technician, boom operator, and so on?

Will there be a studio audience? If so, how will it be arranged? Will the audience interfere with audio operations such as boom microphones or cable runs? Will you have to provide microphones to cover audience response or to pick up audience participants who may speak during the show?

It is extremely important for the audio engineer to be thoroughly familiar with the studio and all of the audio requirements. If the program is scripted, the audio engineer should receive a copy of the script to use to indicate specific audio cues. (See Figure 8-9.) After meeting with the program director, the audio engineer will usually meet with the audio production crew to go over specific job assignments, crew responsibilities, and possible problem areas. (The director of the program does not usually supervise the audio team.)

THE CREATIVE USE OF SOUND Once the audio engineer learns of the specific audio requirements, he or she can begin to plan for the use of sound on the show. Audio planning involves two related considerations: technical and creative. On the technical level we would expect, as a minimum, the quality of the audio reproduction to be technically acceptable. But the audio must also be planned creatively to provide the maximum contribution which sound can make to any show.

As you select equipment and plan its use, you must always keep both the technical and the creative considerations in mind. For example, will you need very precise cuing for sound effects or music? If so, cartridges are the best choice. Are the microphones you selected appropriate for the coverage pattern and the material which must be picked up? If you have a choice of equally good frequency responses and pickup patterns, you might consider how the microphones will look on-camera to determine which one you ultimately choose to use. Are the types of microphone stands you have selected the ones which will give talent the most flexibility during their performance? Are there any ways in which sound can be used to complement the director's visual concept, such as special sound effects, music, or audio-space manipulation?

The point is that with the director preoccupied with so many production details, the audio engineer must serve as both artist and technician. While almost anyone can spot a glaring audio problem, it is only a creative and skilled audio engineer who can provide the many subtle audio elements which do so much to help the program's sound fulfill its indispensable role in the production.

Setup and Rehearsal

Television is a team operation, and each crew member should know his or her specific job responsibilities. In an efficient production operation, each production area—audio, lighting, cameras, and so on—should be working simultaneously so that the setup can be completed in a minimum of time. It is vital for the audio engineer and crew to make the most efficient use of the available time to position microphones, check cable connections, set audio levels, and balance audio sources. Division of responsibility and closely coordinated team work are the keys to a smooth setup.

During the setup period, the audio engineer must not only supervise the activity of the audio crew on the studio set, but must also patch in the correct feeds; identify each fader on the audio console; audition all tapes, records, and cartridges; and check all incoming remote feeds, including videotape and film sound-track levels.

On the studio floor all microphone cables

should be run neatly around the set, taped to the floor or behind set pieces if necessary, and connected to the proper mike outlet boxes along the studio wall. The audio engineer should run through each microphone individually to make certain it has been patched into the correct fader on the console and is working properly. Audio crew members should secure hand and lavalier

FIGURE 8-9 Television Script with Audio Engineer's Cues.
The audio engineer marks a copy of the script before production to enable him or her to operate the audio console quickly and accurately.

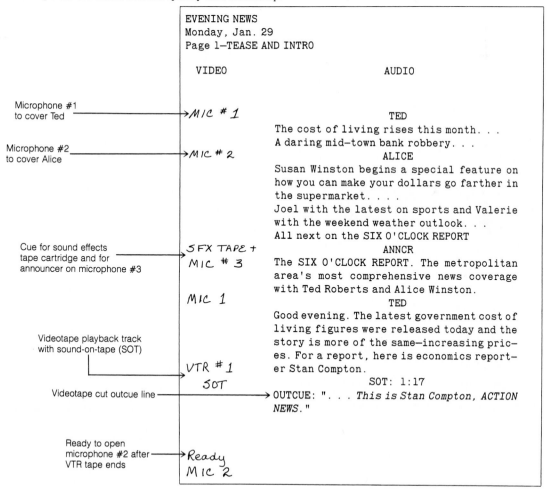

EVENING NEWS
Monday, Jan. 29
Page 1—TEASE AND INTRO

VIDEO   AUDIO

Microphone #1 to cover Ted ——→ MIC # 1

                                               TED
                                The cost of living rises this month. . .
                                A daring mid-town bank robbery. . .
Microphone #2 to cover Alice ——→ MIC # 2   ALICE
                                Susan Winston begins a special feature on
                                how you can make your dollars go farther in
                                the supermarket. . . .
                                Joel with the latest on sports and Valerie
                                with the weekend weather outlook. . .
                                All next on the SIX O'CLOCK REPORT
Cue for sound effects tape cartridge and for announcer on microphone #3 ——→ SFX TAPE + MIC # 3   ANNCR
                                The SIX O'CLOCK REPORT. The metropolitan
                                area's most comprehensive news coverage
                                with Ted Roberts and Alice Winston.
                                MIC 1   TED
                                Good evening. The latest government cost of
                                living figures were released today and the
                                story is more of the same—increasing pric-
                                es. For a report, here is economics report-
Videotape playback track with sound-on-tape (SOT) ——→ VTR # 1   er Stan Compton.
                                                      SOT: 1:17
                                SOT
Videotape cut outcue line ——→ OUTCUE: ". . . This is Stan Compton, ACTION
                                NEWS."

Ready to open microphone #2 after VTR tape ends ——→ Ready MIC 2

microphones in a place where they will not be damaged. Booms must be positioned and checked, not only for audio but also for compatibility with the needs of the program director and the lighting director. Live sound effects should be checked; and audio levels, obtained. The audio engineer must also test all communications systems with the production crew on the studio floor and, if necessary, with the control room crew. Finally, the monitor levels in the control room and studio should be set, checked, and approved by the director or the assistant director.

If there will be a camera rehearsal before the actual production, make a final check on all sound levels, balance, and sound perspective. This is particularly important for unscripted programs where there is little, if any, rehearsal time available for a complete run-through before air.

Production

During the production phase, the audio engineer works from inside the control booth, riding levels, cuing records and tapes, and coordinating the activities of the audio crew on the studio floor. The audio engineer has a television monitor which displays the video portion of the program as it leaves the control room. Some studios also provide the audio engineer with an additional "preview" monitor, which shows the next video source to be put on the air. The preview monitor is the video counterpart to the audio "audition" system in that it displays the next video source before it is actually placed on the air. This is particularly advantageous if the director will be rolling in a film or videotape segment with an audio track or if the program will use a remote video/audio feed originating from outside the production studio.

OPERATIONAL CUES Once the program is underway, the audio engineer receives instructions from the director. The following are commonly used operational cues and their meaning:

Open mike: Audio engineer should open the microphone fader.

Fade in audio: Audio engineer should open the fader to permit the sound level to increase gradually. The speed of the fade-in is variable and should be established during rehearsal.

Fade out audio: The opposite of a fade-in. Audio engineer gradually closes a fader. The speed is variable.

Hit music: Usually refers to music on a prerecorded tape, record, or cartridge. At this command, the director expects the music to come in at full level.

Cut sound: Abruptly cut off a sound. This can refer to a microphone, prerecorded material, a remote feed, or the entire show.

Sneak sound (in or out): Slowly fade in (or out) a sound under existing sound. For example, "sneak music in under announcer" would mean to fade in music slowly under the announcer's voice.

Sound up and under: Sound should come in at full level and then be faded under for mixing with another sound source which will be dominant. This is often used in conjunction with an announcer.

Segue: Follow one sound immediately with another. This usually refers to prerecorded material although it can also apply to microphones on different studio sets. A segue means no silence between two sound sources.

Cross-fade: Simultaneously fade in one sound source while fading out another. This refers to prerecorded material or to live sound. The speed of the cross-fade can vary depending upon the director's command. This is the audio equivalent of the video "dissolve."

Dead-pot (or "backtime"): Prepare a prerecorded music or sound track so that it will end at a predetermined time. If the director wanted the musical theme of a program to end exactly as the program ended, the audio engineer would have to dead-pot the record or tape. This is accomplished by first timing the prerecorded material. Once you know how long the material runs, simply subtract the running time from the time when the tape or record should end. This is the time when the material should be started but with the fader closed. At the appropriate time in the program, the audio engineer fades in the sound, and it will end exactly when the program ends.

For example, a show is scheduled to go off the air at twenty-eight minutes and fifty seconds after the hour. The audio engineer has timed the theme, and it runs two minutes and fifty seconds. Subtract the running time from the program's end time:

```
 28:50  (program's end time)
-2:50   (record's playing time)
 26:00  (dead-pot time when
            record should be
            started)
```

At exactly twenty-six minutes after the hour, the audio engineer must start the record with the fader closed. When the director calls for the music, fade it in. The record will end just as the director fades out the video.

The only major problem with the dead-pot is that the audio engineer must remember to start the sound source exactly on time. The assistant director should cue you, but it is a good idea to

mark the dead-pot start time in your script as a reminder.

Postproduction

For a live broadcast the only postproduction operation is to strike all microphones and audio cables, clear the patch panel, and return the audio control console to normal. Records must be

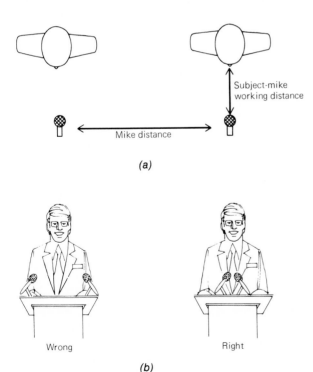

FIGURE 8-10 Microphone Phasing.
Phasing problems occur when an audio signal reaches two different microphones wired into the same audio system at slightly different times. Whenever you use multiple microphones on a production, the mike test should always include a test for phasing. To do this, simply open the fader for one microphone, and read the VU meter level. Then open another fader, and watch the level on the meter. The combined level should increase since both microphones are picking up the sound check. If the VU meter level dips instead of rises, you have a phasing problem. By checking each microphone against all the others, you can ensure that all mikes are in phase. If you find a microphone that is out of phase, reposition it until the phase cancellation effect disappears.

A good rule of thumb to follow in positioning microphones to avoid the phase cancellation problem is to follow the "3:1" rule. As illustrated in (a), the distance between microphones should be at least three times the subject-to-mike working distance in order to ensure proper mike phasing. Naturally, the closer the microphone is to the subject, the closer multiple mikes can be positioned to one another without causing phasing problems. Another source of phasing difficulties is illustrated in (b). When using two directional mikes on a podium, do not separate them since this increases the possibility of phasing problems. Instead, position them as illustrated, close together to eliminate a possible time lag in each mike's picking up the sound from the source.

Preview monitor

Line monitor

Glass window to control room production area

FIGURE 8-11
It is helpful if the audio engineer can see a monitor which shows the line video leaving the control room. In addition, a preview monitor, which indicates upcoming video, is valuable when the engineer must mix in the sound from videotape or film segments and from remote feeds which originate outside the production studio. *(Courtesy: ABC-TV)*

returned to the record library; and tapes and audio cassettes erased so they can be reused.

For productions which are video-recorded for editing during postproduction, audio can play a very important role in the postproduction process. This is especially true when the audio is recorded on a multitrack recorder and is mixed down to create the final monaural and/or stereophonic tracks.

The use of a separate multitrack ATR (audiotape recorder) combined with a video recorder requires the use of SMPTE (Society for Motion Picture and Television Engineers) time code to synchronize the VTR and audiotape machines precisely. On an elaborate musical program, the program's sound may have been recorded directly onto a multitrack ATR, which was run in sync with the VTR which recorded the program's video. In this case, the existing tracks are mixed together; and EQ, reverb, and other audio processing techniques are applied as necessary to each track until the final mix is achieved. If the program's audio was not recorded on a multitrack machine, the first step is to strip the original program audio which was recorded on videotape onto a multitrack machine which is run in sync with the VTR, using SMPTE time code. After the audio is dubbed onto the multitrack ATR, additional music, narration, and sound effects can be added to individual tracks and then the final program track is mixed to combine all of the elements into a completed sound track.

During postproduction it is also possible to *overdub* sections of dialogue which were not well recorded on the original sound track. This is most common on productions which were produced in the field, where poor acoustics or extraneous noise from airplanes or traffic cannot be adequately controlled. The talent is brought into a studio and, watching a video playback of the scene, they deliver their lines to match their lip movements. If a multitrack ATR is used, the overdubbing can be recorded on a separate track, and the dubbing repeated over and over until everyone is satisfied, without affecting the existing audio on other tracks. The audio engineer will most likely have to apply EQ and reverb in order to match the sound quality of the overdubbed dialogue with the original sound track.

Once all of the individual tracks are laid down on the multitrack ATR, the audio engineer and program director work through the mix-down, balancing each track to achieve the final monaural and/or stereo mix.

Since a multitrack mix-down may use as many as sixteen or even twenty or more individual inputs, it is impossible for the audio engineer to manipulate each fader simultaneously. Yet there are times when a complex mix requires the audio engineer to have more than two hands, so a *computer-assisted mix-down* system was developed. Using the system, a microcomputer literally memorizes the position of each fader during a mixing rehearsal. The system can even remember precisely when a fader's position was changed to raise or lower an input source's gain and will repeat the exact move precisely at the necessary time. The system works by referencing off the SMPTE time code which synchronizes the VTR and ATR machines, and once it has "learned" a particular mixdown, it will repeat it exactly unless the audio engineer chooses to override it manually during the mix.

Finally, the mixed multichannel audio track is re-recorded in perfect synchronization onto the original videotape, which completes the audio postproduction process.

For a more detailed discussion of postproduction audio mixing, see the section in Chapter 10 about videotape recording production techniques.

REMOTE AUDIO PRODUCTION

All of the equipment and techniques which we've discussed in this and the previous chapter on audio apply equally to productions which are produced on location as well as those produced inside a studio. The only difference is that the audio engineer's normally difficult job is complicated further by working on a remote location where there are few available controls over the acoustical environment and where there is extraneous noise and the lack of elaborate built-in equipment like that normally found inside a production studio. Nonetheless, producing a high-quality sound track which is technically clean and aesthetically effective is just as important when working in the field as it is inside the studio. In general, location production will be remixed during a postproduction session, so the primary goal is to produce a quality audio track which can be processed, manipulated, and ultimately mixed later on.

In Chapters 12 and 21 we will discuss the specifics of audio remote production for the three common location production situations: electronic news gathering, electronic field production, and multiple-camera remote production.

SUMMARY

The audio engineer's job in production is to plan for the technical and creative use of sound. Building the audiospace is the audio engineer's most important responsibility because this serves as the audience's only reference of aural reality. Two important factors in creating the audiospace are sound presence and sound perspective.

Sound from multiple audio sources must be mixed by adjusting the individual fader controls on the audio console until the proper sound balance has been achieved. First the audio engineer must establish basic audio levels for each audio input. Once this is accomplished, the engineer must vary the signal strength for each source until the mix sounds correct and reflects the proper relationship between the various audio sources.

Prerecording audio material is common in television production. Voice-over, prerecording music tracks, and lip sync are three commonly used techniques. Care should always be taken when recording material for use within the show to make certain the sound perspective and sound presence between the live audio and the prerecorded material match properly.

Music is an important aspect of television sound. Aside from its role as a primary element in entertainment programming, music is used for opening and closing themes as well as for background to help establish a mood or atmosphere. Sound effects are also useful in shaping the audiospace and enhancing the visual image. Effects can be produced mechanically inside the studio, but most are obtained from records or tapes which are available commercially. The recorded sound effect can be modified or altered by editing it on tape or varying the playback speed to achieve a specific effect.

Special electronic effects are used to manipulate the original sound for dramatic or creative purposes. Filters are used to cut off high and low frequencies to replicate the sound of a telephone voice or radio speaker. Equalizers enable the engineer to "color" the sound quality by boosting or attenuating certain frequencies within a sound signal. This manipulation can be used to correct for acoustical problems or to modify the sound for creative purposes. Artificial reverberation enables the audio engineer to add "echo" to the original sound to compensate for a "dry" recording studio or to enhance the sound for aesthetic reasons.

Compression permits the audio engineer to limit a sound source's dynamic range to smooth out variations in level or to permit the engineer to raise a particular sound source so it achieves greater prominence in the overall mix without distorting during occasional high peaks in intensity. Noise reduction systems are employed to eliminate or reduce extraneous noise due to audio processing and multiple generation dubbing.

The audio engineer is a key member of the production team and should attend all major preproduction meetings. Once the audio engineer learns about a show's sound requirements, he or she can begin to plan the audio approach from both a technical and aesthetic standpoint. During setup the audio engineer supervises the activities of the audio crew. During rehearsal and production, the audio engineer works the control console mixing the various sound sources to fashion the program's audio. If the program has been videotaped with the sound recorded on a synchronized multitrack audio recorder, during postproduction the audio engineer can remix the sound, "sweeten" it to improve its quality, and add additional audio such as narration, music, or sound effects.

CHAPTER 9

VIDEO RECORDING

Equipment and Operation

As millions of West Coast viewers tuned in the *CBS Evening News with Douglas Edwards* on November 30, 1956, few could have realized they were watching television history being made. While the news of the day in front of the camera was fairly routine, behind the scenes television took a significant step forward. On that day, CBS became the first network to use the new electronic videotape recorder (VTR) to delay the nightly news broadcast three hours for viewers in the western time zone. The videotape machines could record and replay a program with quality indistinguishable from the original live broadcast. Until the development of the VTR, the only way to delay a live broadcast was a 16-mm *kinescope,* made by filming directly from a television monitor. Not only was the kinescope a clumsy and time-consuming process—the film had to be developed and dried before it could be broadcast—but the fuzzy images it produced left much to be desired. With videotape, programming could be recorded hours, days, even weeks in advance of air time and replayed with pictures and sound quality identical to a live broadcast. (See Figure 9-1.)

We take videotape so much for granted now that it is hard to remember a time when broadcasting live, recording a show on kinescope, or producing entirely on motion picture film were the only production alternatives available. Today it is

FIGURE 9-1
The original VTR machine which recorded the *CBS Evening News with Douglas Edwards,* the first network program ever to be videotaped. *(Courtesy: Ampex Corp.)*

difficult to think of any television program which does not, in one way or another, use electronic video recording. Over half of all programming is recorded and edited on tape. Much news and commercial production take advantage of the speed and portability available with modern VTRs. Even such "live" productions as sports coverage use tape and video discs to slow down and freeze the action for instant analysis. Videotape plays such an important role in modern television production that it has become a primary production tool, as essential as the camera and the microphone.

In this chapter we will discuss how video recording works, the various VTR formats used in television, and videotape editing techniques. We will start with a short overview of the history and development of video recording to help you to put into perspective both the evolution of videotape and how it has come to affect virturally every aspect of television production.

THE HISTORY OF VIDEOTAPE RECORDING

As soon as television production began, it became apparent that some means of storing a broadcast for delayed playback was essential. This was particularly important for the national networks, who broadcast many shows live from New York City and needed some way to delay the program for the three-hour time difference on the West Coast. More importantly, television producers and directors had been looking enviously at their counterparts in radio, who—with the recent introduction of magnetic audiotape recording—had the luxury of taping their shows. Not only could productions be scheduled at more convenient times than the actual air date, but editing the audiotape allowed radio producers to eliminate technical errors and to assemble the best possible performances for the air show. Television broadcasters wanted to record pictures on magnetic tape with the same ease and production versatility available for sound.

Although many attempts were made to develop an electronic method of recording video, not until the Ampex Corporation unveiled its first commercial VTR, at the 1956 meeting of the National

Association of Radio and Television Broadcasters in Chicago did videotape recording become a reality.

Within six months of the initial CBS broadcast in November 1956, all three national networks made videotape recording a permanent part of their broadcast operations. Within a year or so, Ampex had developed a color VTR which continued to expand the production capability of the video recorders. But still something was missing.

Programming had to be recorded "live on tape." This meant that the show was recorded as if it were live, the only difference being that instead of broadcasting the program immediately, it was recorded on tape for later broadcast. Videotape was used as a storage medium but, while this was certainly a convenience, it did not provide the producer and director with significantly greater production flexibility than they already had. It was obvious that if videotape was to realize its greater potential as a production medium, some method of tape editing would have to be developed.

By 1958 the first mechanical tape editing process was introduced. Mechanical editing meant that the editor had to cut and splice the tape physically. At best, it was a crude, imprecise, and time consuming process, but it did allow producers and directors to record their shows in segments, to eliminate errors in production and performance, and to combine the best takes into a composite broadcast master.

By the early 1960s, Ampex had greatly refined the editing process with the introduction of electronic editing. Replacing physical cutting and splicing, electronic editing was a transfer process in which recorded scenes were copied onto another tape at a specific edit point. Editing could now be accomplished more quickly and more accurately, opening the door to the use of tape editing as a production tool itself. There was little that creative directors and editors could not do, given the time and the patience. Such programs as NBC's classic *Laugh-In,* with its rapid-fire cutting between different scenes and shots, were a direct result of the increased editing capability which was then available.

In 1967, the Electronic Engineering Company of California developed a time code system which enabled each individual video frame to be "ad-

dressed" with its own unique eight-digit code. The time code simplified retrieval of various scenes within a tape and permitted editing with reliable single-frame accuracy. The time code system became so valuable and so widely used that in 1970 it was standardized by the Society for Motion Picture and Television Engineers (SMPTE), so that the entire industry would utilize one, completely compatible time code system.

Also in 1967, the first color, slow-motion, video disc was introduced. The disc could record up to thirty seconds of material, which could then be replayed in slow or fast motion or stopped at any single frame for detailed analysis. The first "slo-mo" unit was used by ABC Sports at the *World Series of Skiing,* and it quickly became an indispensable part of sports and special event production.

The early 1970s saw the creative requirements of producers and directors continually pushing tape recording and editing to their limits. The trend-setting series *All in the Family* was the first videotaped situation comedy show to depend heavily on precision editing to actually "build" a show out of different live performances, all of which were recorded on tape. In fact, an average episode of *All in the Family* contained about sixty edits within the thirty-minute show. The editing techniques pioneered and developed for this show established the basic production techniques for all videotaped situation comedies which continue to be used today.

In 1972, the first network, made-for-television "movie," *Sandcastles,* was produced entirely with electronic cameras and edited on videotape. This was made possible by the introduction of computer-assisted editing systems, which combine a computer's vast memory with video production equipment to permit the same sophisticated, precision editing on videotape which had previously been available only with motion picture film.

The same technology which, by the mid-1970s, had reduced the size of radios and televisions and made pocket calculators and digital watches commonplace, also reduced the size of video-tape recorders, making them far more portable. And, with the reduction in the size of color cameras, these portable and reliable VTRs formed the heart of the electronic news gathering (ENG) systems which revolutionized television news. Unlike motion picture film, ENG tape does not require any processing before it can be broadcast. Television news could now use ENG equipment, which combined the portability of film with the immediacy of tape.

At the same time, videotape recorders—which but a few years earlier were available only to the largest stations and production facilities because of their enormous initial cost and the expensive operating maintenance—now became available to schools, community groups, and industrial users. The first broadcast recorders were as large as a kitchen stove and portable only insofar as a Mack truck could haul them from place to place. The newer machines could be carried on a technician's back or shoulder, transported in the back seat of a car, and operated on battery power for completely remote production. The introduction of the digital *time base corrector* permitted the signals from these low-cost VTRs to be transformed into broadcast quality for immediate broadcast or for taping onto conventional studio VTRs.

The introduction of high-quality 1-inch helical VTRs in the mid-1970s ended the dominance of the 2-inch quadraplex machines, which had been the industry standard since video recording began twenty-five years earlier. The new helical machines provided picture and sound quality which equaled, and in some cases even surpassed, that which was possible with quad VTRs, but at much lower cost. In addition, the operational characteristics of the helical VTRs offered such important features as slow-motion, fast-motion, freeze-frame, and greatly expanded editing capabilities which were previously impossible using the older quadraplex format. By the 1980s it was clear that the helical VTR had become the new industry standard for video recording.

Throughout its brief history, videotape record-

ing has made some remarkable advances; we have touched only on the most significant ones here. Today it is at the heart of all television production and is considered a primary tool. When you stop to consider that not long ago just the ability to record and replay a black and white picture on magnetic tape brought incredulous broadcasters to their feet, the commonplace use of videotape today and its broad production capabilities are truly remarkable.

HOW VIDEOTAPE RECORDING WORKS

At a very basic level, videotape works much like the audiotaping process described in an earlier chapter. As with the audio system, video recording operates by storing electronic information on magnetic tape. The tape is made of a plastic base coated with minute particles of metallic oxide. As the video signal activates the tape heads, they function like an electromagnet, creating a series of electromagnetic fields on the moving tape. These electromagnetic fields align the particles on the tape into a specific pattern, which corresponds to the original video information. To play back, the tape travels past the video heads, which "read" the previously aligned patterns on the tape. The action of the heads and the moving tape together produces a series of electrical signals in the head system. These tiny signals are amplified by the VTR's internal circuitry and result in an exact duplicate of the original video signal.

Although the basic idea behind videotape recording is similar to audio recording, in reality the two are very different, due primarily to the enormous amount of electronic information needed to record and to reproduce a television picture. A good audiotape recorder must handle a signal with a frequency range from about 20 to 20,000 Hz. A videotape recorder—in order to record and to reproduce a detailed color image—must deal with a far more complex signal with a frequency range in the millions of cycles per second.

Videotape Scanning

All audio tape recorders operate in pretty much the same way. While there are some differences in tape size, the number of recording tracks, and the speed of the tape, the basic operation of the audio machines is similar. That is not the case, however, with video recording. There are a number of different ways in which a VTR can "write" or align the magnetic particles on the tape. The particular method which is used determines the *scanning* system of the video recorder and describes the VTR's basic characteristics.

We have already said that in order to record and reproduce a picture, a VTR must deal with an enormous amount of electronic information over an extremely wide frequency range. You ought to remember from our discussion about audio recording that, all things being equal, the higher the tape speed, the higher the maximum frequency which can be recorded and reproduced, and the better the overall sound quality. That explains why, under similar recording conditions, a tape recorded at 15 inches per second (ips) will reproduce sound with better quality than one recorded at 7 ½ ips. In order to record and reproduce a satisfactory video signal, engineers have determined that the effective "head-to-tape velocity" (or "writing speed") would have to exceed 1,000 ips. But there lies a problem. Using a stationary video head and rapidly moving tape at about 1,000 ips, we would need almost 500,000 feet of videotape to record just one hour of programming!

The ingenious solution to this problem was to move not only the tape but the heads as well. The combination of rapidly rotating video heads and tape traveling past them at a much more practical speed (around 6 to 15 ips, depending upon the particular VTR format) produced an effective writing speed which exceeded the minimum necessary for a technically satisfactory picture.

All VTRs use the rotating head approach, in which the video head system is mounted on a drum which rapidly rotates as the tape passes by. However, there are two different scanning systems, or ways of tracing the electrical impulses onto the tape. These describe the two basic videotape formats used in television. (1) *Transverse scanning,* which uses the quadraplex or four-head system; and (2) *helical* or *slant track scanning,* which uses either one or two video heads, depending upon the particular helical system.

Quadraplex Format/Transverse Scanning

The quadraplex format was the original recording system introduced in 1956. Despite its high cost for both equipment and operation, it remains widely used throughout the television industry for a number of reasons. First, as the industry standard until the introduction of high-quality helical recorders a few years ago, it was the only video-recording format capable of delivering high-quality pictures and sound. Second, the fact that it was the industry standard meant that there was interchangeability of tapes between production facilities. This guaranteed that a quadraplex recording produced at one facility could be played and/or edited at any other facility which had quadraplex VTRs available.

The quadraplex system uses four separate video heads, hence the name "quadraplex" or, as it is more commonly known, "quad." All quad recorders use magnetic tape which is 2 inches wide and available in a variety of reel sizes and playing time.

The four video heads are mounted on a head wheel which revolves at 14,400 revolutions per minute (rpm) as the tape passes by at 15 ips, achieving an effective writing speed of 1,561 ips. (Although most quad machines can be adapted to record or play at a slower and more economical 7½ ips, the reproduction quality is not as good as at 15 ips. and the faster speed remains the industry standard.)

As the 2-inch tape passes by the revolving head system, an air vacuum gently holds the tape in contact with the heads. Each of the heads scans the tape for a brief instant, and a series of logic circuits switches every head on and off rapidly as it comes into contact with the tape. As you can see in Figure 9-2, the four heads rotate at a right angle to the tape and scan the tape in an up-and-down or "transverse" pattern.

Quadraplex Tape Tracks

The quad VTR organizes all the electronic infor-

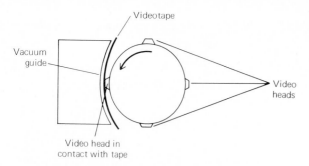

Videotape

Vacuum
guide

Video
heads

Video head in
contact with tape

FIGURE 9-2 Quadraplex Video Heads.
Four recording heads, mounted on a rapidly spinning headwheel, scan the tape in an up-and-down or "transverse" pattern.

mation necessary to record and reproduce a program on the 2-inch tape into a series of tracks. In addition to the video track, which is written by the rotating heads, three stationary heads lay down additional tracks. The tape, then, carries four basic tracks: (1) the audio track, (2) the video track, (3) the cue track, and (4) the control track. (See Figure 9-3.)

AUDIO TRACK The audio track runs along the uppermost edge of the tape and contains all the audio information for the program. Some quad recorders include an optional second audio track to enable you to record sterophonic sound or to add additional audio, such as music, sound effects, second language translation, or narration.

VIDEO TRACK The video track occupies almost three-fourths of the 2-inch tape, and carries all the video information for recording and reproducing the picture.

CUE TRACK The cue track was originally de-

signed as an optional audio track, but its poor frequency response has made that use impractical. It is most commonly used to carry the SMPTE time code information, which is used in postproduction editing.

CONTROL TRACK As the VTR records, it organizes the magnetic particles into a series of individual video frames (remember, 30 fps). Each video frame is marked with a control pulse, which is carried on the control track and used by the VTR to synchronize its playback. You might think of the control track as the electronic equivalent of sprocket holes in motion picture film. The control track is especially important in videotape editing since it indicates the beginning and end of the most basic element in video recording: the video frame.

There is one other head used during recording which does not lay down a video track. This is the *master erase head,* which is the first head to come into contact with the tape as it leaves the supply reel. The erase head wipes the tape clean of previously recorded signals and prepares it to record new information. (See Figure 9-4.)

Operating Controls on Quadraplex Recorders

There are five basic operating controls on a quadraplex VTR: (1) play, (2) stop, (3) fast forward, (4) fast reverse, and (5) record.

RECORDING MODES A VTR can record in any of three different modes: (1) *video and audio,* (2) *video only,* or (3) *audio only.* In the first mode, both video and audio are recorded; in the second, only the video information is recorded and the sound track is left intact. The reverse is true in

FIGURE 9-3 Quadraplex Tape Tracks.
All quad VTRs record and playback, using identical tape track configurations. This ensures total compatibility among all quadraplex videotapes.

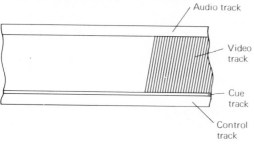

Audio track

Video
track

Cue
track

Control
track

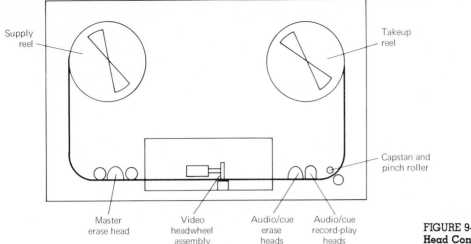

Supply reel

Takeup reel

Capstan and pinch roller

Master erase head

Video headwheel assembly

Audio/cue erase heads

Audio/cue record-play heads

FIGURE 9-4 Quadraplex Head Configuration.

audio only; only the sound track is recorded, leaving the video track intact. The ability to record independently on one track makes it possible to record narration or other audio, such as music or effects, onto a previously recorded video track or to match a new video track to an existing audio track. Matching video to audio is frequently used in news editing and in commercial production.

QUADRAPLEX VTR FORMATS

Quadraplex VTRs are available in two formats: (1) reel-to-reel machines and (2) video cartridge machines.

Reel-to-Reel Quad Machines

Reel-to-reel quads are commonly used for most production and playback purposes. Since the machines are rather large and very expensive and can be temperamental unless the proper maintenance and operating conditions exist, most VTR machines are installed in a special tape room, where the temperature, humidity, and other environmental factors are kept at a constant level for optimum VTR operation. (See Figure 9-5.)

Quadraplex Cartridge Machines

The quad cartridge machines are designed to provide television production with the same operational flexibility in programming short program segments as audio cartridges provide for sound. With more and more rapid-sequence segments, such as commercials, news clips, program inserts, and station IDs being aired on a regular basis, the use of a conventional reel-to-reel machine to handle these complicated chores is inconvenient and inefficient, keeping an expensive quad machine unavailable for use in production, editing or playback functions.

The quad cart systems are designed to eliminate these problems by providing a means of programming short-segment, rapid-sequence events. Each segment is contained in its own plastic cartridge. There are two machines currently available, one manufactured by RCA and the other by Ampex. The machines can hold over twenty individual carts, and some carts can hold a maximum of six minutes of program material.

The carts are loaded in storage bins either in sequence or in random order. As each cart is loaded, you enter the cart's ID number and its bin location into the memory system. Once the machine is loaded, you can program its memory for the proper playback sequence. If you wish, the cart machine can be operated *manually,* calling up each individual cart when necessary, or you can *preprogram* the machine, entering a specific sequence of events (a station break or program

257

commercial break, for example). Once the machine is told to begin the preprogrammed sequence, it will automatically follow it with split-second precision. Finally, if you are working at a studio where the operation is computer automated, you can operate the cart system in the *automated* mode, in which the entire operation (except for loading and unloading) is controlled and executed by the station's central computer system.

Each cart carries a prerecorded identification code which shows the title of the segment or commercial and its running time. The ID code appears on the cart system's playback monitor to indicate what the next program event will be. The ID strip disappears automatically as soon as the cart begins to play back program material.

Playback is virtually instantaneous since it takes the cart machine less than one second to produce a stable, locked-in picture. This makes the use of the cart system ideal for playing back

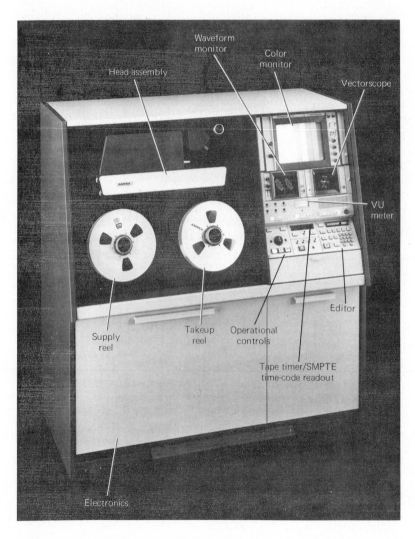

FIGURE 9-5 **Reel-to-Reel Quadraplex VTR.**
(Courtesy: Ampex Corp.)

news stories within a news program, station editorials, program openings and closings, as well as the usual station breaks and commercials. Some stations even use the cart system as an editing device, placing individual program segments on each cart, preprogramming their sequence and duration, and recording the precisely timed, rapid-sequence output onto a conventional reel-to-reel machine.

The compact carts are convenient to store and make cataloging and locating the necessary commercials and other short-segment program material easier than with reel-to-reel tape. (See Figure 9-6.)

HELICAL VTR FORMATS

The most exciting and significant advances in video recording over the past few years have occurred in the helical format. Helical equipment was originally developed as a low-cost recording alternative for industrial, educational, and other closed-circuit operations which could not afford

the high cost of quadraplex equipment. Compared with quad, these earlier helical VTRs were much cheaper to buy and to operate, but these advantages were off-set by some very serious drawbacks. First, the early helical machines did not produce tapes which met the exacting broadcast standards required for over-the-air transmission. Second, the machines were notoriously unreliable, often breaking down completely at a critical time during production or playback. Finally, editing helical tapes was a nightmare and resulted in a very limited postproduction capability. Consequently, broadcast television operations had little use for helical VTRs and regarded them as nonprofessional gadgets to be viewed with considerable disdain.

But the technological advances of the past decade have changed all that. The new generation of helical VTRs offer quality, performance,

FIGURE 9-6 **Quadraplex Cart Machine.** (*Courtesy: Ampex Corp.*)

and reliability at virtually every production level from sophisticated production and editing systems to the most basic home video cassette recorder. The role of the helical VTR in television production is no longer a question. Rather, it is a matter of matching the particular helical machine and format to the various production requirements which can take advantage of the many capabilities that the helical format has to offer. Before we discuss the various helical formats and machines, we will first review how the helical VTR records and reproduces a television image.

Slant Track Scanning

All helical VTRs thread the magnetic tape around a large diameter headwheel drum like a spiral or helix, hence the term "helical" format. Unlike quad machines, where there is only one way to thread the tape for all machines, helical recorders have two different tape wraps which signify important differences in their design. The two are (1) omega wrap and (2) alpha wrap.

OMEGA WRAP   In an omega wrap, the tape is threaded around the headwheel drum in a 180° configuration. The shape of the tape path resembles the Greek letter "omega." (See Figure 9-7.)

ALPHA WRAP   In an alpha wrap, the tape is threaded completely around the headwheel drum. The shape of this tape threading resembles the Greek letter "alpha."

Since the tape is wrapped around the headwheel drum in both the omega and alpha wraps, the video heads align the magnetic particles on the tape in a series of long, slanted lines, hence the term "slant track" scanning. These longer lines permit the helical machine to move the tape past the rotating heads at a much slower speed than is possible with quad. Since helical machines operate with tape speeds varying from about 5 to 12 ips, depending upon the machine model and manufacturer, they require much less tape than a conventional quadraplex VTR.

Segmented and Nonsegmented Formats

There is one other important variation in helical design: the way the VTR writes and organizes the

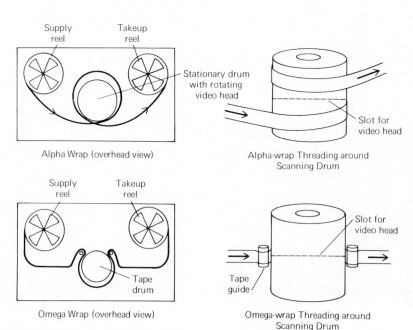

Alpha Wrap (overhead view)

Alpha-wrap Threading around Scanning Drum

Omega Wrap (overhead view)

Omega-wrap Threading around Scanning Drum

FIGURE 9-7   Helical Scanning.
Helical recorders which use the *alpha wrap* format require the tape to be threaded completely around the head drum. The shape of the threaded tape resembles the Greek letter *alpha*, hence the name. *Omega wrap* formats require the tape to wrap only halfway around the tape drum. The threading configuration resembles the Greek letter *omega*.

video information onto the tape. In a *segmented* scanning machine, two video heads are used to write the information on the tape. Each video frame is divided, or "segmented," into a series of video line packages. The operation of the rotating heads is similar to quad—each head is rapidly turned on and off as it comes into contact with the moving tape. In the *nonsegmented* or "continuous field" format, a single video head is used to record all picture information. The primary advantage of the nonsegmented format is that each complete video field is written on every head scan of the tape. This permits accurate slow motion and freeze-framing. In addition, the single head does not need the elaborate switching circuitry which is necessary for the two-head segmented format, and some picture imperfections associated with multihead recording, such as "banding," are eliminated with the continuous field approach.

HELICAL VIDEOTAPE RECORDERS

By this time you may wonder why there are so many differences in the design and operation of helical machines. Although all quad VTRs operate the same way, use the same size videotape, and offer 100 percent interchangeability between tapes recorded on one machine and played on another, this is not the case with helical VTRs. As various manufacturers developed their own, improved helical systems, the industry was faced with a bewildering array of helical formats. This resulted in the inability to establish a single, industrywide format while helical was in its earliest stages of development. The resulting lack of compatibility among various machines was a serious impediment which initially hindered the growth and acceptance of the helical format for some time.

Fortunately, the helical picture has become more organized over the past few years with the introduction of two industrywide formats for 1-inch VTRs and with the widespread acceptance of the ¾-inch cassette for ENG use. Despite these improvements, however, the various open-reel and cassette formats still do not provide for total compatibility, so it is important for you to become familiar with the different formats which are currently available.

It is convenient to organize all helical VTRs into two broad categories: (1) *1-inch video recorders* and (2) *video cassette recorders (VCRs)*.

1-Inch Production VTRs

One-inch production VTRs are the most sophisticated helical recorders, capable of recording and reproducing pictures with a quality equal to or better than those currently available with quad. Yet their initial cost is about half the cost of a comparably equipped quadruplex VTR, and their lower tape speed combined with smaller-sized tape reduces their operational cost. Of particular importance is the fact that the helical VTRs are much smaller in size and lighter in weight than quad machines and, significantly, their ability to provide pictures in slow motion, fast motion, and still frame offer production and editing capabilities which are impossible with the quadruplex format.

One-inch helical machines are available in either of two formats: (1) *Type B* and (2) *Type C* format. Although there are some exceptions, for the most part Type B format has become the helical recorder standard in those countries using the 625-line system while Type C is the most commonly used format in the 525-line countries, including the United States. Remember that a tape which is recorded in either format can only be played on another machine which uses the identical format.

TYPE B FORMAT Type B format uses 1-inch tape threaded in an alpha wrap with a segmented recording approach. In addition to a studio VTR, these machines are available in various battery-powered remote designs. Bosch-Fernseh, IVC, and Philips are among the major manufacturers currently offering Type B format production VTRs. (See Figure 9-8.)

TYPE C FORMAT Type C format uses 1-inch tape threaded in an omega wrap and a nonsegmented recording approach. Like Type B machines, these are available in both studio and

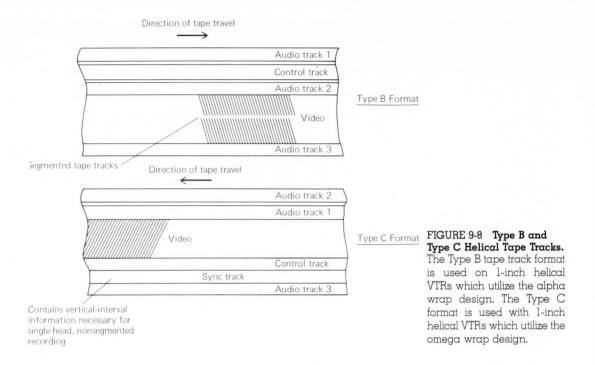

Direction of tape travel

Audio track 1
Control track
Audio track 2
Video
Audio track 3

Type B Format

Segmented tape tracks

Direction of tape travel

Audio track 2
Audio track 1
Video
Control track
Sync track
Audio track 3

Type C Format

Contains vertical-interval information necessary for single-head, nonsegmented recording

FIGURE 9-8 Type B and Type C Helical Tape Tracks. The Type B tape track format is used on 1-inch helical VTRs which utilize the alpha wrap design. The Type C format is used with 1-inch helical VTRs which utilize the omega wrap design.

portable configurations. Sony and Ampex are the two major manufacturers currently producing 1-inch production VTRs utilizing this format. (See Figures 9-9 and 9-10.)

Helical Cassette Recorders

Videocassette recorders have become universally accepted for a variety of production applications because of their low cost, ease and reliability of operation, and excellent picture reproduction. All VCRs use videotape which is completely enclosed inside a plastic cassette. The size and configuration of the tape and cassette depend on the particular cassette format. VCRs are now available in four basic formats: (1) ¾-inch U-Matic, (2) ½-inch VHS, (3) ½-inch Beta, and (4) ¼-inch CVC.

3/4-INCH U-MATIC FORMAT The ¾-inch U-Matic format has become the industrywide standard for ENG and EFP operation. Available in remote and studio configurations, the ¾-inch format VCRs are probably the most widely used video-recording machines in operation today.

The editing capabilities of the ¾-inch machines are now as sophisticated as those found in top-level 1-inch and 2-inch VTRs. In addition, many newer ¾-inch VCRs offer such additional features as high-speed search with a visible picture at up to forty times normal speed in forward or reverse, as well as conventional jogging, shuttle, and freeze-frame operations.

While the widespread acceptance of the ¾-inch format for remote operations is clear, many production facilities are combining the ¾-inch VCR with a time base corrector and using the machine as a primary production recorder inside the studio as well. Although the picture reproduction of the ¾ system can not rival that of the more expensive 1-inch and 2-inch VTRs, the cost savings and production flexibility of the U-Matic format is enormous. (See Figure 9-12.)

1/2-INCH VHS FORMAT Originally designed for in-home use, the ½-inch VHS format has been upgraded for industrial and broadcast production use as well. The most common use for the VHS format at this point is in primary production work at nonbroadcast industrial facilities and for use as

FIGURE 9-9 1-Inch
Production VTR.
(Courtesy: Ampex Corp.)

FIGURE 9-10 1-Inch VTR
Room.
The compact size of the
1-inch VTR permits two VTRs
to be positioned in the space
necessary for a single larger
quadraplex videotape ma-
chine. (Courtesy: Ampex
Corp.)

FIGURE 9-11 **Portable 1-Inch VTR.**
The portable 1-inch VTR permits top-quality recording on location. *(Courtesy: Ampex Corp.)*

a workprint for off-line screening and editing at more sophisticated production facilities.

The new generation of integrated camera/recorders for ENG operation also uses a VHS format, but this is a specially converted ½-inch system which is incompatible with conventional VHS machines.

1/2-INCH BETA FORMAT As with the VHS system, the Beta format was originally designed for

FIGURE 9-12 **¾-Inch U-Matic Video Cassette Machines.**
The studio model *(a)* is used for playback, recording, and editing. The portable version *(b)* is used for video-recording in the field. *(Courtesy: Sony Corp.)*

(a)

(b)

FIGURE 9-13 U-Matic Videotape Cassettes.
The cassette on top is the standard ¾-inch size and is available in play lengths from ten minutes to ninety minutes. The smaller cassette underneath is designed for use in portable VCRs, which are used extensively in news and other remote applications.

home use and has been adapted and upgraded for industrial and broadcast operation. The ½-inch Beta cassette—which is incompatible with the ½-inch VHS system—is also used in its own version of the integrated camera/recorder for ENG use. As with the VHS camera/recorder, the Beta tapes used in the ENG system are recorded and played back in a specially converted system and cannot be used with conventional Beta equipment. (See Figure 9-14.)

FIGURE 9-14 ½-Inch Beta Format VCR.
(Courtesy: Sony Corp.)

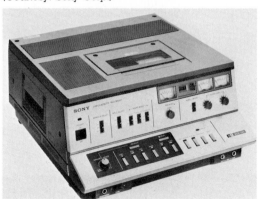

1/4-INCH TAPE FORMAT The newest VCR format utilizes ¼-inch tape inside a cassette for recording and playback. The ¼-inch CVC format cassette is three times smaller than a ½-inch Beta cassette and five times smaller than a ½-inch VHS cassette which enables the ¼-inch VCR to be utilized in an integrated camera/recorder unit that is exceptionally lightweight.

VCR OPERATION The operation of all video cassette recorders is extremely simple. To play back a tape, simply insert the cassette into the VTR machine and press "play." The machine will automatically thread the tape and adjust for optimum playback. If the picture is still not completely stable, adjust the "tracking" control until playback improves. If you see "flagging," or wavy lines, at the top of the picture, adjust the "skew" control until the lines straighten out.

To record, insert a blank cassette and depress "record" and "play" simultaneously. All ¾-inch cassettes are equipped with a record button, which is inserted in the bottom of the cart. The plastic button must be in place in order to activate the record mode. Removing the button prevents accidental erasure of previously recorded material.

Still-Frame and Slow Motion

One of the important and unique advantages of the helical format is its ability to slow-motion or still-frame a picture. This is made possible by the wraparound threading and the operation of the video head system, which can be made to scan the same video line repeatedly to produce still-frame or slow motion. This means that videotape editing can now be done with the same convenience and accuracy which had once been possible only with motion picture film on a viewer. The picture can be slowed down or stopped until the precise edit point is located. Although the slow-motion and still-frame capability of most cassette helical machines is not acceptable for on-air use, it enables the editor to work with videotape in a way which had been possible before only with motion picture film.

The 1-inch production machines offer an additional capability by permitting the operator to slow-motion or still-frame a rock-steady broadcast-quality image. This capability is commonly used in place of the more expensive slow-motion disc for sports, entertainment, and instructional applications. Another advantage of 1-inch VTRs and some newer ¾-inch cassette machines is their ability to shuttle the tape back and forth at up to thirty times normal speed while still producing a coherent picture. This makes locating an edit point even easier than before and increases videotape's production flexibility.

Time Base Correctors

Even the most expensive and technically sophisticated VTR cannot move the tape past the video heads at an absolutely precise and constant speed. Variations are to be expected in any mechanical system. In audio recording, these variations are known as "wow" and "flutter." In video recording, the variations are called "time base errors." A time base error shows up onscreen as picture jitter, "skewing," or color breakup.

Quadraplex machines have long used *time base correctors* (TBC) to correct for minor errors, but the high stability of the quad format required these TBCs to make relatively minor adjustments. Consequently they were designed to offer a very limited range of picture correction. Helical machines—which use a slant track rather than the more stable transverse writing approach—are much more susceptible to time base errors and require much greater correction. Only with the introduction of the *digital time base corrector* in the early 1970s did the wider correction capability necessary for helical recording become available.

If a time base corrector is used, any helical machine can produce pictures which meet the most stringent technical specifications for direct on-air broadcast or for up-dubbing to more sophisticated helical and quadraplex machines for editing and playback. The TBC works by converting the nonstandard output of the helical VTR into a numerical "digital code" which corresponds to the various picture elements in the original signal. Once the video signal has been coded, it is processed, and errors are corrected before it is reconverted to a conventional signal at the TBC output. (For a complete explanation of digital operation, see the section in Chapter 11.)

Time base correctors also permit the integration of ¾-inch or ½-inch program material which was recorded on cassette machines, with sequences that were recorded in the studio on 1-inch or 2-inch VTRs. Frequently, a remote sequence will be recorded in the field by using a ¾-inch VCR. Back in the studio, the material is fed through the TBC and recorded in either 1-inch or 2-inch VTRs depending upon the format which is used for principal recording and editing. This is called a *bump-up* because we transfer material from a lower quality recorder to one of higher quality. By "bumping-up," the director can record on the less expensive and more portable machines and then transfer the raw footage without significant loss of picture quality for subsequent postproduction editing and integration with studio recorded material.

Needless to say, a TBC cannot take a video signal from a $2,000 cassette recorder and produce a picture which rivals that of a top quality helical 1-inch or a quadraplex VTR. But the TBC does permit pictures from a cassette recorder to be used on-line and enables 1-inch production

FIGURE 9-15 **Digital Time Base Corrector.** *(Courtesy: Ampex Corp.)*

VTRs to replay broadcast quality pictures in slow-motion and still-frame. It also permits the nonsynchronous signal from an inexpensive cassette or reel-to-reel VTR to be fed into the program switcher and used along with an array of special electronic effects such as "wipes," "dissolves," and "keys."

VIDEO DISC SYSTEMS

Up to this point we have discussed the magnetic tape systems for recording and reproducing the video signal, but there is one other video recording system used in television production—the video disc. Unlike tape systems, the video disc uses a flat, magnetic disc—similar to those used by data processing computers—to store and retrieve video information.

You will find two types of disc systems in production: (1) the slow-motion disc and (2) the still-frame-storage system.

Slow-Motion Disc

The slow-motion disc was originally developed for use in sports coverage to permit the director to replay the action instantly at a variable speed or to completely stop the action in a freeze-frame.

The disc system uses a large, metallic disc pack, which rotates rapidly as the recording head system sweeps across the disc like a needle on a record turntable. Unlike a standard record, where the information is laid down in a continuous spiral, however, the video disc records its information in a series of concentric rings.

The disc can record about thirty to thirty-five seconds of material at normal speed. Once the storage capacity has been reached, the unit progressively erases the older material and replaces it with the new information. This allows sports productions to record the action continuously, recuing for playback once the event is over.

On playback, the video heads scan the disc and replay the recorded information. Stop-motion is obtained by continually replaying a single video frame, which is one concentric circle on the disc. Slow motion is produced by playing a frame more than once before moving to the next. You can replay action forward or reverse, at normal speed,

faster or slower than normal, or stop the action completely. For example, if you were recording a football play, you might replay the action at normal speed while the quarterback sets up to pass, then gradually slow the action down as the receiver reaches for the ball, freezing the action completely the instant the receiver is tackled. Rapid recuing is possible by a device which marks the start of a specific event and allows the operator to recue to that spot immediately for instant replay.

It is important to realize that the slow motion created by the disc is not really the same as motion picture slow motion. In film, the scene is photographed at a much higher frame-per-second rate so that the action is spread out over more frames than normal. Since television slow motion merely repeats a single frame over and over before moving on to the next, the effect gives the illusion of slow motion although the action is not really presented in any more detail, as is the case with film. This is why some very fast events such as auto racing or ski jumping may show a blurry still-frame when replayed on the disc. However, for most production requirements, the slow-motion effect is more than adequate.

The capability of 1-inch production VTRs to produce slow-motion, (slo-mo), fast-motion, and freeze frames has reduced the reliance on the slo-mo disc for many production situations. Since the open reel VTR can hold up to an hour's worth of recording tape, it does not suffer from the major drawback of the disc, which is its inability to record a sequence that is longer than 30 seconds.

Electronic Still-Frame-Storage Systems

Still-frame-storage systems are designed to replace the use of slides and camera graphics by storing hundreds of individual video frames (you might think of them as individual video "slides") in its memory. Each individual image is assigned a code number and is accessible immediately at the push of a button. (See Figure 9-16.)

(a)

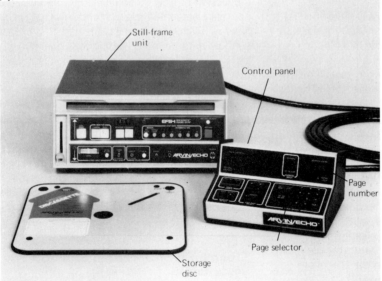

Still-frame unit

Control panel

Page number

Page selector

Storage disc

(b)

FIGURE 9-16 **Still Frame Storage Systems.**
The studio version (a) has an extremely large capacity capable of storing all of the still graphics for a production facility. Each graphic is provided an "address code" permitting instant retrieval whenever necessary at the touch of a button. The smaller unit (b) is portable and stores each of the video frames' on a disc which is capable of holding hundreds of individual frames. The small size of the unit makes it ideal for use on remote productions as well as for in-studio applications. *(Courtesy: Ampex Corp. and Echo Sciences Corp.)*

The operation of the still-frame storage system is simple. To record an image, you simply locate an open frame space in the memory, feed the image into the still-frame system, and press "record." The image is immediately captured, and a code address number is assigned, which corresponds to its memory location. To retrieve the image, you punch-in the address code number, and the image reappears in less than a second. Images can be recorded from any video source, such as film or slide chains, cameras, or videotape.

For example, before covering a baseball game you might use a field camera to record a still-frame of each player, assigning an address number which corresponds to his jersey number. Then, whenever the director needs the "slide" during the game, the still-frame operator punches-in the player's number and the image appears instantly. Most operators use the jersey number for the home team and 100 plus the

jersey number for the visiting team. This way, if there are two players with number "24," the home player is coded as "24" and the visiting player as "124."

The still-frame system offers many advantages. Slide and camera graphics always look fresh and new since they are electronically stored. They cannot be misplaced, lost, or scratched. Nor do you need multiple copies of the same slide for different studios. For those programs where a great many still images are necessary—such as news programs election returns, sports events, and so on—the still-frame disc system can hold hundreds of images, any of which can be put on the air in less than a second. The rapid access of individual frames enables the frame-store device also to be used to create an animation effect by playing back individual frames in a preprogrammed sequence to create the illusion of a moving image. In Chapter 13, Television Graphics, we will discuss the use of the still-frame storage system to create electronic graphics for a variety of production applications.

FIGURE 9-17 **Slow Motion and Still-Frame During Sports Remote.**
The operator is checking the list of call numbers which are used to recall any individual frame from the still-frame system's memory as necessary during the game. The control unit at the bottom left controls the slow-motion and freeze-frame operation of 1-inch VTRs, which are used for instant replay and slow motion. (*Courtesy: Echo Sciences Corp. and ABC-TV*)

VIDEOTAPE EDITING

The ability to edit videotape is one of its most useful and important properties. Without an accurate and rapid method of editing tape, electronic news gathering and electronic field production would be impossible. Editing has become such a key element of the production process that it is difficult to imagine any program which does not utilize it in some way. In order to function effectively in television, you must understand how videotape is edited and the many exciting and creative production possibilities that editing can offer.

Mechanical Editing

When videotape was first introduced, the only way it could be edited was mechanically. That is, the tape was physically cut with a razor blade and spliced together, much as we do with audiotape. But unlike audiotape, which can be cut anywhere, videotape must be spliced by cutting precisely in between the invisible electronic video frames on the tape, or the picture will tear and break up during playback. Not only is mechanical editing time-consuming and inaccurate, but physically cutting the master tape is always a dangerous procedure. Consequently, mechanical editing is never used today except in an emergency or under very rare circumstances.

Electronic Editing

Electronic editing eliminates all of the problems associated with mechanical editing and allows us to take advantage of the many technological advances in modern equipment which greatly expand the editor's postproduction capability. There are a number of different methods which are used to control electronic editing, but the basic process of electronic editing is the same regardless of the control technique and equipment you may be using.

Essentially, electronic editing is a transfer process in which a playback VTR containing the originally recorded material is used to transfer its sound and pictures onto an edit/record VTR,

which assembles the raw footage into a completed sequence or program. You might think of electronic editing as a dubbing process in which the edit/record VTR makes a copy of the playback VTR's program material. However, the *electronic editor*—a device which controls the operation of both the playback and edit/record machines—enables us to control precisely when and where each edit will occur. This is accomplished by determining where the old material on the edit/record VTR will end and the new material playing in from the playback VTR will begin. It may sound fairly simple, but to make the electronic edit look like a clean and precise camera cut requires some complex electronic circuitry. (See Figure 9-18.)

All electronic editors let you choose the kind of edit you can make. There are two kinds of edits which are available: (1) *assemble* edits or (2) *insert* edits.

ASSEMBLE EDITS When the electronic editor is in the assemble mode, the new program material is added to existing program footage on the edit/record VTR beginning at the predetermined "in" edit point. The edit/record VTR will continue recording new material until it is stopped. As it records, it automatically lays down the necessary control track, audio, and video information, beginning with each new edit point. Using assemble editing, you simply add material—or assemble the program—segment by segment.

Since the playback VTR needs control track information from the edit/record VTR in order to

FIGURE 9-18 **Electronic Videotape Editing.**
Electronic editing is a dubbing operation in which the original material from the master tape is transferred to the edit/record VTR. The editing system controls the operation of both machines and switches the edit/record machine from "playback" to "record" at the proper time to make the edit.

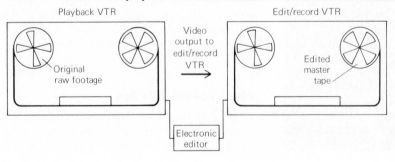

Playback VTR

Edit/record VTR

Video output to edit/record VTR

Original raw footage

Edited master tape

Electronic editor

synchronize its playback to avoid picture break-up, it is important to have some extra control track in the form of a recorded signal at the precise edit point. This is easily done by simply letting the edit/record VTR record about 10 to 15 seconds *beyond* the "out" edit point where you want the old material to stop and the new material begins. When the edit is made, the playback VTR's programming is recorded over the existing information on the edit/record tape, but it will have some control track to lock onto providing a much cleaner edit.

Assemble edits are very convenient because you simply add segment by segment to build a show on the edit/record VTR. However, the lack of a continual control track, which is used like film sprocket holes to guide the VTR's operation, can

result in the picture "tearing" or breaking up momentarily even though you have laid down additional control track at the edit point. (See Figure 9-19.)

INSERT EDITS Insert edits were originally designed to let you add video or audio to an existing program without disturbing the rest of the program material. For example, you might have recorded a scene where a character reads a letter and you want to insert a close-up of the letter, which was shot after the original scene was

FIGURE 9-19 **Assemble and Insert Editing.**
In assemble editing (a), each scene, take, or shot is added on after the previously edited material to assemble the completed master. The edit/record VTR lays down a new control track for each segment. In insert editing (b), the new material is edited onto the master tape by referring to the existing control track. This produces a more stable edit but requires that a control track already exist on the tape in order to make an edit. Insert editing permits audio-only and video-only edits in which only the sound or picture is added to the master tape without disturbing existing program material.

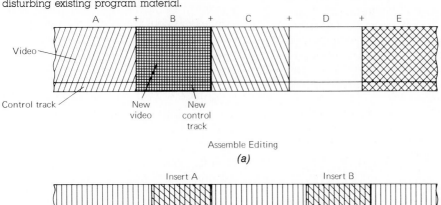

Assemble Editing
(a)

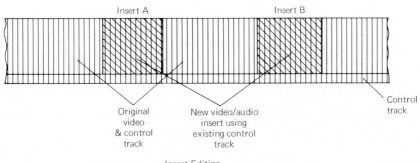

Insert Editing
(b)

recorded. Operating the electronic editor in the *insert* mode, find the correct in and out points for the insert and made a video-only edit using your normal editing procedure. Remember, since the insert edit is being made on a previously recorded tape, you must program both in and out points; otherwise the insert will continue to erase the previously recorded program material.

Insert editing uses the control track, which is already present on the recorded tape. This produces a much more stable edit than is possible with assemble editing, where no control track exists at the edit point. This is why many editors like to make all video edits in the insert mode. But to do this, you must prepare a tape for the edit/record VTR with a complete control track. This is called a "crystal black tape," and it is made by recording the entire tape with "black" from a studio output or a "black burst" generator. The crystal black tape contains the necessary control track to guide the VTR, assuring very stable edits at all times. Using the crystal black tape, you edit in the normal way, but instead of operating the editor in the assemble mode, you operate in the insert mode, indicating both in and out edit points. Remember to leave a video pad after each segment to help you select the best edit point for the next.

The only real disadvantage to the insert editing mode is the need to prepare the crystal black tape. Obviously, a one-hour tape requires you to record one hour of black before you can begin to edit. Many production facilities keep a supply of prerecorded crystal black tapes on hand so that emergency editing jobs or fast-breaking news can be insert edited without delay.

Although all electronic edits are made the same way, by transferring material from a playback VTR to an edit/record VTR, the way we control the editing operation varies, depending upon the type of edit control system which is available. There are essentially three different types of electronic editing control systems, and each requires a different kind of editing control equipment: (1) *manually controlled* electronic editing,

(2) *programmed* electronic editing, and (3) *computer* electronic editing. The manual system is the simplest and least expensive, but it is also limited in its accuracy and editing flexibility. Programmed electronic editing is most commonly used for news editing and many simple editing situations. Computer systems offer the ultimate in editing capability and control but are very costly to purchase and complex to maintain.

Manually Controlled Electronic Editing

In the manually controlled system, it is the tape operator/editor who controls exactly where and when the edit will take place as the playback and edit/record VTR machines are rolling. This method is sometimes referred to as "editing on the fly" since the precision of the edits depends on the editor's reaction time and skill in precuing both VTR tapes accurately prior to the edit.
The process works like this:

1 Find the "in" point where the edit should begin on the edit/record VTR. For example, if the last scene on the edit/record VTR is a close-up of a hand reaching for a pen, we might want the in-point to be just before the hand actually touches the pen. This is the point where the new material should appear.

2 Find the start of the new scene which will be played from the playback VTR. In this case, it may be a close-up reaction shot of a character in the play. The point on the tape you select will immediately follow the in-point on the edit/record VTR.

3 Rewind both the edit/record VTR and the playback tape *exactly* the same length. This is the *preroll cue,* which is used to give both machines enough time to come up to speed and stabilize before the edit is actually made. Usually this preroll time is around ten seconds.

4 Start both machines simultaneously, and watch the edit/record VTR monitor. At the precise edit point, press the record/edit button, which will immediately change the machine from playback to record, making the edit.

5 Let the edit run on past the point where you

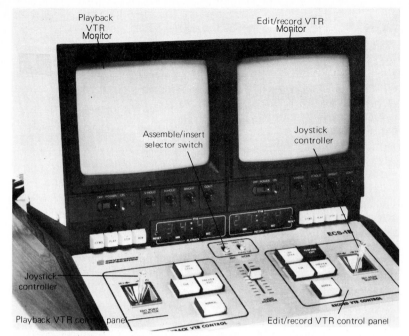

FIGURE 9-20 Edit Programmer.
When used with helical VTRs which offer slow-motion and still-frame capability, editing on videotape resembles film editing. The tape can be run slowly forward and backward until the exact edit point is located on both the playback and edit/record VTRs. The edit programmer controls both machines, so the actual edit is accurate to the desired frame. By viewing the still-framed image on the playback and edit/record VTR monitors, the editor can see how the video transition will appear before actually performing the edit. *(Courtesy: Convergence Corp.)*

wish to make the next edit on the edit/record VTR. This is a video "pad" which is helpful in locating the best edit in-point for the next edited segment.

6 Stop both machines and recue the edit/record tape to check the edit.

7 Repeat the process for each new edit.

You can see that the reliability and precision of the editing depends primarily on the skill and reaction time of the tape editor. If either tape machine is slightly miscued, the edit will be too early or too late. An early edit means that the new scene has erased some of the material on the edit/record VTR, and you may have to record the previous scene all over again. A late edit may look bad because it comes after the originally deter-

mined edit point. These are some of the reasons why manual electronic editing is sometimes, less than charitably, referred to as "punch and crunch" editing. Even with its limitations, however, a skilled director and editor can do some very creative editing using this method, although the accuracy of each edit is always subject to some possible variation and error.

Programmed Control Electronic Editing

Programmed electronic editing eliminates the imprecision of the manually controlled method by enabling the tape editor to preprogram the exact point on both playback and edit/record tapes. There are a variety of edit control devices which are used to interface the playback and edit/record VTRs and to control their operation. Since edit programmers are designed for use with helical 1-inch and cassette machines, their operation takes full advantage of the slant-track format's capability to display an image in slow-motion, fast-motion, and freeze-frame modes.

Edit programmers operate by counting television frames using the control track pulses which are on both the playback and edit/record VTR tapes. As each frame is counted, its precise location is displayed by a series of numbers which appear on a readout device on the edit controller console. Most edit controllers employ a *joystick* controller for each VTR. The joystick gives the tape editor motion control over the playback and edit/record VTRs by permitting the operator to manipulate the movement of the tape in the following ways: (1) forward or backward at normal speed; (2) forward or backward at slow motion or fast motion; (3) forward or backward moving frame-by-frame; and (4) still-frame, freezing the picture completely at the exact frame where the edit is to be made. (See Figure 9-20.)

While each edit programmer has slightly different operating features, they all work basically as follows:

1. Find the "in" edit point on the edit/record VTR by jogging back and forth with the joystick or pushbutton controls until you have located the precise frame where the edit is to occur. This frame is then displayed as a freeze-frame and appears on the edit/record VTR's television monitor, which is usually located on the right-hand side of the edit control console.

2. Follow the same procedure on the playback VTR to locate the frame where the new material is to begin. The freeze-frame image will appear on the playback VTR's monitor, which is usually located on the left-hand side of the edit control console. By looking first at the picture on the edit/record monitor and then at the picture on the playback VTR monitor, you

can see how the video transition will appear once the edit is made.

3. Once both edit points are determined, press the cue button, which automatically backrolls both VTRs the same distance to synchronize precisely their pre-edit roll time.

4. If you wish, you may "rehearse" the edit. In this case, the program controller rolls both VTRs and electronically switches the video image on the edit/record VTR's monitor to show you exactly how the edit will appear without actually making the edit on tape. If the preview edit is unsatisfactory, you simply recue either or both VTR's until you have corrected the problem and produced the desired edit transition.

5. Once you have approved the preview edit, press the cue button again to backroll both machines for the actual edit. By pressing the edit/record button, you start both machines simultaneously, lock them into synchronization, and cause them to make the edit precisely at the preprogrammed edit point.

Edit programmers which count each video frame in order to control the editing operation are the most commonly used editing devices owing to their versatility and low cost. Virtually all of the ENG footage which is produced for news broadcasts are edited on these machines daily, and many editors have become sufficiently proficient at using the edit programmer to do some very elaborate editing jobs as well.

However, the control track counting system is not without its disadvantages. Since the system merely counts the video frames which exist on the tape, it is susceptible to error and is not always frame-accurate. This is rarely a serious problem for most day-to-day editing requirements, and operators usually learn how to compensate for their system's minor variations and idiosyncracies. Another disadvantage of the control track counting system is its inability to locate automatically footage which may be anywhere along a tape. Instead, the editor must visually search through the tape in order to locate a particular cut. Again, this is not usually a problem for most common editing situations, but on some productions which require elaborate editing of many hours of raw footage, it can be time-consuming. Finally, and most significantly, the control-track counting system does not permit the edit-control device to synchronize and control the operation of more than one playback VTR or the operation of auxiliary equipment such as multitrack audiotape recorders, switchers, and special effects generators. In order to utilize these items to achieve some very sophisticated postproduction operations, we must turn to the third type of edit controller, the *computer-assisted* system. But before we can discuss the computer system, we must first examine the standard on which it is designed to operate: the SMPTE time code.

SMPTE EDIT TIME CODE

Although electronic editing will make a perfect edit exactly where programmed, locating the precise edit point can be difficult, particularly on quad VTRs, which can neither be run in slow motion nor still-framed to display individual frames. In order to identify each video frame better and to permit much more accurate editing, the time code system was developed in the late 1960s and later standardized by the Society for

Motion Picture and Television Engineers (SMPTE) in 1970. Standardization means that all time code systems—regardless of the tape format or VTR machine model—read, write, and understand the same "language."

The system works by adding an eight-digit code number to mark each video frame, in effect giving every frame its own unique "address." The time code information is usually carried on the cue track or, on some machines, on a special time code track. The code can be recorded either while the program is being recorded or afterward. The numbers appear on specially equipped monitors at the bottom of the picture screen and on time code readout counters.

The eight code numbers refer to the hour, minute, second, and video frame. Since there are thirty frames per second, the last two digits reach "29" before the seconds move ahead one number. (See Figure 9-21.) Although the time code itself does not function as an editing system, it is the key to all sophisticated electronic editing

FIGURE 9-21 SMPTE Time Code. The time code digits—representing hours, minutes, seconds, and frames—are superimposed over the video image. *(Courtesy: CMX Systems, an ORROX Company)*

because it serves as the basic reference used by both conventional and computer systems to locate, cue, and edit videotape with single-frame accuracy.

Drop Frame Versus Nondrop Frame

The time code is created on the tape by a "time code generator," which can be run in one of two modes: (1) *drop frame* or (2) *nondrop frame.*

For certain technical reasons, if we generate the time code continuously—starting at 0:00:00:00 for the first frame and continuing for each subsequent frame on the tape—we would wind up with a discrepancy between the time code which appears on the videotape and the actual time of day or standard clock time. Consequently, the SMPTE developed an alternative generating mode called "drop frame," which is designed to correct this discrepancy. In the drop-frame mode, each minute—with the exception of every tenth minute—automatically drops two frames in its time code count. The result is a time code which precisely agrees with actual standard clock time.

The reason this is important is because computer systems which use time code can not function with different tapes encoded with different time codes. The different "dialects" of the time code "language" confuse the machine and inhibit its operation. This is why most production facilities have standardized on one mode—usually drop frame—for all video recording. Whichever mode you select, make certain that *all* tapes are recorded with the same time code mode whether it's drop frame or nondrop frame.

Using Time Code in Production

Every videotape which is to be edited must be encoded with its own unique series of ascending time code; that is, the time code should begin with a low number and progressively increase to a higher time code number. In addition, it is crucial to avoid any duplication of similar time code numbers on the same tape or on different tape reels since the time code reader which uses the tape as reference deals only with the eight-digit code and becomes confused if it reads two identical numbers. There are two ways you can use time code: (1) *zero start* or (2) *time of day.*

ZERO START In *zero start* the digits begin from zero at the start of each tape reel or recording, and indicate the amount of lapsed time as the recording progresses. Most recording sessions use zero start since you can use the hour digits to designate different tape reels. The first reel would begin at 01:00:00:00, the second reel at 02:00:00:00, and so on.

TIME OF DAY In *time of day*, the digits run synchronously with a master studio clock, and the numbers reflect the actual time of day. Time of day is a useful method when you want to use the clock time of day to log events during a recording. Recordings of news events, sports, and other productions which require you to locate events within real time are recorded with this method. Producers, writers, and other production personnel do not need access to the SMPTE readout digits to log events for later editing. All they have to do is refer to a studio clock since it is identical to the time code digits on the recording tape.

RECORDING TIME CODE In most videotape systems, the time code is recorded on a secondary audio track or on the VTR's cue track. It is crucial that the time code be recorded at a sufficiently high audio level and without distortion, or the system will be unable to read the code properly, and poor edits or a lack of synchronization between different machines may result. It has been found that time code which is recorded at a level of +3 dB over zero on a VU meter provides the most satisfactory results. It is definitely worth the extra few minutes it takes to make certain that the time code is being properly recorded to avoid having to completely restrip the time code after the original recording is completed.

For productions which are recorded inside the studio, time code is usually recorded at the same time the VTR records the program material. For remote productions which are recorded outside the studio on location, the time code is usually added later, often when a ¾-inch cassette is bumped-up to a 1-inch helical or 2-inch quad for subsequent editing and postproduction. Some newer remote cassette recorders and 1-inch VTRs have a built-in time code generator which automatically lays down the time code as the original program material is recorded. If a tape is produced without time code, either inside the studio or out on location, time code can always be added before editing, but this involves the use of a VTR to record the time code.

PLAYING BACK TIME CODE Although the time code information is recorded and stored on a separate audio track, a "time code reader" is required to enable you to see the numbers during playback. The time code reader can display the figures either on the top or bottom of the television screen along with the rest of the video image, on a separate digital clock-like readout device, or on a CRT unit.

It is also possible to *burn in* time code numbers onto a videocassette "workprint" so that you can watch picture and time code on a playback videotape machine which is not equipped with a time coder reader. This permits you to preview the raw footage on any conventional videocassette machine and to plan your editing by noting the exact time code numbers where edits are to be made. Of course, the burned-in time code can never be used on the master tape since the playback image will include the time code numbers along with the rest of the program video.

Off-Line and On-Line Editing with Time Code

Although the time code itself does not function as an editing system, it is the key to all sophisticated electronic editing because it serves as the basic reference used by both conventional and computer systems to locate, cue, and edit videotape with single-frame accuracy.

Until the development of the time code, all editing was accomplished on the same mastering/broadcast-quality machines on which the program was recorded. On-line editing is costly because it uses expensive, broadcast-quality VTRs and trained technicians to operate

them. Also, machines used for editing cannot be used for other production or playback purposes. The expensive costs, combined with the distracting environment in the VTR tape room, often make on-line editing a pressured and difficult situation for a director and editor.

The introduction of the time code gave editors and directors an alternative to on-line editing. Using an inexpensive helical workprint of the original master tape, which carries identical time code numbers, the director and editor can preview scenes and made edit decisions off-line. The off-line approach offers a number of important advantages:

1 Creative edit decisions can be made in a quiet and pressure-free environment.

2 The simple and easy-to-operate helical machines let nontechnical people operate them, eliminating expensive technician time and giving creative people more "hands-on" control.

3 Most helical machines can slow-motion or freeze the action. By means of the slow- and stop-motion control of the helical VTRs, the specific frame where the edit is to be made can be jogged into position, and its exact frame number noted in an edit decision log.

4 Expensive, on-line VTRs can be more efficiently used for production and playback since long edit sessions are planned first on helical machines.

Using the off-line method, a helical copy (either reel-to-reel or cassette) of the original master tape is made containing the same time code that appears on the original tape. The helical workprint can be recorded at the same time the master tapes are recorded, or a dub can be made after the taping session is finished.

The editor and director use the time code digits, which appear along with the picture on the playback monitor, to develop an *edit decision list*. The decision list is used later as a reference to direct the editing of the master tapes on either

2-inch quad or on helical VTRs. Since all creative editing decisions are made off-line, the actual editing of the master is a straightforward and simple process of editing by the numbers.

If you wish, you can actually produce an edited helical workprint on most off-line systems. This is valuable in complex editing situations, where the editor and director need to see how the cuts actually look on the television monitor. But for most simpler editing jobs, many directors simply use the edit number to develop the written decision list without going through the additional step of actually producing the edited workprint.

Another advantage of off-line editing is on remote productions. Many editing decisions can be made in the field, since the equipment needed to preview and prepare an editing log is very portable. While at NBC, executive producer George Heinemann and his staff pioneered the use of off-line location editing. Each night the footage recorded during the day is rough-edited by the director and tape editor in the production company's hotel. The producer, director, and editor can actually see how the various pieces will fit together, and any additional footage which seems necessary can be shot or reshot the next day while the production unit is still out in the field. The director leaves the location with the program almost "edited." Only some additional trimming is necessary before the master tape is actually edited to conform to the previously edited workprint.

COMPUTER-ASSISTED EDITING SYSTEMS

Computer-assisted editing systems offer the ultimate in sophisticated videotape editing operation. The computer systems combine a dataprocessing computer with television production equipment, enabling the operator to accomplish easily a wide range of complex editing in a relatively short period of time. As more and more productions are being shot "film style" with a single camera/single VTR approach, literally hundreds of edits are necessry to assemble a finished production and only the computer's ability to keep track of this vast amount of information keeps the editor from being overwhelmed by the task.

The systems which are currently available can operate either on-line or off-line, using quadraplex or helical format VTRs or a combination of the two. In addition to controlling the various playback and edit/record VTRs, the computer system interfaces with such auxiliary production equipment as audiotape recorders and production switchers, enabling the operator to perform a variety of complex postproduction functions efficiently and accurately.

While the operation of a computer system is highly complex, the key to understanding how it functions is relatively simple. The basic idea is for the computer to use the SMPTE time code as a means of both controlling the operation of all equipment during the editing operation and to store in its vast memory every edit decision as it is made by the editor or director. What you are doing as you use the computer system to edit is to develop an *edit decision list,* which is literally a list of every edit point; the duration of the transition (if it is a fade, dissolve, wipe, or key); and the "in" and "out" points for each edit sequence. As you go along making your edits, the computer stores all of this information in its memory and uses the finished edit decision list as a guide for the final automatic assembly of the master footage.

Unfortunately, this brief explanation cannot convey the tremendous power of these systems to produce exceptionally complex postproduction operations. A few years ago, computer systems were used only by the major networks and by a few large production facilities which had editing requirements sufficiently complex to take full advantage of the capabilities that the computer system can offer. However, today it is not uncommon for many medium- and small-scale production facilities to use computer systems for a variety of complicated productions. If you have a very elaborate editing job, you might consider renting the use of a computer system from a nearby production facility which offers computer editing. While the cost is more than for conventional editing, the system's enormous capability can more than offset the increased expense both in time saved and in the quality of the final product.

Computer Editing Operation

The operation of a computer editing system is highly complex, and the average editor requires a considerable amount of time working with the system in order to master its full capabilities. Although some newer computer systems are equipped with "smart" control consoles in which the computer literally takes you step-by-step through the editing operation, for the most part

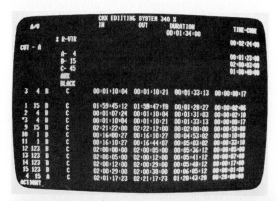

computer systems are operated by trained editors who will follow the director's instructions in assembling the program. However, it is important for you to know the basic operations of the computer system so that you will understand how to take full advantage of the editing capabilities it has to offer.

The system is controlled by a specially designed keyboard and a CRT terminal which are used to communicate with the computer and to control the operation of the VTRs. Using the keyboard, the editor enters editing information into the computer, instructs the system where and when to perform an edit or transition, and enables the operator to control each VTR and any auxiliary equipment such as multitrack audiotape recorders. (See Figure 9-22.)

The computer system can be used either on-line or off-line. If you are editing on-line, you will actually be editing the master tape as you go along by making edit decisions and executing

FIGURE 9-22 Computer Editing.
The editing information as it appears on the CRT readout. Each series of time code numbers represents an edit point. All editing operations are controlled by the keyboard unit, which enables the editor to operate all videotape machines, as well as to enter information into the computer system. *(Courtesy: CMX Systems, an ORROX Company)*

them at the same time. The real advantage of the computer system is to work off-line, however, because this enables you to produce a helical cassette "workprint" and then to have the computer automatically assemble the master footage to conform to the edited workprint.

The basic operating procedure works like this:

1   Assign each playback VTR and audio recorder a specific letter or number to be used by the computer during the editing procedure. For example, "R-VTR" would be the edit/record VTR while "A-VTR" and "B-VTR" would be the two playback VTRs.

2   Preview your scenes or takes by entering the time code address for the start of each scene or shot. The computer will automatically search and cue up for playback.

3   Determine your edit points. This can be accomplished in a number of ways: You can do it "on the fly" as the VTR plays back at normal speed by simply pressing a button when you see the point at which you want the edit to occur. The system will automatically note the appropriate time code numbers of the video-frame at the edit point and store them in its memory. Another method is simply to enter into the keyboard the specific time code numbers where you want the edit to take place. Of course, to do this, you will have had to preview the tape earlier, but if you have edited "on paper" by noting specific time code numbers for frames where edits are to occur, it is a simple matter to instruct the computer to make the edit at the exact frame you want it. Finally, you can use the jog and shuttle capabilities of helical VTRs and cassette machines to move the tape back and forth until you've located the specific frame for the edit point. Again, when a button is pressed, the computer will automatically note the correct time code numbers and enter them into its memory.

4   Preview the edit. The computer will automatically preview the edit or transition (fade, dis-

solve, wipe, and so on) so that you can see how it will actually look. If you are not satisfied with the edit, you can "trim" the edit point forward or backward until you have determined the precise location for the edit. Again, all corrections to the edit point are automatically updated in the computer's memory.

5   Create the edit decision list. Once you have completed your editing, the computer assembles all the information you have entered into an edit decision list. The list is the computer's copy of your edited workprint and uses time code to indicate every edit or transition, its location, and its duration. However, before you can use the edit decision list to conform the master tape to the edited workprint, the edit list must be "cleaned" to process the list for automatic assembly. The cleaning is called "list management," and it is done by a computer program which scans through the list and searches for any discrepancies or errors. The cleaning program also organizes the events in the edit decision list to provide maximum efficiency during automatic assembly of the master footage based upon the edit decision list you have created in preparing the workprint.

The edit decision list can be printed in a hard copy log which indicates every transition according to its time code numbers. In addition, the computer prepares either a magnetic floppy disc or a machine-readable punch tape, which is used later to control the automatic assembly of the master footage according to the instructions on the edit list. The advantage of the floppy disc over punch tape is the vast amount of information conveniently and safely stored on disc as compared with the more cumbersome punch tape.

6   Automatic assembly of the master tape. Once the edit decision list is on a machine-readable medium (either floppy disc or punch tape), you are ready to conform the master footage to the edited workprint to create the final, edited version. The operator must load the necessary master reels onto the various VTRs, set up the edit/record VTR, and instruct the computer to follow the edit decision list to assemble the master version. At this point, no

further manual control is necessary as the computer controls the operation of the VTRs, editors, and auxiliary equipment such as audiotape recorders and program switchers to produce the edited master tape.

Although cut-only edits are made extremely simple by the computer system, its enormous capabilities become apparent when you need to make such other transitions as dissolves, fades, wipes, supers, or chroma keys, with two or more playback VTRs at the same time. It is a relatively simple matter to instruct the computer as to the start and end points of the transition and how long the transition should run in video frames. The computer will control all precuing, playback, editing, and switcher operations, assuring you precise accuracy regardless of the complexity of the edit or transition.

If a computer-assisted editing system is used, there is little a creative director or editor cannot do. The system enables you to record an entire program with a single camera, making hundreds of edits within the show with single-frame accuracy in a fraction of the time it would take using other editing methods. The computer system also allows you to perform a variety of electronic special effects, such as video animation—having a stage fill up with props, sets, and performers apparently without benefit of a stagehand—or

making one performer into a duo, trio, or quartet, all singing and dancing together in perfect synchronization.

Audio Editing with Computer Systems

Up to now you may have had the impression that videotape editing deals only with pictures. That is not really the case, however, and the use of SMPTE time code, combined with computer-assisted systems, has given directors and editors far more flexibility to deal with sound as well.

The heart of the audio editing system is, again, the SMPTE time code, which is used as a reference to keep the multitrack audio recorder operating in perfect synchronization with the videorecorders.

On many complex programs, the audio is recorded directly onto a multitrack audiotape recorder (ATR). One of the ATR's tracks is assigned to record the identical time code used by the program VTR. Once the audio is recorded on the multiple tracks, the audio engineer can mix-down the sound to either monaural or stereo in a

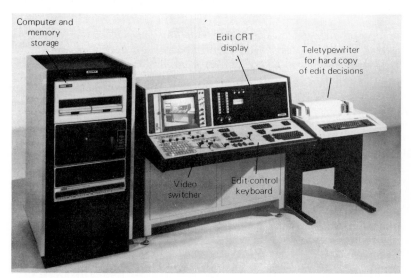

FIGURE 9-23 **Computer-Assisted Editing System.** The computer system includes a keyboard controller, CRT display unit, computer memory and storage device, and a printer terminal for producing hard copy of edit decision lists. *(Courtesy: Ampex Corp.)*

leisurely postproduction session without the pressure of an entire studio full of crew and cast members waiting around until the right audio mix is obtained. If the program requires additional sound effects, music, narration, or a laugh track, these can also be added and mixed in during the postproduction phase. A major advantage of this approach—aside from the high-quality sound it can produce—is that musical performers appearing on the show can be involved in the final mix of their own performance. This is particularly important to many rock musicians who are concerned about technical quality and want to replicate their recording "sound" on television.

Once the multiple tracks are mixed-down, the final monaural mix is transferred onto the audio track of the quadraplex or helical production VTR in perfect synchronization. If the program will be broadcast or distributed in stereo simulcast, the multitrack mix can be used to create two stereo tracks.

Multitrack techniques are also valuable on dramatic programs, where there is a great deal of postproduction dubbing and mixing. Usually a workprint is made by off-line editing the helical copy of the original recording. Using this as a guide, the editor, director, and audio engineer can assemble the sound track on a multitrack recorder, which, again, runs in sync with video via SMPTE time code. Sound effects, music, dialogue overdubbing, and any other audio can be added and mixed in at this time. Once all the tracks are mixed-down to a monaural or stereo version, they are laid back on the quadraplex or helical master tapes for broadcast or distribution.

The SMPTE time code, along with the computer editing system, can also be helpful in smoothing out the audio after you have cut and edited a show. For example, most videotaped situation comedy shows ultimately build the broadcast show from a variety of taping sessions, each with a different audience. The edit points may naturally contain an audio mismatch since one audience may respond differently to a joke from the way another does. Using time code, the audio is lifted off the master tapes, remixed, and "sweetened" to smooth out rough sound transitions between edits and laid back onto the edited master tape, producing a smooth-sounding track with a constant audio level.

SUMMARY

Electronic videorecording is a primary production tool used (1) to record a program for delayed playback; (2) to edit a show, eliminating errors or poor performances and combining the best takes and segments into a final edited master; (3) to manipulate time and motion with slow motion and still-frame techniques; and (4) to play back previously recorded material into a production.

There are two basic video-recording systems: (1) quadraplex, which uses a transverse scanning process, and (2) helical, which uses a slant-track recording format. Quadraplex machines offer complete tape interchangeability, but the 2-inch tape is expensive and does not offer slow-motion or still-frame capability. Helical recorders are available in 1-inch, reel-to-reel tape machines as well as in cassettes which use ¾-inch, ½-inch, and ¼-inch tape. One of the primary advantages of the helical machines is their ability to reproduce pictures in fast motion, slow motion, and freeze-frame, which can be used as both a production tool and during postproduction editing.

A third electronic recording format is the video disc. The slow-motion disc enables pictures to be replayed in slow-motion or freeze-framed, and the electronic still-frame storage system holds hundreds of individual video images—or electronic "slides"—in its memory; each image is instantly accessible at the push of a button.

The fact that videotape can be edited is one of its most significant features. On-line editing refers to the making of edits on the same equipment on which the program was recorded. Off-line editing refers to the making of a helical cassette "workprint" and using SMPTE time code to first edit the tape on less costly equipment before employing expensive and heavily used production machines to make the edits on the final master version.

Most sophisticated production editing uses either programmable edit controllers or computer-assisted editing systems. Programmable controllers are used mainly with cassette machines and control the editing operation by counting each video frame. Computer-assisted systems utilize the more precise SMPTE time code, which not only identifies the edit points on the videotape recorder, but also enables auxiliary equipment such as a multitrack audiotape recorder to be synchronized with the videotape equipment.

CHAPTER 10

VIDEO RECORDING

Editing and Production

Techniques

Videotape and video recording offer a wide range of enormously creative and versatile production possibilities. As with any production tool, however, it is effective only if you have carefully planned for its use. Just as the artist selects a particular paint, brush, and canvas in order to achieve a specific purpose, so too must the television producer and director select the right video recording technique for a particular production situation.

In this chapter we will discuss the different videotape production techniques which are available to you, and explain when, how, and why to use them effectively in a variety of production situations.

VIDEOTAPE PRODUCTION MODES

Once you have decided to produce your show on videotape, you must determine which videotape production mode you will use to record the show. There are four different techniques available: (1) live on tape, (2) recording in segments, (3) recording with single camera/single VTR, and (4) recording with multiple cameras/multiple VTRs. Each approach has its share of advantages and disadvantages, and your decision to use a particular method or a combination of techniques will depend on the type of show you are producing, the available production and editing facilities, and the size of your production budget.

Live on Tape

Recording a program *live on tape* means that you are, for all intents and purposes, producing a "live" show, using multiple cameras and switching between them from the control booth as the show proceeds. The only difference between a truly "live" show and a "live-on-tape" program is that instead of broadcasting the program as it is produced, the show is recorded on videotape for delayed broadcast. There is usually little, if any, postproduction editing on a live-on-tape show. The entire program is produced straight through, and only the most serious performance or production errors are edited after the taping is completed.

WHEN TO USE LIVE-ON-TAPE APPROACH

Consider the following factors in deciding whether or not to produce your show live on tape:

1 *Does the program need postproduction editing?* An interview show, for example, needs no real postproduction editing, and the flow of a live program provides the continuity necessary for an interesting and effective interview. Shooting such a program in short segments with constant starts and stops would destroy the interaction between participants. Of course, you might decide to let the interview run overtime and edit it later, cutting out irrelevant or uninteresting parts.

2 *Can the program be easily produced without editing?* Some programs are produced just as easily without postproduction editing. A game show, for example, as well as some other routine format shows, such as a daytime "home" show or some "demonstration-how-to" programs can usually be produced straight through without any need for later editing.

3 *Do you have the facilities, time, and budget necessary for postproduction editing?* Editing costs money, requires available editing facilities, and takes time, delaying the production's final completion date.

Recording in Segments

Recording a program in segments combines the

advantages of multiple-camera production with the flexibility and creative control of postproduction editing. If this approach is used, the entire program is broken down into relatively short scenes or segments. Each segment is recorded by multiple cameras, and the director switches between them from the control room, as with any conventional program. During postproduction, the segments are edited and assembled into the composite program.

The advantage of the segment approach is that it lets you break down a large, complex production into smaller individual segments, which are much easier to deal with one at a time. Since both performers and production crew concentrate on only a small piece of the larger program, they can devote their full attention to the particular segment being rehearsed and taped. Once the scene is approved by the director, everyone can turn to the next segment without concern for those sections which are already "in the can." In this way, a one-hour program becomes a series of shorter shows, each of which is rehearsed, polished, and videotaped before the next segment is produced.

Another advantage is that scenes, or segments, can be recorded out of sequence for the convenience of the production. For instance, say you are producing a one-hour drama which begins with a courtroom scene and winds up at the end of the show on the same courtroom set. If you were broadcasting live or recording live on tape, the entire set would have to remain standing and lit so you could use it at the end of the show. However, if you record the program in segments, you can record all the scenes involving the large courtroom set and the extras who must appear as spectators at the same time. Once all courtroom scenes are recorded, the extras can be released and the set taken down to make room for other sets needed in the production. In the same way, if your program has the host and hostess open and close the show on the same set, recording both the opening and closing at the same time eliminates the need to reset and light the same area twice.

The segment approach also lets you tape many segments on one day, which can be used later on a number of different shows. The long-running PBS series for children *Sesame Street* uses the segment approach to optimum advantage. For example, on one day the production unit may shoot a series of segments involving animals and a trainer from the local zoo. These segments are recorded, logged, and filed away where they are available for a month's worth of programs without the need to arrange for the animals to be brought to the studio on a more frequent basis. You can shoot all a program's introductions and closes on one day, all the puppet sequences the next day, and so on. Actually, the production unit never really produces an entire show from start to finish in one day. Instead, segments are continually shot—according to the program's scripts, which are written in advance—and the broadcast program is a composite of different segments assembled in postproduction.

Recording in segments can be a very efficient production approach, provided you have planned the taping sessions carefully. You must be sure that you have developed a shooting schedule well in advance and that the sequence of scenes or segments is arranged to take maximum advantage of your facilities and talent. And, of course, you must be certain that all the necessary segments have been recorded. It is most embarrasing to learn in an edit session, after you have completed production, that you forgot to record a particular segment!

ISOLATED CAMERA An ingenious variation on the segment approach is to use one of the multiple cameras as an *isolated camera.* The concept was borrowed from sports production, where it was developed to provide the viewer with another angle of the action for the instant replay. Any of the cameras can be "isolated," or assigned to its own VTR or to a disc recorder, while the director uses the remaining cameras to cover the play. In a nonsports production, the "iso" camera is used in much the same way. One of the studio cameras is used to feed an independent VTR, which continuously records whatever the camera photographs. The particular camera assigned to "iso" can be changed by the switcher during the program. In this way, the director cuts the show normally by switching between all the available studio cameras, while another VTR independently records the output of the designated isolated camera at the same time. The isolated technique is frequently used in situation comedy or musical programs, where its additional footage comes in handy during the postproduction editing session. It can also be used as insurance to make certain that the most important element of a program is on tape regardless of how the director cuts the show from the control booth. For example, on an interview program, the camera which is trained on the guest can be designated the "iso" camera. Regardless of how the director cuts the show from the booth, there will always be extra footage of the guest in case a camera cut is missed or to provide additional material for reaction shots. Similarly, in producing an industrial demonstration program, the camera which is covering the demonstration in close-up can be "iso-ed" to ensure that additional footage will be available should postproduction editing be necessary.

When to Use the Segment Approach

1 *Does the program lend itself to taping in segments?* Some shows are simply better produced straight through than in segments. Many others lend themselves more readily to the segment approach, particularly shows which do not require the continuity and flow of a complete performance from start to finish. If the program will not be hurt by constant starts and stops and if segment taping is more efficient, it should be the production choice.

2 *Would it be more convenient to produce the show in segments?* The horror stories from the days of live television about performers racing across the set, changing their costumes, and just making (or missing) their next cues are no longer applicable when you tape in segments. In addition, your studio may not be large enough to accommodate all the sets you will

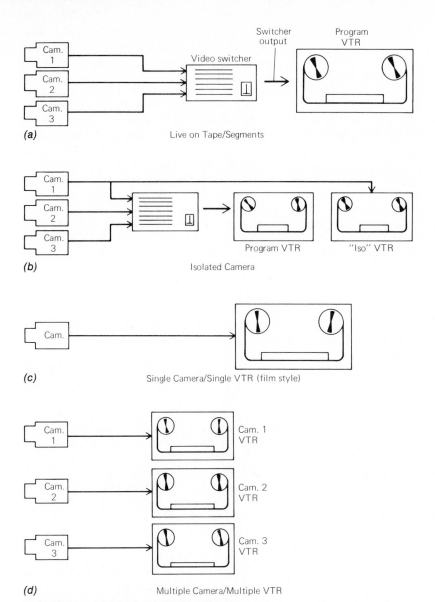

FIGURE 10-1 Videotape Production Modes.
Producing live on tape or in segments *(a)* uses multiple cameras with the director selecting each camera shot as the production proceeds. A variation of the approach uses an isolated camera *(b)* in which one camera feeds both the program switcher and a separate VTR to provide additional footage for editing. The single-camera/single-VTR approach *(c)* is similar to motion picture film production, in which each new camera angle requires a separate setup. The multiple-camera/multiple-VTR approach *(d)* uses a separate VTR to record each camera's continuous output. All camera shot changes are done in postproduction editing.

need at the same time. Finally, it may be less expensive to produce in segments, particularly if you will need a large number of performers, extras, or musicians for only a part of the production. If you recorded the show in its on-air sequence, all the performers would have to be on hand constantly, even for those parts of the show where they were not needed. Taping in segments permits you to schedule talent only when they are actually necessary; this is most efficient and economical use of rehearsal and production time.

Recording with Single Camera/Single VTR (Film Style)

Unlike the two videotape techniques just discussed, which use the conventional multiple-camera approach, the single camera/single VTR method is closer to most film production than to television. This method is commonly used for electronic news gathering and electronic field production, but it has many other production applications as well.

In this approach, a single video camera is used to shoot the entire program, shot-by-shot. Every camera shot is an individual setup with its own precisely planned lighting, audio, camera angle, and performer blocking. Since the production is concerned with only one shot at a time, the director and cast can concentrate on a specific part of the show, sometimes only a line or an actor's reaction. At the same time, the technical crew can tailor the lighting, audio, and camera movement for each individual shot without compromising, which is often necessary in multiple-camera production.

Of course, shooting a scene many times with a single camera takes longer than shooting a scene once with multiple cameras. The precision it provides both the production crew and the performers, however, makes it a very valuable technique when the situation warrants. The production method in video is identical to the shooting technique used for motion picture film except that an elec-

tronic camera and VTR are used in place of a film camera. An obvious advantage over film is that the director, cast, and crew can immediately replay every recorded take without waiting for the next day to see the "rushes." Most field productions are shot with this approach since it simplifies the personnel and equipment necessary when using multiple cameras operating out of a large remote van. The single camera and VTR can be carried in a stationwagon or minivan. The smaller vehicle is unobtrusive—particularly when compared with a remote van—easy to maneuver and park, and its equipment can be unloaded, set up, and loaded back into the van or wagon in minutes, enabling the unit to shoot at a number of different locations on the same day.

Because every shot is a separate setup, each transition from shot to shot on the final master must be made with a video edit. This makes the postproduction phase quite involved. At the same time, however, complete postproduction editing brings to television the same creative control and precision once available only with film.

In order to use the single-camera technique, you must first break down the entire shooting script into small scenes and then further into individual shots or setups. You juggle the shots into the most efficient shooting sequence, which is to be followed by the director and production crew during taping.

Generally a scene is taken first with a master shot. The *master shot* is a fairly wide angle which contains all the main action. The same scene must later be replayed while the camera is repositioned to shoot from a new angle, concentrating on smaller elements such as close-ups, reaction shots, and so on. Of course, if you are certain that a scene will require only one shot or setup, then the master shot is unnecessary.

Shooting shot-by-shot means you must be very conscious of *continuity*. These are the small details in camera and actor movement, position, and the placement of sets and props, which must all be kept in their proper order to show the viewer a logical sequence of events once the individual shots are edited together. In other words, a cigarette should not miraculously get longer from one shot to the next or appear first in one hand and then another; nor should an actor be looking toward the right in one shot and toward the left in the next.

When to use the Single-Camera Approach

1 *Does the production require the additional precision in setup and editing which the single-camera approach can offer?* Such productions as commercials, technical demonstrations or training tapes, and musical, dance, or dramatic shows often call for the maximum in production control. If you are working with special effects, where the proper camera position is essential to develop the most effective use of the videospace, the single-camera method is a good choice.

2 *Is the single-camera approach preferable to multiple-camera methods?* Sometimes simple productions are more suited to the multiple-camera approach than to the single-camera method, particularly if you are in the studio where multiple cameras are easily available, and there are enough to cover the action. Remember that it is usually faster to shoot the scene or segment once with a number of cameras than to shoot the same scene a number of times with one camera. Also, some productions cannot be carefully controlled, started, and stopped on cue to accommodate each camera setup. Certainly a sports event is one example of a program which generally requires multiple cameras in order to be covered effectively.

However, the single camera technique is sometimes preferable to the multicamera methods, particularly on remote productions. Single-camera setups can be done quickly and efficiently with a relatively smallcrew. This is one reason so many commercials, documentaries, and even dramatic shows are shot with a single camera on location.

Recording with Multiple Cameras/Multiple VTRs

The fourth production approach combines the flexibility and efficiency of multiple-camera shooting with the precision and accuracy of the single-camera/single-VTR method. Using the multiple-camera/multiple-VTR approach, you use two or more cameras on the production but, instead of feeding all the cameras into the program switcher, where each shot is selected by the director

during production, each camera is connected to its own VTR. Just as with the isolated technique, each VTR continuously records everything its camera photographs. The result is simultaneous footage of the same scene from a variety of camera angles.

The approach was originally developed for film by Desi Arnaz when he and his wife, Lucille Ball, were planning the classic *I Love Lucy* series. Prior to that series, all filmed comedies were shot with a single camera and without a studio audience. The performers and director had to estimate how much time to leave between jokes for the audience "reaction," which was later added to the sound track. Arnaz and Ball wanted to film before a live audience to help the performers deliver their lines and gauge their timing to a real audience's reaction. The three-camera technique Arnaz pioneered is still used today for film-produced comedy shows and has been widely adapted for electronic recording as well.

The technique works like this: the director blocks out performer moves and camera shots, as in any conventional production. During taping, the performance proceeds straight through, as each camera continuously sets up and shoots its preplanned shots from positions marked on tape on the studio floor. Usually the production is divided into acts, and each act is shot in its entirety, unless a set change is required.

Once the shooting is over, the director and editor use the footage from each camera/VTR to assemble the show. Their edit decisions are made in a relatively pressure-free environment, which permits experimentation and improvement rather than "on the fly" from the control booth during the actual production.

The advantages are obvious. The performers can develop and build their characters through an entire act without the starts and stops which are necessary with single camera production. The live audience is indispensable in guiding the performers' delivery and timing. Yet the director and editor need not worry about editing the show until the production is completed.

Although three cameras are the usual complement for most multiple-camera/multiple-VTR productions, any number of camera/VTR systems can be used, depending upon the production requirements. For example, on location, a difficult stunt or special effect might be recorded with two camera/VTR systems to ensure that the difficult-to-repeat stunt will be captured on tape from a variety of angles.

When to use the Multiple-Camera/Multiple-VTR Approach

1 *Do you need to cover continuous action while maintaining the capability for precision postproduction editing?* Many comedy and music shows are vastly improved when performers can work before an actual audience. The interaction of actor and audience can often spark a great performance, but this means that the musical number or play must run straight through with a minimum of interruptions. If your production will benefit from this continuity, the multiple-camera/multiple-VTR approach is a good choice, particularly if you will require careful editing later in postproduction.

2 *Do you have the necessary technical facilities?* While the technical preparations are neither exotic nor expensive, the technique does require as many independent VTR machines as there are cameras. This may cause a scheduling problem at those facilities where the number of VTR machines is limited. In addition, your postproduction editing will be rather involved and expensive. Unless you have adequate editing facilities available and the time and money necessary to undertake the editing, you might think about an alternative technique which will require less sophisticated editing equipment and less postproduction time.

Combining Videotape Production Approaches

Although we have had to talk about each production approach individually, you should not get the idea that only one technique can be used on any show. In fact, you might find that a combination of two or more approaches is best for your production situation. For example, say you are shooting an instructional or information show which involves mostly interior scenes with a few location segments. You might decide to shoot all the interior studio scenes with the segment approach and to shoot the location scenes with the single-camera/single-VTR method.

The point is that every show presents its own, unique set of production problems. Your choice of production technique should favor the one which will give you the best combination of creative control and production efficiency.

USING VIDEOTAPE IN PRODUCTION

Unlike some other production areas such as audio or lighting, where a single individual is usually responsible for the planning and execution, videotape and video recording involve a large number of television team members. The producer and director must determine the basic production approach, the tape editor must understand how the program will be edited, and the production staff must prepare for the use of videotape. Although the actual recording and editing of a show take place during the production and postproduction stages, the planning and preparation for the most effective use of videotape begins long before that, during the earliest stages of preproduction, and continues throughout every phase of the production process.

Preproduction Planning

As the program concept is developed and the script prepared, the director and producer must make a number of important production decisions involving the use of videotape.

DETERMINE THE VIDEOTAPE PRODUCTION APPROACH The first decision you must make is which production approach or combination of approaches you wish to use. The decision should be based on the various points just discussed, including the type of production, the level of precision editing required, the production budget, the technical facilities capability that you have, and the amount of production and postproduction time available.

SHOOTING SCHEDULE The shooting schedule is used to organize the sequence of taping during your production time. If you are shooting live on tape, the shooting schedule will be the same as the program's operational schedule and rundown sheet. If you will be shooting in segments or with a single camera, you must carefully develop a shooting schedule which will take into account each camera setup, the location, the set, the performers who are necessary, and the order or sequence in which you want to record them.

SCENE BREAKDOWN SHEET A scene breakdown sheet is a handy way to organize the thousands of details you need to keep track of when planning your shooting schedule. On the one shown in Figure 10-2, there is a place to indicate scene and shot numbers, performers, sets, props, miscellaneous items involved in the shot, and any other necessary information. Particularly if you are shooting in segments or out of sequence, organizing your script into different scenes and using the scene breakdown sheets will help to eliminate confusion or errors in production.

Some producers and directors like to write each scene and shot on a separate index card and spread them out on a large table or on the floor. Then they shuffle the cards around in different orders to find the most efficient shooting sequence. Regardless of how you develop your shooting schedule, be sure you have included all the scenes and shots and have estimated the amount of production time for each as accurately as possible.

VTR SCHEDULING FACILITIES You will have to

be certain that someone on the production staff has scheduled VTR facilities prior to your production date. You must estimate the amount of taping time you will need, any playback facilities necessary, and the amount of recording tape you want on hand for the production. Most facilities like to record a production simultaneously on two VTRs: one serves as the "master" copy; the second, as a "protection" copy in the event there are technical problems with the master recording. If you are using a protection VTR, be sure to double your tape estimate. Finally, remember to schedule recording and playback facilities only for the times you will actually need them. You do not need recording VTRs during your camera blocking and rehearsal period, for example, so be sure to schedule them only during those times when they will actually be used.

TAKE SHEETS Take sheets are indispensable in production and postproduction, as they identify each cut on the video recording, its running time, and whether or not it is usable footage. Prior to the production, a series of take sheets should be prepared, and a production staff member should be assigned to keeping track of all recording information on the take sheet. One copy of the take sheet should be kept in the VTR reel box; and another, in the program script file. (See Figure 10-3.)

SMPTE TIME CODE If you will be using SMPTE time code, you must decide whether to use the counter to reflect *time of day* which is the actual clock time, or *zero time,* in which the counter is started at zero at the top of every tape reel.

Time of day is convenient when you are recording an event in normal time, such as sports or news, where the actual time of day is used to index the action for later editing. Zero time is most useful when you are recording material that has no relation to clock time, but you want to control

BREAKDOWN SHEET

SHEET NO. _____

PROD NO. _____   TITLE _____

LOCATION _____

DAY NITE	EXT INT	NO PAGES	SCENE NO.	DESCRIPTION	NO.	CAST	LINES	COSTUME NO.

SHOT NO.

SET PIECES	ATMOSPHERE	EQUIP.

CARS – LIVE STOCK	SPECIAL EFFECTS	

FIGURE 10-2 **Breakdown Sheet.**
The breakdown sheet organizes all shots, performers, and props for every scene.
The producer and director use breakdown sheets to plan the most efficient
shooting schedule.

"Good" takes circled

Approximate starting time for each take (if using, time code)

For reference during editing

Running time of program material

PRODUCTION TITLE "BIOLOGY"

PRODUCTION NUMBER 5-32

VTR REEL # 60-108

VTR DATE JAN 29

SCENE/SHOT	TAKE	GOOD	NG	TIME	NOTES	Hr	Min	Sec	Fr	
7-A	1		X	:18	CAM OUT OF FOCUS	00	02	20	10	
	2		X	1:35	GLARE ON DEMO CLOSE-UP	00	04	15	18	
	③	X		1:45	ENDS ON WIDE SHOT		07	15	20	
12	①	X		2:10			10	55	25	
25	1		X	1:25	POSSIBLE CU FOR INSERT		14	15	00	
	②	X		2:38			18	30	15	

FIGURE 10-3 **Take Sheet.**

the time code numbers for index and logging purposes.

If you will be using time code, your take sheet should include the start and stop time code numbers for each cut. This will simplify locating the different cuts during playback and editing.

VIDEOTAPE PLAYBACK INTO THE SHOW

Often you will need to "roll in" or play back previously recorded tape segments into the show you are producing. To be certain that the right cut appears at the right time, prepare a cue sheet for the VTR operator, which identifies each cut on the insert play back reel and its proper sequence in the show. Every segment on the reel should have been recorded with an identification slate. Using your cue sheet, the VTR operator can match the slate information with the cue sheet. This means that even if the segments are not in order on the insert reel, they will appear in correct order during the show. Checking both the cue sheet and the ID slate will eliminate such errors as cuing the wrong take or cut, particularly on commercials or news clips, which often look alike without the slate and cue sheet information.

Setup and Rehearsal

During the studio setup, as the final touches are made before rehearsal, the VTR record and playback machines should be readied for use. If you are using a VTR to play back an insert into the

show, be sure that the assistant director, director, or production assistant has gone over the playback instructions with the VTR operator to eliminate confusion between the control booth and the VTR room.

The VTR room should also play back a test segment to the production control room to make certain that picture and sound levels are feeding properly. At the same time, the control room should send the recording VTR color bars and audio test tone, through the switcher and audio control console to check for outgoing levels and technical quality. It is better to learn about an incorrectly patched audio line or video problems at this stage than during the production.

Production

Since the production process will vary somewhat from show to show, depending on the kind of production and production technique being used, it is difficult to give you a single account of how the production process works. However, we can discuss a number of important functions which are standard to all recording sessions regardless of the production or the production technique. These include preparing video leader, playing back recorded material into a show, and some of the common problems which can develop during a taping session.

VIDEO LEADER Video leader contains all the technical and program information necessary to properly align the VTR machine and to identify the program or cut. The video leader includes: color bars, audio test tone, identification slate, and, in some instances, program cuing information.

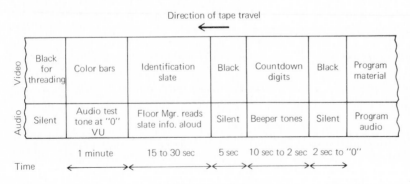

Direction of tape travel

FIGURE 10-4 **Videotape Leader.**
Reading from left to right, the tape leader should first have black for threading, color bars, the slate, and so on. The time line indicates approximately how long each element should be recorded.

FIGURE 10-5 **Slate.**
The slate is recorded prior to program material and is used to identify the program or various scenes and takes. *(Courtesy: Children's Television Workshop)*

COLOR BARS AND TEST TONE   Most VTR operators like to have at least one minute of color bars (or a monochrome test signal if recording in black and white) and audio test tone recorded at 0 VU at the top of the tape for playback alignment purposes. The tone and "bars" should be fed through the control room switcher and audio console just before actual recording.

SLATE   A visual slate is a must whenever you make a recording. Take sheets or box labels can be lost or mixed up, but the proper slate, recorded on the tape, is a reliable safeguard against playing back the wrong cut or a false start. The slate should contain the following information: (1) the name of the program, (2) the segment or series name and production number, (3) the recording date, (4) the air date (if known), (5) scene and take numbers, and (6) any special instructions. Most production facilities use an erasable slate, and it is the floor manager's responsibility to make certain that the correct information appears on it before each recording. If you have a character generator (CG) available, you can use the CG to create the slate information and leave the floor manager free to concentrate on the activities on the studio floor.

The slate is recorded immediately after bars and tone. To be sure that the slate is kept on screen long enough to be read, it is a good idea to have the floor manager read the slate aloud into an open mike. The aural information recorded is a double check for the VTR operator, when the tape is cued for playback or editing. (See Figure 10-5.)

PROGRAM-CUING TECHNIQUES   In order to play back a videotape into an existing program precisely or to begin a program playback on time, some method of accurately cuing the tape is necessary. A common visual cue is to use electronic or mechanical digits. These are numbers which appear on the screen in one-second intervals from "10" seconds to "2" seconds. At the same time each digit appears, a 400-Hz tone—sometimes called "beepers"—is automatically triggered and recorded on the sound track. The final two seconds are left blank and without audio to prevent accidental airing of the cue leader. Before a scene is recorded, the director calls for the digits and beepers which are recorded on the tape. Precisely at "zero," the director cues talent and the segment begins.

If such an electronic cuing device is not available, you can make a cheap substitute by splicing together a number of segments of 16mm film academy leader. This is the cuing leader which counts down in one-second intervals from "11" to "2." When you are ready to begin a recording, roll

FIGURE 10-6 Videotape Recording Procedure.

Time to Start of Segment	Video	Audio	Commands	Results
2:00	Black	None	DIR: "Roll tape for recording." VTR: "Tape is recording and locked."	Instructs VTR operator to start VTR recording. Machines are ready to record sound and picture.
1:30	Color Bars	0–db tone	DIR: "Bars and tone."	Color bars and audio tone are recorded for approx. one min.
0:30	Slate	FM reads slate into open mike	DIR: "Take slate."	Slate is recorded as FM reads information aloud into open mike.
0:10	Cuing Digits	Beepers	DIR: "Start digits and beepers."	Visual and audio cuing information is recorded just before program material begins.
0:02	Black	None	DIR: "Black and ready to fade up on——. Ready to cue talent."	Two seconds of black are recorded as a safety pad to prevent cue markers from inadvertently appearing on air.
0:00	Fade in	Audio Up	DIR: "Open mike. cue talent. fade in——"	Segment or program begins.
At end of show or segment	Fade to black	None	DIR/AD: "Ready to stop tape . . . stop tape."	Tape is run in black for a few seconds as a safety pad. VTR should be stopped on command, and VTR operator spot-checks and OKs recording before cast and crew are released.

the leader through the film chain, and punch it up on the air. If you want audio cue tones to accompany the numbers, have the audio engineer pulse a tone each second as the numbers change. Just as with video digits, the numbers stop at "2" seconds. At this point, fade to black, count down two seconds to "zero," and fade up both video and audio.

Figure 10-6 shows the various operations which must be done before recording a program or segment. The time column shows the approximate time before the segment begins, the video and audio columns show what is happening in each of the respective areas, and the command column shows the directions given by the program's director or assistant director in the control booth and what is heard, via the talk-back system, from the VTR room. The last column describes what is happening and why.

Remember that the cues are only as accurate as the director is in starting the program's audio and video precisely on time. If you start the show sooner or later than the "0" mark, the cuing system will be off, which can result in a miscue during playback.

STOP-START RECORDING If you are recording in segments or with a single camera, you will naturally be stopping and starting the tape each time you begin a new take. In this case, it is very important for the take sheet and slate to have corresponding information. Make sure that whoever is filling out the take sheet reminds the floor manager to change the slate at the start of each new scene or take.

To make locating different cuts easier, particularly if you are not using time code numbers, have the audio engineer add tone, or "beepers," right after the floor manager reads the slate aloud. During fast forward or rewind, the tone makes a distinctive sound. By counting the number of tones you can determine how many cuts you are into the tape.

Once you have recorded a take you *think* is usable, you may want to view it to be *sure* it is usable before going on. Often actors and crew can improve their performance by watching the playback. But beware of the danger of spending too much valuable studio time viewing playbacks and running out of production time at the end of the day.

VIDEO PAD Whenever you are taping for editing —either in segments or with a single-camera/ single-VTR—be sure to record a *video pad* at the start and end of each take. The pad is simply an overlap of the action, which will give you more edit room during postproduction. You may find that the particular edit point you had planned to use looks wrong after you have previewed the edit. If you overlapped the action, you can use the video pad to locate a better edit point. Overlapping the action also helps performers gain the momentum necessary to reach the proper intensity level to match their preceding performance.

If you have enough tape stock available, most directors would encourage you to tape everything except the first few rehearsals. Even though the scene may not look ready, you might get that perfect performance or an unexpected shot that is a one-time occurrence. If you have it on tape, you can use it later in editing; but if it was never recorded, it is lost forever. There are some risks in extensive taping, however. Sometimes actors and crew adopt the attitude, "If I make a mistake, so what? We'll just take it again." That sort of approach can result in sloppy technical production and poor performances. Also, too many retakes can drain the cast and crew, resulting in a worse recorded performance than you had achieved in earlier rehearsals. It is important for the director and producer to be aware of these problems and to weigh the value of taking the scene "just one more time" against the fatigue and carelessness which may result when you have honed the cast and crew past the fine edge to a dull and lifeless performance.

CONTINUITY Recording a show in segments, particularly if you are taping out of sequence, means that you must be particularly careful about continuity problems. Costumes, props, actor positions, and set pieces must be kept constant, so there is a natural progression when the different shots or segments are finally edited into the composite program. Even small details which went unnoticed during the production can be ultimately distracting to the audience.

Usually the production assistant assigned the job of supervising continuity keeps a careful record of the dress and costumes, props, sets, and actor and camera positions for each shot or setup. Some production units use inexpensive Polaroid cameras to make a snapshot of each scene for continuity reference.

COMPLETING THE RECORDING Once you have completed recording a program or segment, go to black and continue recording for an additional fifteen or twenty seconds. This black pad serves as a safety cushion in the event the program switcher does not cut to another video source immediately after the tape playback material is finished. Finally, be sure to let the VTR operator know to "stop recording." Many VTR operators are instructed to let the machine continue recording, even if black is on the line, until they are given specific instructions to stop.

VIDEOTAPE PLAYBACK OF RECORDED MATERIAL Often you will need to play back a previously recorded segment, such as a commercial, news spot, or program segment, into a production. In order to make good use of such insert material, the director, assistant director, and VTR operators must coordinate their operations so the transition into and out of the taped segment is executed smoothly.

PREROLL CUING Since most VTRs need a certain amount of running time to lock and stabilize picture and sound, the VTR operator must run the tape backward for a few seconds before the start of the actual program material. The amount of preroll you will need depends upon the VTRs in your studio and the facility's standard operating policies.

Each of the insert segments should already have cuing leader recorded prior to the program material. Using the leader or the VTR's tape timer, the operator will cue each cut to the agreed-upon preroll time. For example, if you will use a ten-second preroll cue, the tape will be rewound ten

seconds before the first sound and picture at the start of the program material.

Once you know what your preroll time will be, you can coordinate with talent where in the script to begin the roll cue. If the show is fully scripted, locate the word which is ten seconds (if that is your roll cue time) before the end of the last sentence. This becomes the roll cue, and, once the word is said by talent, the director calls for the VTR operator to roll tape. The talent should finish speaking ten seconds later just as the stabilized VTR playback begins the program material.

If your program is mostly ad-lib, without a fully written script, you may want to write out the tape lead-in on a cue card so that you and the talent will know precisely how long the tape introduction will run. On some very spontaneous shows, such as news programs or sports coverage, you might use a "bumper" before the tape playback. A bumper is a title card or slide which appears between the end of the talent's lead-in and the start of the pretaped segment. For example, on most sports coverage, the bumper is the game score, which is punched up prior to the tape roll. As soon as the director cuts to the bumper, the VTR operator is instructed to roll tape, and the director cuts or fades to the VTR playback as soon as the picture and sound appear.

Figure 10-7 will give you an idea of how the videotape playback procedure works.

Postproduction Editing

The complexity of the postproduction session depends mainly on the production approach you used to record your program. Shows produced live on tape or in long segments will usually need little, if any, postproduction editing. However, shows produced in short segments, with a single camera or with the multiple-camera/multiple-VTR approach will require an extensive postproduction edit session since almost every transition must be made with a video edit.

SCREENING THE FOOTAGE The first step in postproduction is to screen all the material you have recorded. Using the take sheets from the production, start to view those scenes which were marked either "good" or "maybe." Those which are definitely "no good" usually need not be

FIGURE 10-7 Videotape Playback Procedure.

Script	Director's Commands	AD's Commands	Results
WE'LL BE BACK WITH OUR SPECIAL GUEST IN (ROLL Q) A MOMENT BUT FIRST, LET'S PAUSE FOR SOME IMPORTANT COMMERCIAL MESSAGES	Ready to roll VTR Roll tape	Five .... four .... three .... two	VTR operator is ready. Floor manager gives hand signals corresponding to AD's countdown.
VTR COMML #1 (1:00)	Fade to black Fade up on tape	one	Floor manager counts down to zero and gives "cut" sign. Switcher fades out on picture and up on VTR, as audio engineer cuts studio mikes and brings up VTR sound track on audio console. AD begins stopwatch as VTR segment begins.
		Thirty seconds to end Twenty seconds.	Floor manager readies studio.
	Ready on the floor Ready to cue talent and fade up on Camera 2.	Ten seconds ... Nine ... eight ... seven .... six .... five .... four ... three ... two	Floor manager begins five-zero silent hand cue.
	Fade to back Cue talent; fade up on Camera 2.	one	Floor manager finishes countdown and cues talent on director's command, as switcher fades to black and fades up on CAM 2. Audio engineer fades out VTR sound track and fades up studio mikes.

viewed unless you think they might contain a good shot or two, which may be usable if edited into another take.

Locating the particular cut on the tape for each take or segment will go quickly if you have indicated on the take sheet either the approximate time into the tape of each cut or the SMPTE time code numbers. In addition, the audio beeper cue tones, which precede each cut, make a distinctive sound as the tape is run fast forward or reverse. Counting each audio blip as you hear it will tell you how many cuts into the tape you are.

If possible, use a helical copy of the original footage for previewing and screening purposes. Even if you will be editing on-line, the convenience of actually operating the playback machine as you view each segment will make the viewing session go a lot faster. In addition, the stop-frame and jogging capabilities of many helical machines allow the director and editor to evaluate their edit decisions frame by frame.

If you will be editing your master tape with SMPTE time code and a computer-assisted editing system, you should have requested a helical cassette copy of the footage with burned-in time code. This will permit you to screen the footage with time code numbers on any conventional VCR and monitor. Noting takes or potential edit points by time code will make your on-line editing go much faster.

Once you have viewed the various takes and decided which ones you will use, watch the program material again, this time making some rough editing decisions. Many directors suggest you keep a fresh copy of the shooting script to use as an editing script, marking all edit decisions and possible alternative cuts in case the edit you planned looks wrong onscreen.

If you are working on a production where there are a number of different takes of the same scene, such as a multiple-camera/multiple-VTR show or a production which taped both the "dress" and "air" shows, as in many situation comedies, it is a great convenience to have a couple of helical playback decks available. This allows you to compare two different takes directly or to preview quickly edit points as you watch. If extra machines are unavailable, simply go through the script, taking a few lines of dialogue or action at a time, look at all the takes you think look good, and determine which complete take or combination of different takes can be used to build the sequence.

EDITING ON-LINE If you will be editing on-line, that is, actually editing the master tape as you go along, you will generally start at the beginning of the program and proceed through to the end, scene-by-scene and shot-by-shot. If you have used SMPTE time code during screening, you should have already prepared a rough edit decision list indicating the various edit points and their time code numbers. If you are working with a computer-assisted system, you will actually be making the final edits as you move along through the show. Of course, if you are editing manually—without time code or computer—you will have to preview each segment to locate and identify the edit points, using your take sheet and the VTR tape timer as a guide and watching the playback monitors for the precise edit point.

EDITING OFF-LINE If you are editing off-line, you will be working with a helical cassette copy of the master footage in order to prepare time code instructions for later assembly. There are a number of editing options available for off-line editing, depending upon the complexity of the editing job and the equipment you have available.

EDITING ON PAPER The simplest off-line approach is to use a helical cassette with burned-in time code and edit "on paper" by simply noting the time code numbers for each edit point. Using the jog and pause controls, you can stop the tape in freeze-frame in order to copy the exact time code numbers where each edit is to take place. Many directors find that simple editing tasks go quickly and efficiently by using this method.

EDITING A WORKPRINT ROUGH-CUT When you are dealing with a more complex production, however, such as a single camera shoot or where multiple camera/multiple VTRs are used, the paper-and-pencil approach is usually not satisfactory because the individual shots are too short to give the director any real sense of visual continuity. In this case it is preferable actually to produce an edited workprint "rough-cut," which is used to guide the editing of the master tapes on-line. If you have a sophisticated computer-assisted, off-line system available, the computer will not only help you to make the edits, but will also keep each edit decision in its memory, eliminating the need for the director to engage in complicated bookkeeping for each time code edit point. However, you can edit footage without such a computer system and still provide a completed rough-cut with the necessary time code information to guide the automatic assembly operation of the computer system during on-line assembly.

To do this, all you need is a helical copy of the master footage with burned-in time code and a conventional helical edit system such as the kind used for ENG editing. You simply edit together the program without regard to the time code (since you aren't using a time code reader, you cannot use the time code for edit control anyway). Once you have completed the rough-cut to your satisfaction, it is a relatively simple job to note the time code numbers for each segment and edit point and develop the edit log which will be used to control the automatic assembly of the master tapes on-line.

Conforming the edit decision list or the computer-assisted workprint to the actual master editing is a completely technical process, and, quite often, neither the director nor the producer is present during the on-line edit session. Often the assistant director will coordinate the master tape editing since all creative decisions have already been made.

COORDINATION SESSIONS The usual VTR editing system of a playback and an edit/record VTR limits you to cut-only edits. If you wish to use any other type of video transition—dissolve, super, wipe, fade, or even a chroma key insert—you must arrange for a VTR coordination session.

A *coordination session,* or "coord," means that you play two or more playback VTRs through a program switcher and then into the edit/record machine. The transition or key is executed with the program switcher exactly as if you were using

two or more studio cameras during a regular production.

As an example, let us say we want to dissolve from a comedy sketch into a close-up of a guitar as a singer begins a song. The first thing you must do is to prepare two separate playback reels: an A reel and a B reel. The A reel, or "roll," as it is often called, contains the last shot of the comedy sketch, and the B roll contains the close-up shot of the guitar as the song begins.

Next, decide how long you want the dissolve to last. It is a lot easier if you are using time code since you can describe the length of the dissolve in terms of video frames. Say we want a dissolve which lasts about a second and a half. That would be forty-five video frames. Using time code makes it a relatively easy matter to precue both the A and B tapes so that, when they roll, there will be a forty-five-frame overlap between the end of the sketch on VTR A and the start of the song on VTR B. Both machines play through a switcher, and the edit/record VTR records the transition as you make the dissolve between A and B machines.

The same technique is used for any multiple-image effect, such as wipe, split screen, or chroma key. The important thing is to determine how long you want both scenes to overlap and to precue the tapes so you can produce the effect through the switcher for the desired length of time.

While time code makes coords run more smoothly, it is not necessary in order to do one. If time code is unavailable, you will probably have to experiment a few times until you find the right combination of preroll cuing between the different tape machines to produce the effect you want.

Computer-assisted systems make coord sessions relatively painless since the computer controls the entire operation—the cuing and rolling of up to six playback VTRs, the operation of the program switcher and the transition or effect, and the operation of the edit/record VTR—according to the preprogrammed information in its memory. (See Figure 10-9.)

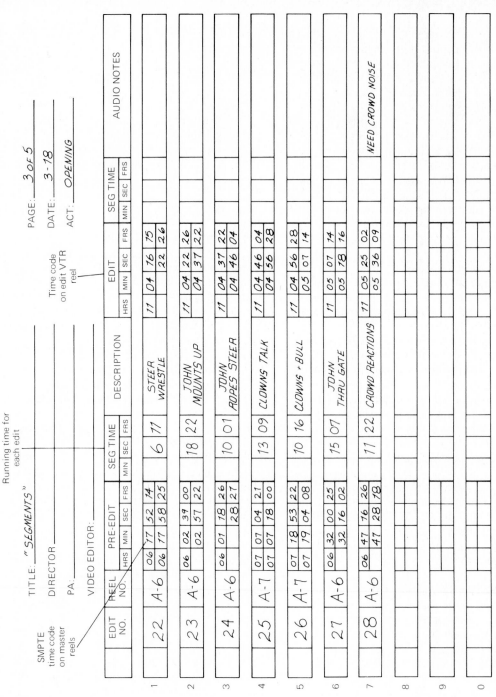

FIGURE 10-8 Time Code Editing Sheet.

By using time code, a program can be edited on paper, using off-line equipment. Once the edit points are indicated by time code numbers, the actual on-line editing of the master can proceed quickly.

Digital video-effects controls

CMX keyboard

Video monitors

Computer-editing CRT display

Audio-control console

Video switcher

FIGURE 10-9
Postproduction Room.
For elaborate postproduction operation, a postproduction facility enables editing, coordination sessions, and audio sweetening to be performed without the need for an expensive studio control room. *(Courtesy: ABC-TV)*

AUDIO MIXING AND SWEETENING One of the most significant production advantages of a computer-assisted editing system is its ability to synchronize a multi-track audiotape recorder (ATR) to permit you to work independently with the program's audio. By using an ATR which is controlled by the same SMPTE time code which is used for VTR operation, it is possible for the director and the audio engineer to add additional audio during postproduction, to correct or enhance the sound quality of the original audio, and to "sweeten" the sound track by remixing the audio to produce the best possible sound.

What is especially important about postproduction audio mixing is the fact that it eliminates the need for the program's director to concentrate on the audio portions of the program during production. At the same time, postproduction audio mixing enables the audio engineer and director to devote full attention to the audio requirements of the production without the pressure of a studio full of cameras, crew, and talent waiting while the sound mix is attended to.

For maximum postproduction audio flexibility, a multitrack audiotape recorder is necessary. Most production facilities are using 16 or 24 track ATRs, although 8 track and 32 track machines are occasionally used as well. Obviously, the greater the number of audio tracks, the more flexibility the audio engineer and director have to work with

during postproduction. We've already discussed how the ATR is synchronized with the operation of a VTR using time code in Chapter 9. Now we will see how to use the equipment during the postproduction mixing process.

The first step is for the program director and the videotape editor to build an edited workprint, using an off-line computer system and a helical cassette machine. The actual sound which was recorded during the production is used as an audio reference but with little regard at this point for how well the audio edits might sound. What is of primary concern are the video transitions since audio problems can be corrected during the postproduction audio sessions. Once the edited video workprint is completed and approved, work can begin on the audio track by using the SMPTE time code as a reference guide.

If the original program audio was recorded only on the production VTR, the sound track must be dubbed to one of the tracks on the multichannel audio recorder. At the same time, one of the ATR's tracks are used to record SMPTE time code so that the audio track can be played back on the ATR in perfect synchronization with the video. On some productions the original program audio may have been recorded on a multitrack recorder at the same time the videotape was recording picture. In this case, both VTR and ATR will already have time code and can playback in perfect sync.

Now the director and the audio engineer must select and assemble all the additional audio material they will need for the production. This includes voice-over narration, dialogue which is to be dubbed in place of existing audio, sound effects, and music. For each segment in the program where the director decides that additional audio is necessary, the segment must be timed to determine how much audio will be required to cover the video. For example, a car-chase sequence might need sound effects of squealing tires, sirens, an automobile engine, and music. Timing this sequence off the original VTR time code shows that it runs thirty-seven seconds, so at least that much audio is gathered from various sources—live recordings, sound effects tapes and records, and so on—and recorded onto conventional ¼-inch audiotape.

Once every additional sound is assembled and recorded onto audiotape, we are ready for the next step: the *audio laydown* session. At this point, we record or "lay down" each of the individual sound elements onto one of the respective audio tracks on the multichannel ATR at the appropriate point in time where it will be needed during the audio mix. To do this, we must develop a *track log,* which will assign each discrete track to a particular sound function. For example, in this session we decided that Track #1 would hold the original dialogue which was dubbed from the VTR to the ATR. Tracks #2, #3, and #4 carry sound effects, Tracks #5, #6, and #7 contain music, Track #8 holds voice-over narration, and Track #15 carries the SMPTE time code signal. The reason we assign three tracks to sound effects and music is to permit the audio engineer to cross-fade from one music or sound effects track to another. We have left a number of tracks blank to permit us to record the final mix-down which will become the program's audio track (See Figure 10-10).

During the layback session we use the SMPTE time code to locate the exact point on the ATR where we will need each sound sequence. Every audio segment was prepared with additional ma-

terial before and after its time code number to provide a safety pad in case we want to fade the sound in or out. Also, we record every track at full level since the mixing and balancing of the sound will come later.

By the end of the audio layback session, each track contains sound material corresponding to the video action at the particular time in the program. For example, say a car-chase sequence begins at 00:18:27:09 in time code and ends thirty-seven seconds later at 00:19:04:09. On the multitrack audiotape, the sound effects and music for the chase sequence are located on their respective tracks directly between the two time code numbers.

At this point, we are ready for the final audio mix. As the director and the audio engineer go through each sequence of the show, the various sound tracks are mixed and balanced. Reverberation and equalization are applied where necessary to improve the sound quality or to provide the proper sound perspective. Cross-fades are timed and executed until they deliver the exact effect desired by the director. Usually, the director and editor will work with only a short sequence at a time, perfecting the sound mix until it is just right and then recording the completed mix on one of the empty tracks of the ATR. In this way, the final audio mix for a program is assembled piece by piece, much like assemble editing of video. Working this way in postproduction, the director and the audio engineer can concentrate on the sound and work with various mixes and sound balances to achieve the perfect combination.

The *audio cue sheet* is used to indicate the various levels for each fader on the control console and the point where fades and other audio cues should begin and end. Following the audio cue sheet, the audio engineer records the final mix as the program proceeds. The output of the audio console is fed onto the empty track on the multitrack audio recorder. If a stereo mix is necessary, two tracks are used to record the stereo mix from the audio board.

Once the audio mix is recorded, the master production tapes are first edited to conform to the time code or the edited workprint for video only, and then the mixed audio track or tracks are added, using the SMPTE time code to assure audio and video synchronization.

FIGURE 10-10 Audio Mix Track Log.
When editing and mixing audio with a multitrack audio recorder, a cue sheet such as this is used to identify the material on each of the audiotape recorder's tracks and to indicate to the audio engineer exactly how to mix each of the separate tracks in order to build the final program audio. In this example, the blank tracks will be used to record intermediate mixes, which are later remixed until the final program audio is produced.

AUDIO MIX TRACK LOG

Prod. No. **31-8**   Title "**THIEF**"   Director **HENRY**   Audio Eng. **ALLISON**

AUDIO TAPE TRACKS

1	2	3	4	5	6	7	8	9	10	11	12	13	14	15	16
ORIG. VTR AUDIO	SFX "A"	SFX "B"	SFX "C"	MUSIC "A"	MUSIC "B"	MUSIC "C"	V.O. NARRATOR							TIME CODE	
17:32:18 →	CUT #6 CAR CHASE 18:27:09 →	SIRENS 18:52:09 → 18:59:10	GLASS CRASH 18:55:16 → 18:57:21	"FRENZY" CUT #2 18:27:09 →			18:27:09 → 18:59:10								
	→ 19:04:09			→ 19:04:09	"ESCAPE IN NIGHT" CUT #5 FADE X 19:00:12 →		19:04:09 → 20:35:19								
→ 20:35:19					FADE OUT 20:35:19										

Admittedly, such an elaborate postproduction audio procedure is time-consuming and expensive and requires sophisticated equipment. However, it is being increasingly used for a variety of productions since it offers enormous creative and production flexibility and ensures a top-quality final production.

EDITING ENG FOOTAGE Editing ENG news footage utilizes the same approach as editing any single-camera/single-VTR production. The editor must literally build a story by editing and sequencing a number of different video takes since every cut is, in fact, a video edit. However, the special nature of news makes ENG editing a somewhat different procedure in that the timeliness of the material requires editing decisions and the editing process to be completed quickly and efficiently in order to take maximum advantage of the speed which ENG offers the news production team. In Chapter 12, "Electronic News Gathering Production," we will discuss in detail some of the special approaches which are commonly used in editing ENG tape footage.

SUMMARY

A program can be produced with videotape in four ways: (1) live on tape, (2) recording in segments, (3) using a single camera/single VTR, and (4) recording with multiple cameras and multiple VTRs. Each of these production approaches has its own set of advantages and disadvantages, and the particular production requirements will determine which approach or combination of techniques will provide the most creative control and production efficiency.

During the preproduction phase, the director and producer must plan a shooting schedule. This is particularly important when shooting in segments or with a single camera since nonsequential shooting can often result in increased production efficiency.

During recording sessions, video leader must be laid down at the top of each tape. This usually includes color bars and tone, an identification slate, and cuing information. Usually a combination of audio beepers and video countdown digits is used to roll in VTR program material accurately. Most VTRs require a preroll to run up to speed, and this must be carefully coordinated between the director, talent, and the VTR operators.

During postproduction editing, the program can be edited either on-line —using production recorders and making edits as you go along—or off-line—using a helical workprint to make all editing decisions. Coordination sessions are necessary when you wish to use more than one playback VTR for such transitions as dissolves, wipes, supers, and chroma keying. During postproduction, the audio can be remixed and sweetened to improve its quality and to add additional sound, such as narration, dubbing, music, and effects.

CHAPTER 11

THE SWITCHER AND SPECIAL ELECTRONIC EFFECTS

A television show is the result of many different pictures and images, which are combined to create the videospace. The *production switcher* is the equipment which enables us to select from the many different camera shots, videotape playbacks, slides and film, character generators, and remote feeds, which are available. Yet controlling which video source will go on the air is only the beginning of the switcher's production capabilities. The switcher, combined with its *special effects generator,* can shape a picture's size and dimensions, combine a number of video sources to produce a composite image, and, with the addition of more sophisticated equipment, produce a visual reality which exists only in the videospace.

In this chapter we will first discuss the functions and operation of the switcher and special effects generator and then will cover some of the special electronic effects used in production.

THE PRODUCTION SWITCHER SYSTEM

The production switcher is really a multipart system which includes the program switcher, special effects generators, and an array of television monitors, which display both incoming video as well as the switcher's output.

Control Room

The production switcher is located inside the control room, where the program's director and other key personnel control the production operation. (See Figure 11-1.)

PRODUCTION CONSOLE In most control rooms, the switcher is mounted on a long desk called the "production console." Seated at the production console are the switcher operator, the program's director, the assistant director, and such other production personnel as the producer and the production assistants.

THE SWITCHER OPERATOR The crew member responsible for operating the production switcher is known variously as the "switcher" or the "technical director" (TD). Although, strictly speaking, the term "technical director" refers to the senior technical crew member responsible not only for operating the switcher but for all the technical aspects of a show, the terms "TD" and "switcher" are often used interchangeably.

At some studios there is no TD or switcher assigned, and the program's director is expected to do the switching. Since the operational and personnel situations vary so widely from studio to studio, we will follow our usual practice of referring to the crew position in terms of its production

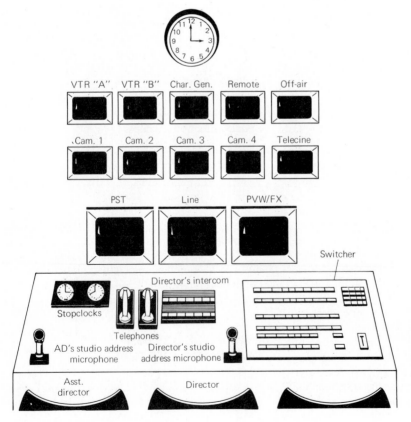

FIGURE 11-1 **Studio Control Room.**
The layout of a typical studio control room. Although control rooms may vary slightly, the basic equipment and how it is positioned are relatively standard.

role. Whether the person operating the switcher is a technical director, switcher, or the program's director, the functions and responsibilities associated with the job remain the same.

The switcher sits in front of the switcher control panel where he or she can operate all the controls and see the various monitors which face the control console.

THE PROGRAM'S DIRECTOR The director sits in the middle of the production console next to the switcher. From that position, the director can see all the video monitors and can communicate directly with the switcher operator.

On the production console, in front of the director, are a number of communication systems. A microphone and switch are a part of the studio address (SA) system, which enables the director to speak to the entire studio floor using overhead loudspeakers. During rehearsal and production, when mikes are live in the studio, the headset private line (PL) is the director's main communication link to camera operators, the floor manager, and other key production personnel.

MONITOR DISPLAY Facing the production console are a number of black and white television monitors, which display the picture output of every video source available to the production. There is a separate monitor for each camera, VTR playback machine, character generator, telecine camera, remote feed, and so on.

In addition to the smaller black and white monitors, there is a larger monitor (color, if the studio is so equipped) labeled "program," or "line." This monitor shows the final output of the video switcher, and indicates which picture is being fed to the station transmitter or being recorded on videotape. In some control rooms there are additional larger monitors labeled "preview," or "preset." We will explain how these monitors are used later on.

A *tally light* on each monitor lights up whenever that video source is fed on the air. If two or more sources are combined to produce a composite picture (such as a superimposition), a tally light will show on each of the monitors whose pictures are involved in the effect. The tally lights are important because they remind the director and

crew which of the many picture sources is being fed out of the studio at any time.

Production Switcher

The production switcher plays a central role in control room operations. At first glance, some of the more elaborate switchers, like the one in Figure 11-2, can look terribly intimidating, having literally hundreds of buttons, switches, and controls. Actually, all switchers operate in pretty much the same way, and once you understand the concept of switching, it is easy to learn to use any switcher, regardless of how simple or complex it is.

The many different models of switchers available make it impossible to provide step-by-step operating instructions. Instead, we will use a "typical" switcher layout to illustrate the basic principles of switching and the switcher's production capabilities. First, we will start with the simplest switcher sysem possible and then will add on to our basic model as we progress from simple cuts to more elaborate transitions and effects.

Program Bus

Every video source must be represented by its own, individual button on the switcher console. If, for example, a studio had three cameras, a character generator, a film chain, and a videotape playback machine assigned to it, the switcher would require a minimum of six input buttons plus another for video black. Actually, most production facilities install switchers with a larger number of inputs than they actually need to permit the addition of more sources as they expand the operation or acquire new equipment.

The single row of buttons you see in Figure 11-3 is called a *bus*. Regardless of how many buses a switcher may have, each bus is identical in that it has the same number of buttons, arranged in the same sequence and corresponding to the same video sources.

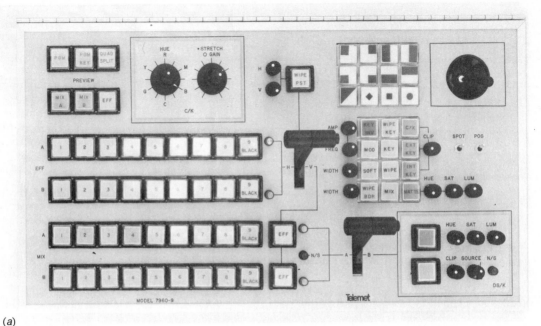

(a)

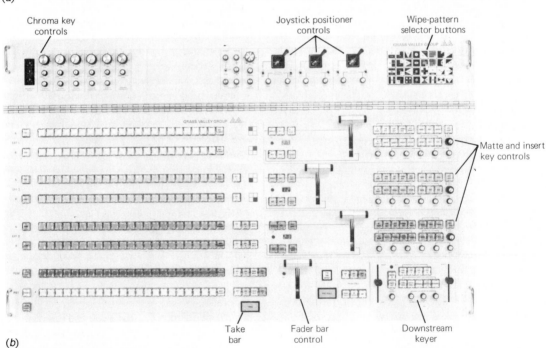

(b)

FIGURE 11-2 Video Switcher.
(a) Example of a small video production switcher; (b) A large-scale production
switcher. (Courtesy: Telemet, a Geotel Co. and The Grass Valley Group, Inc.)

BLACK		CAM 1	CAM 2	CAM 3	FILM	VTR	CHR GEN

FIGURE 11-3   **Program Bus.**

The function of the program bus is to determine which video source leaves the switcher to go out over the air. You might think of the program bus as similar in one respect to the master fader on the audio control console. Whatever video source is punched up on the program bus is the image which appears on the line monitor and leaves the control room to be recorded or broadcast. Of course, our simple switcher makes this appear quite obvious because there is only one bus. However, later on, as we add more buses to the switcher to permit us to do more, the program bus still functions the same way. Regardless of what else may be punched up on any other buses on the switcher, unless it has been routed through the program bus, it will not appear on the air.

CUTS   A cut is an instantaneous switch from one video source or camera shot to another. The cut is the most direct method of switching between sources and is usually the least noticeable transition between shots. The program bus illustrated in Figure 11-3 allows us to make cuts by simply punching up whatever video source is desired. If you want to see Camera 1 on the air, press the button labeled "CAM 1." If you want to see film on the air, press the button labeled "FILM." We can start our show in BLACK and end it the same way, but the single bus in the figure will permit us only

to cut between the various sources and video black. In order to produce more complex transitions, we require an additional pair of buses.

Mix Buses

As you can see in Figure 11-4, we have added two identical buses with a fader bar control between them. You will also notice that we have added another button next to the PROGRAM bus. We will now look at each of these new additions individually.

PRIMARY AND SECONDARY SOURCES   In the first example, when we had only a single row of video inputs, our switcher contained only *primary sources*, that is, those video inputs which actually produce a picture or image, such as cameras, VTRs, character generators, and so on. The button labeled "MIX," which we have just added to the switcher, is a *secondary source* because it represents the output of the two MIX buses below. You will remember that unless a video source appears on the PROGRAM bus, it will not go out over the air. The secondary source button, labeled "MIX," enables us to send the output of the MIX buses through the PROGRAM bus as though it were any other video source. In a moment you will see why secondary sources are important

FIGURE 11-4   **Program and Mix Buses.**

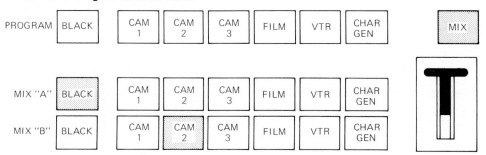

311

when we wish to do more complex switching and transitions.

MIX BUSES The two buses labeled MIX "A" and MIX "B" consist of the identical primary sources on the PROGRAM bus. As you can see, CAM 1 appears in the same position on all three buses, FILM appears in the same position, BLACK appears in the same position, and so on. This eliminates any possible confusion when working with any of the switcher buses since the same video source appears in the same location on every primary source bus.

FADER BAR What makes the MIX A and MIX B buses unique is the fader bar control positioned between them to their right. Actually, the fader bar consists of two handle controls, which are *ganged,* or locked together, to travel as if one. The fader bar is, as its name suggests, a potentiometer, such as those used on audio control boards or lighting dimmer boards. Using the fader, you can vary the video output of either bus over a continuous range from 0 to 100 percent signal strength.

Whenever the fader bars are in the upward position, adjacent to the MIX A bus, that bus is activated. Whichever video source is punched up on MIX A will feed the PROGRAM bus and, assuming MIX is also punched up on PROGRAM, will go out over the air. Whenever the fader bar is

in the bottom, or MIX B, position, the MIX B bus is activated. If, for instance, the fader control is adjacent to MIX A, only sources punched up on MIX A will appear on the line monitor. Regardless of which source is punched up on MIX B, none of the MIX B sources will appear because the bus is inactive at the time.

By now it should appear obvious that the fader bars are the key to producing all "mix" transitions between video sources. Using the two MIX buses and the fader, we can produce three basic transition effects: the *fade,* the *dissolve,* and the *super.*

FADE A fade is a gradual transition from black into a picture or from a picture into black. In order to produce a fade, we require two video sources: (1) the image source, and (2) video black. For example, assume we wish to fade in from BLACK to CAM 2:

1 Be sure MIX is punched up on the PROGRAM bus, so its signal will feed the line monitor and studio output.

2 Punch up BLACK on MIX A.

3 Preset CAM 2 on MIX B.

4 On the director's command, move the fader handle from MIX A to MIX B. The program monitor will show the fade-in as you move the fader bar controls toward MIX B. The speed of the fade is variable, depending on how quickly or slowly you move the fader controls (See Figure 11-4.)

DISSOLVE A dissolve is a simultaneous fade-in of one picture while fading out another. The

FIGURE 11-5 **Switcher Set for a Dissolve.**

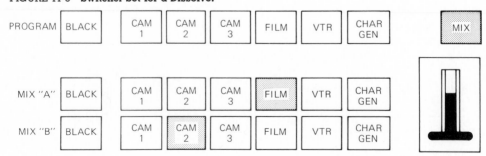

dissolve is produced exactly like a fade-in or fade-out but, instead of using BLACK, we preset another video source. For example, say we are on CAM 2, and the director wishes to dissolve to FILM (See Figure 11-5):

1 The fader handle is already on MIX B with CAM 2 punched up. This is the source going out to the program line.
2 Preset the upcoming source—FILM—on MIX A. Since the bus is inactive, it will not show on the line monitor.
3 On the director's command, move the fader bar from MIX B to MIX A. As it begins to move you will see the FILM picture begin to appear faintly on the line monitor. By the midpoint of the fader bar's travel, both CAM 2 and FILM have transparent images, which share the videospace equally. As the bar moves closer to MIX A, the FILM source becomes visually prominent and CAM 2 looks progressively weaker. At the end of the fader bar's travel as it reaches the MIX A position, FILM accounts for 100 percent of the signal strength, and the CAM 2 picture on MIX B totally disappears as the MIX B bus becomes inactive. As with the fade, of course, the speed of the dissolve is variable and dependent on how quickly you move the fader bar from one position to the other.

It makes no difference which MIX bus you start from and which you end on; just remember to be sure to preset the proper video source *before* you begin to move the fader. Even the slightest fader movement will show whatever is punched up on the other MIX bus.

SUPER A superimposition, or "super," is simply a dissolve that has been stopped midway. Usually a super is produced at the midpoint in the fader bar's travel between the two buses, but this can vary depending on the picture brightness of the two sources. A darker picture will require more signal strength than a lighter one. Most supers must be set by watching the effect as it is produced on the video monitor and setting the fader

bar at the position which offers the best visual mix. While you are in the super position, you can *undercut,* or change, either video source on MIX A or MIX B, while the super effect is on the air.

The command for a super is: "Super CAM 2." If you want the supered image taken out, the command is: "Lose super." If you want to keep the supered image and lose the previous source, the command is:"Take the super through to CAM 2."

OPERATING PROGRAM AND MIX BUSES
Using the simple switcher we have illustrated so far enables us to produce four transitions or effects: cuts, fades, dissolves, and supers. Many programs can be switched entirely on the two MIX buses with the secondary MIX button depressed throughout the entire show on the PROGRAM bus. For cuts, simply cut between sources on either MIX A or MIX B. For fades, dissolves, and supers, use the fader controls and preset the upcoming source on the inactive bus.

USING SECONDARY SOURCE CONTROLS

There are some situations, however, when you will need to utilize both the MIX buses as well as the PROGRAM bus. For example, say you are on FILM and the director wants to cut to a super of CAM 1 and CAM 2. Using the switcher we have illustrated the only way to do that is:

1 As soon as possible, punch up FILM on PROGRAM bus. Since FILM is already on the air via the MIX bus, the audience will not see any change, but the switch does free you to work with the MIX buses while FILM feeds PROGRAM.
2 Preset your super, using MIX A and MIX B busses and the fader handles.
3 On the director's command, punch the MIX *secondary* button on the PROGRAM bank. On the air, the switch will show a transition from FILM to a super of CAM 1 and CAM 2.

Preview Bus

In the example just mentioned, where the director wished to cut to a preset super, it would certainly be helpful to see what the super looked like before it went on the line. Without a preview ability we would have to take our chances that the effect which was set up blind would be acceptable. To enable us to look at an effect or, in fact, at any video source before it is punched up, most switchers have a PREVIEW bus, which is used to feed the large monitor labeled "preview."

As you can see in Figure 11-6, we have added a PREVIEW bus to our switcher. Both the primary and secondary buttons on the PREVIEW bus are identical to those on PROGRAM. Now we can preview any single video source or any MIX effect by punching it up on the PREVIEW bus before taking it on the air. For example, if the director wanted to cut from FILM to a super of CAM 1 and CAM 2 that we mentioned earlier, the switcher could punch up MIX on the PREVIEW bus and adjust the fader bar to produce the exact video mix the director wanted. Once the director approves the super, it can be put on the air as soon as it is called for.

Mix/Effects System

Although the MIX and PROGRAM buses are capable of producing the basic transitions and effects we need for the simplest of shows, modern television production demands far more video flexibility. At the very least, we ought to be able to produce composite effects using multiple picture sources. To do this, we must add a *mix/effects* system to our switcher.

A mix/effects system consists of: (1) a pair of mix/effects buses and fader bar control, and (2) a special effects generator. Using the system, we can increase our switching repertoire to include wipes, split screens, and keys. Before we discuss how to produce these effects, however, let us take a look at our new, expanded switcher. (See Figure 11-7.)

You will notice that we have added two additional buses labeled "MIX/EFFECTS 1" and "MIX/EFFECTS 2," a fader bar control between them, and a special effects panel with a variety of buttons and controls. Our secondary source array has also increased to include an EFFECTS button, which now appears on PROGRAM, PREVIEW, and importantly, on both MIX A and MIX B buses.

WIPES A wipe is a video transition in which one picture is literally wiped off the screen by another. Most switchers have a variety of selectable wipe patterns, and some advanced switchers even have the capability of adding custom wipe patterns. Although the actual operation may vary from switcher to switcher, all wipes are produced using the M/E buses, the fader bar controls, and the special effects generator (SEG).

FADER BAR CONTROLS The fader controls for

FIGURE 11-6 **Switcher with Preview Bus.**

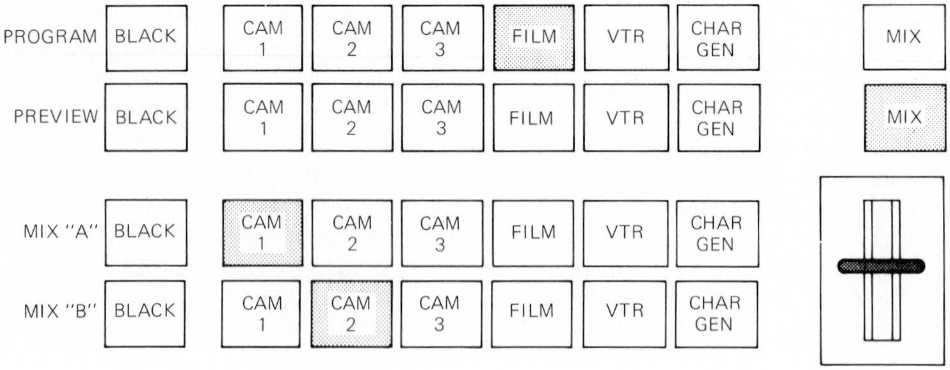

MIX/EFFECTS systems are used to control the direction and speed of a wipe. They work somewhat differently from the normal MIX fader control just discussed, and they commonly offer three operational modes: normal, reverse, and normal/reverse.

Normal Moving the fader always produces a wipe movement in the direction indicated on the pattern selector button. The fader's position makes no difference; the wipe always appears in the direction indicated whether the bar is in the MIX/EFFECTS 1 or MIX/EFFECTS 2 position.

Reverse Moving the fader control always produces a wipe movement in the opposite direction indicated on the pattern button.

Normal/reverse The fader bar handles work like a normal mix control. The movement of the pattern is associated with the movement of the fader relative to the M/E buses. When the bar is moved toward M/E 2, the wipe spreads out; when the bar moves toward M/E 1, the wipe closes in.

WIPE PATTERN INSERT A wipe consists of one image being wiped off the screen by another. Most switchers use M/E 1 and M/E 2 to preset the existing image and the video source which is

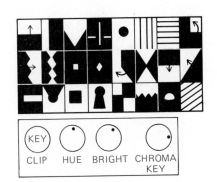

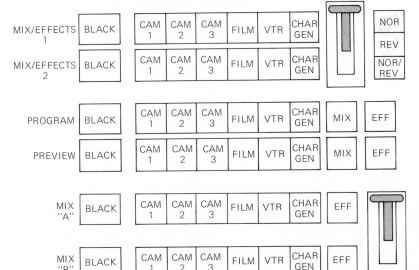

FIGURE 11-7 **Switcher with Mix/Effects Buses and Special Effects Generator.** The series of buttons at top left are for selecting specific types of wipe patterns.

inserted into the pattern and wiped on screen. You can also wipe to or from BLACK by punching up BLACK instead of a video source with a picture. If you have a color matte generator in your switcher, you have the option of replacing black with a color. The hue, saturation, and brightness levels of the inserted pattern are all adjustable with controls on the SEG.

MODULATION For special visual effects the borders of a wipe pattern can be "modulated," or made to vary in size and shape along the edges. The rates of movement and undulation are variable.

BORDERS A wipe can be produced with borders around the edges of the pattern by switchers which have bordering capability. The borders are described by their names: hard wipe, soft wipe, and edge wipe.

A *hard wipe* is the conventional wipe available with all SEG-equipped switchers. The edges of the wipe pattern are hard and distinct, producing an obvious separation between the scenes.

Switchers equipped with *soft wipe* permit you to blend or soften the edges of the wipe pattern to produce a variety of interesting effects. The amount of softness or blending along the edges is variable, and the diffused effect produces a very gentle transition. Some directors use soft wipes in place of dissolves for a slightly different visual impression.

The *edge wipe* generates an obvious border around the outline of the wipe pattern. The width and color of the border are adjustable from the SEG panel. The border around the wipe edges produces a very obvious and distinct separation between scenes and is valuable when you want to tell the viewer that there is a significant difference between the scenes being wiped across the screen. For example, some directors use the edge wipe in going from live action on a sports event to the instant replay. The edge on the wipe's borders makes the distinction clear between what

is live and what is being replayed (See Figure 11-8.)

ROTARY WIPE Switchers equipped with rotary wipe capability enable you to rotate the wipe as the pattern sweeps across the screen. The rate of speed of the rotation is variable from a minor turn to a continuous spin. Of course, all of the wipe pattern features such as border edging and wipe positioning can be combined with the rotary effect.

SPLIT SCREEN A split screen is basically an incomplete wipe. As with a wipe, the video sources must be selected on the M/E buses, and the split is controlled with the M/E fader bar handles. Any pattern wipe can be a split screen, although the horizontal, vertical, diagonal, and square-hole splits are the most common. Because a split screen is a variation of a wipe, all the border effects available for wipes are also available to you in producing the split-screen effect. Split screens are very useful when you wish to show a subject or scene from a variety of different angles or viewpoints at the same time.

JOYSTICK POSITIONER It is possible to vary the position of a split-screen pattern, placing it anywhere within the screen raster, by using the joystick control found on most SEG panels. For example, you can produce a square-hole split screen and then position it high, low, to the left, or to the right, where it best suits your purposes. The positioner is often used in conjunction with a *spotlight effect*. The spotlight slightly darkens the entire picture except for a pattern which is generated like a wipe or split screen. Combining the spotlight with the joystick control enable you to use the effect like an electronic pointer, to highlight or accentuate any part of the screen.

QUAD SPLIT A sophisticated variation of the split screen is the quad split, in which up to four different images can appear simultaneously on the screen. The size of each quadrant is variable, and the vertical and horizontal edges can be positioned independently to produce a staggered pattern or a symmetrical one. Quad splits require special equipment and a switcher large enough

FIGURE 11-8 **Wipe and Split-Screen Effects.**

to accommodate four individual sources, but the effect can be quite dramatic. (See Figure 11-9.)

INSERT AND CHROMA KEYING The term "keying" refers to a family of video effects in which one video signal is electronically cut out or "keyed" into another. Unlike a super, in which two (or more) video sources share the total signal

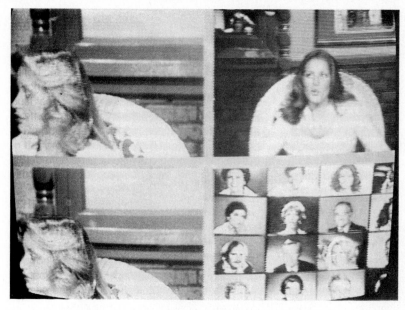

FIGURE 11-9 **Quad Split.** The borders enhance the separation between the images. *(Courtesy: Vital Industries, Inc.)*

strength, producing a faint and transparent image, keying permits all sources to reproduce at 100 percent intensity since the key source literally cuts out a video hole for itself in the background picture.

There are three types of keys: insert keys, matte keys, and chroma keys. All three have become standard in television production and, because of their flexibility, offer almost endless possibilities for every type of program.

INSERT KEY Insert keys are used to display words, lettering, and other graphics on a composite picture which includes the background image and the keyed information. The source of an insert key can be a graphic photographed by a studio camera, a slide or film from a telecine, or the output of a character generator.

The insert key operates by using the brightness level of the key source to trigger the effect. As the camera's output is routed through the special effects generator, an electronic "switch" senses when the scanning beam strikes a high-luminance portion of the key source picture. At a particular level of brightness, or intensity, the switch changes from the input of the background to the insert key source. That is why graphics used for keying work best with white lettering

against a black or dark background. This provides enough contrast for the keying system to sense the brighter letters and trigger the key effect.

Clip control The switcher operator uses a clip control to adjust the level, or "threshold," at which the keying action will occur. The correct clip setting is the point where it produces the cleanest insert onscreen without tearing or bleeding around the edges of the key. Clip adjustments must be made each time a new key is set up since variations in subject brightness and camera shading will affect the clip level. The output of the mix/effects system you are using to prepare an insert key should always be checked first on a preview monitor to enable you to adjust the clip level before punching up the key effect on air. The video engineer can contribute to a good insert by using the camera's shading controls at the CCU to increase the picture contrast between the white lettering and the dark background.

Producing the insert key An insert key requires two video sources: the key source—from a camera, telecine, or character generator—and a background source, which can be any video picture, video black, or a color, if your SEG is equipped with a matte colorizer. Usually the insert key is produced using the M/E buses, one for the insert key source and the other for the background video. Depending on your switcher's capabilities, an insert can generally be "popped" in or out instantly, faded in or out, or wiped onscreen or off. As with the superimposition, you can *undercut* an insert, selecting different background video sources while maintaining the key, or *overcut* the key, changing the key source over the same background picture.

Key reversal Switchers equipped with a key-reversal feature enable you to produce insert keys with black lettering on white backgrounds as well as with the more conventional white lettering on black.

MATTE KEY In an insert key, the hole cut out by the key source is filled in with its own video information. In a matte key, however, the hole can be filled in with a variety of video sources, or manipulated to produce a number of special effects. The sources commonly used to fill, or "matte in" are color, borders, shadow, or outline.

Color A graphic produced with plain white lettering against a black background can be electronically colorized, using a color generator and matte keying. In fact, color can be generated within any insert key from such sources as cameras, film chains, and character generators. The hue, saturation, and luminance are all adjustable from the SEG control panel.

Border A border feature enables you to outline the insert to make the lettering or graphic stand out more prominently on screen. The black border outline around the lettering is especially helpful when you must key an insert against a light background.

Shadow A variation on the border effect is the drop-shadow, which is produced by moving the outline down and slightly to the right. The on-screen image appears as if the lettering were illuminated by a hard, directional light source.

Outline In the outline mode, the inserted lettering appears only as an outline, as though punched out of an electronic stencil. The background scene shows through the lettering.

CHROMA KEY The use of chroma key has become so widespread that what was once considered an exotic special effect is now as commonplace as a dissolve or a fade. It is difficult to think of many electronic color productions which do not use chroma key at one time or another. Unlike an insert or matte key, which permits only the keying of graphics, with chroma key any object or subject can be inserted into a background picture to create a reality that exists only in the videospace.

As its name suggests, chroma key utilizes a specific color to trigger the keying operation. Although most keyers enable you to select virtually any hue as the "key color," blue and green have been found to work best for most production situations. This is because the two hues are farthest from the colors of normal skin tone, and produce cleaner looking keys with a minimum of technical imperfection.

The operation of a chroma key closely parallels insert and matte keying in many ways. However, instead of sensing brightness level variations, the chroma key triggers the keying operation whenever it detects the chroma key color in the source picture. The most familiar chroma key production example is the newscaster in front of a chroma key blue or green set. During the key operation, whenever the scanning system senses the key color of the studio set, it switches the video to the background picture. In the case of Figure 11-10, the background video is a slide from the telecine.

Needless to say, the chroma keyer cannot distinguish between "important" and "unimportant" parts of any picture. Anything in the key-source picture which approaches the hue and saturation of the key color will disappear and be replaced with the background video source. This is why the color of talent's clothing, the lighting and the colors used in the foreground set, and the color of on-camera graphics must all be carefully considered whenever you intend to use chroma key in production.

Chroma keyers are available in two different formats with very different production capabilities: (1) RGB chroma keyers, and (2) encoded chroma keyers.

The RGB keyer requires three separate and discrete red, green, and blue video signals from the camera in order to produce the key effect. The advantage of RGB keyers is their superior keying quality. However, since the RGB keyer must use separate signals, its use is limited to cameras which supply RGB video; it cannot be used to produce keys with such composite video sources as VTR playbacks or remote feeds.

The *encoded chroma keyer* uses a composite video signal, enabling it to key from any video source including VTRs, various remote feeds, and non-RGB cameras. However, we pay for this increased production capability with a key that is neither as clean nor as sharp as the one produced with the RGB format. Most switchers offer encoded keyers as standard with RGB as an option. In this case, you can select the keying format which will give you the best results for

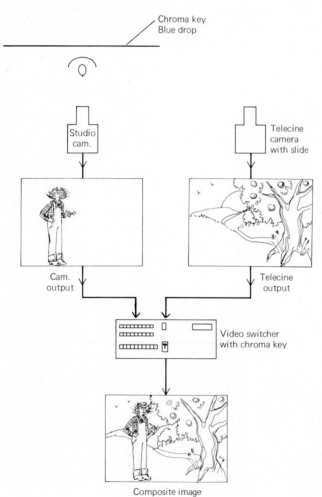

Composite image

FIGURE 11-10 Chroma Key.
The outputs of two video cameras are combined electronically by the switcher and special effects generator to produce a composite image.

every key situation. When you must key only studio cameras which feed RGB, use the RGB mode. When you must key in composite video, use the encoded keyer.

Hard keys A hard key, like a hard wipe, is the oldest and most familiar keying mode. A hard key produces a very distinct, hard edge around the keyed subject, as though the insert were cut out with an electronic cookie cutter. Unfortunately, the hard edge is susceptible to picture imperfections such as image "tearing" or "crawling" along the edges of the key. That is because the hard key operates on an all-or-nothing principle. The key system either triggers the foreground key subject or inserts the background video source. The on-or-off operation gives the insert a "keyed-in" look, and minute objects such as hair, or subjects without a well-defined edge, tend to key poorly.

Soft key The soft key has eliminated many of the problems associated with hard keying because it does not operate on the all-or-nothing principle. Instead, the soft keyer samples the key source subject and the background video for both color and brightness value. The electronic switching operation is then controlled proportionately, based on both the color and brightness values in

the composite shot. This results in a soft outline around the key insert, which blends naturally into the background picture. The degree of softness is variable, so you can produce the most realistic-looking effect with a little experimentation. (See Figure 11-11.)

Soft keyers not only make conventional chroma keys look better on screen, but they also permit the otherwise impossible keying of such transparent objects as glass, fire, smoke, and water. Since the soft keyer uses the proportion of color and brightness in the foreground and background images to produce the composite effect, it is sometimes variously referred to as a "linear," "proportional," or "luminance" keyer.

Shadow key One of the easiest ways to detect a chroma key composite is the lack of shadow falling from the keyed subject onto the background. Conventional hard keyers either could not capture the shadow at all, or, if they did, the shadow created bleeding and tearing, which ruined the key. These problems are solved with the shadow keyer, which captures the shadow detail

THE SWITCHER **321**

FIGURE 11-11 **Soft Key.** Notice how it is possible to see through the glass vase insert to the background video. Soft key permits keying transparent or translucent objects and can also produce a realistic shadow when necessary. *(Courtesy: The Grass Valley Group, Inc.)*

in the foreground key picture and casts it naturally onto the background scene, producing a very realistic and lifelike composite in the videospace.

Using chroma key in production The effects possible with chroma key are limited only by your imagination and ingenuity, especially when using the new soft and shadow keyers, which produce exceptionally clean and natural-looking composite pictures. Chroma keying is used routinely on news and informational shows, on sports programs, and on many entertainment shows. With chroma key, the viewer can be shown a reality which exists only in the videospace: a boy and girl flying over a city skyline on a children's show; a dancer whirling across the walls and ceiling of an otherwise normal room; a demonstrator actually walking through a car engine in an instructional program.

By experimenting with chroma key, you can find many more imaginative ways to use it in different types of productions. Transitions are one effective way to employ chroma key. Any object on the set that is painted the chroma key color can be revealed while the next program scene is inserted in the "window." If the camera on the key source zooms in and increases the amount of chroma key color in the shot, the background picture becomes increasingly larger until it completely fills the screen. Colored window shades, balls, pages in a book, doors, and other set and prop materials can be easily used in this way.

The most important factor in producing a clean key is to pay careful attention to such production details as lighting, staging, performer wardrobe, and camera angle. Even the best and most expensive keyers cannot work to their potential unless the backgound chroma key set is evenly lit without hotspots or random shadows. It has been found that encoded keyers require more illumination on the colored background than RGB keyers to produce a clean key effect. If at all possible, check your lighting, costumes, and other production variables on the keyer before you begin rehearsals.

MULTIPLE MIX/EFFECTS SYSTEMS All the effects we have just discussed—wipes, split-screens, and keys—are possible using a single mix/effects system, although not necessarily at the same time. However, modern television production requires the ability to produce a number of special effects simultaneously or to execute various transitions between composite video effects. For example, if we want to wipe from a chroma key composite with a graphic insert to another chroma key with another graphic, we will require a number of individual mix/effects systems in order to create the two effects and the transition between them.

The reason large production switchers appear so complex and intimidating is because they contain multiple mix/effects systems to provide greater operational flexibility. You might think of the two, three, or more mix/effects systems on a switcher as being like having several small switchers all contained in one switching console. The output of any one of these mix/effects "mini-switchers" which, by itself, is capable of producing an elaborate composite video effect, can serve as an input to a second mix/effect "switcher" with the same operational capabilities. So a composite image produced by one mix/effects system—let us say it is a split-screen produced by two cameras—can be entered into another mix/effects system where a performer on a third camera is chroma-keyed over the original split-screen. If we have a third mix/effects system available, we can take the composite split-screen/chroma key image and by entering it into the third mix/effects system produce another video effect, say, an insert of a graphic across the lower third of the screen. You can keep on going like this until you've exhausted your supply of mix/effects systems.

Switchers with multiple mix/effects systems operate on a "cascading" principle with the video flow starting at the mix/effects system at the top of the switcher and working its way down through each subsequent mix/effect system on the board. At any point the output of one mix/effects system can be "reentered" into the switcher by using the *secondary source* buttons which control the output of each MIX/EFFECTS bus. While all of this may sound complicated, we really aren't doing anything more than we did when we used one

Individual Video Sources

"Downstream" Building of Composite Image

Wipe to split screen

Chroma-key talent over split screen

Insert matte lower-third graphic over composite

JOHN DOE
Reporting

JOHN DOE
Reporting

FIGURE 11-12 Illustration of Video Cascading to Produce a Complex Composite Image.

mix/effects system. The only difference is that we can continue to reenter the composite image produced by one system into another, take that newly created video and enter it into a third system, and so on, with each new mix/effects system adding another layer of video manipulation to the image. (See Figure 11-12.)

Multiple mix/effects systems allow such complex production effects as a dissolve from a single source to a scene consisting of a chroma key insert over a quad split. It is also possible to wipe a chroma key background from one source to another while maintaining the keyed subject and a third graphic on screen during the background transition. The more M/E systems available, the more elaborate and sophisticated your switching capabilities become. However, this also increases the complexity in working the switcher and requires an operator who can keep track of the many video sources and transitions and correctly preset them so you do not exhaust your mix/effects systems before completing a composite or transition switching sequence.

Preset Buses

A number of switchers use a PRESET bus, which is located directly over or under the PROGRAM bus. The function of the PRESET bus, as its name suggests, is to enable the TD to preset the next shot or effect. For example, assume the director has CAM 1 on the air and plans to cut next to CAM 3. Using the preset system, the switcher operator punches up CAM 3 on the PRESET bus as soon as the director says "Ready three." As the CAM 3 button is pressed on the switcher, Camera 3's picture automatically appears on the large preset monitor, which is located next to the line monitor. This enables the director to glance from one monitor to the next to see how the two shots will cut before actually making the switch on the air.

FLIP-FLOP CONTROLS   When the director calls for the switcher to "Take 3," the switcher could easily punch up CAM 3 on the PROGRAM bus. However, if the PRESET bus system is used, there is a more convenient way to make the switch, and this involves what are called "flip-flop" controls.

A flip-flop circuit is designed automatically to exchange, or flip-flop, the video sources which appear on PROGRAM and PRESET whenever the take bar switch is pressed by the video operator. In the example above, the switcher has CAM 1 on

the PROGRAM bus and has preset CAM 3 on PRESET. Now on the director's command to "Take three," the switcher presses the take bar, which instantly cuts to CAM 3 on the line monitor. At the same time, the video sources flip-flop, so CAM 3 now lights up on the PROGRAM line and CAM 1 appears on both the PRESET bus and on the preset monitor.

There are two reasons for presetting: first, it enables the director to see what has been preset and prevents the switcher from pressing the wrong button on the air in the excitement of a production. It is harder to make a mistake when you have seen the correct picture on the preset monitor, and you have to press only a single take bar, rather than to locate one button along a row of identical buttons on a conventional primary source bus. Secondly, it has been found that a great deal of cutting within a show often occurs between the same two cameras. Using the flip-flop system, the switcher can cut back and forth between two cameras, or two other video sources, without having to constantly punch each up on the PROGRAM or other primary source bus. This is especially helpful to a busy switcher, who must cut the show while preparing for upcoming effects or transitions.

If the transition were a mix requiring the fader bar handles, the flip-flop system would work the same way. It makes no difference in which position the fader bar handles are located. A movement of the fader in either direction automatically fades out the PROGRAM source and fades in the PRESET source. At the completion of the fade or the dissolve, the video source on PROGRAM and PRESET automatically flip-flop.

Naturally, there are times when it is impossible to preset every shot before punching it up on the air. In this case the switcher would simply do most of the cutting directly on PROGRAM or on a MIX bus and use the PRESET system when convenient.

Downstream Keyer

As its name suggests, the downstream keyer enables you to insert or key a video source over the output of the PROGRAM bus "downstream," at the point just before the video signal leaves the switcher. The advantage of the downstream keyer is added production capability because it can produce an insert key without using any of the switcher's mix/effects systems. This frees all of the mix/effects systems for the production of more elaborate effects, such as wipes, or for chroma keys and still enables the switcher to produce a final video insert such as a lower-third graphic over the composite image.

Downstream keyers are used frequently on news programs, where a great deal of mix/effects work is already done on the video signal, such as a chroma key insert of a news graphic or live picture with its own caption insert behind the camera shot of the newscaster. Another lower-third graphic insert is easily produced with the downstream keyer without affecting any of the other mix/effects presets. Downstream keyers are very easy and convenient to use, especially in situations such as news programs, instructional shows, or sports events, where a great deal of graphic material must be quickly inserted over a composite image which is produced with one or more mix/effects systems. Most downstream keyers are equipped with a master fader which enables you to fade the entire output of the switcher either up or to black.

Programmable Switchers

As production switchers become more versatile, they also become increasingly complex. Although modern switchers are capable of virtually limitless effects and transitions, the ability of the switcher operator to keep track of the various operations can be the most serious limitation in the switching system. For example, a thirty-second commercial may utilize multiple keys, inserts, wipes, and split-screens, all of which must be preset and executed flawlessly in less than half a minute.

To help the switcher operator produce these effects and transitions, a *computer-assisted programmer* is available which is interfaced with the switcher. These "smart switchers" are able to memorize a complex sequence of switching events and execute them precisely on command. To program the memory, the switcher operator simply runs through each sequence of events, one step at a time, as they are to appear on air. Once the sequence is in the computer's memory, a single press of a button will produce the prepro-

grammed series of switches and transitions. Obviously, a number of complicated sequences can be programmed in advance, each being assigned a different cue number. Then the switcher operator can call up any one of the cues whenever necessary throughout the production. An added bonus is the ability of the smart switcher to interface with a computer-assisted videotape editing system to produce frame-accurate postproduction switches and transitions.

SPECIAL ELECTRONIC EFFECTS

At one time anything more than a straight cut was considered a special electronic effect. Now, modern television production uses wipes, inserts, and various chroma keys so often that calling them special effects is like calling pizza a gourmet food. In this section we will deal with some of the electronic effects which are, indeed, special, either because they require elaborate auxiliary equipment or because the effects are less frequently used in most production situations.

Camera and Switcher Effects

A series of special effects can be produced either with the camera and its CCU controls or in conjunction with the video switcher. They include video feedback, debeaming and solarization.

FEEDBACK Video feedback is produced whenever a camera is turned on a monitor which displays the camera's own picture. The resulting electronic circle, or "loop," produces feedback as the camera sees itself over and over. Video feedback is the picture counterpart to the audio feedback squeal we hear whenever a microphone is held too close to its monitor speakers. You can demonstrate video feedback for yourself by punching up a studio camera on the line and then turning the camera on a television monitor which displays the camera's shot. The hall of mirrors effect you have created—seeing a monitor in a monitor, in a monitor, to infinity—is feedback. Varying the camera-to-monitor position by zooming, trucking, panning, or tilting and also adjusting the camera shading and lens iris opening will create an endless series of cascading visual images.

To produce a picture with the feedback, you must use another camera, which photographs the subject, and then mix the primary subject camera with the video feedback camera. It is advisable to produce feedback pictures on the best quality

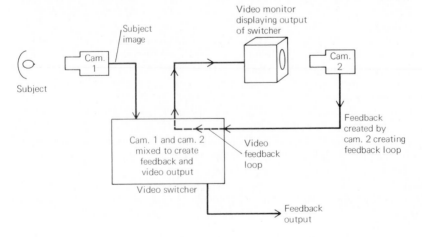

FIGURE 11-13 **Video Feedback.**
The picture above was produced by photographing Camera 1's image by Camera 2 and feeding the image back on itself.

monitor you have since the effect is created in part with the monitor's picture and its resolution and color quality will influence the final image.

Video artists have long used feedback as a basic way to build and manipulate the video-space. Each time you experiment with feedback you will likely find new and exciting possibilities, especially if you combine the feedback effect with other special video effects, such as keys, wipes, and supers. (See Figure 11-13.)

DEBEAMING AND SOLARIZATION Debeaming refers to reducing the scanning beam's voltage strength as it reproduces the video image. The result is a highly contrasted image with stark whites, blacks, and colors.

You can achieve a similar effect, without disrupting the camera controls, by using the insert keyer. Although the keyer is intended to be used mainly for graphics, shooting a normal subject and then routing the signal through the insert keyer will produce a picture with stark whites and blacks. Since the keyer works only on brightness levels, you will not see the entire subject, but only a surreal impression, depending on how it is lit and the brightness values of clothing. Varying the

lighting will, of course, make a major difference in how the key appears on screen. Adding matte key color to the effect is yet another way to manipulate the image. The effect is sometimes called "solarization" because the image produced looks like the photographic process which goes by the same name. (See Figure 11-14.)

Digital Video Effects (DVE)

The use of digital technology has made it possible to expand our production capabilities far beyond what we have come to expect as the state of the art. We have already mentioned digital technology in terms of digital audio, the time base corrector, and the digitally controlled camera. Remarkable as these devices may be, it is in the area of video signal manipulation and processing that the full impact and potential of digital technology and its application to production can be fully appreciated.

FIGURE 11-14 **Solarization Effect.**

DIGITAL OPERATION   In order to understand digital equipment and its use in production, we first need a very brief introduction to the basics of digital operation. You'll remember that a conventional television signal is an *analog* signal because it uses the variation in electrical current to represent or reproduce an "analogy" of the image which is photographed. From our discussion of the scanning and reproduction process in Chapter 2, you should recall that it is the variation of the electrical current in the pickup tube which corresponds to the brightness values of the subject before the camera. The voltage level which is produced by the pickup tube and associated electronic circuitry within the camera varies across a theoretically infinite range of values in order to reproduce accurately the subject's image on the television screen. In effect, what we are dealing with is an electronic *waveform* which is a continuously varying electronic signal.

A *digital* signal, on the other hand, does not use current variation or waveforms but instead converts the analog signal into a series of numerical code numbers, one number for each element of the picture.

In order to process video signals digitally, we must first convert the analog waveforms into a digital signal. This is accomplished by "sampling" the analog signal as it enters the digital converter and dissecting each sample of the original picture into any one of 256 brightness levels. A code number is assigned to each sample according to its brightness level and its position on the television raster. The higher the brightness value, the higher the digital code number. In this way, the entire video signal is broken down into a series of computer-type binary code numbers which represent the original picture. Of course, the sampling and conversion process must be done very rapidly since a new video field occurs sixty times a second, and a complete new video frame is produced thirty times a second. Once the analog video signal has been *digitized,* or converted into digital code numbers, it can be processed, manipulated, and then reconverted back into its original analog waveform at the output of the digital device.

Why go through all this trouble? Because using digital signals, as opposed to analog signals, offers us a number of significant advantages. First, digitizing enables us to store, process, and retrieve an enormous amount of video information within a relatively small memory area. Data processing computers have long used this concept, and now such television production equipment as the digital video effects unit or the still-frame-storage device use digital techniques to hold and manipulate an enormous amount of complex information. The greater the memory capability, the more we can manipulate the digital signal in a variety of highly sophisticated ways. Also, because digital television equipment relies heavily on computer-type technology, we reap the continuing advantage of increasingly more powerful equipment at lower cost.

Secondly, once a digital signal is digitized, the equipment works only with the numerical code numbers and not with the actual electronic signal itself. This means that there is virtually no degradation in picture quality regardless of how much we process and manipulate the digital signal. The ability to manipulate video with "transparent" processing is extremely important and permits us a tremendous amount of production flexibility without the problem of deteriorating picture quality. In fact, some digital devices, such as the time base corrector and the video noise reducer, actually produce a better quality picture leaving the device than the original signal which was entered into it before any processing or manipulation.

Finally, and perhaps most importantly, digitizing the signal enables us to manipulate the video in ways which are impractical or impossible using analog techniques. Digital video effects (DVE) units can produce video effects which were possible earlier only by using film and the optical printer, a costly and time-consuming process at best. When we use the DVE, it takes only the press of a few buttons and controls to transform an image in countless ways instantly. Digital video effects have become so widespread that it is difficult to think of any area of production where they have not been applied to produce exciting and effective visual techniques.

With this simplified explanation of the digital process in hand, we can now look at a number of digital devices and see how they can be used in production.

FRAME SYNCHRONIZER The electronic "clock" which synchronizes all the video signals in a system and keeps them running together is the *sync generator.* As television production ventures out farther and farther from the studio, problems in synchronizing various wild feeds with inhouse video sources have become increasingly difficult. ENG and EFP feeds are routinely microwaved back to the station from all over town. Cameras are run from boats, planes, helicopters, blimps, and automobiles. Satellite feeds are becoming increasingly commonplace as both broadcast and closed-circuit systems establish temporary interconnection networks for special events. Until the introduction of the frame synchronizer in 1974, synchronizing these various remote feeds was handled in a variety of ways, none of which was completely satisfactory. The digital frame synchronizer (frame sync) unit has eliminated these problems and has also introduced a number of unique and exciting production techniques.

The frame sync unit is designed to accept any "wild" (nonsynchronized) feed from any source, convert the signal to digital bits, synchronize the signal with the in-house video, and then read out the processed signal in analog form, where it enters the program switcher like any other video source. The result is a signal that is completely synchronous and can be keyed, wiped, or dissolved in the switcher. If you have seen any programming on the major, national networks, especially remote news and sports events, you have seen television signals which were digitally processed by a frame synchronizer.

In addition to synchronizing a remote video source, the frame synchronizer offers two valuable production options: (1) freeze-frame and (2) image compression.

FREEZE-FRAME The frame synchronizer works by storing a complete frame of video and then reading it out in step with the in-house sync generator. The device is also programmed to hold the last complete video frame and continually repeat it in case a new signal does not arrive from the remote source. For example, a signal being microwaved-in from a moving car or helicopter may pass through a blind spot where transmission is temporarily interrupted. The frame synchronizer simply reads out the last complete frame over and over until microwave reception is restored. Since the freeze-frame effect can be produced intentionally with a control switch, this provides a new way of freeze-framing video without using a slow-motion disc. Freeze-frame is equal in quality to stop-motion on film, and easily surpasses the quality of stop-motion from a helical VTR.

IMAGE COMPRESSION Since the frame synchronizer reads out the processed video signal in synchronization with the in-house sync, if we vary the rate at which the stored signal is read out, we can reduce the size of the complete picture in the raster. For example, if the unit is told to read out the stored signal twice as fast as normal, the result is an image with half the normal vertical height and half the normal horizontal width. In other words, an image which is one-quarter normal size. Using a joystick control, we can position this compressed image—which is the complete video picture squeezed down to one-fourth normal size—anywhere on the screen.

The advantages of image compression are obvious. Camera operators no longer need compose a split screen or chroma key shot in the camera viewfinder. A normal, full-frame action image can be electronically shrunk in size and positioned behind the newscaster. Since the compressed image is always in sync, the newscaster at the studio can have a two-way, live talk with a reporter in the field, and the compressed background image can then be faded, wiped, or popped off the screen whenever necessary.

DIGITAL VIDEO EFFECTS (DVE) UNIT The digital video effects (DVE) unit is a remarkable device which greatly expands the operational potential of the frame synchronizer. The DVE is integrated with a full-capability program switcher providing the operator with enormous flexibility in manipulating and modifying the video image. What is especially important is that the manipulation of the image takes place instantly, in real time so you can see exactly how the effect appears in the videospace and modify it until you are satisfied with the result.

Over the past few years, a number of manufacturers have produced DVEs under a variety of trade names. Rather than discuss each one in detail, we have designed the following description to acquaint you with the wide range of potential possibilities which a DVE unit offers. While some of the effects mentioned may not be available on every unit, for the most part DVE systems provide basically the same type of video manipulation capabilities regardless of the individual manufacturer. (See Figure 11-15.)

CONTINUOUS IMAGE COMPRESSION The DVE goes the frame synchronizer one better by enabling the operator to compress a full-frame picture *continuously* until it literally disappears. The effect appears as if we electronically zoomed out on the picture, which, of course, we have. Naturally, we can stop the compression at any point and then position the compressed image anywhere on screen by using the joystick posi-

tioner control. This technique is commonly used in news programs where a compressed picture of a graphic or a live camera feed is positioned over the shoulder and to the side of the on-camera newscaster.

Picture compression and positioning are also used to build a graphic electronically by combining the compressed and positioned image along with additional graphic material, which is usually provided by an electronic character generator. You have seen this effect often in news, in sports, and on many instructional programs.

One of the greatest advantages of image compression is the ability to shoot a scene in full frame and then compress and position it later during postproduction so as to vary its size and position until it is perfectly integrated with other program elements such as the on-camera talent, graphics, and so on.

CONTINUOUS IMAGE EXPANSION The reverse of the compression technique is continuous picture expansion, which, again, appears as though we are electronically zooming in on a picture. This effect permits the director to enlarge an image much as a photographer might enlarge a negative in the darkroom to eliminate areas along the edges of the frame which are unimportant and emphasize better the relevant parts of the picture in the videospace. Some DVEs permit up to eight times normal image expansion although after four-times expansion the image becomes very grainy.

The possibilities of image expansion are tre-

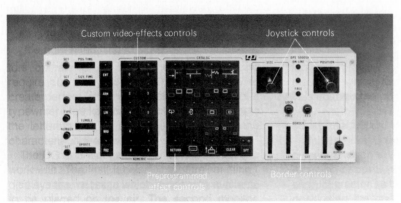

FIGURE 11-15 Digital Video Effects Control Panel. The various buttons enable the switcher operator to call up any preprogrammed and custom-programmed video effects. The joystick controls enable the image to be positioned anywhere on screen and also control the compression/expansion of the video image. *(Courtesy: MCI/Quantel)*

(a)

(b)

(c)

FIGURE 11-16 **Image Stretch/Expansion Effects.**
Picture *(a)* shows the original image before digital
manipulation; *(b)* the image is rotated along the
Z-axis; *(c)* the image is stretched and reduced in size.
(Courtesy: Ampex Corp.)

mendous and offer the director new production
capabilities which were never before available.
The director can literally move around inside a
normal frame, highlighting a portion of the picture
by expanding it and then positioning the newly
expanded image anywhere within the videospace. This has been used to great advantage on
news and sports coverage where certain scenes
are recorded on videotape, replayed in slow
motion, and then expanded to provide the viewer
with an even closer and more vivid look at the
action.

IMAGE STRETCHING EFFECTS The DVE makes
it possible to expand or compress either the
horizontal or vertical dimensions of the image
independently. This means you can literally re-
shape the picture's aspect ratio although some
distortion becomes evident after a certain amount
of stretching. (See Figure 11-16.)

IMAGE ROTATION EFFECTS Using a DVE, you
can take an image and rotate it within the videospace about all three axes. When we rotate an
image on the X-axis, we create a "tumble" effect
similar to viewing an object rolling over from front
to back. Rotating about the Y-axis creates a
"spinning top" effect. Rotating along the Z-axis
creates a depth perspective which gives a three-
dimensional look to the image. Some sophisticated DVEs enable you to control the depth of the
perspective so that, as an object passes from one
plane to another, you control the amount of warp-
age or "keystone" effect.

On most DVEs these sophisticated rotation
movements can be preprogrammed and stored in
the system's memory so you can re-create the
exact movement and number of rotations which
you rehearsed precisely on cue.

TRANSITION EFFECTS Among the various transitional effects possible with the DVE are the
push-off—in which one image literally pushes
another off the screen—and the *video split*—in
which the image is literally split apart, providing

space for a new image behind it. The *page flip,* as its name suggests, appears as though the new image is flipped across an existing picture as if you were turning the pages of a book. There are also a family of *matrix wipes,* in which small squares of the screen randomly disappear to be replaced by a new image.

All these transitional effects can be varied and modified by using different controls on the DVE and the switching console, providing literally dozens of different transitional effects. (See Figure 11-17.)

ADDITIONAL DVE EFFECTS Since the DVE operates much like a digital computer, manufactur-

ers are continually developing new software programs which enable the operator to produce countless new and interesting effects. For example, the *startrail* effect leaves a decaying trail of starlike dots as the image moves across the screen. This can be combined with rotated lettering to create a striking graphic image which can, in turn, be inserted over another picture. The *multifreeze* effect produces an apparently neverending series of multiple images much like those we see in a hall of mirrors in an amusement park, where the original image is repeated to infinity. *Mirror-imaging* produces the mirror-reversed image of a subject and can be used to create realistic-looking reflections. (See Figure 11-18.) *Bouncing* moves the image like a Ping-Pong ball across the screen from edge to edge until it is in its predetermined position in the videospace.

All of these—and other effects too numerous to mention—can be modified or combined to create

(a)

(b)

(c)

FIGURE 11-17 **Digital Video Effects.**
The most sophisticated DVE systems enable the picture to be manipulated in countless ways. Picture *(a)* shows the original image. Picture *(b)* shows the image which is wrapped around itself into a cylinder. The "wrap" is variable and appears in real time in the videospace. Picture *(c)* shows a "page turn" effect in which the original image appears on the back of the "page" and a second image appears on the front of the "page." The page flip is variable in real time. *(Courtesy: MCI/Quantel)*

FIGURE 11-18 **Mirror-Image DVE Effect.**
(Courtesy: Vital Industries, Inc.)

entirely new visual images. The biggest problem in using a DVE is in knowing when to stop manipulating the image because it is so effortless to create dozens of entrancing and visually startling effects. The secret, of course, as with any item of production equipment, is to apply the technology to the primary objective of effectively communicating to the viewing audience. (See Figure 11-19.)

CHROMA KEY TRACKING Combining a DVE unit with a chroma key special effects generator

FIGURE 11-19 **Bounce and Multifreeze Effect.**
The original image is compressed and then bounced across the videospace. The multifreeze effect enables the switcher operator to produce a trail of images which decay at a previously determined rate. *(Courtesy: Vital Industries, Inc.)*

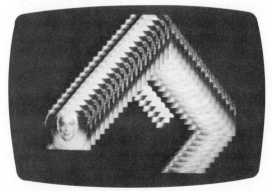

enables the DVE system to compute automatically the size and position of the chroma key cutout and to fit the compressed picture into the chroma key window. As the size of the window increases or decreases owing to the camera panning, tilting, or zooming on the shot, the size of the compressed picture insert varies proportionately so that the proper perspective between the foreground subject and the compressed keyed image in the background is maintained at all times. This eliminates one of the most irritating production problems associated with chroma key and produces a key effect which looks extremely natural and realistic.

POSTPRODUCTION WITH A DVE The DVE is especially useful for postproduction work as well as during actual production. We have already mentioned how a director and a graphic artist can use picture compression and positioning to integrate a camera shot or still-frame picture or slide with graphic material to build an electronic graphic. But this is only one of the many production capabilities offered by DVE in post production.

The primary advantage of the DVE is that the director can use any normally produced full-frame camera shot, film, slide, or videotape and, by modifying its size, shape, and position within the videospace, integrate it perfectly with other production elements. Without the ability of the DVE to vary the full-frame image's size and position, these elements would have to be shot with the final matte insert position in mind, and no last minute changes in picture positioning would be possible, greatly reducing the director's postproduction flexibility.

The image expansion capability is also useful in correcting minor compositional mistakes such as a boom microphone or some other unwanted element which appears on the edges of a camera shot. By replaying the videotape and slightly expanding the image size, we can completely eliminate the offending item from the edges of the videospace without having either to reshoot the scene or settle for an imperfect shot.

The transitional capabilities of the DVE are also

a natural for postproduction work. If two or more videotape recorders are played through the switcher and DVE, highly effective co-ord transitions can be made and recorded on another VTR machine. The ability to interface the switcher and DVE with the VTRs through the use of SMPTE time code enables the edit computer to accurately control all DVE operations and ensures perfect transition effects during a co-ord session.

SUMMARY

The production switcher is used (1) to select the particular video source which is to appear on the air and (2) to manipulate the picture's size and shape and/or to produce composite images with multiple sources. The switcher is commonly located in the control room facing a bank of video monitors. Each monitor continuously displays the output of a potential video source—cameras, VTR machines, telecine, character generator, and so on.

Each video source is represented by a separate button on the switcher, which has rows of buttons called buses. Even the simplest of switchers includes a PROGRAM bus and a pair of MIX buses with a fader bar control. Using these buses and controls, the switcher operator can produce cuts, fades, dissolves, and superimpositions. To produce more elaborate effects, a switcher must be equipped with a special effects system, consisting of (1) a pair of MIX/EFFECTS buses with a fader bar and (2) a special effects generator (SEG). The special effects system enables the switcher to produce such video effects and transitions as wipes, split screens, and keys.

A wipe consists of one video source literally wiping another off the screen in a preselected pattern. A split-screen is a wipe stopped midway through, so two or more pictures share the videospace simultaneously. Wipes and splits can have hard edges, with an obvious distinction between the two pictures, or soft edges, where the edges blur slightly for a less obvious distinction. Borders and color can be added to the edges of the wipes and split screens on some switchers.

There are three types of keys: insert keys, matte keys, and chroma keys. The first two are used normally to insert graphic material into a background picture. The insert key simply cuts itself into the background; the matte key enables the switcher operator to fill in the cutout space with a color with special borders, or with a drop-shadow.

Chroma key utilizes a particular color to trigger the insert operation. Any portion of the key-source camera shot containing the preselected color will be replaced on-air by the background picture. A hard chroma key produces a sharp, defined edge to the insert. A soft key produces a softer blending between the key and background, producing a more realistic and natural composite effect. Soft keyers also permit keying transparent objects, and some enable the switcher operator to include the foreground subject's shadow in the background composite.

Special electronic effects include those possible with the camera and the switcher and those which require auxiliary equipment. Camera and switcher effects include feedback, debeaming, and solarization. Video feedback is produced when a camera is trained on a monitor displaying the camera's output, creating a continuous video loop. Debeaming is the gradual reduction of the camera's electronic output, using the CCU controls, and produces a stark, high-contrast image. A similar solarization effect is possible using the insert and matte key controls on the switcher.

Digital technology permits the processing and manipulation of video signals in a variety of unique ways. The digital operation requires a conventional analog video signal to be converted into digital code numbers before being processed, manipulated and then reconverted to analog at the output. A digital frame synchronizer assures perfect sync when using remote video feeds, and also permits still-framing and image compression.

The digital video effects (DVE) unit enables the operator to manipulate the image by continuously expanding or compressing it and to modify its size, shape, and position in a variety of ways.

CHAPTER 12

ELECTRONIC NEWS GATHERING
(ENG) PRODUCTION

Not long ago, any television production which took place outside the confines of the studio was considered a rather extraordinary event. This was due, in large part, to the nature of the available production equipment, which was designed primarily for in-studio use. Producing on location required a large crew to transport literally tons of bulky studio equipment to the remote site. As a result, field production on a regular basis was limited to the largest stations and the national networks, which could afford the costly mobile units, associated production equipment, and the sizable manpower necessary to conduct a remote.

The introduction of lightweight and relatively inexpensive cameras and 3/4-inch video cassette recorders in the early-to-middle 1970s changed all of this forever. The new equipment was truly portable and needed only a one- or two-person crew to bring it to the location and to operate it during production. Unlike motion picture film, which required a time-consuming processing step, videotape footage which was recorded in the field could be replayed and edited immediately. The new equipment and the associated production capability which it offered resulted in an entirely new kind of television production: *electronic news gathering*, or ENG (sometimes referred to as "electronic journalism").

Virtually overnight, 16-mm motion picture cameras were replaced by portable color cameras and battery-powered VCRs. Not only could a recorded tape be replayed instantly, but combining the video camera with a portable microwave unit also enabled sound and picture of a news event to be sent directly from the field back to the home studio and broadcast live as the event unfolded before the viewer's eyes.

ENG equipment and techniques not only revolutionized the way we cover daily news events, but also resulted in a significant change in nonnews remote productions as well. Television production in the 1980s is no longer primarily studio based, and virtually every production is likely to be produced—if not totally, then at least in some part—outside the studio on location.

One of the most immediate applications of the ENG approach was the development of *electronic field production*, or EFP. EFP uses many of the same equipment and techniques of ENG but applies them to nonnews production situations which can be more carefully preplanned and which require somewhat higher production values than is necessary for news coverage. Not only could these productions be produced relatively inexpensively in the field, but shooting on location also enabled the producer and director to take advantage of the very special flavor which only remote shooting can bring to a production. Entertainment programs; commercials; instructional, educational, corporate communication programs; and the magazine-format feature shows which appear on virtually every local station and on many cable channels are only a few examples of the application of EFP remote production techniques.

Even large scale *multiple-camera remote* productions, which use mobile vans complete with a miniature control room, took advantage of the trends in miniaturization which the modern technology offered. Today's mobile units offer the producer and director in the field, virtually the same production capability as can be found inside a conventional studio control room.

There is no question that remote production has become an indispensable and essential ingredient of television production. What is important for you to understand is how to approach a remote production and how to select the best equipment and techniques to match each production situa-

tion. In this chapter we will discuss electronic news gathering, which is a special type of remote production since news coverage, by its very nature, offers a unique set of production challenges. In Chapter 21, we will discuss the other types of remote productions, which, although they take place outside the studio as does ENG, utilize somewhat different production equipment and techniques.

A final word: Any remote production is, first and foremost, a television production and utilizes virtually all of the various techniques which we discuss in each chapter of this book. Rather than repeat what we have already said about equipment operation and production applications, we will concern ourselves in the chapters about remote production with how to apply what you already know about the equipment to the special requirements of a remote production situation. For more information about any specific equipment and its operation, refer to the particular chapter for details.

ELECTRONIC NEWS GATHERING (ENG)

Electronic news gathering refers to the family of production hardware and techniques which are used to cover the news on a day-to-day basis. Although ENG production uses equipment and techniques which are, in most cases, virtually identical to those used inside the studio, the very nature of "news" means that these approaches must be applied in a very special way. News coverage cannot be preplanned in advance, there is little time for equipment setup, and rehearsal is impossible owing to the spontaneous elements of a news event. An effective ENG production team combines the aesthetic and technical abilities which are necessary for any conventional television production with the capability to think and act quickly in response to rapidly unfolding situations. In the following sections of the chapter, we will discuss the use of production equipment and techniques specifically as they apply to the coverage of news events.

ENG EQUIPMENT AND OPERATION

When we talk about ENG production, we refer to a single portable camera which is connected directly to a portable video cassette recorder. Since both camera and VCR are powered by batteries, the camera operator has complete mobility to cover the news event. Usually, the ENG crew records program material on a videocassette in segments, and the various segments are edited later into a complete news story. Another capability of many ENG systems is to microwave live pictures and sound of an event back to the studio for immediate broadcast.

The primary goal of ENG operation is to cover fast-breaking news events, so the equipment is designed for maximum operational flexibility. Not only the camera and VCR, but also audio and lighting equipment used for ENG should be selected with portability and efficient operational capability in mind. Consequently, we usually do not demand ENG units to produce material with production values as high as we might expect from programming produced inside the studio or by EFP or large scale remote operations. That is not to say that ENG units must not produce programming with a minimum level of technical and aesthetic standards. There is no point in producing material which is so poor that it cannot be edited or broadcast, or which does not enable the viewer to see and hear clearly what is taking place.

ENG Mobile Unit

At its simplest, the ENG mobile unit can be an ordinary car, which the camera operator, sound engineer, and reporter pile in with their equipment and race around town. Although some smaller stations take that approach, the use of a van or

minibus, which has been specially designed for ENG operation, offers some important advantages over a car. The van is large enough to hold all the equipment and personnel, small enough to drive, maneuver and park easily, and able to carry a microwave system to permit both cable-free camera operation and mobile-unit-to-studio transmission for live broadcasts from the field. (See Figure 12-1.)

POWER ENG units run primarily off batteries, and the ENG van must have some way to charge the batteries while the unit is in the field. An internal generator, which is powered by the van's engine, is an effective way to make the vehicle independent of cable current. The generator supplies all the needed power whenever conventional ac power is unavailable. Most ENG units can be powered from common, utility, 120-volt ac, 20-amp current.

COMMUNICATIONS A two-way radio provides the necessary communications link between the ENG unit in the field and the news and engineering departments back at the studio. For on-site communications, many ENG units equip their crews with a couple of inexpensive walkie-talkie units, which permit communication over some distance between the camera crew covering the event and the engineer located inside the ENG van.

ENG Camera

We have already covered the various hand-held and shoulder-mounted cameras which are normally used for ENG operation, in Chapter 2. The cameras selected for ENG must be battery-powered, light enough to be handled easily and sufficiently durable and rugged in design to withstand daily operation. In addition to the camera head, the ENG unit should carry a tripod, shoulder brace, lens shades and filters, and any test equipment necessary to set up and align the camera in the field.

ENG CAMERA OPERATION Most ENG crews consist of one or two technicians. In a two-person crew, one person is responsible for operating the camera while the other handles the VTR, audio, and any supplemental lighting.

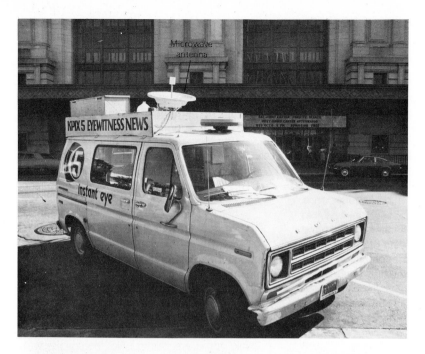

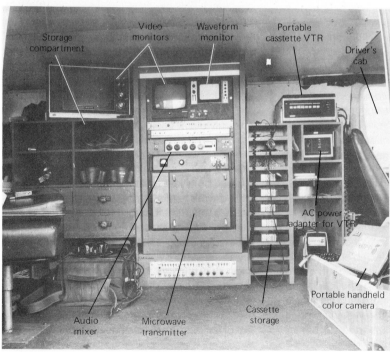

FIGURE 12-1 ENG Mobile Unit.
The medium-sized van is easy to drive and park. The microwave antenna atop the truck is folded for travel. Interior of the mobile unit contains microwave transmitter, production equipment, and supplies. *(Courtesy: Farinon Electric and KSTP-TV)*

The most important requirement for an ENG camera operator is to be able to react quickly to unpredictable events. Yet you should avoid the temptation to shoot all ENG coverage hand-held. First, the camera is never as stable as when it is mounted on a tripod, and, secondly, you restrict the use of telephoto focal lengths when you hand-hold owing to excessive picture jitter on a long lens. Of course, there are situations where a tripod cannot be used, but whenever you have the time and opportunity to use one, the extra work is more than offset by the improved picture quality.

As soon as you arrive at the location and set up your camera, make certain you adjust the cam-

FIGURE 12-2
Although a mobile van permits optimum production convenience, the small size and portability of ENG equipment enables an ordinary car to substitute for the ENG van. The open sunroof provides an effective camera position for the camera operator.
(Courtesy: RCA)

era's white balance before you begin recording. Many ENG crews keep a white cardboard square permanently attached to the bottom of the newscaster's clipboard, so it is a simple matter to set the white balance quickly.

Avoid shooting subjects against a very white background or against an open sky. The flesh tones of the subject will turn very dark and reproduce poorly. Remember that an "auto-iris" always adjusts exposure for the brightest portion of the picture. This should be the principal subject, not the background area.

If you have the time, make a quick test recording of both sound and picture and replay the tape through the camera's viewfinder while you or the VTR engineer listens to the audio playback. Even though you may have made perfect recordings only a few minutes earlier at another location, it does not take much for the cable connecting the recorder to the VTR to break or short out and the test recording will confirm that both camera and VCR are functioning properly.

While you are shooting ENG footage, remember that it will be edited later into a complete story, so make certain that you shoot sufficient cutaway shots to permit flexible editing. Shoot some wide, establishing shots and some over-the-shoulder footage from behind the speaker so that the editor can cut between two statements by using the cutaway footage without creating a distracting jump cut. Another common mistake in ENG shooting is too much camera movement, including excessive zooming. It is difficult to edit a number of shots in which the camera is constantly moving about the scene, so try to keep camera movement and zooms to an absolute minimum.

ENG Audio

Under most circumstances, keep ENG audio as simple as possible. Reporters usually use a hand-held microphone for standuppers and for interviews. On longer interviews, where sharing a single hand mike is inconvenient, you can have the reporter and subject wear lavalier microphones, but you will need a portable mixer to combine the two mike inputs.

Sometimes a shotgun mike is mounted atop the ENG camera to permit one operator to handle both audio and camera at the same time. This is

FIGURE 12-3 **ENG Crew.**
Most ENG crews consist of two persons. One operates the camera while the other operates the videocassette recorder, monitors audio, and sets up lighting when necessary.

not a particularly good idea, except in an emergency, because the resulting sound quality is never very good. A camera located far from the subject can zoom in for a close-up, but a microphone mounted on the camera must pick up the subject from too great a distance. The resulting sound is usually poor, lacking in sound presence, and includes too much extraneous noise on the track.

Windscreens are an indispensable microphone accessory which should be permanently mounted on microphones whenever the reporter must use the mike outdoors.

You should always monitor the audio through a headset as you adjust levels and while you are recording the sound. This is the only way you will know exactly how the sound is being picked up by the microphone and recorded on the tape. As soon as you have finished a recording, spot-check the tape to make certain that you have recorded sound before you leave the location.

Whenever possible, try to record some "room noise" before you leave a news location. Room noise is simply the background or natural sounds of a location which are recorded without anyone

speaking directly into the microphone. The room noise is useful for covering audio edits or to provide some background to voice-over narration, which may be recorded later back at the studio. To record room noise, simply hold the microphone toward the general area of the event, and let the VCR record the sound for ten or fifteen seconds.

ENG Videotape

The 3/4-inch videocassette recorder has become the industry standard for ENG videotaping. The VCR can be powered by its self-contained batteries for portable field operation or plugged into ac outlets using a special adapter when you are working indoors.

In addition to the VCR itself, make sure you bring along necessary tape stock, spare VCR batteries, extra cables, and a shoulder strap or backpack to permit carrying the VCR once you've arrived on location.

Most ENG cameras enable the operator to start and stop the videotape by using controls contained on the camera itself. Before you begin recording, make certain that these controls are working properly and make a test recording which you can play back through the camera's viewfinder.

Whenever you start a videotape recording, leave enough time before you cue the talent for the VCR to engage the tape and to bring it up to speed. A couple of seconds at the top of a recording and a few seconds after you have completed recording will ensure that the tape editor will have good quality program material.

Before you start each recording session, make certain to check the battery power supply. If you have any question about how much power is remaining, replace the old batteries with a fresh unit. Weak batteries can result in either camera or VCR failure at a crucial time during the recording.

It's important to label every cassette and cassette box after recording so they don't get mixed up or inadvertently erased. The safest approach is to affix pregummed labels to *both* the cassette itself and to the cassette box. If you label only the box, you might mix up cassettes and boxes, so take the extra minute to label both clearly and you will avoid possible confusion later.

ENG Lighting

Most ENG lighting is kept as simple as possible with the primary objective to provide the necessary illumination for the camera to produce technically acceptable pictures. The best all-around lighting kit for ENG includes two open face spotlights and a floodlight, along with aluminum light stands and a few gaffer-grip clamps for use in locations where light stands can not be set up. Do not forget to include plenty of ac power extension cable, barndoors, and scrims in your light kit.

Since you may have to shoot at locations where no ac power is available, a battery-powered light unit is a useful accessory to carry in the ENG van.

By its very nature, news production does not always allow you time for careful lighting, and there are times when all you can do is mount a lensless spot atop the camera. But it is a mistake to dismiss how important good lighting can be to quality pictures, and when the time is available, a few minutes of attention devoted to lighting can dramatically improve the quality of your tape.

A single light source aimed straight at the subject creates the unflattering home-movie effect which in-studio lighting is designed to avoid. Whenever possible try to light a subject from at least two different angles to bring out depth, texture, and detail in the picture. Using the conventional key, back, and fill positions for the instruments will produce the best lighting. If you lack the time or space for a three-instrument setup, use at least two lights: one as a key and the other as a combination fill and kicker. Light can also be bounced off a wall or ceiling to create a soft, even illumination, but make sure the bounce surface is painted white. Colored areas will reflect the same color light onto the subject and will upset proper color reproduction of the subject's skin tone.

When shooting outdoors, avoid positioning a subject directly in the sun. The tremendously wide variation in brightness level between light and shadow areas will exceed the camera's acceptable contrast range and will produce a poor-quality picture. Shady areas with indirect lighting are usually ideal because their even illumination eliminates wide brightness differences on the subject. If you must shoot in the sun, set-up the camera so the sun is behind you. If you are shooting outdoors at night, position the subject in front of a background which is already illuminated. If you must also light the background, have the subject stand as close to it as possible, so the subject's spill light will fall on the background area. Unless a portion of the background is sufficiently illuminated to be photographed by the camera, the subject will appear to be standing against a completely black void.

Live ENG Microwave Operation

Live ENG coverage of a news event is made possible by using *microwave* transmission. Microwave—the point-to-point radio transmission of video and audio signals—has been used in television for years, mainly as a method of transmitting network programming to local affiliate stations. Recent technological advances in miniaturization have produced microwave units which are small enough to be installed inside an ENG van and used for news gathering operations. The microwave enables the ENG crew to cover fast-breaking news live, as the event happens, by feeding sound and picture from the remote back to the studio. The microwave system also permits ENG crews to relay videotaped segments of a news story back to the station where it can be edited and broadcast while the ENG unit goes on to cover another assignment in the field.

There are two types of microwave links which are useful for ENG operation: (1) mobile-unit-to-studio, and (2) camera-to-mobile-unit. In both cases, the microwave system requires an uninterrupted line of sight between transmitter and receiver in order to operate.

MOBILE-UNIT-TO-STUDIO LINK ENG mobile

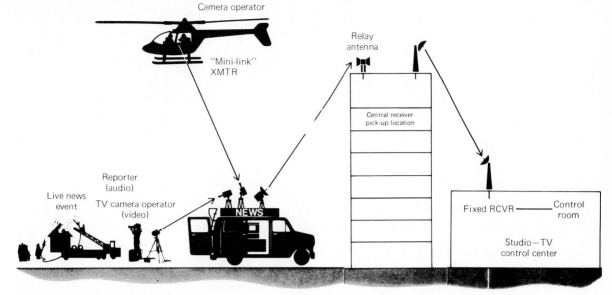

Camera operator

"Mini-link" XMTR

Relay antenna

Central receiver pick-up location

Live news event

Reporter (audio)

TV camera operator (video)

NEWS

Fixed RCVR — Control room

Studio—TV control center

FIGURE 12-4 Microwave Links.
Camera-to-mobile unit links permit wireless transmission of audio and video. Camera can be in a stationary position or in a mobile configuration as in a boat, car, or helicopter. The signal output from the mobile unit is fed to a relay system located atop a tall building or to an antenna within the mobile unit's operating area. The signal is then amplified and retransmitted to the station for recording or live broadcast. *(Courtesy: Farinon Electric)*

units equipped with a microwave system use the dish antenna atop the truck to radio signals back to the studio. Since few stations own a tower high enough to permit a line of sight between the receiving dish at the station and the ENG mobile unit's coverage area, a *microwave relay* system is used to send signals from the field to a repeater unit atop a high-rise building and from there to the station itself. In most cities, the relay receiver dishes are mounted on the tallest, centrally located building which is available.

To help the ENG crew quickly position the truck and its microwave antenna dish for the best transmission, some stations have found it useful to develop a log book of recommended truck positions at various news locations around town. Since many stories occur in the same standard locations (City Hall, downtown hotels, sports arenas), the log book saves time by informing a crew member who is unfamiliar with the location of the best place to park the mobile unit. When a direct line of sight cannot be established such as in a downtown area surrounded by tall buildings,

sometimes a signal can be bounced off one building and ricocheted to the receiving tower. (See Figure 12-4.)

CAMERA-TO-MOBILE-UNIT LINK When a direct cable run from the camera to the ENG mobile unit is impossible, a small battery-operated microwave unit can be used to establish the necessary link. This makes it possible for the ENG unit to cover live events where the mobile unit cannot get as close to the action as the camera can. As with larger microwaves, a direct line of sight must be established between the camera's transmitter and the truck's receiver. Most portable units permit an operating range of about one mile and, depending on the size of their batteries, anywhere from four to eight hours of continuous operation. (See Figure 12-5.)

Needless to say, the same portable microwave unit can be used for other types of remote productions as well, to link cameras which are positioned where cable connections to the mobile unit are either impossible or impractical. Portable microwave units make it possible to obtain those dramatic shots from a blimp or a helicopter and are indispensable in providing the type of coverage viewers have come to expect on sports, news, and entertainment remotes.

ENG PRODUCTION

The ENG production operation works somewhat differently from other kinds of remote productions owing to the special nature of news coverage. For that reason, it is difficult to discuss ENG production during each of the four production stages as we can other kinds of remote operations. Since it is obviously impossible to predict every news event, there is little in the way of elaborate preproduction planning or rehearsal which can be directly applied to ENG operations. Also, most ENG units consist of a limited one- or two-person crew which handles all of the technical operations and a reporter who functions as a "producer-director" on the scene.

Nevertheless, there are a number of important considerations which relate specifically to ENG production, and we can organize them into two of the production stages: preproduction planning and production. Since there is little rehearsal and because postproduction editing is usually com-

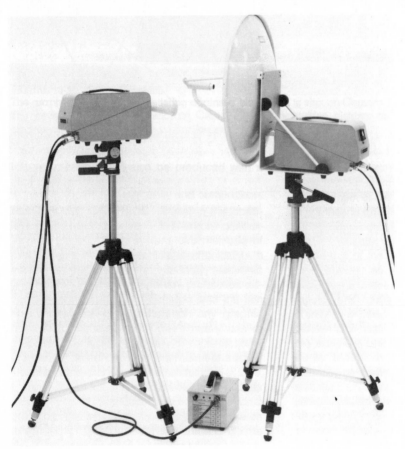

FIGURE 12-5 **Portable Microwave Unit.**
The horn on the left provides a 1½-mile range while the larger dish provides a 26-mile range. Battery pack is on the ground.
(Courtesy: Farinon Electric)

pleted back at the studio, we won't concern ourselves with these stages here.

Preproduction Planning

While it may not be possible to preplan every ENG event, there are some basic planning details which should be taken into account before the unit leaves the studio for the day's work in the field. Many ENG crews develop a standard checklist of equipment which they review prior to leaving the studio. The list in Figure 12-6 is an example of an ENG checklist.

In most towns and cities, the police department issues a "press pass," which permits reporters and crews to cross police lines in order to cover a story. Each crew member should have his or her own pass and should make certain that he or she carries them to every location.

Another indispensable item for the ENG van, which can be easily overlooked, is a complete road map of the local areas which you will be covering. You never know when you will be dispatched to cover an event on a moment's notice, and you may need to consult a map to help you find your way.

Although news can happen anywhere at any time, there are usually a number of locations which the ENG unit will cover on a regular basis. City Hall, the state capitol, major hotels in the area, the airport, the local courthouse, and major sports or events arenas are all examples of locations where news is regularly covered. Over time ENG units become familiar with these locations and develop routines for parking the van, setting up equipment, and so on, which makes for a more efficient operation. Some stations develop an ENG *logbook*, which contains essential information about regularly covered locations which is kept on board the ENG van and available for instant reference. The book includes information about the location of public telephones, ac power outlets which are working, names of official contacts, the best location to park the car or van, access-ways to news locations, and so on. Whenever an ENG crew visits a new place which is not included in the logbook, the information about the new location is added so the book is continually updated. It may sound trivial, but only if you have

ever spent twenty minutes trying to find an open door to an enormous sports arena while the event you are suppose to cover takes place without you, can you begin to appreciate how important such information can be to an efficient ENG operation.

Production

Most ENG units consist of one or two people to handle the technical operation of camera, VCR, audio and lighting, and a reporter who also functions as a "producer-director." The reporter and crew work closely in establishing camera positions, camera shots, and the approach which will be used to cover the event. After a certain amount of time working together, the two usually require little formal communication between them, and this is a real advantage when breaking news events which do not permit extended conversation between reporter and crew are being covered.

What is most important in ENG operation is for the crew to remain flexible so they can cover unexpected situations which might develop. The weakest equipment link is the cable which connects the camera to the VCR because it restricts the camera operator's movement and is subject to damage and shorting-out if any stress on the cable and connectors develops as the camera operator's movement outdistances the VCR operator's ability to keep up. Of course, if you are working with an integrated video camera/ recorder, the camera operator is completely self-contained and has no cable constraints on his or her movement around the scene.

The exact production sequence for an ENG story usually depends upon the event which is being covered. Normally, the ENG crew will shoot interviews or the statements of the major participants first. Once these *sound bites* are recorded, the news reporter will prepare an opening and closing "standupper" which introduces the piece, sets up the sound bites, and provides a closing summary to the story. Whenever possible, the

FIGURE 12-6 **ENG Equipment Checklist.**

CAMERA
Camera
Lens
Lens Shade
Battery Pack/Belt and Spares
Tripod
Lens Cleaning Tissue
White Balance Card

AUDIO
Microphones (Handheld and Shotgun)
Microphone Cable
Headset
Portable Audio Mixer

LIGHTING
Openface Spotlights
Openface Floodlights
Portable Floorstands
Gaffer grip clamps
AC Power Cable
Scrims
Barndoors
Gloves
Battery for Portable Light

VCR
Videocassette Recorder
Spare Cable/VCR Cable
Batteries and Spares
Headsets
Cassettes
AC Power Adaptor

MISCELLANEOUS
Police Permits for Crew
Roadmap
Flashlight
Pads and Pens
Gaffer Tape
AC Extension Cable
Walkie-Talkies
Video Monitor/Receiver
Stopwatch

camera operator should shoot a number of cutaway shots which can be used to provide smooth visual transitions for editing the story back at the studio.

During the editing process, the various sound bites, cutaways, and standuppers will be edited and sequenced to produce the completed story. The length of any of these segments and the overall running time of the completed story depend upon the relative importance of the event and the amount of time which is allocated to the story in the final news program lineup.

CUTWAYS AND REVERSE ANGLES It is impossible to overemphasize the importance for the ENG camera operator to shoot the event with videotape editing in mind. Especially if you will not be editing the piece yourself, remember to provide as much cutaway material as possible for the editor to use in assembling the story. The cutaway can be a wide shot of the scene, a close-up of the reporter responding to the speaker, close-ups of other cameras and reporters, and so on. You can't provide too much cutaway footage, and anything which the editor doesn't use is simply erased, but it is impossible to shoot more footage once you have left the news scene.

A "reverse angle" is simply the opposite angle of the main subject. For example, whenever you shoot an interview, you will most likely position the camera over the shoulder of the reporter so you can get a close-up of the subject who is being interviewed. Unfortunately, the single camera does not permit you to photograph the reporter, in close-up, while she or he asks questions. To provide this footage, once the subject's interview is over, position the camera in the reverse angle —behind the subject's back although the actual subject is often gone by this time—and have the reporter ask each question again while you shoot the reporter in close-up. For both journalistic as well as editing reasons, it is important that the reporter repeat each question identically to the way it was asked during the actual interview. The reverse angles footage enables the editor to cut in a close-up of the reporter asking each question as the news story is assembled in the edit room.

CAMERA MOVEMENT Another important point to remember during ENG production is to avoid excessive camera movement or zooms. News is exciting, and the natural inclination for the camera operator is to pan and zoom about a scene much as our eyes might do normally. Although lots of camera movement may seem perfectly natural to you while you are taping the news event, once you view the footage in the edit room, you will probably discover that to your dismay there is so much camera movement that it is difficult for the viewer to become oriented to what is taking place, and the constant movement makes editing difficult.

Live ENG Production

One of the most important capabilities of ENG is the ability to broadcast live from a remote location. In order to do this, you must establish a two-way communication link between the home studio and the ENG unit in the field. Although the two-way radio is sufficient for engineering purposes, it is not useful for a live feed because it does

not permit easy cuing or two-way conversations on the air between the field reporter and the anchorperson in the studio.

A convenient way to establish the necessary communication link between talent in the field and the home studio is to use a battery-powered TV receiver and a portable FM/TV audio receiver. The TV monitor is simply tuned to the station's broadcast frequency and placed out of camera range, where it can be viewed by the reporter. The reporter wears an earpiece connected to the FM/TV audio receiver, which is tuned to the station's audio frequency. By listening over the radio earplug and watching the television receiver, the reporter can see and hear everything that is going out over the air. As soon as the reporter hears the cue and sees the monitor picture switch

FIGURE 12-7 **Live ENG Broadcast.**
The reporter uses a conventional television receiver tuned to the station for video cues and an FM radio with TV audio band connected to an earpiece for audio cues. The monitor helps to cue the reporter for voice-over narration of tape segments which are inserted by the station. The audio enables the reporter to carry on a live two-way conversation with talent back at the studio.
(*Courtesy: WSB-TV*)

to the live ENG feed, he or she begins the report. By listening to the program audio, the reporter can also engage in a two-way conversation with the anchorperson who is back at the studio. (See Figure 12-7.)

Postproduction Videotape Editing

Every ENG story which was videotaped in the field in segments by the single camera/single VTR approach must be edited into a completed story before it can be used. In Chapters 9 and 10, we discussed the videocassette editing equipment which is commonly used for ENG editing. In fact, editing an ENG story is no different from editing any nonnews single camera/single VTR production in that every cut must be made by videotape editing during postproduction. What makes ENG editing somewhat special, however, is the fact that news footage must be edited quickly and efficiently in order to prepare the material for use on the day's news program.

Virtually all ENG footage is recorded on helical cassettes. While the 3/4-inch U-matic is by far the most widely used format, a number of operations are recording and editing on 1/2-inch format. In either case, the news footage can be edited in its original format, and most stations actually broadcast their news segments directly from the edited cassette by playing the VCR through a time base corrector.

LOGGING THE ENG TAPE Unlike commercial, industrial, or entertainment production, where you can preview your recorded takes at a fairly leisurely pace, news production demands that quick decisions be made immediately and often under serious time pressures. To edit ENG footage effectively, someone should provide the editor with an accurate log of each recorded cassette's contents so that the producer and tape editor can locate important cuts quickly for previewing and editing.

If you are using SMPTE time code, the time code is either recorded in the field or recorded

simultaneously as the producer and/or editor screens the raw footage. By noting the time code number for each important cut on the tape, you will be able to return to a particular segment quickly when you edit the piece. If the story is microwaved into the studio from the field, the time code is usually laid down as the microwave feed is recorded.

Many news operations prefer to use control track time code editing since laying down the time code requires no additional time. In this case, you should have made a log of the tape's contents, noting approximate reel-times where important footage is located. For instance, if an important sound-bite interview was recorded about twelve minutes into the tape, the editor will punch 00:12:00:00 into the edit programmer, and the tape will automatically run forward. Shuttling and jogging the tape back and forth will locate the exact part of the tape where the sound bite begins. A well-prepared log is particularly important in news operations since quite often the crew and reporter who shot the story are out in the field covering other news while the editor and news program producer cut the tape for broadcast.

EDITING THE ENG NEWS STORY Most news spots contain four types of footage: (1) *actuality*, which is coverage of the event as it occurs; (2) *interviews*; (3) *standuppers*, where the reporter or correspondent talks directly to the camera; and (4) *cutaways*, which are wide-angle or reverse angle shots of the scene and are used as transitional footage to eliminate jump cuts or an audio mismatch.

While each news piece is often slightly different, most follow a fairly basic format. First, the reporter gives a brief orientation or background, to the story. This is often done as a standupper, the reporter talking directly to the camera, using the news event as a background. Next, various sound bites are used, including actuality footage, or interviews. Finally, the reporter ends with another standupper, summing up the story.

Many ENG editors like to lay down the sound track first, going from the original recorded cassette onto a prerecorded crystal black tape. Then they go back and execute video-only edits, building the visuals to match the existing sound track. Sometimes, the reporter's opening or closing standupper is eliminated entirely if there are enough good actuality visuals to take the place of the "talking head." Additional voice-over narration can be added later if necessary. Of course, you can also simply assemble various segments, beginning with the first actuality or the reporter's standupper and continuing through until the end of the news spot.

Since many stations are broadcasting their news spots directly from cassette, a number of ingenious and time-saving shortcuts have been developed to speed up the editing process. At WSB-TV in Atlanta, ENG footage is edited in the following sequence:

1 The editor uses the written log, which was prepared previously, to locate important scenes within the cassette tape. All editing is done using time code numbers.

2 The producer, editor, and sometimes the reporter make the final decisions on which segments to include in the spot and their proper order.

3 A prerecorded cassette containing film academy leader and video black is used for a visual cue before the actual program material. A crystal-black cassette is inserted into the

edit/record VTR, and the academy leader and video black are transferred to the blank cassette.

4 The news story is assembled and edited according to the predetermined sequence. Sometimes the audio track is recorded first and the visuals are matched later. Other times, the sequences are assembled in order, depending upon the type of story and the available footage.

5 Another cassette, which is prerecorded with video black, is used to edit-on black at the end of the story. This is a safety pad in case the program director does not punch out of the taped segment quickly enough on the air.

6 The edited tape is labeled with the story and content, date, running time, and the in and out cues. The labeled cassette is sent by messenger to the VTR room for air.

ENG VIDEOTAPE PLAYBACK As we mentioned earlier, most operations play back the

FIGURE 12-8 **ENG Editing.** ENG footage is edited with two video cassette machines and a program edit control unit. The monitor on the left shows the editor the output of the playback VCR. The monitor on the right shows the output of the edit/record VCR. (*Courtesy: KSTP-TV*)

edited cassette without "bumping up" to 1-inch or 2-inch VTR formats. To do this, the playback VCR must feed through a time base corrector and then into the studio control room, where the playback VCR appears as one of a number of video inputs on the production switcher.

Since the edited tape should have been edited with some visual countdown leader, it is a simple matter of cuing up the tape and still-framing it with the pause control at the appropriate number. A three-second roll cue is common, so all you do is play back the cassette until you see the number 3 and press the pause control. The freeze-frame allows the cue leader number to appear on the VCR playback monitor in the studio control room and indicates that the tape is cued and ready. At the correct time, the director calls for the VCR operator to roll tape, and three seconds later picture and sound appear and the tape is punched up on the air.

SUMMARY

Electronic news gathering production utilizes a portable television camera connected to a battery-operated video cassette recorder. The ENG mobile unit can be as simple as a conventional car, or a specially modified van or minibus which contains all of the production equipment and a microwave unit which enables sound and picture to be transmitted directly back to the home station for recording or live broadcast.

The ENG camera is a portable camera which operates off of battery-supplied power. Although ENG coverage frequently requires the camera to be hand-held, a tripod should always be used when the situation permits to eliminate jitter and permit shooting at longer lens focal lengths. When shooting ENG footage, always record some "cutaway" footage to enable the tape editor to bridge edits.

ENG audio production is usually simple and straightforward. Hand-held or lavalier microphones are frequently used although shotgun mikes are used when it is impossible to position the microphone directly in front of the subject. Audio recording should always be monitored through headsets to ensure quality. Before leaving a news location, the engineer should record some "room noise" to aid in audio editing of the videotape footage.

The ENG news material is recorded on either 3/4-inch or 1/2-inch video cassettes. The VCR is battery operated, and fresh batteries should always be kept on hand to avoid running out of power at a crucial time during recording.

ENG lighting requirements are generally straightforward although when the situation permits, a number of portable lensless instruments should be positioned to illuminate the subject using the conventional three-instrument lighting approach.

Live sound and pictures of a news event can be transmitted back to the home station by using a portable microwave unit, which is installed in many ENG mobile vans. In addition, smaller microwave units enable the ENG camera crew to work at a distance from the mobile van without having to run cable between the camera and the mobile unit.

ENG production is somewhat different from most other television production situations owing to the nature of news coverage. Although few events can be preplanned, the ENG crew can utilize a "logbook" which includes pertinent information about each news site that is regularly visited by the ENG crew. During a recording, the ENG crew should remember to keep editing in mind and to shoot the event in a way that provides the editor with enough footage to assemble an effective news story. This includes the use of cutaway footage, reverse angles, and a minimum of random camera movement or zooming.

ENG footage which was recorded in the field must be edited into the completed news piece. The first step is to log the tape and note either the time code or counter numbers where each important segment appears. During actual editing, the various taped cuts including sound bites, standuppers, and cutaways must be edited and sequenced into a complete story. Most operations utilize the edited cassette for playback during production. The playback VCR is run through a time base corrector which enables the cassette to be integrated with other video elements through the production switcher in the control room. The use of the pause control on the playback VCR makes it possible to preroll the tape to any point and freeze-frame it until called for by the program director.

CHAPTER 13

TELEVISION GRAPHICS

What makes a program successful? Certainly a lot depends on such major production elements as the performers, the camera shots, the audio, and the direction. But there are many other elements, not all obvious perhaps but nevertheless important, which contribute to a production. One of these is the use of television *graphics*—the various titles, photographs, lettering, illustrations, and diagrams that appear in virtually every television program. Graphics generally serve a very utilitarian function: to convey information to the viewer. Yet when graphics are carefully designed and imaginatively used in a production, they are capable of doing much more.

Think about how a program opens. Usually there are program titles, which give you some idea of the show's content and its tone. Certainly the titles which appear before a suspenseful murder mystery look much different from the titles which introduce a situation comedy. The look of the graphics can enhance and complement a show and interest the viewer in what is to come. To do this, graphics must be aesthetically pleasing to look at and, at the same time, effective in communicating their message to the audience.

You can begin to develop an eye and appreciation for graphic design by watching your local television stations carefully. Does each station have its own, unique graphic image? What do the different station logos or identification symbols look like? Do they match the station's "personality?" Take a close look at the same kind of programming as it is produced on different stations; local newscasts are a good example. How are the graphics designed and used on each station? Are some more effective than others? Why?

In educational and industrial communications, graphics frequently play a very significant role, since these productions often deal with complicated ideas and processes. Well-designed graphics can help to clarify program content and to focus the audience's attention on the most important elements in the presentation.

In short, effective television graphics should be designed to:

1   Convey information clearly and directly.
2   Establish a show's overall mood and tone through the graphic style.
3   Help to present facts, concepts, or processes visually so that the viewer will better understand and appreciate the program content.

GRAPHIC DESIGN

When you start to design a graphic there are a number of technical and aesthetic factors to consider. The *technical* considerations are those necessary in adapting the graphic to the capabilities and limitations of the television medium. The *aesthetic* or creative considerations are the various elements which contribute to make the artwork both interesting to look at and also effective in communicating your message to the viewer.

Technical Design Considerations

Technical considerations always come first and foremost because they determine how a graphic

will appear in the videospace. No matter how imaginative and creative a graphic may be, if it does not reproduce well over the television system, it cannot be effective.

ASPECT RATIO   All television screens, regardless of their size, are designed in a 3:4 *aspect ratio*. This means that all pictures must be framed in proportions of three units high to four units wide. Graphics which are designed for television must also fit this ratio, or should be adapted to it. (See Figure 13-1.)

If you find you must use a graphic that is not in the 3:4 ratio, there are a number of options open to you. First, you can choose to come closer with your camera and fill the screen with only a portion of the visual. If you must show the entire graphic, you will have to mount the artwork on black art cards and accept the fact that the broadcast picture will show black borders along the edges of the screen. (See Figure 13-2.)

Fortunately, as the 3:4 ratio is about the same size as a 2 × 2 inch 35-mm slide, there are no serious format problems in shooting slides for use in the film chain. However, slides fit the aspect ratio only when they are projected horizontally. If

FIGURE 13-1   **Aspect Ratio.**
All television graphics must be designed in the proportion of three units high to four units wide.

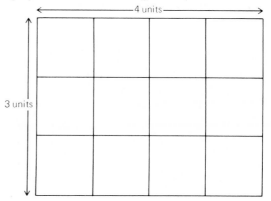

3:4 aspect ratio
of TV screen
image

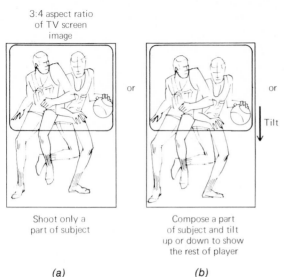

or

or

Tilt

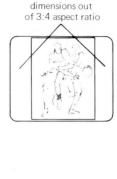

Empty area
due to subject
dimensions out
of 3:4 aspect ratio

Shoot only a
part of subject

(a)

Compose a part
of subject and tilt
up or down to show
the rest of player

(b)

Shoot subject smaller
so entire player fills screen
and accept the lack of
picture material on either
side of the screen

(c)

FIGURE 13-2
If a graphic cannot be produced in the 3:4 aspect ratio, the director must either shoot only a portion of the graphic (a), shoot a portion and pan the camera to reveal the rest of the graphic (b), or shoot the entire graphic and accept border edges without graphic material (c).

you use a vertically framed slide in the slide projector, you cannot avoid showing black edges on the sides of the picture, and you also risk cropping important visual information at the top and bottom of the television screen.

A graphic that has been properly framed into the 3:4 aspect ratio enables you to make the most efficient and effective use of the videospace.

SCANNING AND ESSENTIAL AREAS We mentioned in an earlier chapter that the picture you see in the camera's viewfinder contains much more information than what ultimately appears on the audience's television screen. That is because the outer edges of the picture are lost as the video signal travels through the television system. To make certain that the audience sees the most important visual information, you must be sure it is contained within the *essential area*—the portion of the picture that will definitely appear on all television receivers.

Figure 13-3 shows a graphic card that is divided into two areas: the *essential area* inside is the "safe" area that will be seen by everyone. The next larger area is the *scanning area*, which contains information that is photographed by the television camera but which may not appear on all audience receivers. Since many members of the audience will see all or part of the scanning area,

it should contain some visual information but none of the essential words or illustrations which must be seen in order to understand the message.

PICTURE RESOLUTION The limited ability of the television system to reproduce fine picture detail is an important factor in the planning and preparation of graphics. Lettering styles should be selected for the best television reproduction. Avoid light, ornate styles with fine lines and delicate serifs. The resolving power of the television system simply cannot handle the fine detail, and your lettering on screen may be unreadable. Instead, choose medium-sized type styles which are simple and direct in design. Most of the "sans serif" typefaces with good visual weight and clarity reproduce well on television. (See Figure 13-4.)

You must also consider resolution limitations in planning the use of photographs or illustrations. Photos should not contain extremely small details, which will blur or become invisible on the television screen. Very complex illustrations, such as maps or technical diagrams, must be simplified and redesigned for television use.

If in doubt about a type style or illustration, preview the graphic on camera to make certain it is legible. How it looks in the videospace is, of course, how the audience will ultimately see it.

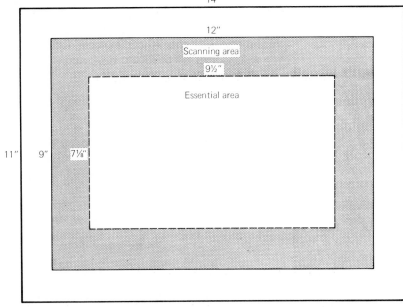

14"

12"

Scanning area

9½"

Essential area

Edge of
graphic
card

11" 9" 7⅞"

FIGURE 13-3 Essential Area.
The gray shaded area will be cropped by the technical characteristics of the television reproduction system. All essential graphic information should be designed to fall into the central "safe" area to ensure that viewers will see it.

CONTRAST RANGE AND GRAY SCALE From our earlier chapters on cameras and lighting, you already know that the television system works best when it photographs a scene consisting of a contrast ratio which is about 20:1. In addition, the television camera cannot reproduce pure white and pure black accurately, so we work with television white (about 60 percent reflectance) and television black (about 3 percent reflectance). This also applies to graphics.

The television gray scale is especially helpful to graphic artists for a number of reasons. First, a well-designed graphic needs good contrast between foreground lettering and background ele-

FIGURE 13-4 Typefaces for Television.
The resolution limitations of the television system require simple, bold typefaces. Avoid light, ornate styles with fine lines or delicate serifs, which are difficult to reproduce on screen.

ABCDEFGHIJKL
MNOPQRSTUV

ABCDEFGHIJKLM
NOPQRSTUVWXYZ

ABCDEFGHIJKLMNOPQRSTUVWXYZABCDEFGHIJKLM
abcdefghijklmnopqrstuvwxyz 1234567890

AOPQRSTUV aopqrstuv $5678

Not suitable

AXYZABCDEFGHIJKL
aqrstuvwxyzabc $567

APQRSTUVWXYZAB
abcdefghijklmnopqrst

AZABCDEFGHIJKLMNO
aijklmnopqrstuv $12345

AOPQRSTUVWXYZ
apqrstuvwxyz $567

Good choices for TV graphics

355

ments. Most graphic artists suggest maintaining at least two gray scale steps between them. Avoid pure white or pure black. The only exception to this rule is the preparation of "super" graphics where white lettering or line drawings are prepared against a black background. The super graphic is electronically superimposed or inserted into another picture.

Good contrast within the proper range will not only make your graphics look and reproduce better, but it will also improve the legibility on screen.

COLOR   Color is naturally an important production element in the design and preparation of graphic material. You will recall that a color depends on three factors: (1) *hue*, which is the actual color such as red, green, or blue; (2) *saturation*, which is the intensity of the pure color or how much white has been mixed in to dilute the pure color; and (3) *brightness*, how light or dark the color appears. Brightness is particularly important in designing graphics because we are concerned both with black and white reproduction as well as with color. Colors must vary not only in hue and saturation, which naturally register on a color receiver, but also in different brightness values—the only way to indicate color differences in monochrome. Take two colors such as red and green, for example. They may be completely different in hue and saturation but, because they have essentially similar brightness values, there is no apparent difference or contrast between them in a black and white reproduction. (See Color Plate 8.)

Not all colors reproduce equally well on television. Even the most expensive and sophisticated color camera has difficulty in reproducing red and orange hues accurately. If possible, avoid using large areas of deeply saturated reds which may reproduce off-color or distort the picture into a series of horizontal lines, called "banding," which run across the television screen.

SCALE   The size of the television screen is rela-

tively small, and the videospace can be easily cluttered with excessive detail. Television graphics should be designed as simply as possible, and graphics are better with less information than with too much. If in doubt, leave it out.

Aesthetic Elements in Graphic Design

After you have met the various technical requirements of graphic design, you can begin to tackle the aesthetic elements. Among the aesthetic or creative details you should consider in graphic design are style, composition, the use of color, and the most effective way to convey information.

STYLE   All the elements that go into a graphic—the lettering, the illustrations, the color scheme, the layout—make up its visual style. A program's graphics should be designed to complement the show's style and approach. A contemporary music show might benefit from graphics that are designed in a highly stylized and abstract way. A more straightforward and traditional graphics style might best enhance a news or public affairs show. A drama set in Victorian England might use graphics that are designed to reflect the historical period depicted on the program. (See Figure 13-5.)

A particular graphic style is frequently used by a station or production company in their logo or identification symbol. The CBS "eye" is one of the most effective and enduring logos. While the style has been changed over the years to reflect modern taste, the CBS logo still provides instant identification, recognition, and impact whenever it is used. Graphic style can also unify a series of individual shows. For example, all CBS news programs use the same style lettering to give them a unified and integrated graphic "look." (See Figure 13-6.)

Before work begins on a production's artwork, the graphic artist and the program producer must discuss the concept of the show, the program's overall tone and objectives, and how the graphics are to be used. This will help the artist to design a graphics style that will complement the production. Of course style must also be functional. You should not necessarily use old English lettering in an instructional program on Chaucer's poetry.

(a)

(b)

(c)

FIGURE 13-5 Style.
The style of the graphic should reflect the overall tone of the program, the station's image, or the show's approach. These graphics are examples of various styles which were designed to complement the message, program, or station image. *(Courtesy: KPIX Group W San Francisco; Phyllis Essex–WPLG-TV; Ann Aiken–WTOP-TV)*

The lettering style, while admittedly "English," is unsuited for TV reproduction. The result may be words that prove to be unreadable. The idea is to set the mood and tone with a style that is still effective in communicating with the audience.

COMPOSITION AND LAYOUT The same basic rules that apply to picture composition for cameras hold true for graphic design as well. A well-composed graphic organizes the various elements into meaningful and related units, conveys the necessary information to the audience, and presents a visual design that is aesthetically pleasing and interesting to watch.

It is a good idea to organize different graphic elements into a series of distinct groupings. Graphics may contain a number of different kinds of groupings or units; some are pictorial, and some are lettered. By experimenting with various ways of organizing the videospace, you will come up with those combinations that work best. For example, it has been found that written information is read and understood most easily when it is organized into related units. The program-promotional graphic in Figure 13-7 groups the information into three units: the show title, the picture, and the air date and time. Do not forget to consider aspect ratio, essential area, and the other technical factors when you compose your graphic.

Graphics are frequently most effective when they are arranged or composed in a balanced or symmetrical fashion. This does not mean that lettering or illustrations must always be centered in the screen. Rather, the overall visual impression which the graphic makes should be one of stability and order. You might decide to run written commercial copy flush left along one edge of the screen and a picture of the sponsor's product on the other side of the picture. The result is a balanced visual that is interesting to look at and still easy to understand. (See Figure 13-8.)

Of course, there are times when unbalanced or asymmetrical compositions can be very effective. Graphics are by definition two-dimensional. Inducing a depth dimension by running lettering or

FIGURE 13-6 Logo.
A logo is an identification symbol which is immediately identifiable in the audience's mind. The CBS eye is one of the most effective and enduring logos although it has been changed and updated over the years to present a more contemporary look. *(Courtesy: CBS)*

pictures along the diagonal or giving the illusion of depth through perspective are useful techniques. The imbalance and instability will add a feeling of tension and immediacy to the graphic, drawing the viewer's attention more closely to the screen. (See Figure 13-9.)

Your layout and composition will depend, in

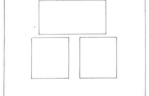

FIGURE 13-7 **Composing Graphic Information.**
This graphic was designed with the visual information organized into three
groupings as indicated in the diagram. *(Courtesy: WAGA-TV)*

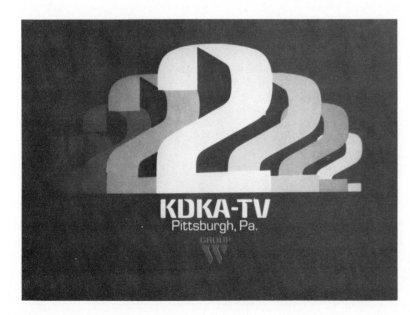

FIGURE   13-8   **Balanced
Graphic Deisgn.**
*(Courtesy: KDKA-TV, Group
W Pittsburgh)*

FIGURE 13-9 **Unbalanced Graphic Design.**
Depth is induced in this electronically generated
graphic through its design and by running the type
on the diagonal. *(Courtesy: ABC-TV)*

part, on how the director intends to use the
graphic in production. For example, a graphic
that is to be "supered" during the show must be
composed so the combination of foreground
super and background picture produce a well-
composed shot.

USING COLOR IN GRAPHIC DESIGN Color
can be a vital element in graphic design if used
well or astonishingly ineffective if used poorly. To
make the best use of color, you must understand
some of its basic principles.

One fundamental color principle is the fact that
colors are influenced by their surroundings. For

instance, the greater the light intensity that shines
on a color, the brighter and more saturated the
color appears. Another important point is that a
light color against a dark background looks larger
and even lighter than a light color against a light
background. Take a look at the two squares in
Figure 13-10. The size and brightness value of the
two circles are identical, but the darker background makes its circle appear even larger and
lighter than its identical twin. The same idea is
easily applied to lettering and layout design.
Important foreground elements ought to be lighter
in color and brighter than the background so that
they will stand out more prominently and attract
the viewer's attention.

When you are working with color, keep in mind
the need for good contrast in hue, saturation, and
brightness. Viewers watching in monochrome,
where only brightness values register, require a
distinct change in the gray scale in order to
perceive any difference between two colors. But
good brightness contrast will make your visual
more interesting and effective in color too.

Colors also play a part in how the audience
perceives a subject. Some experts claim that
colors influence our judgments of size, weight,
and temperature and even affect our psychological state of mind. Introductory art classes make
the point that reddish colors appear "warm" and
bluish colors appear "cold." We tend to associate
certain colors with different activities. Reds are
linked with warmth, fire and tension. Blues are
associated with cold, steel, and ice. According to
Zettl,[1] colors can be viewed as being either

[1]Herbert Zettl, *Sight Sound Motion*, Wadsworth, Belmont, Calif.,
1973, p. 70.

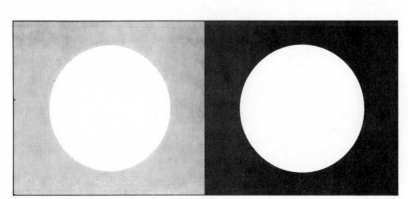

FIGURE 13-10
Both circles are identical in
size, but the light-colored circle against the dark background appears larger and
more dominant. The proper
use of brightness and color
contrast will help to lend visual dominance to the most
important parts of a graphic.

"high-energy" or "low-energy" colors. High-energy colors are reds, oranges, and yellows, particularly when they are deeply saturated. Low-energy colors are cooler blues, browns, and purples, especially when they are less saturated. According to the theory, high-energy colors tend to involve the audience more directly and with greater impact than do low-energy colors.

Of course, these color rules are not necessarily true at all times, and they often vary, depending, in large part, on a particular color's surroundings and the relative brightness and saturation of adjacent shades. A normally warm red may appear cool in comparison to an even warmer red nearby. Our intuition does tell us, nevertheless, that certain colors just seem to go better with certain subjects or emotions. Lighter colors look happier and give an uptempo feeling. Darker colors suggest a somber, dramatic, moody atmosphere. The graphics for a daily morning show would most likely be designed in bright and cheery reds and yellows rather than in dark browns and purples. Yet the darker colors might be just right for titles for a Shakespearean tragedy since they match the program's mood and tone.

DESIGNING FOR EFFECTIVE COMMUNICATION When you come right down to it, a graphic's basic function is to impart information to the viewer. No matter what the particular message is—a program title, the sponsor's product, a station's call letters, a key point in an instructional presentation—the graphic must be designed for the most effective communication possible.

Graphics rarely stay on screen for very long, so they should be designed to make an immediate impression on the viewer. Sometimes images that are frequently associated with a subject can help. A news story about money might use such common symbols as a dollar bill or coins. If you need a long time to absorb everything of importance in a graphic, it will not be very meaningful to the audience. Graphics communicate best when they are clear, simple, and direct.

Illustrations are particularly important in many shows, but they must be carefully designed and prepared for maximum communication impact. You will most likely have to adapt existing illustrations to television's demanding specifications or develop new ones completely from scratch. Ironically, the less material contained in a single

illustration, the more effective the information communicated to the audience. That is because removing excess detail makes your important points or facts stand out more prominently.

Even the use of such simple graphic techniques as a performer writing on a blackboard should be carefully planned and prepared. During production is no time to learn that your talent's handwriting makes a doctor's prescription look like block printing. Talent-produced lettering must be clear, sharp, large, and legible. It should also be organized into distinct groupings which will fit the aspect ratio.

The graphic's moment of truth comes when it is photographed on camera. A graphic may look fine to the naked eye, but if it is not clearly legible on screen it cannot possibly be effective.

GRAPHIC LETTERING TECHNIQUES

All television graphics consist of two basic elements: *lettering*, which includes all text material, and *illustration*, which includes picture material such as photographs, maps, drawings, and charts. Of course, lettering and illustration are often combined into a composite graphic such as a map with key cities identified or a photograph of a newsmaker with her name and title supered on a lower third of the screen.

There are three basic methods used to produce lettering for television graphics: (1) electronic, (2) mechanical, and (3) photographic.

Electronic Character Generator

The electronic character generator (CG) has become an indispensable graphics tool due to its enormous production versatility and its ability to produce perfect lettering instantly. With a wide range of character generator models available for every price/performance range, the CG has become as standard a production device as a camera or microphone.

CREATING LETTERING All character genera-

(a)

Keyboard controls

Cursor controls

Memory controls

(b)

tors operate pretty much the same way and require absolutely no artistic talent or ability to create perfect lettering. The device consists of a typewriter-type keyboard which is used to create the lettering and to control the position of the characters on the screen.

The CG operator generally has two monitors to work with. The first, labeled "program" or "line," displays the message which is currently available to be placed on the air. The second monitor, labeled "preview" or "setup," displays another message which the CG operator is preparing. This enables the operator to set up one message while another is available to the program director during production. (See Figure 13-11.)

Using the keyboard, the operator types out the message which appears on the "preset" screen. A *cursor*, or position marker, shows the location of each character as it is typed onto the screen. By positioning the cursor anywhere on the screen, the message can be placed wherever the director and CG operator wish it to appear. For example, the temperature of key cities across the country can be positioned by using the cursor to correspond to a map of the nation, which is photographed by a studio camera or off a slide on the film chain. Using the cursor, the CG operator can position and insert the temperature of key cities. The director simply supers the CG output over the picture of the map to produce a composite graphic.

CHARACTER MANIPULATION Once the message has been typed on screen, it can be manipulated by the CG operator in a number of ways. Most CGs have an automatic centering feature which accurately positions the message in the center of the screen. In addition, CGs can *flash* individual words or phrases, can *color* both the lettering and/or the background, and can *outline* the lettering with *edging* or *drop-shadow* to make the characters appear more prominent when they are supered against another background picture.

It is also possible to move the lettering across the screen at a variable speed. *Crawling* is the movement of the message horizontally from the right side of the screen to the left. *Rolling* is the vertical movement of the lettering from the bottom of the screen to the top. (See Figure 13-12.)

Many character generators have their own video mixers built in. This permits the CG operator to insert lettering into the video signal which leaves the production console switcher. The only problem is that the director must be able to cue the CG operator when to insert and remove lettering, and during a complex production it can be difficult for the CG operator both to prepare new lettering and to insert material precisely on cue. In most studios, the output of the character generator is handled as another video source like a studio camera or film chain. In this case, the CG appears as an input on the program switcher and on its own monitor in the studio control room. This permits the switcher or technical director to control inserts while following the program director's cues and leaves the CG operator free to concentrate on preparing upcoming graphics.

MEMORY STORAGE AND RETRIEVAL The real value of the CG is not only its ability to produce lettering instantly, but also its capability to store previously prepared material in its memory to be recalled instantly when needed for the produc-

FIGURE 13-12 **Character Generator Lettering.** Character generator lettering can be produced in a variety of type styles. Edging, border, and drop shadows increase the lettering's readability. The CG operator can also colorize individual letters, words, or sentences. *(Courtesy: Thomson-CSF Laboratories, Inc.)*

tion. All character generators include some internal memory which permits the operator to store a number of previously prepared "pages" and to recall them instantly when necessary. This internal memory is usually limited, however, so to increase the storage capacity of a CG, a number of accessory memory units are available. The floppy disc storage unit is the most flexible and powerful memory system as it permits hundreds of individual "pages" to be stored and retrieved virtually instantly. A less costly alternative is the use of a modified audio cassette recorder which stores prepared material on an audio cassette. The problem with the audio cassette system is that its memory capacity and retrieval speed are limited compared with the disc system, but its lower cost makes it an appealing alternative for production applications where instantaneous memory retrieval is not crucial.

The CGs memory operation works as follows: Before a production the CG operator is provided with a list of every lettered graphic necessary for the production. These might include the opening title and credits, lettering material needed during production such as lower-third super identification titles, and the closing credits. As the CG operator prepares each individual graphic or "page," it is assigned a memory number and entered into the CG memory system. For instance, the opening credit might be assigned memory position #01, the super title of the program host's name assigned #02, and so on. During the production, the CG operator simply punches in the appropriate memory number of each individual page, and the previously prepared graphic appears exactly as it was originally typed and composed, available for the program director to use on cue.

CREATING CUSTOM CHARACTERS Basic character generators usually provide one or two different character fonts or typestyles, but most of the advanced and sophisticated models have literally dozens of typestyles from which you can choose. In addition, custom lettering can be

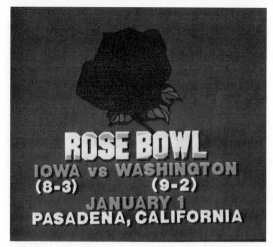

FIGURE 13-13 **Custom CG Characters and Logos.** The rose logo was produced especially for the production and stored inside the CG memory where it could be recalled instantly and positioned anywhere on screen. *(Courtesy: ABC-TV)*

developed and entered into the CG memory to give a production a distinctive "look."

Some advanced CGs also permit you to develop your own custom characters such as "logos," which are the letter/picture graphics used on a station's ID or on an athletic team's trademark signature. The custom character is created by first preparing a high-contrast, black-and-white artwork of the logo. The artwork is photographed by a small black and white camera, which is connected to the CG's memory system. Once the CG has "grabbed" or memorized the logo, it can be recalled from its memory position just as you would retrieve any other message which is stored in memory. In addition, you can vary the size of the logo, position it anywhere on screen to integrate it with lettering, and colorize it as with any other character. (See Figure 13-13.)

The operational speed and memory capabilities of character generators have made them highly versatile graphics tools. Difficult and involved graphic situations such as election returns or the production of instant news or weather bulletins are made faster, easier, and less expensive with an electronic CG than with alternative mechanical or photographic methods. The character generator has also made possible such special graphics applications as captioned programing for those with impaired hearing and the English subtitles for foreign language films.

Mechanical Lettering Methods

Although the electronic character generator is an extremely flexible device, there are times when lettering must be produced through mechanical means. The most commonly used mechanical lettering techniques are transfer lettering, the hot press, and punch-out lettering devices.

RUB-ON TRANSFER LETTERING As its name suggests, rub-on lettering uses special self-adhesive letters which, when rubbed, are transferred from a plastic carrier sheet to the artwork. Transfer lettering is available from a number of manufacturers under such brand names as "Letraset," "Presstype," and "Instantype." The lettering is available in dozens of sizes and type styles. Although the size you select depends on the size of your artwork, the most popular television graphic sizes are 24, 36, and 48 point type sizes. Most lettering styles are available in either white or black. Black is used for lettering against a light background; white, for lettering against a dark background.

To use the rub-on lettering, first, align the appropriate letter on your artwork and gently rub across it with a ballpoint pen, a pencil, or a burnishing tool. Once the entire letter has been completely rubbed over, carefully remove the plastic carrier sheet, leaving the letter transferred onto the artwork. Be careful that you rub the lettering gently and evenly to avoid cracking or peeling. If the lettering does crack—which can happen on older sheets which have lost their elasticity—fill in the cracks with either a black felt-tip pen or white ink, depending upon the lettering's color. (See Figure 13-14.)

Rub-on letters can be applied to almost any dry surface including illustration or matte cards, photographs, transparent acetate cels, or glass. Acetate cels are particularly convenient when you are lettering information to be used with a photo or illustration. By lettering onto the cel sheet, you can lay the lettered cel over the illustration and vary the position of the lettering without permanently marking the background picture.

FIGURE 13-14 Rub-on Transfer Lettering.

Rub-on lettering is relatively inexpensive and convenient to use. Since you need no special equipment, you can produce graphics almost anywhere. But the technique does take time, patience, and an eye for the proper spacing of letters and words.

HOT PRESS The hot press is a printing machine that uses metal type to produce lettering. Words are composed into a "chase," or type holder. The type is inserted into the chase backward in a mirror image of the actual message. Both the chase and type are preheated to about 250° and locked into the heated hot press. The artwork is positioned on the base, and a piece of specially colored hot press foil is placed over the artwork with the shiny side facing up. The press is brought down into contact with the foil, which allows the hot type to burn the foil onto the artwork in the shape of the lettering. You can produce white, black, or colored lettering depending on the color of the foil that you use.

Many stations and studios have their station logos or other frequently used graphic symbols specially made into hot press type. This makes it convenient to reproduce the logo whenever needed.

The hot press can letter almost any graphic surface including matte board, acetate cels, or photographs. Its primary advantages are speed of operation and the ability to make several copies of the same lettering with ease.

(a)

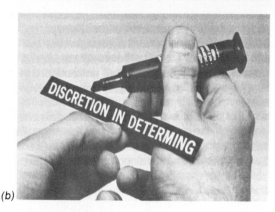

(b)

FIGURE 13-15 **Varityper Headliner.**
The lettering (either black on white or white on black) must be trimmed and mounted on an art card. When preparing a strip of type for a super card (b), run a black felt-tip pen around the edges to prevent any white from being photographed by the camera, which can ruin the electronic insert effect.

Photographic Lettering Methods

A machine such as the Vari-Typer Headliner produces lettering through a photographic process. The "type" is actually a large plastic wheel with lettering on it. Each wheel is selected for its type style and size and inserted into the machine. In order to produce words, you turn the wheel to the appropriate letter and press the "print" button. The letter's image is produced on specially prepared photographic paper, which is inside the machine. The paper is automatically processed and emerges a few minutes later through a slit on the side of the Headliner. The sprocket holes along both edges of the paper strip must be trimmed off before mounting the lettering on your artwork. (See Figure 13-15.)

Aside from various types, styles, and sizes, you have a choice of producing either black lettering on a white background or white lettering on a black background. If you use the latter for super graphics, be sure to blacken all four sides of the trimmed paper by running a felt-tip pen across the edges. This prevents the white edges from picking up any light and ruining the superimposition or electronic insert.

The disadvantages of the hot press are its high initial cost for both the machine and the type and the limited number of different type styles you have available as compared with other lettering techniques.

PUNCH-OUT LETTERING An alternative to the rub-on and hot press mechanical systems is a machine which produces words by punching out letters from a roll of plastic. It works much like a hand-sized label marker and produces clean lettering which can be mounted on matte cards, transparencies, or photographs.

GRAPHIC ILLUSTRATION TECHNIQUES

Illustrations include all nonlettered information such as photographs, drawings, maps, and charts.

Photographs

Television is a visual medium and television graphics natually rely heavily on pictures and photographs. If you have the opportunity to take the pictures yourself or to assign someone to do original photography, you can control for all of the usual graphic requirements. The photos must be composed into the 3:4 aspect ratio, they should have sufficient contrast in both color and monochrome, and they should follow the recommended guidelines for picture detail, resolution, and essential area. At times, you will not be able to take your own pictures, and you may have to rely on a file of existing pictures, on wire service photos, or on cutouts of pictures from books, magazines, or even newspapers.

COPY STAND   The copy stand is a device which enables you to photograph original artwork or to reproduce pictures from books or magazines. You can buy a professionally-made stand, or you can

design and construct a simple one yourself. You will need a flat surface for the artwork, a long upright column with an adjustable screw mount so you can vary the camera's height, and a light source to illuminate the artwork evenly. Some stands are constructed as double-deckers with a top level made of glass and a second level below. This lets you use cels on the top level and helps to induce depth and dimension between foreground and background elements. (See Figure 13-16.)

A 35-mm reflex camera is commonly used since the film is inexpensive, easy to process, and can be made directly into black and white or color slide transparencies for the telecine slide projector. A number of 35-mm cameras accept a special television cutoff mask inside the reflex viewfinder to help frame and compose the artwork within the essential and scanning areas.

PHOTO PRINTS   If you are producing photo prints to be used on studio camera cards, make the prints on nonglossy, "matte" paper, which will not glare under studio lights. A good working size

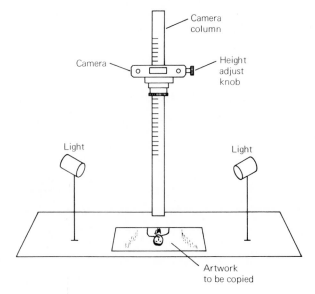

FIGURE 13-16   **Copy Stand.**

for prints is 8 × 10 inches. Prints much smaller than that are difficult to frame in the camera and do not permit steady movement of the camera across the photo to animate the picture.

MOUNTING PHOTOS AND PICTURES Photo prints and cutout pictures can be mounted on illustration board in a number of ways. The easiest and most convenient is to use spray adhesive or rubber cement. Rubber cement is inexpensive and has the advantage of being almost "goof-proof." As soon as the rubber cement dries, excess cement is easily removed by rubbing it gently with an eraser. Rubber cement solvent can be used to remove pictures and photos from an art card, allowing you to recycle the card for future use.

A more elaborate mounting procedure uses the *dry mount press*. To use this device the picture and mounting board must be preheated inside the press. Using a hot tacking iron, adhere a sheet of dry mount tissue to the back of the picture. Trim off the excess mounting tissue into the shape of the original photo or cutout. Finally, cover the picture with plain paper to protect it, and insert the picture with the mounting tissue into the press for ten to twenty seconds at about 225°. Remove the mounted graphic and let it cool.

The dry mount press takes much longer than adhesive spray or cement but does a much neater and more permanent mounting job. (See Figure 13-17.)

PHOTO LIBRARY A photo library is a valuable resource for pictures and illustrations. Some studios and stations make it a practice to save old books and magazines so the graphic artist can cut out different photos as necessary. You can either cut out pictures to be used individually, paste them up into a collage, or trace the picture for an outline or silhouette which is cut out from different colored art paper. Of course, you must be careful not to use copyrighted material without permission. It is also a good idea to file old

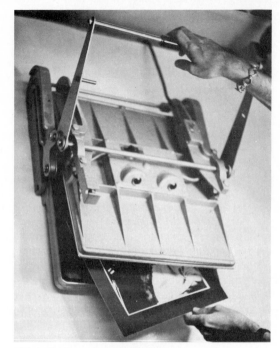

FIGURE 13-17 **Dry Mount Press.**

pictures and photographs from past productions since you never know when they may come in handy or can be modified for another purpose.

Many broadcast stations keep photo files of subjects especially for news programs. Stock photographs of commonly used places and subjects such as city hall, the mayor, city officials, and national figures are filed and indexed so they are available quickly when needed.

Maps and Diagrams

Various technical illustrations such as maps, charts, drawings, and graphs are often used on television productions. The trick though is to prepare an accurate diagram or illustration of a technical subject while still keeping it simple and readable for the television screen. This means you will usually have to adapt the original illustration or prepare one completely from scratch.

MAPS Maps are one of the most often used illustrations particularly on news and documentary programs. Conventional maps are impossible to use on television because they contain far too much fine detail. Maps must be redesigned for simplicity, and you may have to eliminate or refine some unessential details such as coastline indentations, roadways, or rivers in order to present better the important information that the viewer needs.

All art departments should have a number of atlases and map directories on hand since you never know what sort of map you may need, especially for news events. The easiest way to produce a map is to trace the outline of an area from the original map. Then transfer the outline onto colored art paper and cut to shape. The color of the land area can vary, but representing land masses as lighter in color and brightness than water, which is shown in darker shades, is common practice. Once you have cut out the map's shape, mount the cutout on a contrasting color/brightness background, and draw in important landmarks. You can use transfer lettering or the hot press to label significant cities or locations. Remember to keep the map simple. The point is to orient the viewer, not to give explicit directions.

DIAGRAMS AND CHARTS Charts and diagrams are helpful in explaining, clarifying, or simplifying ideas, issues, or processes. Once

FIGURE 13-18 **Diagrams and Charts.**
This graph was produced by an electronic character generator. Note the simplicity and clarity of the design. *(Courtesy: Thomson-CSF Laboratories, Inc.)*

again, clarity and simplicity are essential in producing an effective diagram. Avoid using an illustration taken directly from a book or other printed material. Aside from copyright infringements, these diagrams are usually unsatisfactory because they were not designed for television reproduction. Instead, it is up to the graphic artist to adapt the basic idea of the chart or diagram for television use. The same is true when preparing graphs. Only the barest amount of information should be included in the legend along a graph's horizontal and vertical axes.

Sometimes you may wish to produce a white line drawing against a dark or black background for maximum contrast and legibility. The white line drawing can also be used as a super graphic against a background picture.

Combining Illustrations and Lettering

Most television graphics combine lettering and illustrations. The lettering can be an integral part of the picture, such as on a map or chart, or it can use the illustration primarily as a background, a common technique in program titles. One way to integrate titles and illustrations is to insert or superimpose electronically the lettering from one graphic source on a picture from another. For example, you can prepare a title on the character generator and insert it over a plain picture on a slide or camera card.

Preparing a combined lettered-illustrated visual into a single graphic relieves the director of the need to coordinate two video sources during the show and gives the graphic artist more control over the integration of picture and lettering. First decide where you want the lettering. If the lettering is to be superimposed on the picture, try to find a dark area where you can use light lettering which will stand out more prominently on screen. Of course, you can also locate a light background area and use dark lettering. If you are producing artwork to be shot and used as slides, a good

(a)

(b)

FIGURE 13-19 Using Cels to Build a Graphic.
In this sequence, the graphic artist prepared a series of cels to build the final graphic. *(a)* Silhouette of a woman is laid over a background; *(b)* another cel with the target sight and gun is prepared and laid over the original; *(c)* lettering is added to complete the final graphic. The composite can be either photographed directly by a studio camera, made into a 35-mm slide, or photographed and stored in a still-store memory system. *(Courtesy: Ken Dyball)*

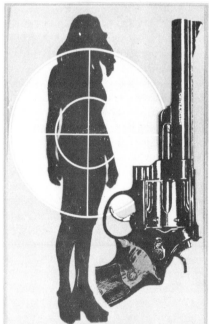

MURDER IN FOREST HILLS

(c)

method is to prepare your lettering on a transparent acetate cel and lay the lettered cel over the photograph or illustration. You can move the cel around to find the best location for the lettering. A cel overlay should not be used on studio camera cards because you risk the possiblity of glare from the shiny acetate. (See Figure 13-19.)

If you wish to letter directly on a photographic print, a simple technique is to prepare black lettering on a cel and place the lettered acetate over a piece of blank photographic paper. Sandwich the cel and printing paper between two glass plates, and place it under the photographic enlarger. Make the print as you normally would, but increase the exposure time by three or four seconds to compensate for the lettering. After the paper is developed, white lettering will appear burned into the photograph where the cel overlay was positioned. Of course, this approach makes the lettering a permanent part of the photo print so be sure to plan for the location of the lettering before making the composite print.

A simple way to integrate both pictures and copy is to prepare lettering on the art card with either mechanical or photographic methods and then mount a picture or illustration on the card next to the lettering. This technique is frequently used for station identification slides or in promotional graphics for upcoming shows.

DIGITAL ART SYSTEMS

The most significant and exciting advance in television graphics has been the development of the *digital art system*, which completely eliminates the need for an artist to use paper, pencil, inks, or photographic techniques in the production of television graphics. The digital art system combines the flexibility of digital video processing with the power of a computer memory to enable an artist to create television graphics complete with illustrations and lettering totally electronically.

Although there are some minor differences in the operation of a digital art system depending upon the manufacturer and model, most of the systems work in much the following way. The graphic artist works with an electronic stylus and an electronic tablet. As he or she draws on the tablet, the artwork appears instantly on the television monitor directly in front of the artist. Using a variety of techniques, the artist first produces an outline of the image which is either drawn completely in freehand, is traced from an existing photograph or drawing, or uses any of the pre-drawn shapes such as circles, rectangles, and ellipses, which are electronically modified as necessary. Once the outline is complete, the artist fills in the outline with a virtually infinite number of colors and shadings to produce the completed illustration. If lettering is required, a built-in character generator is used to add text material to the illustration producing a complete composite graphic. (See Figure 13-20.)

The completed graphic is then stored in the computer memory, where it can be recalled

whenever necessary. It is also possible to create an animation effect by recalling a stored graphic brushstroke by brushstroke, just as the artist drew it originally. (See Color Plate 10.)

The digital art system can also use existing camera-ready photographs, maps, charts or drawings, which are photographed by a camera and captured by the system memory. Once the picture has been captured, the artist can integrate it with electronically produced illustration and lettering to create a composite graphic.

The advantages of the digital art system are tremendous. Original artwork can be prepared quickly without any intermediary mechanical or photographic steps. Changes or revisions can be made quickly, or the artist can produce a number of versions and select the best one without having to completely redraw the graphic for each version. Maps, charts, and diagrams can be quickly modified for TV and prepared for production by tracing the outline with the electronic stylus and then adding color and shading as desired. With all the artwork stored inside the computer's memory, the graphics cannot be misplaced or damaged, and it is impossible to produce a slide with a fingerprint smudge. (See Figure 13-21.)

Naturally, the cost of these systems is enormous, which means they are not likely to be readily available except at those networks, stations, or production facilities which have graphics requirements sufficiently complex to justify the high costs of obtaining the equipment. But as with all modern production equipment, it is only a matter of time before technological advances result in economies and price reductions which will make the digital art system available to a wide group of production users. We need only remember that just a few years ago the ever-present electronic character generator was considered so sophisticated and expensive that its use was limited to a handful of production facilities. It is likely that the trend toward totally electronic graphics production at virtually every level of television production will be a reality in the near future.

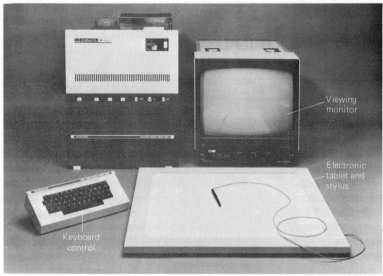

(a)

Art viewing area

Artist

Palette

(b)

FIGURE 13-20 **Digital Art System.**
The system consists of a keyboard, electronic touch tablet with pressure-sensitive stylus, and viewing monitor which displays both the artwork that is being produced and the "menu" of possible art effects. The insert (b) shows what the artist sees on the viewing monitor. The top shows the artwork and illustrates how lines, color, and brightness can be varied by the amount of pressure applied on the electronic stylus. The bottom area shows the artist's "palette," which is used to "mix" various colors and shades. (Courtesy: MCI/Quantel)

TYPES OF GRAPHICS

Graphics can be classified according to how they are used in production. Graphics which are located off the studio set and appear onscreen from an unidentified source are called *offstage* graphics. Those graphics which do appear onstage in front of the camera or which form an integral part of the program's set are called *onstage* graphics.

Offstage Graphics

Offstage graphics include the camera or title card, slides, the crawl, super slides or cards, and such electronic graphic techniques as the character generator and still frame storage devices.

CAMERA CARDS A camera card is a graphic card that is placed on an easel or card stand so the television camera can photograph it. Camera

FIGURE 13-21
An example of artwork created entirely by a digital art system. The actual artwork was animated by using the memory capability of the digital art system. Animating art with a digital art system is far less time-consuming and less expensive than conventional animation techniques. *(Courtesy: ABC-TV)*

cards are variously called "flip," "studio," or "title" cards.

Camera cards vary in size from studio to studio, but the 11 × 14-inch card is very popular, particularly because it is exactly one-quarter of a standard 22 × 28-inch illustration board. No matter which dimensions you choose for your camera cards, it is important that they are all kept the same size. Not only will this keep your aspect ratio and essential and scanning area calculations consistent for all artwork, but a uniform size will prevent a smaller graphic from getting lost in a stack of cards and make it easier for the camera operators to frame sucessive shots.

Each card should be indexed and numbered with a small piece of masking tape. The numbered tabs, which are staggered to run along one edge of all the cards, help to identify each graphic and make the job of pulling cards during the show much easier.

One of the major advantages in using camera cards is the opportunity to "animate" the graphic by panning, tilting, and zooming your camera across the card. If you use two or more cameras, each with its own set of camera cards, and move

the cameras over the graphics while cutting, wiping, dissolving, and so on between them, you can produce some interesting and effective visual essays. A camera card also makes it easier to recompose a poorly framed graphic and gives the director more flexibility in positioning the graphic within the camera shot. (See Figure 13-22.)

SLIDES Slides are 2- × 2-inch, 35-mm transparencies, which are projected from the telecine slide chain. Slides can be used as both a principal photographic medium and a production medi-

FIGURE 13-22 Camera Cards.
The tabs, which are made of masking tape, help keep the cards in order and make it easy for the floor manager to pull cards during production.

Single Camera Card with Tab

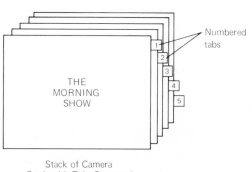

Stack of Camera
Cards with Tabs Staggered
Along Edge

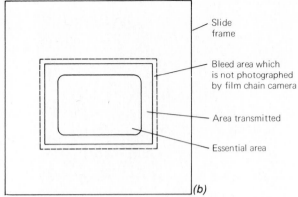

Slide frame

Bleed area which is not photographed by film chain camera

Area transmitted

Essential area

(a)

(b)

FIGURE 13-23 **Thirty-five Millimeter Slides.**
(a) 35-mm slide in holder; *(b)* template indicating cutoff and essential area of a 35-mm slide.

um. In the first case, pictures are photographed with slide film, and the slides are used in place of camera cards. This eliminates costly prints, which are also time-consuming. If you know you will be shooting film for use in the film chain, shooting directly on slide film can save you a lot of time and trouble. In the second example, slides are used as a production medium to eliminate the need for a studio camera to shoot any graphics. In this case, all of the artwork—program titles, lettering, identification supers, and so on—is photographed on slides, and then all graphics are loaded onto the slide chain for production convenience.

A 35-mm slide's horizontal format closely approximates the 3:4 aspect ratio but has a slightly greater width. You must be careful to leave enough room on both edges of the slide so you will not lose important visual material from either end. If your 35-mm camera accepts a TV cutoff template in the viewfinder, by all means use it both on and off the copystand. Otherwise, remember to frame your slides slightly looser than usual, particularly at both sides of the frame. The template in Figure 13-23 is the exact size of a slide and shows the excess side areas which are not reproduced on television, plus the scanning and essential areas. You can trace it and make a cardboard template to use in evaluating your slides for television.

Slides are processed either at the studio or sent to a professional laboratory. Professionally processed slides usually return in cardboard holders,

but it is a good idea to remount the slides in a more permanent plastic holder with glass panels. These holders completely enclose the slide to prevent it from getting scratched or dirty, and the glass keeps the film from buckling or warping under the heat of the slide projector.

There are a number of production advantages to using slides instead of camera cards. First of all, slides are easy to work with and convenient to store and catalogue. Unlike art cards, which take up lots of storage space and are easily damaged, properly stored slides will retain their picture quality for a long time. Replacing all camera cards with slides releases a valuable studio camera from any graphics responsibilities, giving the director greater flexibility in shooting the set. Once you have loaded slides into the telecine projector, there is little chance of a slide getting lost or out of proper sequence, two potential disasters when using a large number of studio camera cards.

But slides have their share of disadvantages too. Once the slide is mounted and loaded in the projector, you cannot animate the picture or change the position of the artwork. This is very critical when you are supering a slide over another picture or inserting the slide into a chroma key window on the studio set. Since the slide is fixed in place, the only way to change the relative position of the super slide against its background is to vary the background camera shot.

CRAWL The crawl is a long, continuous strip of

paper which is mounted between two drums or rollers. The rollers are turned either manually or electronically to move the paper past the camera. The crawl is traditionally used to display a list of production credits. Crawls are usually produced with white lettering on a black background so you can super the credits over another shot. There is no reason, though, why you cannot make a crawl with black lettering and art work on light paper. (See Figure 13-24.)

Crawls are also convenient for displaying long lists of information such as a cooking recipe or text material. If you are preparing a crawl with mechanical or photographic lettering methods, it is convenient to prepare the lettering on individual paper strips, which are then mounted onto the long crawl sheet. Producing the names on separate strips will not make a mistake in spelling or positioning a major catastrophe since you need do over only one strip and not the entire crawl.

When you prepare a crawl it is important to maintain the aspect ratio and to group all lettering within the essential area. Long names or credit lines should be broken down into smaller groups if at all possible. Otherwise the camera must pull back to include the overly wide lines, reducing the size of the other lettering and making the crawl difficult to read.

SUPER CARDS AND SUPER SLIDES Super cards and super slides refer specifically to cam-

era cards or slides which contain white lettering and/or line drawings against a black background so the graphic can be supered or inserted into another video source. Supers are very useful when you want to show a number of visual elements simultaneously. To use a familiar example, you can see a baseball player at the plate and at the same time read his vital statistics. Supers are commonly used to introduce program titles, to identify speakers on news clips or panel shows, or to display key words or phrases in instructional programming. (See Figure 13-25.)

A frequently used super is the *lower third*. This is an identification super which is positioned, as the name suggests, in the lower third of the television screen. If you are producing lower-third slides you must carefully determine the lower-third position by using an essential area template. Remember that a slide super cannot be repositioned much once it is in the slide projector. Producing lower thirds on a camera card gives you more latitude, since the camera operator can tilt up or down to vary the super's position on screen. A good rule of thumb for preparing super visuals is to position the lowest lettering line about 1½ inches from the bottom of the essential area.

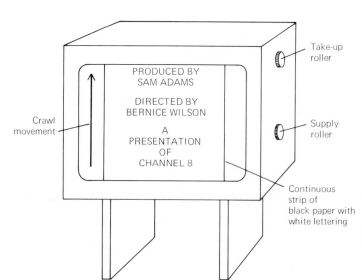

FIGURE 13-24 **Mechanical Crawl.** The mechanical crawl uses a pair of rollers with a long paper strip to display a long series of credits or other graphic information. Crawls are powered either electronically with variable speed motors or mechanically by hand.

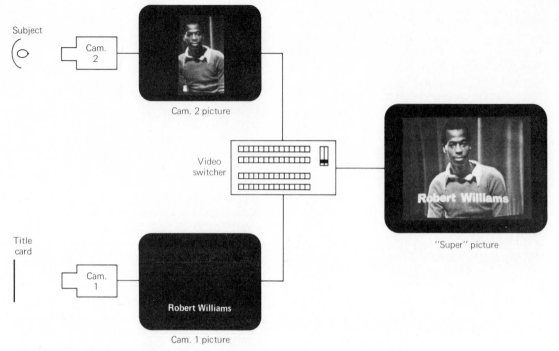

Subject

Cam. 2

Cam. 2 picture

Video switcher

"Super" picture

Title card

Cam. 1

Robert Williams

Cam. 1 picture

FIGURE 13-25 Super Graphics.
The graphic card with white lettering against a black card is shot on Camera 1.
The background subject is shot on Camera 2. The two video pictures are
electronically combined in the production switcher to create the super.

Naturally, supers can also be produced with a character generator. (See Figure 13-26.)

ELECTRONIC GRAPHICS Earlier in this chapter we discussed the electronic character generator and the digital art system, two methods of producing television graphics electronically. In addition to their speed and versatility, electronic graphics offers another important production advantage: they free studio cameras and the film chain from having to photograph any graphic material. On shows which make heavy use of graphics—such as news or instructional programs—this can be the equivalent of adding another camera or two to the production.

There is another electronic device which is frequently used to produce and display electronic graphics: the *still-frame-storage* unit. As it was described in Chapter 9 on videorecording, the still-store device enables the operator to "capture" and store a complete video frame in its memory so it can be recalled and punched up on

the air whenever necessary during the production. The still-store unit can store and display a camera shot or a slide, or it can capture and store a composite graphic which is made up of two or more video sources such as a camera picture and a character generator super.

The use of the still store is becoming increasingly popular as a graphics device because it enables the director to create composite graphics quickly and easily and have them available instantly as needed during a production. For example, on coverage of a football game we can use the still-store unit to produce a number of electronic "slides" combining a photograph of a player with his name, number, and some relevant statistics. First we shoot the player in close-up with a conventional television camera. If you are using a DVE, you can compress his picture and position it wherever you want on the screen..This is usually done prior to the game using a camera, or you can use slides or photographs if they are available. Then we use the character generator to

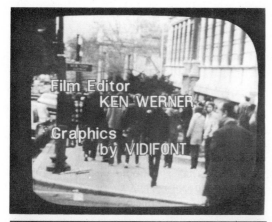

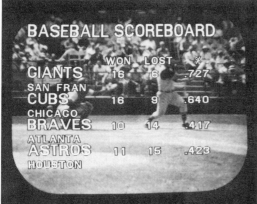

FIGURE 13-26
A super is useful whenever the director wants to show the viewer the graphic material along with a background subject. *(Courtesy: Thomson-CSF Laboratories, Inc.)*

display the player's name, number, and relevant statistics. The CG lettering is inserted over the picture of the player, and the composite image is captured by the still store device memory. The process is repeated for every player until we wind up with a library of electronic slides, which are available to the director during coverage of the game. (See Figure 13-27.)

The same technique can be used to build a news graphic by combining a photograph or a still frame from a camera or videotape with CG text. Once in the still-store memory, the newly created graphic is instantly available to the director during the production. If this approach is used, it is possible to create and store a number of complex graphics combining illustrations and let-

tering quickly and relatively easily, an important factor on news and sports productions where quick graphics turnaround is essential.

Onstage Graphics

Onstage graphics are used directly on the studio set. These include chroma key windows, rear projection units, hand or desk cards, and various graphic set pieces.

CHROMA KEY WINDOW This is one of the most frequently used onstage graphics. The "window" is an area of the set which is painted a chroma key color, usually blue or green. Using the program switcher, any video source can be inserted into the window. It is used mostly in news shows, but has application in countless other production situations. (See Figure 13-28.)

The most important factor in preparing graph-

FIGURE 13-27 **Electronic Graphic.**
The graphic was produced and composed entirely electronically by a character generator. Even the helmet logo was "captured" and stored by the CG. The frame at the left is a blue area which will be used to chroma-key a close-up of the player into the space to produce the final graphic. Then the graphic will be "captured" by a still-store system and retained in memory until called for by the operator. *(Courtesy: Thomson-CSF Laboratories, Inc.)*

FIGURE 13-28 **Chroma Key Window.**
The area behind the newscaster is painted a chroma key blue, which permits combining any graphic from a camera, slide, still-frame unit, or character generator to produce a composite shot.

ics for the chroma key window is to preplan the positioning and layout of the material so it will fit into the limited "window" area. You have to know in advance on which side of the screen the director plans to use the key. Precise registration of picture material is not overly critical with camera cards. A properly designed card with sufficiently wide borders permits the camera operator to vary the graphic's position in the shot to match the window. The original graphic location is very critical with slides, however, since they are fixed and cannot be repositioned.

REAR SCREEN PROJECTION An alternative to the chroma key window is to use rear screen projection (RP). This consists of a translucent screen and a powerful slide or overhead transparency projector positioned behind the screen. The size of an RP screen can vary from a small window a few feet wide to enormous background screens. (See Figure 13-29.)

The visual source can be either a 35-mm slide projector or an overhead transparency projector. Slides are easier to work with, but the overhead transparency is popular because you can animate the visual by overlaying a number of cels, each with different information, onto a single background. A more elaborate overhead setup involves using two projectors with a variable lighting control between them. The operator can dissolve from one projector to another or use both simultaneously for a composite effect. This is one

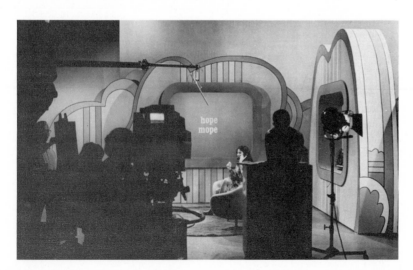

FIGURE 13-29 **Rear Screen Projection.**
The words are projected onto the screen from a slide projector located behind the screen at the rear of the set. *(Courtesy: Children's Television Workshop)*

way to produce the often-used "splat" on news programs—a highlight, arrow, or symbol which appears on a graphic such as a map or diagram to draw attention to an important area.

HAND AND EASEL CARDS Sometimes you may want your performer to handle graphic cards onstage. The hand card is a conventional camera card that is usually held by talent. If used on an easel or stand, it is called an "easel card."

The advantage of a hand card is that talent can completely control the appearance and timing of the graphics and can easily point out specific features. The problem is that some performers find it hard to hold the cards steady, and close-up shots greatly magnify any shaky or unsteady handling. To minimize this, instruct your talent to prop the bottom of the card on a solid object, like a desk, an arm of a chair, or even his or her lap, to help hold it steady. Also, be sure to tell the talent which camera will be used to shoot a close-up of the graphic.

GRAPHIC SET PIECES Graphics are frequently used as set pieces, making them an integral part of the setting and its surroundings. The set piece can be as simple as a blackboard or drawing easel or as complex as the mammoth tally boards used for television coverage of national elections. You are undoubtedly familiar with the most com-

mon graphic set piece—the ever-present weather map which is used on virtually every station's newscast. The weather map is simply a set piece designed so talent can either draw on it or attach words and symbols to it, the better to explain the weather forecast visually. The same basic graphic idea can be adapted to a variety of production situations. (See Figure 13-30.)

BLOWUP PHOTOS These are photos or illustrations which are enlarged to poster size and mounted on thin plywood boards. The boards can be hung from the studio lighting grids with heavy-test fishing line or mounted on spring-loaded "polecats" and positioned around the studio floor. Blowups are useful for adding depth and dimension to the studio set and for integrating graphics with oncamera talent.

MOUNTING BOARDS Although these have been adapted from the classroom, some television programs can benefit from their use. They consist of a mounting device—usually magnetic pieces or a hook and loop—which permits talent to attach graphic material directly to the board.

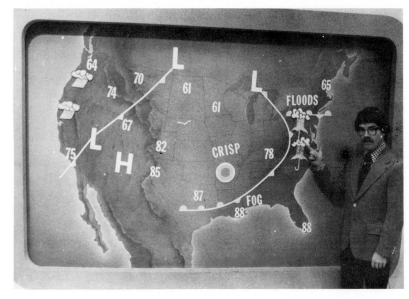

FIGURE 13-30 **Graphic Set Piece.**
The weather board is a commonly used graphic set piece. *(Courtesy: WXIA-TV)*

Other common set piece graphics are name plates, which identify participants on panel shows and logos or lettering for desks or set areas.

USING GRAPHICS IN PRODUCTION

Most graphics planning and preparation are handled during the preproduction stage. During setup and rehearsal, the graphics are organized in the studio and the slides are loaded in the film chain. During production, graphics are integrated into the show. Although graphics are rarely used during postproduction, there are times when they are added to a previously recorded show during videotape editing. In this section we will discuss all graphics-related activities as they occur in each of the four production stages.

Preproduction

At some point in the preproduction stage, the producer and director must meet with the graphic artist to discuss the graphic needs for the program. The producer explains the show's overall mood, tone, and objectives to the artist so he or she can begin to work up a style and approach for the graphics. Artists will usually prepare a few rough sketches to give the producer and director some idea of how the graphic artists conceive the graphic look. Once the style decisions are made and agreed upon by the producer, director, and graphic artist, the show's graphics are then developed.

One of the most frequent complaints from graphic artists is that they are never given sufficient advance time to design and prepare a show's graphics properly. Of course, on news programs it is not always possible to know far in advance which graphics will be necessary, but there is little excuse for a late graphics request on other scripted shows, especially those which will require elaborate or time-consuming graphic preparation.

The graphic artist also needs to know money and equipment availability. Some graphic techniques require outside work, and the budget must have money allocated for such activities. Equipment availability is important because if you know you will have the use of a character generator, for example, the artist will have little or no lettering to do.

Here is a quick checklist of some of the graphics considerations which should be discussed during preproduction:

1 *How much time is available for graphics? What is the due date? This is usually rehearsal day and not necessarily taping or air date.*

2 *How are the graphics to be used in the show? Is it simply a matter of opening titles and credits? Will the show require elaborate graphics as are often necessary in informational or educational programs?*

3 *What offstage graphics are necessary?*
   Titles for the program?
   Super cards or slides?
   Any special illustrations, such as maps, drawings, or graphs?
   Should graphics be delivered on camera cards? On slides? From a still-store unit?
   In the case of chroma key or super slides, how will the director use them? Is there a special videospace area where they should be positioned?

4 *What onstage graphics are necessary?*
   Will you use chroma key? Rear screen projection?
   Are any special onstage graphics, such as blowups, required? These may have to be sent to a special lab to be produced, and this can take time.
   Are special set piece graphics necessary? For example, identification panels for speakers, mounting devices such as magnetic panels, or integrated set and graphic pieces? These set plans should be coordinated with the set designer and the lighting director.

5 *Can stock photos or file graphics be used, or will all graphics have to be specially prepared?*

6 *Does the graphic artist have all necessary names and spellings for lower third identification supers and for the credit crawl?*

Be sure to double-check the correct spelling of names, places, and the proper use of official titles (for example, "Dr. John Doe," "Senator Robert Smith").

Setup and Rehearsal

SETUP By studio setup time all of the graphics —camera cards, slides, set pieces—should be available to the show's production team. The tabbed and numbered camera cards are taken by the floor manager or a floor assistant and placed on the proper camera graphic stands, while a production assistant or the telecine operator loads the slide chain.

The graphic floor stands are lit, along with the rest of the set, by the lighting director. Camera card stands are usually illuminated best with a fresnel spotlight that is flooded out, scrimmed, and positioned at a steep vertical angle. The angle prevents most reflections and glare from reaching the camera lens. If a glossy photo or graphic is used, dulling spray may help to reduce the glare or tilting the card forward may change the angle of light reflectance. Onstage graphics must also be lit carefully, especially large mounting board panels or blackboards. The lighting must be arranged so the talent's arm or hand will not throw distracting shadows across the graphic set piece.

The camera shooting the camera card must be positioned at a direct, 90° angle from the card stand. If the camera position is not perpendicular, the graphic will *keystone*, producing a distorted image on screen. (See Figure 13-31.)

Slides should be loaded properly into the projector and run through once to be sure they are all in proper sequence and that none are upside down or backwards.

If you are using a character generator, preprogram as many graphics as possible during the setup period. Each separate "page" or graphic display should be stored in the CG memory, and a sheet should be drawn up listing all memory numbers and their graphic content. This sheet will help you to retrieve any graphic immediately when called for by the director. For sports events, use the player's number as the memory address. For example, player number 38's lower third graphic would be stored in memory address

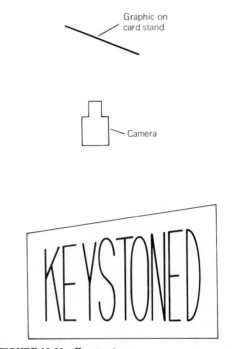

FIGURE 13-31 Keystoning.
If the camera is not directly perpendicular to the card on the graphic stand, the resulting picture will show keystone distortion.

number "38." To distinguish between two players with the same number, use the plain number for the home team and a 100 prefix for the visiting team. For instance, "38" would be the home team player and "138" the visiting player. In the case of a news program, all of the show's titles, lower third ID supers, and sports and weather information should be preprogrammed. All you need to do during the show is to punch in the correct memory address to retrieve the graphic. Of course, existing messages can still be updated instantly while the show is on the air.

If you intend to build electronic graphics by using a still store device, you must plan sufficient facilities and crew time during setup to permit you to produce all of the graphics which will be necessary. Remember that this will require use of the switcher console, so make certain that engineering is aware of your plans and has cameras, videotape recorders, and the CG available for use. If you will be preparing still-store graphics for use during sports coverage, use the same memory number system we described earlier so that you can store and recall each player's composite graphic by using his jersey number.

REHEARSAL During the show's rehearsal period, the director will start to integrate the graphics within the production. Even the simplest graphics should always be rehearsed. Programs which rely heavily on visuals may have to spend even more rehearsal time coordinating all of the various graphic elements.

If a floor assistant or floor manager will pull camera cards, the director must decide how the floor assistant will receive the cue to pull to the next card. Some directors prefer to give a specific graphic "pull" or "flip" cue. Others have the floor assistant automatically pull the card as soon as the camera's tally lights go off, indicating that the director has cut to another camera. Whichever way you decide, be sure to keep it consistent throughout both rehearsal and production.

Rear screen and chroma key operations usually require additional rehearsal since there is little margin for error. RP operators should have a specially marked copy of the program script to follow, particularly when they will have to animate a visual on cue.

Camera operators must have an opportunity to rehearse their "break" from the studio set to the graphic stand and back. If you wish to have the camera animate the graphic, be sure to rehearse each camera's moves just as you would any other sequence of camera shots. Cameras should be alerted if they will have to shoot close-ups of onstage graphics so they can plan their moves and shots accordingly.

Do not slight graphics during your rehearsal period. Integrating supers, slides, RP, camera cards, and electronic graphics into a show takes practice and coordination among a number of production and crew team members. By the end

of the dress rehearsal, every graphic that will be used in the show—including the opening titles and the closing crawl—should have been rehearsed at least once.

Production

During the production, slides and electronic graphics usually present the least difficulty. Slides which were preloaded and checked are very reliable. Electronic character generators and still frame storage units are preprogrammed, and the operator's biggest job is often simply to call up previously prepared graphics from the memory systems. Of course, the operator should also be prepared to change existing copy if necessary. (See Figure 13-32.)

During the production, the director should refer to each studio camera card by its number, not by its description. Describing the card is easy if it is a picture of the president of the United States, but what happens if you are directing a show on an exotic subject which neither you nor the crew knows much about? A picture of one chemical molecule can look a lot like another.

Visual essays can be very effective by using two or more studio cameras and by cutting, dissolving, or wiping between them as the cameras move across the pictures. Remember, though, if you are trying to establish a mood or illusion with the graphics you cannot use poorly composed or formatted graphics that are outside the normal aspect ratio. If your shot reveals the mounting board of the camera card or shows the sides of a vertical slide, you will ruin the illusion of reality which you are trying to create in the videospace.

Some graphics are more effective when you add information step by step on screen. This can enhance the viewer's understanding of the information and increase the graphic's visual impact. Most mechanical animation of this sort is done with super graphics since the black background will disguise the various animation tricks used to achieve the effect.

A *pull-slot* is simply an art card that is sandwiched between two plain cards. A slot in the front card permits you to reveal a portion of the graphic by pulling an overlay strip out slowly. A similar technique involves cutting a slot in the

Still-store monitors

CG program monitor

CG preset monitor

Still-store control unit

Character generator

FIGURE 13-32 **Electronic Graphics in Production.** The operator controls a character generator as well as a still-frame store unit during the production of a news show.

shape of the artwork or the drawing that is to be animated. A piece of black masking tape is attached to the rear of the graphic, covering the slot, and a light source is directed on the back of the graphic. As the masking tape is pulled away, the light source will shine through the opening and reveal the graphic image. A little experimentation with light levels and camera angles will show you the combination that works best.

The crawl which is used to end many shows can be used as a pad to vary the show's overall timing. Slowing down the speed of the crawl will stretch out the show's time; speeding up the crawl will shorten it. Be careful, though, of either slowing down or speeding up the crawl excessively. If your time problems result in a crawl moving at a snail's pace or so fast that it is an unreadable blur, you ought to find some other way to adjust the program's timing.

Postproduction

Graphics can be added to a videotaped program during the postproduction editing stage. This is done either by editing the tape or by performing a super or insert of the graphic as the videotape plays through a program switcher.

In the first instance, a graphic is added by performing a video edit. Usually this will be a "video-only insert" since you can keep the original sound track and alter just the video portion of the show. For example, while reviewing a recorded tape, you realize that instead of seeing the talent describe an object, it would be better to show a diagram of the object as the talent continues talking. By executing a video-only edit at the appropriate point, you can insert the new graphic while retaining the original sound track narration. Once you have shown enough of the graphic, punch-out of the edit and return to the original video.

Supers or inserts are accomplished by playing back the recorded tape through a program switcher as you produce the composite effect. For example, you may have recorded a panel discussion and realized that the viewer would follow the discussion better if the speakers were identified from time to time. If you are editing the show, you can play segments through a switcher, create the insert or super, and record the composite video output on another videotape recorder.

A DAY IN THE LIFE OF A TELEVISION GRAPHIC ARTIST

To give you some insight into how graphics are integrated into the overall production process, we offer a brief account of how one graphic artist goes about preparing the art material for use on a nighly news program which is broadcast by a local television station. Although it details the activities of only one operation, the approach and techniques involved can be found at virtually every production facility and television station.

When the graphic artist and her assistant arrive in the morning, they are confronted with a pile of slides, camera cards, and photographs which were the graphics used on last night's newscast. The first order of business is to return each slide and each photograph carefully to its place in the graphics library so that all will be easily available when needed again. The graphic artist also keeps a listing of the electronic "slides" which were stored in the station's still-store unit's memory system. Each day she evaluates any new electronic slides which should be saved because they are likely to be reused in the future. Electronic slides which aren't reusable can be erased to provide storage space for new slides and material. Having sent that information to the CG/still-store operator, the artist begins working on some new projects.

During the early morning hours, the graphic artist and her assistant work on new graphics which will be needed for upcoming feature stories or for other programming on the station. They also work on developing new "stock" graphics which will be kept on file in the graphics library and available for future use.

By noon the news program's producer calls a staff meeting to begin to develop a tentative story rundown for the evening's show. The graphic artist attends the meeting and takes notes on the show's graphic requirements, noting each graphic into one of three categories: (1) "stock" graphics, which are already on file in the library; (2) graphics which are on file but require some modification; and (3) graphics which must be prepared from scratch. A story about the President's trip to Europe will need only a stock slide, a story about a threatened strike of school teachers can use an existing slide about "strikes" but with some modification, and a story on the economy will require the preparation of a new graphic showing the rate of inflation.

Once the meeting is over, the graphic artist returns to the Art Department to begin work on the night's show. Her assistant begins to pull all the existing graphics from the library which may be on slides, camera cards, or already on memory file on the still-store unit. Meanwhile, the graphic artist begins sketching out ideas on the new slide which must be prepared and those which will require some modification of existing material. A number of slides will be "built" electronically by integrating lettering from the character generator with camera card photographs and then storing the composite image in the still-store unit's memory.

By midafternoon, completed art material and slides are sent to the studio and to telecine while the artist goes to the production studio to supervise the building of electronic graphics. The chart which was prepared about the inflation rate is placed on a camera card stand and photographed by the studio camera while the CG operator punches up the necessary lettering. Once the two video sources are mixed in the program switcher and appear properly composed on the "line" monitor in the control room, the graphic artist nods approval and the still-store unit's memory button is punched, capturing the newly created electronic "slide" into memory. The CG/still-store operator notes the memory number of the graphic on his cue sheet so he will be able to retrieve the graphic when called for by the program's director. A number of additional electronic "slides" are built in this fashion by combining studio cameras, the character generator, and some special effects from the program switcher until the day's list of slides are completed and stored in the still-store unit's memory.

The graphic artist returns to her office to watch the run-through of the news program's graphics and film material on a closed-circuit monitor. By the time the news program is broadcast, the artist's attention has already turned to the next day's art requirements. As she clicks off the light to the Graphics Department and leaves the building, she is thinking about a new graphic for a feature series on energy conservation that she will work on tomorrow.

SUMMARY

Television graphics include all the lettering, illustrations, photographs, titles, and drawings which are used on a television production. Effective television graphics will (1) convey information clearly; (2) establish a show's mood and tone through the graphics' style; and (3) help to present facts, concepts, or processes visually so the viewer will better understand and appreciate the program content.

There are two major considerations—technical and aesthetic—in the design of graphics. *Technical* considerations include all the factors necessary to prepare a graphic within the technical specifications and limitations of the television system. These include adherance to aspect ratio, the layout within the scanning and essential areas, picture resolution limitations, and proper use of contrast, gray scale, and color. *Aesthetic* considerations are the creative elements which make a visual both interesting to look at and effective in communicating its message to the audience. These factors include style, composition, and the use of color in design.

Television graphics consist of two basic elements: (1) lettering and (2) illustration. Lettering is produced electronically, mechanically, and photographically although the electronic character generator is the most commonly used lettering device. Illustrations include the use of photographs, drawings, maps and charts; the production of television slides; and the integration of lettering and illustration. Digital art systems enable the artist to produce illustrations and lettering completely electronically without any mechanical or photographic means.

Graphics are classified according to how they are used in production. Those which are located off the studio set and appear onscreen from an unidentified source are offstage graphics. These include camera cards, slides, super cards and slides, the crawl, electronic character generators, and electronic still-frame-storage units. Onstage graphics are those which either appear with talent in front of the camera or which make up an integral part of the studio set. These include hand or easel cards, chroma key windows, rear screen projection systems, and graphic set pieces.

Graphic design and preparation occur during the preproduction stage. The graphic artist must be told both the program's objectives and how the graphics will be used, so that the artwork can be designed to best enhance and complement the show. Good graphics take time to prepare and ought to be requested early enough so they can be designed and produced properly.

Graphics must always be rehearsed along with all other production elements. By dress rehearsal, every graphic that is to appear in a show should have been rehearsed at least once. Although graphics are most commonly integrated during the actual production, they can also be added to a videotaped program in postproduction editing.

CHAPTER 14

FILM FOR TELEVISION

Film is used in television in a variety of ways. Many network entertainment programs, commercials, and public service spots are produced and distributed on film. Film clips are frequently used within a program to show locations or activities which cannot be re-created inside the studio. News coverage and documentary productions have relied on film since the earliest days of television.

The development of portable electronic cameras and videotape recorders, combined with the sophistication of videotape editing systems, has resulted in tape replacing film as a primary production medium. Videotape stock is less expensive than film stock, can be reused, and does not require the costly and time-consuming step of processing before it can be edited and broadcast. For the most part, film is no longer used as a primary production tool at most television production facilities. However, it is still used extensively as an efficient means of distributing such program material as syndicated programming, commercials, and such special program insert material as video-music.

In this chapter we will discuss film as essentially a distribution medium. Our emphasis will be on how to use previously produced film material in a television production. For more information on the techniques of film production, see the reading list in the back of the book.

FILM FORMATS

Motion picture film is available in a wide variety of *formats*. The format describes the film's physical characteristics including (1) the size, (2) the packaging, and (3) whether it is silent or sound film.

Film Size

Film size is determined by measuring the width of the film in millimeters (mm). Film sizes range from 8 mm, used in home-movie equipment, to the gigantic 70 mm, which is used on some wide-screen feature films. In television, only three formats are used: 35 mm, 16 mm, and Super 8 mm.

35 MM The 35-mm format is used in the production of prime-time entertainment programs as well as for many nationally produced commercials. The size of the format produces exceptionally beautiful pictures but, since only the networks and a handful of the largest stations have 35-mm projection capability, it is rarely used for most videofilm production. (See Figure 14-1.)

16 MM The industry standard both for local television film production and for projection is 16 mm. It is much less costly than 35 mm, yet the various professional film stocks available in 35 mm are also available in 16 mm. Secondly, 16-mm production equipment is completely professional and comparable to the quality of 35-mm equipment, but is much lighter and more portable.

SUPER 8 MM Super 8 mm is a format used primarily for amateur home movies. However, a number of production facilities keep a Super 8 mm projector on hand in the event some special film arrives in the smaller, nonprofessional format.

Sound Film

Film which does not carry any sound information is called *silent* film and is available on double-perforated film stock. Film which does have a sound track is called *sound* film, usually abbreviated SOF (sound on film). There are two kinds of SOF: (1) optical and (2) magnetic. Sound film has perforations on only one edge since the sound track occupies the other edge.

OPTICAL SOUND TRACK The optical sound track is produced photographically and used on all standard-release prints. If you look closely at the track running along one edge of the film, you will see a wavy line or variations in the density of the strip, depending upon the particular process which was used.

Inside the projector is a sound head which holds a photoelectric cell and an "exciter lamp." The film is threaded so the optical sound track passes between the exciter lamp and the photocell. The light falling on the cell varies according to the pattern on the sound track, creating an

FIGURE 14-1 Film Sizes.

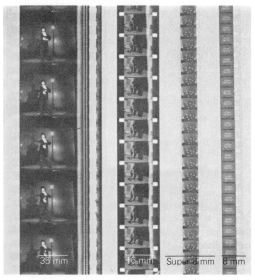

electric current which, when amplified, reproduces the original sound.

MAGNETIC SOUND TRACK Magnetic sound on film uses specially prepared film stock with a sound stripe running along one edge of the film. The stripe consists of the same iron oxide particles used in audio recording, and the recording and reproduction processes are identical to those used for audio tape. A tiny recording head inside the camera selectively magnetizes particles on the sound track during filming. Inside the projector, an identical head "reads" the track as the film runs, and generates a tiny electrical signal which, once amplified is an exact reproduction of the originally recorded sound.

Virtually all of the film which is previously produced and distributed on film, uses an optical sound track. However, there are some special situations when film is delivered with a magnetic track, so it is useful to have a projector which can accommodate both sound formats. (See Figure 14-2.)

SPLICING FILM

In order to combine various pieces of film, we must physically splice the film together. One of the most common videofilm production tasks is to assemble a "show reel" which includes all of the film elements for a program or production. Other times, film may tear and must be repaired in order to run through the projector. To do this, we must be able to edit film together.

Film Editing Equipment

The amount and kind of film editing equipment which a production facility has will depend on how they use film. Facilities which utilize film as a production medium will have a great deal of sophisticated equipment. At other facilities which only use film as a playback medium, only a few basic items of equipment, such as a splicer, rewinds, and viewer, are usually available. In the following section, we will describe the most commonly used items of film editing equipment.

FILM TABLE AND LIGHTBOX While almost anything can be used as a film editing table, a professionally manufactured table offers you the best editing environment. The table is large enough for you to work comfortably, stands on a solid base which will not move or sway as you use the film rewinds, and includes a lightbox which is set flush in the middle of the table. The lightbox consists of a bulb inside a reflecting well which is covered with a translucent plastic. It is used for close inspection of individual film frames. (See Figure 14-3.)

REWINDS Rewinds are crank-operated shafts

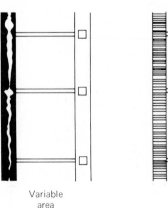

Variable
area

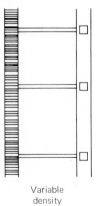

Variable
density

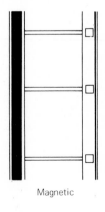

Magnetic

FIGURE 14-2 **Film Sound Tracks.**
Optical sound tracks are produced as either variable-area or variable-density tracks. In the mag-stripe track, the stripe appears on the left side of the film; the smaller stripe running along the opposite edge is a "balance stripe" which is used to equalize the thickness of the film and permit even winding.

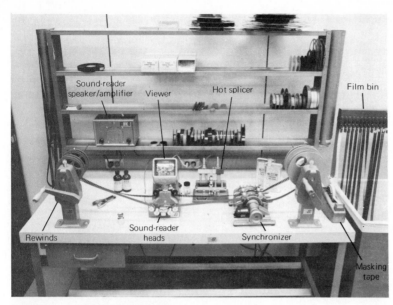

FIGURE 14-3 Film Editing Table.
A well-equipped film editing table for sound and silent film editing. Note the shelves for storing reels, the hole punch for marking film, and the small cement bottle, which is used to hold just a day's worth of film cement. Fresh cement is poured from the large cement storage cans each workday. The film bin to the right is handy for holding small strips of film until they are needed. *(Courtesy: Eastman Kodak Co.)*

used to wind film forward and backward. Typically, a pair of rewinds is mounted on the editing table, one at each end. The rewinds' shafts are very long, permitting you to wind a number of different film reels simultaneously—a picture reel and a reel of full-coat sound, for example. Some rewinds are double-deckers, which are handy for holding different reels of film where they are easily accessible. A tension brake on one of the rewinds is a convenient accessory which helps to control the film's winding speed.

FILM VIEWER   Once the film is placed on the rewinds, it is passed through the film viewer which displays a bright image on the face of its screen. The film viewer is positioned in the middle of the editing table between the rewinds. You can draw the film through the film viewer at whatever speed you wish, slowly moving it back and forth until you locate the exact frame for the edit.

SYNCHRONIZER   The synchronizer is a sprocketed-wheel device with a footage and frame counter attached. It is used to time film and to match different reels of film. Each sprocket wheel, or "gang," as it is commonly called, holds a single piece of film. Most synchronizers are either four-ganged or six-ganged, and all the wheels turn simultaneously. Once film is locked into a gang, it will stay in sync with other reels in other gangs whether you move the film forward or backward, at any speed. The synchronizer is used to synchronize sound and picture film and to prepare multiple reels for laboratory work or projection.

SOUND READER   The sound reader contains a magnetic and/or optical playback head and an amplifier/speaker. It is designed to reproduce sound from magnetic striped, full-coat, or optical track film. The sound reader is located close to the film viewer and synchronizer so you can listen to the sound track as you screen the picture.

REELS   A large supply of film reels is essential for a well-equipped editing table. Reels come in two varieties: (1) basic reels in many different sizes, and (2) split reels or flanges. The *split reel* is actually two metal flanges which screw together around a central plastic core or hub. Once you have wound the film tightly onto the plastic core, you can unscrew the two flanges and store the film-wrapped core. This saves lots of space since film on cores is much easier to handle and to store on the editing table than a number of bulky and cumbersome reels. (See Figure 14-4.)

FILM SPLICERS There are two kinds of film *splicers*, which are used to join two pieces of film: (1) tape splicers and (2) cement splicers. Tape splicers use a special Mylar tape to splice two pieces of film together. Cement splicers actually "weld" two pieces of film with a special film cement. We will discuss how to use both splicers later in this chapter.

ADDITIONAL EDITING TOOLS AND SUPPLIES
Along with the equipment just mentioned, a well-equipped editing bench will have the following: cotton gloves for handling film, scissors and single-edged razor blades for cutting film and tape, colored grease pencils and felt-tip pens for marking and writing on film and film leader, a magnifying glass for close work, rolls of thin masking tape for fastening film to reels, film cleaner and a plush pad for cleaning dirty film, and a supply of film leader including black, opaque, and academy leader.

FIGURE 14-4 **Split Reel.**

Another handy item is a *frame count ruler*, which is marked off in seconds and frames so you can measure a piece of film against it to estimate its running time. If you lack a commercially made ruler, you can make one by laying down a strip of masking tape and counting off twenty-four frames (one second) and then intervals of two, three, five and ten seconds. It is then easy to lay film against the tape to see how much footage in frames and seconds you are working with.

Many film editors also like to work with a *film bin*. This is a canvas sack which is mounted beneath a series of thin nails or clips running along an overhead frame. Each clip is numbered so you can temporarily mount individual shots and let the film run into the canvas basket.

Film Editing Methods and Techniques

There are two ways to splice film together: with *tape splices* and with *cement splices*.

MAKING TAPE SPLICES Tape splices are by far the easiest and most convenient way of joining together two pieces of film. The least expensive tape splicing method is to use perforated, precut transparent tape. To use this method, the film is seated in a splicing block, which holds the two pieces in alignment. The perforated splicing tape is then laid over the splice point and pressed firmly to remove any air bubbles. The problem with press-on tape is the long time it takes to make a splice, an important factor when many splices are needed or when time is limited.

GUILLOTINE TAPE SPLICER A far more desirable method uses the "*guillotine*" tape splicer, which is designed to join film quickly and accurately. The advantage of the guillotine splicer is that its hinged design incorporates two cutting blades which automatically trim the excess tape to prevent snagging during projection. (See Figure 14-5.)

Here are the steps to follow in using a guillotine splicer:

1   Locate the splice point in your footage. Use the straight, perpendicular cutter located on

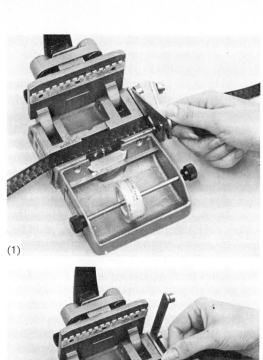

(1)

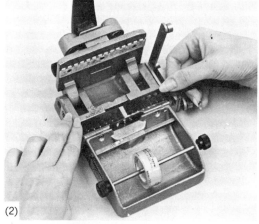

(2)

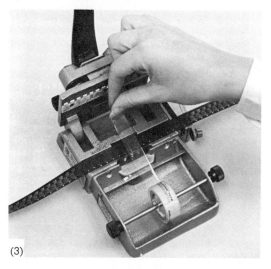

(3)

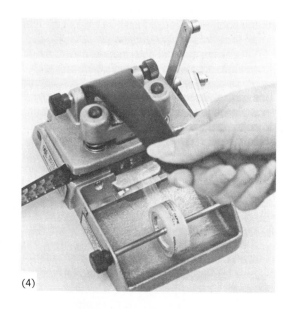

(4)

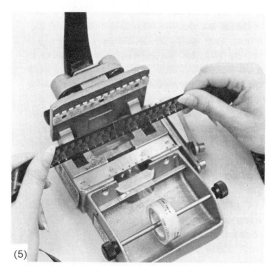

(5)

FIGURE 14-5 Editing with a Guillotine Tape Splicer.
(1) Cut film with cutting bar. (2) Align film on splicing block. (3) Pull splicing tape across the splice point. (4) Press down on handle to trim splicing tape. (5) Remove film and inspect the splice.

the right-hand side of the splicer to cut the film on the frame line. The diagonal cutter is used only for splicing full-coat magnetic film, never for splicing picture footage.

2　Align the two pieces of film in the splicing block. The protruding registration pins will hold the pieces in place. The ends should butt together but should *not* overlap.

3　With the film properly seated in the splicer, pull a piece of splicing tape across the splice point. Press the tape down firmly on the film to remove any air bubbles and to make good contact.

4　Press firmly on the handle to bring the hinged top of the splicer in contact with the film. Press down so the cutter blades will have enough force to trim excess tape from both edges of the film.

5　After you have completed the splice, turn the film over and repeat the procedure on the other side to ensure a strong splice.

6　Inspect the splice to make certain that the tape is neatly trimmed on both edges and the sprocket holes are clear of tape. If necessary, carefully trim away any excess tape with scissors.

If you are splicing magnetically striped film, trim the tape so it will not cover the magnetic striping. Otherwise, the sound will be momentarily interrupted as the tape passes over the sound head in the projector.

The guillotine splicer's speed is only one of its advantages. Tape splices also tend to be more reliable when running through a telecine projector than some cement splices which, if hastily or improperly made, can open during projection. Also, the tape splicer makes a "butt splice" so that there is no overlap in the film. This means you do not lose a frame of film at each edit, as you do with cement splices. Tape can also be peeled off easliy to open a splice if you dislike the edit. However, tape splices tend to jump occasionally

in the projector, stretch after repeated passes on a flat-bed editor, and are unsuitable for use in preparing original footage for laboratory printing.

MAKING CEMENT SPLICES　A cement splice literally welds two pieces of film together. This makes a permanent bond which is actually stronger than the film itself since the cement splice overlaps the film at the splice point. In order to make a cement splice, you will need a splicing block and film cement. Film cement is highly volatile and tends to lose its adhesiveness when exposed to the air. For this reason, most film editors store a large can of film cement and pour just a small amount into a cement bottle each day. Film cement is relatively inexpensive, and the extra cost involved in being conservative is far less than the possibility of splices opening during projection due to weak or stale cement.

The cement splicing block is available in a variety of shapes and sizes. Since the cement bonding is speeded up if the temperature of the splicing block is higher than room temperature, some splicers are made with tiny heating elements which keep the splicing plate warm at all times. These "hot splicers" should always be left plugged in since they take a considerable amount of time to warm up to the proper temperature. However, keep them closed to prevent the heat from warping the plates.

To make a cement splice follow these steps:

1　Find the edit points in both pieces of film and cut the film with scissors, leaving an extra frame on both ends at the splice point. These will be removed later during the splicing.

2　Open the two hinged plates on the splicer. Place one piece of film in the left-hand side and the other in the right-hand side. The film must be positioned with the *emulsion* or dull side facing *up*. Registration pins on the bottom plates will hold the film in position. (See Figure 14-6.)

3　Close both upper plates and lock them to the lower block. Now each side can be raised and lowered independently, yet the film will remain securely in place. By gently raising and lowering first the right side and then the left, cutting blades on the sides of each plate will trim off

(1)

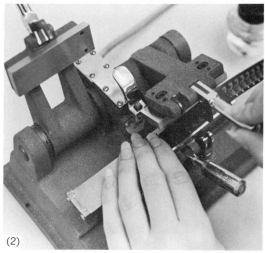

(2)

(3)

(4)

(5)

FIGURE 14-6 Editing with a Cement Splicer.
(1) Place film on splicing block pins. Lock into place.
(2) Place right-hand side of film on right side and lock
in place. (3) Scrape emulsion cleanly off protruding
edge of film. (4) Apply cement. (5) Lock down right
side of splicer to bring both edges of film into contact.
(6) Clean excess cement with cloth. (7) Inspect the
splice by gently flexing the film.

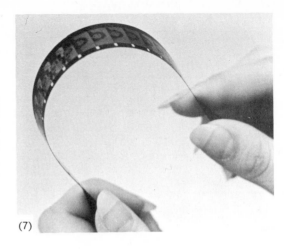

FIGURE 14-6 **Editing with a Cement Splicer. (Continued)**

the excess frames, producing the correct overlap for the splice.

4 Raise the right-hand side completely. The left-hand side will expose a small piece of film protruding from the right edge of the plate. This is where the splice will be made. Since cement bonds only to film base, you must remove the emulsion from the film on the left-hand side. If your splicing block has a built-in scraper, gently slide the scraper back and forth until you have removed all the emulsion. Be careful not to scrape too deeply lest you gouge the film base or damage the edges or sprocket holes. Instead of a scraper, you can use a single-edged razor blade.

5 Inspect the scraped area to make certain it is completely free of any emulsion. Then apply enough cement to wet the splice area, but not so much that it will ooze out once the two films are pressed together.

6 Immediately after applying the cement, swing the right-hand hinged side down and clamp it into position, bringing both film pieces into contact. Film cement dries very quickly, so the faster you bring the two pieces of film together, the better the splice.

7 You must wait at least twenty to thirty seconds for the splice to weld on a cold splicing block and about half that time on a hot splicer. These are slightly conservative estimates, but it is better to wait a few seconds longer than to

produce a weak splice that may open on the air.

8 After the splice has welded, open both upper plates and wipe the film clean of any excess cement with a soft cloth. Remove the film and inspect your splice. Test it by flexing the film into a "U" shape. The splice should hold without buckling or bending.

Some important tips on *cement splicing*:

Fresh cement is essential for a strong splice. Always pour just a small amount into your editing table bottle and cap the large can tightly.

Never add new cement to old in a work bottle. Start with fresh cement every day.

If you are cement splicing magnetic striped film, you must clean off the magnetic striping at the splice point. You can do this easily by moistening a soft cloth with some film cement and gently wiping away the stripe.

Keep your splicing block clean and well maintained. Dirt or misalignment will result in weak splices that may break during projection.

FILM CHAIN

The *film chain* or *telecine* contains all the equipment necessary to reproduce film and slides on television. A usual telecine system consists of (1) the television film camera, (2) the multiplexer, (3) a film projector, and (4) a 35-mm slide projector. (See Figure 14-7.)

Television Film Camera

The television film camera is designed especially for motion picture and slide reproduction. The camera—either monochrome or color—operates exactly as a studio camera, but is designed to respond quickly and to compensate automatically for changes in the brightness level as the film runs or the slides are changed.

Needless to say, only a color film camera will reproduce films and slides in color. Many color facilities have maintained their older black and white film chains for monochrome art work, such as super slides, and for projecting black and white motion picture film.

The output of the telecine camera is fed into the control room production switcher where it is used

like any other video source, alone or in combination with other video for a special effect.

Multiplexer

The multiplexer is a box with a series of internal mirrors and prisms which allows you to feed a number of different picture sources into a single television film camera. Without the multiplexer, we would need a separate television film camera for every film and slide projector—an obviously expensive and wasteful situation. Using the multiplexer, as many as three or four film and slide projectors can feed one camera, although not all at the same time.

The multiplexer, camera, and projectors must

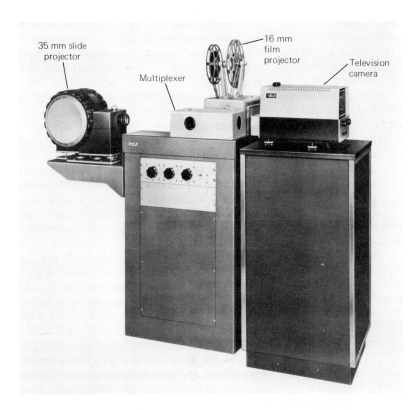

35 mm slide projector

Multiplexer

16 mm film projector

Television camera

FIGURE 14-7 **Telecine.**

all remain in critical alignment, which is why most film chains are built on a "film island." Usually a heavy metal plate is bolted to the floor, and the multiplexer, camera, and projectors are firmly attached to the plate where vibration will not affect the delicate alignment of the various components.

Film Projector

A well-equipped telecine has at least one film projector in the system. This is usually a 16-mm projector since the 16-mm format is the industry's standard for program and commerical distribution.

Most 16-mm film projectors are designed to accommodate reels ranging in size from 50 to 1,200 feet. Some projectors accept giant-sized 4,000-foot reels, which provide almost two hours of continuous programming. Threading the film is usually done manually, although some projectors have an automatic threading feature. (See Figure 14-8.)

SOUND HEADS The 16-mm projector will usually have two heads, one for magnetic sound and one for optical sound. Some projectors have sensing devices which automatically match the proper sound head for the film track. The output of the sound head is patched into the audio control console and regulated by the audio engineer as with any other sound source.

CUING AND PREROLLING All professional film projectors can be *still-framed* to show a single frame continuously without burning or damaging the film. This lets you cue film to a predetermined point, stop the projector, and display a frame on the film chain's monitor in the control room. Seeing the still-frame on the monitor shows everyone in telecine and the control room that the film is cued and ready for air.

Projectors capable of "instant starts" allow you to project sound and picture immediately since the film comes up to playback speed in less than one second. All you have to do is roll the film to the first frame of program material, still-frame it, and start the projector on cue. Older projectors do not have this instant-start capability and require around three or four seconds of rolling time for sound and picture to get up to speed. In this

FIGURE 14-8 **16-mm Film Projector.**
(*Courtesy: Eastman Kodak Co.*)

case, a special cuing leader—called "academy leader—is used to cue the preroll.

FILM CARTRIDGE PROJECTOR The film cartridge eliminates the need for threading the projector since each cartridge, which holds up to two minutes of film time, is automatically threaded and projected on cue. The film cartridge machine is a convenient way to play back a series of short film spots, and the cart machine can be preprogrammed to execute a completely automated film seqence flawlessly.

SUPER 8-MM PROJECTORS Super 8-mm projectors can be fitted to feed into a multiplex unit. Although commercials and programming are not distributed on Super 8 mm, the economy and convenience of the Super 8-mm format makes it popular for production and playback at smaller stations and production facilities. While many Super 8-mm projectors have dual speeds, 18 frames per second (fps) and 24 fps, all film shot for television should be filmed and projected at the standard 24 fps to assure proper reproduction on the telecine camera.

Slide Projector

All 2 × 2 inch, 35-mm transparency slides are normally projected on the film chain slide projector. The slide projector is typically equipped with dual slide drums. As you load the slides, you alternate drums so that all even-numbered slides are on one drum and all odd-numbered slides on the other. An internal mirror flips back and forth between the drums to let you change slides on the air without interruption.

Most slide projectors have a capacity of about thirty-six slides, which is usually sufficient for normal productions. If you will need to project more than the number of slides that can be loaded, you must either use two film chains or assign a crew member to the telecine room to change slides in the drums as they are used during the show. Some advanced slide projectors have a "random-access" feature, which enables you to locate and project any loaded slide instantly.

Slides are a very convenient way to display static visual information since they do not require

a studio camera to cover an easel card. Loaded slides are also less apt to be lost or mixed up during the production than a pile of camera cards. Of course, if you are working with only one telecine chain, you must avoid calling for a slide and a film simultaneously. If you want to super a title over a film, you cannot use the slide chain if the film is running. An old monochrome telecine can be helpful here since most super and title slides do not require a color camera and color reproduction.

Film Chain Operation

The film chain is not located in the studio or control room area since the noise and activity related to its operation would be a serious distraction. The telecine room can be placed almost anywhere; sometimes it is not even on the same floor as the production studios. The program communications PL system is usually connected to the telecine so that the director and control room team can communicate with the telecine operator and vice versa. Some facilities also provide remote controls for the film and slide projectors adjacent to the program video switcher in the control room. Using the remote controls, the program's director or the switcher can start and stop the film projector or change slides from the control room while the show is on the air.

USING FILM IN PRODUCTION

In this section we discuss how to use film in a television production. This includes the proper ways to handle and store film, prepare a show reel for projection, and roll in film footage within a television show.

Film Handling and Storage

Film is a relatively strong and durable material, but it must be handled and stored properly to maintain its picture and sound quality. Film should be stored in a clean, dry, and dustproof environment. Smoking in film-handling areas should be discouraged.

You should check your film reels periodically to be certain that they are not damaged or bent. If the flanges are warped, the reels can seriously damage your film. Also check the condition of your viewer, splicer, synchronizer, and projectors periodically to make sure they are clean and working properly. During projection and viewing the sprocket holes in the film are under a great deal of strain. Dirt or misalignment in the projector can impair the quality of the film or, worse yet, tear it.

FILM CLEANING There are times when film gets so dirty it must be cleaned. To do this, place the reel on one of the rewinds on the editing bench and thread it to a clean take-up reel. Hold a soft plush cleaning pad, which is moistened with film cleaning solution, inside the palm of your hand. As you slowly run the film to the take-up reel, fold the pad gently over the film so you remove dirt from both sides. The idea is to apply a slight, even pressure to dislodge dirt, yet not so much force that you scratch the film's emulsion. If the pad gets dirty, stop the film and turn the pad to find a clean area. If you are cleaning film with a magnetic stripe, avoid touching the striped edge with the pad since it may dissolve the magnetic striping and destroy the sound.

Film Cuing Devices

In order to roll in a film clip into a program we need some means of cuing the film accurately in the film chain's projector. The most commonly used cuing device is *academy leader*, so named because it was developed by the Society for Motion Picture and Television Engineers Academy. It is sometimes called "countdown" leader because it shows a clocklike animation as numbers count down from "8" to "3" seconds.

Academy leader is spliced at the start or "head" of each film clip. The leader numbers stop at "2," and you must splice in exactly two sec-

onds (forty-eight frames) of black leader between the "2" frame and the first frame of program material. This is so that numbers won't show on-air if you punch up the film too early. Remember that academy leader is a cuing device, not threading leader. Be sure you have enough film leader in front of the academy leader to permit the telecine operator to thread the machine properly. (See Figure 14-9.)

Once the academy leader has been spliced into the film, instruct the telecine operator to cue the film to whatever number you wish. For example, if you want a seven-second roll cue, the telecine operator will place "7" in the gate. When you give the roll cue, the film will start, and seven seconds later the program material will appear.

Preparing a Show Reel

It is usually most convenient to prepare a composite reel incorporating all of the film sequences to be used on a particular production. For example, all commercials, filmed spots, news reports, and so on for a daily morning talk show might be assembled onto a single "show reel." Once the clips are assembled in order with the proper leader between them, it is easy to roll in every film segment from a single telecine chain. On some shows which make heavy use of film, such as nightly news programs, you may prefer to isolate the news footage on one telecine and all commercials on another. This gives you more flexibility to change the order of the news spots while keeping the preset commercial rotation.

Many stations prepare a show reel for the entire day's worth of filmed commercials. All the commercials are interspersed with academy leader and spliced in order onto one or two large 1,200-foot reels. All commercial breaks are then run from a single projector.

To prepare a show reel you must use the rewinds, viewer, and splicing block.

1　First, run off about 5 or 6 feet of opaque leader onto the head of the take-up reel. The telecine operator needs this leader footage to thread the machine.

2　Splice in a cut of academy leader. The higher numbers should be on your right-hand (take-up reel) side since the film is designed to

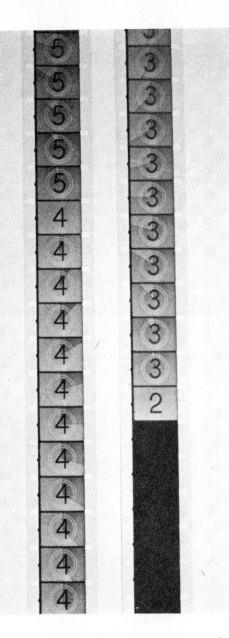

count down as it is projected. Immediately after the "2" frame, retain two seconds (forty-eight frames) of black leader.

3 After the last frame of black leader, splice on the beginning of the program film cut. Roll the film onto the take-up reel.

4 After the last cut of the program material, add some black leader to serve as a safety pad in case the director switches out of the film late.

5 Attach another piece of academy leader, and repeat the entire process until all of the film cuts are on the take-up reel. Of course, if one segment segues directly into another film segment, you can butt splice the two together or separate them with a short piece of black leader. When you are finished, all the film footage will be wound on the take-up reel "tails-out." Simply rewind the footage onto the supply reel to ready it for projection.

If you are using an instant-start projector, you can choose whether or not or use countdown leader. Many directors still like to use academy leader and a short preroll since it gives them a

FIGURE 14-9 Academy Leader.
Academy leader is used to cue up film before each new segment of program material. After sufficient threading leader, splice on academy leader, making certain to leave the two seconds of black before the start of the film program material. At the end of the program material, add additional black leader, and start with another cut of academy leader.

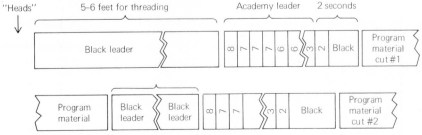

few seconds of rolling time to be certain that the projector is working properly before switching from studio to film. If you choose not to use academy leader, then splice on black leader before and after each program cut and cue the film by either still-framing the last frame of black leader or the first frame of program material.

Film Playback

The ultimate purpose of all television film is, of course, to use either as an insert on a larger show or as an entire program itself. Either way, playing film back on the air uses the same basic methods and techniques.

FILM PLAYBACK IN A SHOW During a program's preproduction, the director, producer, or a production assistant must list all film segments, their running times, whether each cut is silent or SOF and additional helpful information, such as opening and closing audio cues. This information is given to the film editor who will build the show reel, and it is also incorporated into the program's rundown sheet.

During rehearsal someone (director, assistant director, producer, or production assistant) must preview the show reel in the studio control room to make certain that every film cut is correct and in proper sequence and that the proper leader has been spliced before and after each segment. During this time also, the video switcher and the audio engineer should satisfy themselves that the film's audio and video feeds are properly patched in and ready for the production.

PREROLL CUES Preroll cues vary and depend primarily on the studio's operating policy, the telecine equipment, and the director. Cues are usually anywhere from three to seven seconds. Assume that we will be using a five-second preroll cue. This means that the telecine operator will stop the film projector with the number "5" in the gate. You will see a still-framed "5" on the film monitors in both the telecine room and the pro-

gram control room. This indicates that the film is cued and ready.

Once you know the preroll time, you can coordinate with talent exactly where in the program script the roll cue will be given. If your show is fully scripted (such as a news program or an instructional show), find the word in the script that will occur exactly five seconds before you want the film sequence to start on-air. This is the *roll-cue*: once the word is spoken, the director will call for the telecine film projector to roll. Five seconds later, just as the talent finishes speaking, the film segment should appear.

To help the talent and production crew prepare for the transition from studio to film, the director or assistant director should count down aloud the time—in seconds—from the start of the film roll until the beginning of the film segment. As the countdown is heard in the control room and over the PL communication system on the studio floor, the floor manager gives a silent, hand countdown for the talent. This countdown helps talent to gauge the speed of his or her film introduction and to stop just as the film begins.

If your program is ad-libbed, you may want to write out the film introduction on cue cards. This way, both the director and the talent will know precisely how long the introduction will run and where to cue the film roll.

TIMING THE FILM CUT As soon as the film segment begins, the director, assistant director, or production assistant should start a stopwatch, clocking the film's running time. Periodically, the timekeeper should call out how much time remains until the end of the segment when the director must switch back from film to studio. When ten seconds are left, the timekeeper should count down each second over the PL. Hearing this countdown, the floor manager gives a hand countdown from five seconds to air. The last two seconds are done silently in case the audio engineer opens the studio mikes early. As soon as the director cues the floor manager, the talent should be given the "go" cue to begin the action as the director switches back to the studio floor.

Figure 14-10 will give you an idea of how a film segment is rolled into a show and of the various activities that must take place.

FIGURE 14-10 Film Playback Procedure.

Script	Director's Commands	AD's Commands	Results
A FIRE DOWNTOWN	"Ready to roll film"		Director readies telecine for film preroll.
INJURES TWO AND *Roll Q*			
CAUSES (SEVERE)	"Roll film"	"Five . . . ."	Hearing roll cue, the director calls for film roll. At same time, AD starts countdown as the floor manager gives silent hand signals to the talent.
DAMAGE. FOR A REPORT		Four . . .	
FROM THE SCENE, HERE IS		Three	
JANE WINTERS.	"Take film"	Two . . . One"	As talent finishes film intro, director tells switcher to take film. Audio engineer brings up film's sound track and cuts studio mikes.
(FILM/SOF 1:38)			AD starts to time film with stopwatch and calls out time remaining periodically.
		"Ten . . . nine . . . . eight . . . seven . . . . six . . . five . . . four . . . . three . . . two . . . . one"	At ten seconds to end of film, AD begins countdown. Floor manager readies studio floor and gives silent hand countdown to talent so audio engineer can open studio mikes. Director readies switcher and audio.
	"Ready to cue talent"		
	"Ready to take Camera 2"		
	"Cue talent. Take 2"		Director cues talent and instructs switcher to take Camera 2. Audio engineer cuts film track audio and program resumes from studio.

MASTER CONTROL FILM PLAYBACK At most broadcast stations, someone is assigned to master control to supervise not only the programming that is fed out over the air but the telecine operations as well. These operations include playing back film and slide commercials during station breaks and rolling-in filmed spots during breaks in live or syndicated programming. The procedure for precuing is exactly the same as with studio operations. Instead of following a script, however, the assistant director assigned to master control must follow the station's traffic log to learn where and when to roll film. The most common error in master control is projecting the wrong commercial or film spot. You must be sure to check the slates on programs and commercials to make certain that the proper film is loaded on the correct machine. At many stations the actual rolling operation is computerized, and the station's central computer automatically controls each telecine chain, rolling in commercials and spots on films, slides, and videotape.

USING FILM AND VIDEOTAPE IN POSTPRODUCTION

The enormous postproduction editing flexibility which modern videotape editing systems offer is being applied to film in many production facilities. Original footage is produced on film which is processed in the conventional manner. Instead of physically editing the film, however, the processed footage (and sound if shot with synchronized sound) is immediately transferred to videotape. From that point, the program material is edited on videotape, and all of the video and audio postproduction techniques which are normally used for video are available. In effect, this provides the director with the best of both worlds.

SUMMARY

The film format most commonly used in television is 16-mm film. The 35-mm format is used by networks and some larger stations, and the smaller Super 8-mm format is used at some smaller production facilities.

Film is either silent or sound. Sound film contains either an optical or a magnetic sound track which runs along one edge of the film.

The basic equipment used to edit film consists of a film editing table, rewinds, viewer, splicer, sound reader, and synchronizer. Splices can be made either with tape or with cement. Tape splices are quicker and easier than cement and can be easily opened to change an edit. Cement splices take longer, but they are more permanent and stronger than tape.

The telecine—or film chain—contains all of the equipment necessary to reproduce film and slides. A film chain consists of (1) the television film camera, (2) the multiplexer, (3) a film projector or projectors, and (4) a 35-mm slide projector.

Film is commonly cued with academy leader, which contains numbers that countdown in seconds from "8" to "3". Usually a show reel is prepared for each production; it contains all the filmed segments, separated by academy leader. This makes it convenient to roll in all segments, on cue, from a single telecine.

CHAPTER 15
SET AND STAGING DESIGN

Sets and staging are used to create the physical environment in which a show takes place. On most shows the audience's first impression of the program comes from the set. A set which is well designed and effectively staged instantly communicates the show's intentions, tone, and atmosphere to the viewer. It literally "sets up" the audience for the production.

By *sets* we refer to the scenery, curtains, properties, and furniture which appear on-camera. *Staging* refers to how these various elements are designed, arranged, and integrated within the studio to create the mood or atmosphere for a production and to provide the performers with a working environment. While large budgets and elaborate production facilities make set design easier and more accessible, the most important ingredients are imagination and ingenuity, neither of which is the sole property of the networks or large production facilities. Sets and staging are important visual elements which contribute in building the videospace. They deserve as much care and attention as any other facet of a production.

THE SCENIC DESIGNER

The team member responsible for set and staging design is the *scenic designer*, sometimes called the "art director" or "set designer." At larger production facilities, where a great many sets must be designed for various productions, the scenic designer is occupied solely with the job of set design and construction. At smaller facilities, where there may be less call for the design of new sets on a regular basis, the job of scenic designer is often combined with that of the lighting director since the two roles overlap in many areas. At even smaller facilities the program's director or producer may be the person who is expected to design a set, frequently using stock set pieces which already exist. Even if you do not plan to become a scenic designer, an understanding of the basics involved will make it possible for you to communicate with whoever is designing the sets and to understand better the many possibilities which sets and staging offer to a production.

According to Otis Riggs, a veteran art director at NBC, a good scenic designer is someone who combines a heightened visual and aesthetic sense with the skills of design, interior decoration, and drafting. It is the scenic designer's job to convert the writer's script, the director's approach, and the producer's budget into the physical environment in which a show takes place.

A television scenic designer is bound to work on many varied productions. A comedy or drama offers obvious design challenges, but so does the set for a nightly news show, a weekly interview program, an instructional series, or a children's show. Every production presents the designer with a unique set of requirements and problems, and the designer must come up with fresh and interesting designs which will "work" for the production. This means the set must (1) establish the proper atmosphere and environment and (2) facilitate the technical production operation so that the director, crew, and the performers can work within it comfortably.

The scenic designer is one of the key members of the production team. Sets and staging are not only important elements in building the videospace and the audience's perception of the show, but from a technical standpoint the set and staging interact closely with many other production operations such as lighting, camera and performer blocking, and audio.

FUNCTIONS OF SETS AND STAGING

Sets and staging should be designed to serve four basic functions:

1 *To provide the background and physical environment for the action.* At its most fundamental level, set and staging design provides the scenery, furniture, and props for the performers to work with.

2 *To set the time and place and to establish the mood.* The set should tell the audience something about the time and place in which a show occurs. In dramatic productions, this means the specific location, the time of day, and the chronological period. For example, a set might be designed to suggest a bedroom at nighttime in Victorian England. Or a kitchen set might have to establish a kitchen in the daytime in contemporary America.

The mood or atmosphere of the show is suggested by the script, determined by the producer and director, and interpreted into a physical reality by the scenic designer. Set design can suggest happiness, sorrow, loneliness, tragedy, impending doom, fantasy, or any number of different emotional tones. Even sets for nondramatic shows should strive to establish an overall atmosphere. The set for a news program for instance, might be designed to convey the feeling of sincerity, responsibility, and efficiency to the viewer.

3   *To give the show a unique style which unifies its visual elements.* Style is the visual treatment or the "look" of a production. It is used both to unify the visual elements of the show and to enhance its mood and tone. A set's overall style could be designed to be sleek and contemporary, comfortable and homey, businesslike and efficient, or lavish and extravagant, to mention only a few examples.

4   *To work as an effective production element which complements the overall show.* The set must "work" for the performers, for the director and crew, and for the viewing audience. This is probably the most elusive function to put into precise words, yet it is also the most important. If a set fails to do its multifaceted job for a production, the entire show can suffer. A well-designed set is conceived with the performer's needs in mind, provides sufficient shooting opportunities for the director and cameras, and still establishes the necessary environment to enhance the audience's appreciation of the show. While an elaborate and expensively produced set is rarely enough to overcome a weak script or poor direction, even the best written, performed, and produced show works under a serious handicap when using an unimaginative or inappropriate set and staging design.

ELEMENTS OF DESIGN

The various elements which make up set design consist of (1) style, (2) composition, (3) line and texture, (4) contrast, and (5) color.

Style

It is convenient to divide set design into three

stylistic approaches: (1) neutral, (2) realistic/representational, and (3) stylized/abstract.

NEUTRAL   The simplest and most basic style approach is neutral, which can consist of as little as empty space. Two television variations on a neutral setting are cameo and limbo. As explained in Chapter 6, limbo refers to a light gray or colored background behind the foreground subject which suggests apparently infinite space. Cameo is a black background, a seemingly endless void. Both of these effects are created in large part through careful lighting.

A neutral set obviously focuses the audience's interest on the foreground subject since there is little if anything in the background to distract its attention. But since the neutral setting has no real depth perspective, the picture may appear flat and uninteresting. Some designers use hanging pictures, logos, silhouettes, or abstract designs in the background to break up the background monotony. The advantage of a neutral set, aside from giving the foreground subject prominence in the videospace, is its ease of setup and its very low cost. (See Figure 15-1.)

REALISTIC/REPRESENTATIONAL   This stylistic approach attempts either to produce a realistic setting or to represent a normal reality on-camera. Most dramatic and comedy programs use a realistic set, which is usually designed as a boxlike, three-walled setting. Many nonfiction productions, such as news shows or instructional presentations, use a representational set in which the overall environment is used to represent the program's content and objectives. (See Figure 15-2.)

STYLIZED/ABSTRACT   A stylized or abstract setting attempts to suggest a particular reality (or unreality) through the careful selection of sets, scenery, props, and staging. Unlike a realistic/representational set, which is designed to be as accurate in detail as possible, a stylized/abstract approach uses a minimum of detail and often fragments the set by including only certain ele-

FIGURE 15-1 **Neutral Setting.**
(Courtesy: WSPA-TV)

(a)

(b)

FIGURE 15-2 **Realistic/ Representational Setting.** Realistic sets can be used for dramatic programs *(a)* or for informational programming such as the set for ABC-TV's *Good Morning America* *(b)*. *(Courtesy: ABC-TV)*

ments and exaggerating others. Stylized/abstract settings are usually *open sets* without the solid three-wall background which is found in most realistic approaches. (See Figure 15-3.)

Before we leave the topic of approaches to style, a word of caution. These three categories are used as illustrative examples of basic design approaches and not as hard-and-fast categories. Obviously, many settings borrow from each approach, and the lines of distinction often overlap. Exactly when to use one approach over another depends on the show's objectives, the director's approach, the production facilities and budget, and the scenic designer's concept of the particular set and staging design which will best serve all these factors.

Composition

Effective visual composition is as much the responsibility of the scenic designer as it is the responsibilty of the program's director. As with a theatrical set, the scenic designer must produce a setting which looks balanced and unified in the videospace. Unlike the stage, however, where the full set appears across the entire stage at all times, television dissects the full set into smaller segments as the director combines various camera shots and angles to build the videospace. This means the television scenic designer should

have some idea of the director's overall shooting plan, so that all background areas will look well composed and framed from any shooting angle.

As an example think of a simple two-person interview show. The opening shot from a center camera will be used to open and close the show and from time to time provide some visual variety. But the director is more likely to concentrate on medium or close-up shots of the host and guest, which are photographed from side camera angles. In this case, the scenic designer should consider three separate set compositions: the overall wide shot, the background on the guest's medium or close-up, and the background on the host's medium or close-up. (See Figure 15-4.)

Line and Texture

Line refers to the set's overall shape, form, and use of depth and perspective. Texture refers to the physical quality of its surface. Naturally, the two are closely related and both can be used to the designer's advantage in shaping the audience's perception of the videospace.

FIGURE 15-3 **Stylized/ Abstract Setting.**
(Courtesy: Berkey Colortran)

Wide shot
(Cam. 1)

Close-up
(Cam. 2)

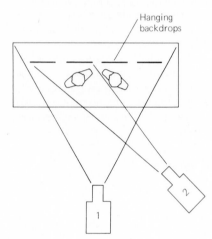

Hanging
backdrops

FIGURE 15-4
The entire set, which is pictured as Camera 1's wide shot, is not the only way the set will appear in the videospace. More often, the director will shoot the subjects in close-up, as pictured in Camera 2's shot. The set should be designed to be equally effective for both shots.

LINE The shape, structure, and form of a set should provide a sense of unity and composition while, at the same time, communicating the show's atmosphere or mood to the viewer. Line is an effective way to do this. A realistic setting will probably use normal lines and perspective since the set must appear normal. A stylized or abstract set, on the other hand, might use distorted lines and exaggerated perspective to produce a fantasy or surreal impression.

Since the television medium is two-dimensional, scenic designers frequently use line to enhance the viewer's perception of depth. We react to larger objects as being physically nearer and to smaller objects as being physically distant, so the scenic designer can use these perceptual processes in adding the impression of depth and dimension to a set. Lampposts or telegraph poles can be made to appear as though they are

receding into the distance by making each one slightly smaller than the one which precedes it. The use of perspective lines, which are painted or drawn on the studio floor, is another commonly used technique to add depth and dimension to the studio.

TEXTURE Texture can be applied physically either by building the set with a three-dimensional surface or through the use of painting and lighting. Needless to say, nothing is as effective as physical depth on a setting to enhance depth and to give the lighting director some texture to work with. The surface quality of scenery also affects the viewer's perception of color and brightness. A smoothly textured surface looks brighter and more colorful than a rough surface which cuts down reflection and appears darker and less vibrant on screen.

Contrast

As we have mentioned in earlier chapters, the television camera operates with a contrast range of approximately 30:1. The familiar ten-step gray scale is a useful reference in selecting the proper brightness values for scenery, furniture, and props. Avoid using pure white or pure black, both of which fall outside the acceptable contrast range as represented by the gray scale. Peak whites should reflect no more than 70 percent light reflectance; blacks, no less than 3 percent light reflectance.

Despite the fact that the designer should try to keep brightness values from exceeding the limits of the contrast range, a set without sufficient contrast produces a dull and lifeless image, which is equally bad. A minimum of two steps on the gray scale is necessary to create a recognizable difference in contrast on-camera. Sets should be designed within the full contrast range using appropriate highlights and shadow to add visual variety and interest. Of course, lighting plays an important part in setting the overall contrast. As a general rule, the background set should appear about two-thirds as bright as the foreground subject. Needless to say, the foreground-to-background brightness relationship will vary sometimes because of special circumstances, but generally as the background brightness increases, it tends to draw the viewer's attention away from the foreground subject.

Color

Color plays an obviously important role in production because it adds depth, dimension, and realism to the videospace. But the use of color must be carefully planned and intelligently used to be most effective.

COLOR REPRODUCTION It is the scenic designer's job to understand how the television camera translates colors, basing the use of color primarily upon its rendition in the videospace. The most accurate way to evaluate a particular color is to see it on-camera under the same set and lighting conditions which will be used on the show. As this is not always possible or practical,

you will need some guidelines to follow in selecting and using color for television.

As you already know, color depends on the interaction of hue, saturation, and brightness. Colors which are very light or highly unsaturated reproduce poorly on color television. The colors tend to wash together into a uniform white or light gray. At the other extreme, dark colors, which are very low in brightness value, also tend to merge together into dark gray or black. These extremes should be avoided because the camera simply cannot distinguish between variations in color at either end of the brightness and saturation scale.

Certain hues also tend to reproduce poorly on-screen because of the limitations of the camera picture tubes and the phosphors on the face of a television receiver. Reds, oranges, and magentas can be troublesome especially when they are highly saturated. This is not to say that you must never use these colors, but that you must apply them carefully and avoid them on large areas of the background.

COLOR PERCEPTION It has been found that most people react to various colors in similar ways. Scenic designers can use this perceptual phenomenon to their advantage in the selection of color schemes for television productions. Here are some basic guidelines in the expressive use of color:

Warm colors, such as reds, yellows, oranges, and browns, look larger and physically closer than cool colors like blues, cyans, and greens.

A bright color against a dark background appears larger and more prominent than it really is. Conversely, a dark color against a bright background looks smaller than it really is.

A dark color appears stronger when positioned before a dark background.

Strong, highly saturated colors look visually heavier and more solid than light, unsaturated pastels. This can be used to the designer's advantage by balancing a large pastel-colored area with a physically smaller but psychologically heavier-looking area of saturated color.

When identical colors are used, a smoothly textured surface looks brighter and lighter than a rough surface, which makes colors appear to be darker and more saturated.

Colors reflect onto different surfaces. If you paint the desk tops of your news set orange, they will reflect orange light onto talent's face, producing an unflattering color effect. Avoid unintentional color shifts due to color reflection either by separating the surfaces or by using a neutral shade which will not upset the proper color tones in the shot.

A subject's color is influenced by the color quality (color temperature) of the illuminating light source. Always match colors under the same 3200°K color temperature, which is the standard for all color studio lighting instruments.

DEVELOPING A COLOR SCHEME A designer's color scheme involves the colors used on all of the background scenery, furniture, and props. On some elaborately costumed productions, the wardrobe designer must also coordinate color schemes with the scenic designer.

To work out a show's color scheme, you will find it helpful to refer to the color wheel in Figure 15-5. Notice that the three primary colors—red, blue, and green—appear opposite their respective complementary colors. It has been found that those colors which appear opposite each other on the color wheel look best when they are combined in the same shot. All human skin tone is a variation of yellow hues, which is one reason why blue is such a popular background color in television. You need not use the exact opposite color incidentally. Variations in brightness and saturation within the same basic family of hues will work just as well. For example, browns and oranges look just as good as yellow against a blue background.

It is also crucial for the scenic designer working

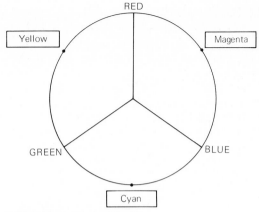

FIGURE 15-5 Color Wheel.
The complementary colors opposite their primary colors match well on screen. Matching colors need not be pure hues but variations on the primary-complementary color combination.

in color to consider how the colors will reproduce on a black and white receiver. You must not select only colors which produce a pleasing color scheme, but also those with sufficient brightness contrast to show up in monochrome. Red and blue are contrasting hues but, unless their brightness values are at least two gray scale steps apart, they will appear as identical on a black and white set.

Colors play an important part in creating the atmosphere and mood of a setting. Obviously, the color scheme for a nightly news show or for a serious interview program will be different from the color scheme for a Saturday morning children's show. The colors on the news set should convey a responsible, businesslike atmosphere and must never compete with the newscaster for the viewer's attention. The colors on the children's show, however, ought to be bright and sparkling suggesting an upbeat, lively environment.

CYCLORAMA

The cyclorama, or "cyc," is a muslin, canvas, or scrim fabric designed to appear as a seamless continuous backdrop. The cyc runs around the edges of a studio suspended from a curtain track. Most studios curve the cyc in a U-shape around

FIGURE 15-6 Cyclorama.
The cyclorama is a cloth scrim which is stretched tightly along the edges of the studio. Cycs offer very flexible staging opportunities. In the photograph, a stylized wood cutout is hung in front of the cyc. Cloud projections complete the setting. *(Courtesy: Group W)*

at least two or more corners to produce a horizonless vista which suggests greater depth. (See Figure 15-6.)

To produce the smooth, continuous surface, the cyc must be stretched taut from the ceiling to the floor by applying tension on the cloth from below. Most cycs use a series of ties at the bottom edge which are attached to a pipe that runs along the floor of the studio in parallel with the curtain tracks above. To set up the cyc, the studio crew secures each tie to the cyc pipe, constantly stretching out the material to remove all wrinkles until it looks completely flat. Another method of maintaining tension is to run a pipe in a skirt pocket on the bottom of the cyc. The pipe's weight pulls on the cyc and creates enough tension to smooth out its surface. When the cyc is not needed, it can be quickly untied from the bottom pipe, bunched together, and pulled along its traveler track until it is out of the way.

Cyc material is typically a neutral gray or off-white, which, if similar in color and brightness to the studio floor, can be used to create a seemingly endless vista on screen. A curved set piece unit called a *ground row* is used to blend the floor and cyc together. The ground row is positioned in front of the cyc where it hides the pipe and any cyc lights which are positioned on the floor to illuminate the bottom half of the cyclorama. (See Figure 15-7.)

A good cyclorama is an extremely flexible, all-purpose scenery element for a variety of reasons:

1 The cyc produces a spacious, horizonless background which enhances the illusion of depth even in smaller studios.

2 The cyc can be "painted" with colored light to produce an endless series of background effects. The use of projected light patterns offers additional variations.

3 A cyc complements many different types of production situations and design styles. Inexpensive cardboard, wood, or styrofoam patterns, logos, or lettering can be hung in front of the cyc to break up the flat background and provide visual variety. Silhouette effects as well as many different foreground/background lighting effects are easily produced using the cyc.

CURTAINS AND DRAPERIES

Studio curtains are draperies which are hung from traveler tracks running along the edges of the studio. The traveler track enable the curtains to be quickly positioned anywhere along the studio edge or to be bunched together and moved out of the way when necessary. Pleated curtains are a simple and inexpensive background element, but unless they are imaginatively used in combination with other set pieces, they can look dull and uninteresting on screen.

Black Velour Curtains

An indispensable drapery for any studio is a black velour curtain, which is used to produce a cameo setting. Although the cameo effect is created primarily through the use of lighting, any background reflectance can ruin the cameo effect of infinite darkness. Especially in smaller studios, where it is difficult to provide much separation between the foreground illuminated area and the dark background, a black velour curtain will absorb much of the ambient light and enhance the cameo effect.

Chroma Key Drops

A chroma key drop is simply a large piece of chroma key blue or green fabric which is hung

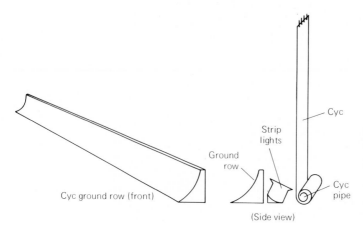

FIGURE 15-7 **Cyclorama Ground Row.**
The ground row is used to enhance the horizonless vista produced by the cyc. It can also be used to conceal the cyc pipe and cyc strip lights, which are used to illuminate the bottom half of the cyc.

behind a subject that is to be inserted into another shot. Because the drop is lightweight and easy to put up and take down, it is a highly flexible scenery item. Chroma key drops are very useful on remote productions, too, since they are lightweight and can be hung virtually anywhere to enable talent (such as a sports commentator) to be keyed into another picture.

SCENERY AND SET PIECES

In this section we will cover the major scenery and set elements and the techniques used in television staging. These include flats, twofold and threefold units, risers, as well as some special purpose staging materials and techniques.

Flats

A studio set is constructed by joining together a number of individual units or *flats*. The size of the flats will vary depending on the size of the studio and other production considerations. A 4-foot width is a convenient size, which allows an individual to carry the flat without assistance. Special flats are also made in larger and smaller widths. The height of the units depends on the grid height of the studio. Smaller studios with lower ceilings use 8- to 10-foot flats, studios with higher ceilings or, where the director wishes to shoot up from a low angle, may use 10- to 12-foot high flats. (See Figure 15-8.)

A flat consists of four parts: (1) the wooden frame, (2) the face of the unit, (3) the hardware used to join the flats together, and (4) a support device.

FRAME The frame of a flat is made of 1- × 3-inch or 2- × 4-inch lumber, cut to size and then joined together with corner blocks or *keystones* made of ¼-inch thick plywood.

FACE The face of a flat can be made either from such hardwall materials as composition board, paneling, or plywood, or from softwall material using a canvas fabric which is stretched over the wooden frames.

Hardwall flats are the most commonly used set pieces in television because they are durable;

enable the designer to add three-dimensional texture to the face of the flat; and permit pictures, mirrors, and other set dressings to be nailed or tacked directly to the face of the unit. Hardwall flats are heavier and more expensive than softwall flats, however, which may be a disadvantage at smaller production studios. *Softwall* flats are frequently used for backing pieces; the outdoor scene which you see through a set window is a typical example.

HARDWARE There are a number of ways to join together individual flats into a complete set. One of the simplest is the *lashline technique*, which uses hardware cleats attached to the frame of every unit. A rope is tied securely around the cleats joining the two flats together.

Another approach is the use of *metal fasteners* which consists of a pin and hinge unit. The hinges are attached to the wooden frames, and a pin is used to secure the two flats together. The disadvantage of the metal fastener is that the positioning of the hinges on every unit must be accurate enough so they will fit together properly when assembling the flats in the studio.

Clamps are probably the most convenient and strongest method of joining the two flats together.

FIGURE 15-8 **Parts of a Flat.**

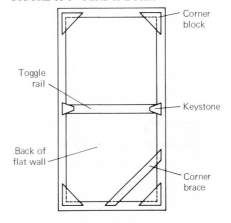

Corner block

Toggle rail

Keystone

Back of flat wall

Corner brace

The C-clamp is an adjustable device, which is tightened on the edges of two frames to secure the units. (See Figure 15-9.)

FLAT SUPPORTS Once flats are joined together, they must be supported so they can stand solidly in place. This is accomplished with the use of braces, or *jacks*, which are attached behind the flat to the wooden frame. The jack is attached either with hardware, such as a pin and hinge, or simply joined with another C-clamp. The bottom of the brace is secured with a sandbag or stage weight. Bracing each individual flat is not usually necessary, but braces should always be applied at such stress points as doorways or windows, on such heavy flats as fireplaces and archways, and at set corners. (See Figure 15-10.)

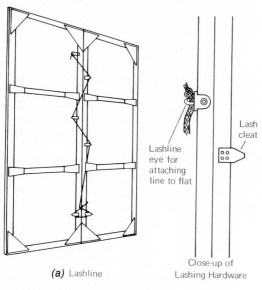

Lashline eye for attaching line to flat

Lash cleat

Close-up of Lashing Hardware

(a) Lashline

FIGURE 15-9 Fastening Flats Together.
The lashline *(a)* uses a strong line, which is wrapped around hardware cleats to join the flats together. Metal fasteners *(b)* consist of a pin and hinge combination. C-clamps *(c)* are a strong and convenient method of joining flats together.

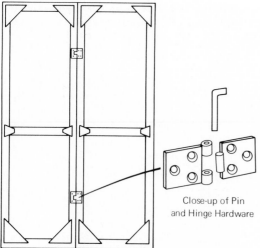

Close-up of Pin and Hinge Hardware

(b) Metal Fasteners

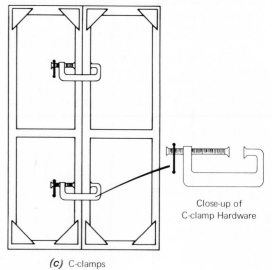

Close-up of C-clamp Hardware

(c) C-clamps

FIGURE 15-10 Bracing Jack.
The jack is attached to the rear of the flat with hardware or a C-clamp and secured by using stage weights or sandbags. *(Courtesy: ABC-TV)*

Flat Construction

The construction of a simple flat is easy, and the necessary supplies and tools are readily available. More elaborate flats containing doorways, windows, and wall pieces may require more advanced carpentry skills.

1 Establish the dimensions you will use and the number of flats you will need. Order either 1- × 3-inch or 2- × 4-inch lumber for the frames.

2 Decide on either hardwall or softwall flats. Hardwall flats are preferable for most television settings. Determine what the face of the flat will consist of: wood paneling, composition board, or plywood. Paneling should be se-

lected for its color, brightness, and appearance on television. Avoid very light or dark colors and overly shiny panel surfaces which may reflect light and create camera or lighting problems. Composition board or plywood is easily painted or wallpapered depending on your needs.

3 Cut the frame lumber to size. Butt the corner joints together using a carpenter's square to ensure a perfect right angle.

4 Position a keystone or corner block piece with the plywood grain running across the joint. This produces the strongest support. Drive the nails partially in and check the shape of the butt joint. If all the corners are tight and square, sink the nails completely. (See Figure 15-11.)

5 Turn the frame over to ensure that no nails are protruding from the face side. Nail the hardwall face material onto the constructed frame.

6 Paint the face, or wallpaper, as necessary. If wallpaper is too light or reflective, it can be

FIGURE 15-11 Corner Blocks and Keystones.
Both of these important structural elements should be positioned with the plywood grain running across the joint.

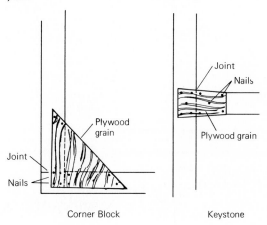

Corner Block   Keystone

toned down by spraying or spattering a darker tone over the area. Spattering is done by tapping a wet brush against your hand or arm, covering the entire area with tiny droplets. The droplets are invisible on camera, but they reduce the brightness and color saturation of the background.

7   Attach connecting hardware to the rear of the frame.

Twofold and Threefold Units

Twofold and threefold units are simply hardwall flats which are permanently joined together with a number of hidden hinges. The advantage of a twofold or threefold is that the unit is completely self-supporting provided the outer wings are not opened too far. A number of neutral twofold and threefold units are convenient to have on hand because they can be used in a wide variety of production situations and can be quickly set up and taken down. Dressing the standard unit with different set trimming will create a unique unit with little cost in money or time. The disadvantage of twofolds and threefolds is their large size and heavy weight, which makes them more difficult to maneuver than ordinary flats. Some studios add casters on the bottom of the units to permit the floor crew to wheel them in and out of the studio more easily.

Risers

Risers are large platform sections made of plywood and 2- × 4-inch lumber, which are used to raise a set above the normal studio floor level. Risers are a very convenient and flexible staging element for the following reasons:

1   Risers elevate the height of seated talent to a reasonable camera pedestal operating height. On an interview show, for example, placing the talents' chairs on risers enables the camera operators to work their cameras at a more comfortable and convenient height while still photographing the performers at eye level. (See Figure 15-12.)

2   Risers add depth and dimension to a set, especially when shooting the entire set on a wide shot.

3   Risers can be joined together in modular fashion to create a multileveled set for dramatic, dance and musical productions. (See Figure 15-13.)

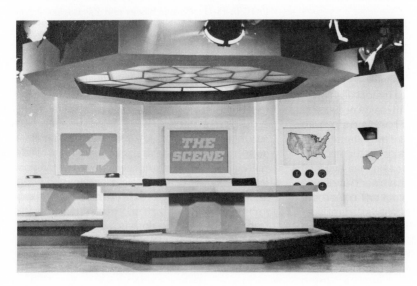

FIGURE 15-12   **Set on Riser.**
(Courtesy: WFBC-TV)

FIGURE 15-13
Risers can be cut to specific shapes and used to create a multilevel setting.
(Courtesy: KRON-TV)

Risers are easily constructed with a large sheet of ½-inch or ¾-inch plywood and a frame of 2- × 4-inch wood planking. Common riser sizes are 4 × 4 feet and 4 × 8 feet. A few smaller risers about 2 × 4 feet and 2 × 8 feet are useful as steps. Risers are usually painted black, although they are most often used covered with carpeting, which not only provides a nicer appearance, but eliminates the sound of footsteps as performers walk across the hollow platform.

Miscellaneous Set and Scenery Items

Although drapes, flats, folding units, and risers are the most commonly used staging and scenery items in television, there are a number of additional techniques which are convenient and effective for a variety of special production situations.

POLYSTYRENE The use of the synthetic plastic material, sometimes called "styrofoam," has become increasingly popular for various set pieces and trimming because of its low cost advantages and production flexibility. Polystyrene is available

from lumber, art supply, and theatrical supply houses in blocks or boards at a reasonable cost. The blocks can be sawed, cut, sculpted, or embossed using a variety of techniques. The large block lettering in the newsroom set pictured in Figure 15-14 was made of polystyrene blocks. Each letter was first penciled on the block and then carefully cut out using a jigsaw.

The foam can be etched or embossed with a paint thinner or solvent such as acetone. Brushing, pouring, or spraying the chemical onto the foam creates a chemical reaction which eats away the surface of the plastic. This should be done in a well-ventilated room, however, since the fumes created by the chemical reation are toxic and must never be directly inhaled.

Polystyrene can be painted with a latex-based paint. Never use oil-based paints because they contain a chemical which will eat away the surface of the plastic.

SEAMLESS PAPER Seamless paper is available on large rolls in a variety of colors and surfaces. The paper makes an ideal neutral background for close-up demonstrations and for product shots. The roll can be mounted on a wall and the paper

curved onto the floor to create a horizonless
background.

POLECATS A polecat is a spring-loaded pole
which is designed to be braced between the
studio floor and the ceiling or overhead lighting
batten. Large posters, pictures, graphics, and
sometimes even flats can be attached to the
upright polecats. A number of polecats with various graphics attached can be strategically
placed about a neutral set to add depth and
visual variety to a normally flat background area.

HANGING SET PIECES Hanging pictures,
graphics, abstract designs, logos, or lettering is
an inexpensive and visually interesting way to
dress a set. Polystyrene, cardboard, and thin
plywood are excellent materials for this purpose,
and both white and black high-test hanging line is
available from theatrical supply houses. For relatively light loads, heavy-test fishing line may serve
as a ready substitute for theatrical line.

MODULAR SET UNITS A number of manufacturers produce modular set units which are mold-

ed out of a polystyrene-like base. It is possible to buy bookcases, molded paneling, and even a custom-designed, vacuum-formed background based on a program's logo or a station's call letters. Though these units are not inexpensive, they are lightweight, flexible, and look good on camera.

FLOOR TREATMENT Most studio floors are a tile or linoleum surface colored off-white or gray. Floors can be painted with a water-soluble paint, which permits floor decoration without interfering with camera and equipment movement. Depending on the production requirements, the floor can be painted a solid color, decorated with abstract designs or patterns, painted with perspective lines to enhance depth, or patterned to suggest concrete, cobblestones, or wooden flooring. (See Figure 15-15.)

Another method of decorating the floor is to use colored tape in various widths for perspective lines, checkerboarding, or abstract patterns. The tape will not usually interfere with performer or equipment movement, but may become worn after a few hours of rehearsal.

Carpets and rugs are often used to cover not only risers but the studio floor on some sets. The advantages of carpeting are a lowering of floor

reflectance, the ability to treat the color and texture of the floor, and an enhancing of the realistic setting. The biggest problem with carpeting is that it interferes with camera movement. Floor assistants must be assigned to lift the corner or edge of the rug whenever a camera must move in closer to the set or performers.

GOBOS In television, a *gobo* is a foreground frame through which the camera shoots the background. Gobos can be used in a variety of abstract or realistic ways, and they are effective because they naturally enhance the depth in a picture by utilizing distinct foreground and background depth planes. (See Figure 15-16.)

PROPS AND FURNITURE

Properties, or "props," and furniture are the additional items which complete a setting, provide talent with the necessary items they must work with during the show, and add detail to the basic background.

FIGURE 15-15 **Floor Treatment.** The studio floor has been painted with a stencil pattern to simulate a cobblestone street. *(Courtesy: PBS)*

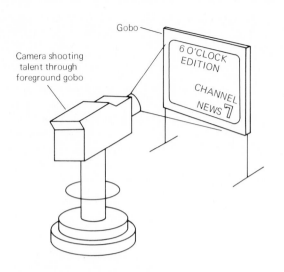

Talent in background

FIGURE 15-16 **Television Gobo.**
The camera shoots talent through the foreground frame.

Props

Props are those items which appear on camera and are neither scenery nor furniture. These include lamps, pictures, curtains on windows, books, dishes, food, and so on. Props are usually classified as either (1) set props or (2) hand props.

SET PROPS All of the varied items which are used to set the atmosphere, provide detail, and simply produce a pleasing background setting are set props. For example, a kitchen scene might need curtains on the windows, cookbooks, pots and pans, dishes, flowers along a window sill, and a picture on the wall to complete the setting. Some set props must be practical; that is they have to actually work. In our kitchen set, the stove, oven, and sink must all be practical since the performer will need to operate them during a show.

HAND PROPS Whatever properties are needed by the performer in the course of the performance are hand props. These include coats, umbrellas, books, guns, knives, telephones, drinks, food, newspapers, and so on. Quite often, talent and the director may request a hand prop which is not necessarily indicated in the script, but which they feel would be useful for the performer to handle during the show.

ACQUIRING PROPS There are a number of standard props which all production studios ought to keep permanently in stock. Pictures, books, lamps, curtains, telephones, dishes, glasses, and eating utensils are required so often that it makes sense to have them on hand. Special props, which are necessary for a particular production, are usually acquired from second-hand shops, local stores, and from individuals or firms willing to lend the production a property in return for an on-air credit. The latter technique is often used, especially when the needed props are expensive and difficult to obtain. For example, a local sports shop might loan a complete line of skiing equipment and accessories for a skiing series in return for an on-air credit after each program.

Always consider the time, place, and chronological period when working out props. A 1940s period drama cannot use a modern, push-button telephone. Since television is a close-up medium, props which will be shot tight must also look as realistic and detailed as possible.

Furniture

All studios require a minimum amount of stock furniture which is always available and can be used for many different production situations. Sometimes a permanent series, such as a weekly interview show or a nightly news program, will purchase its own furniture expressly for the production. But in most other cases, the scenic designer will be expected to use the studio's stock furniture, which is why furniture must be purchased carefully so it will offer a great deal of set and staging flexibility.

Here are some guidelines to follow in buying or acquiring furniture for production:

1 *The furniture should be durable and well made.* Stock furniture gets constant handling (and abuse) and must be able to stand up to it without obvious damage. The fabric should also be strong enough to withstand constant use without becoming easily torn or soiled.

2 *Furniture should be easy to handle.* Furniture should not only be durable but it should also be sufficiently lightweight to be carried back and forth without difficulty. Heavy or bulky furniture pieces are usually more trouble than they are worth unless a production specifically requires them.

3 *The furniture must look good on camera.* Consider how the furniture will appear on camera under studio lighting conditions. Very dark or very light colored furniture may not reproduce well. Overly busy fabrics can prove to be distracting on screen. Polished chrome on modern furniture looks great in-person but may cause flaring problems under studio lights.

The furniture must make performers look their best. An overstuffed chair can virtually swallow up performers. Chairs which look or feel uncomfortable should also be avoided.

4 *Avoid chairs which swivel or rock.* Many people—especially nonprofessionals—have a tendency to rock back and forth during a show which can drive the director, the camera operators, and the audience crazy. In most cases, you will be better off selecting stationary chairs which are comfortable but do not allow aimless movement.

5 *Select furniture which can be mixed and matched together.* While you cannot easily combine starkly modern, contemporary furniture with French Provincial, there are a number of styles which lend themselves to many

different set arrangements. Of course, there are times when a production requires a very special period setting or furniture style. For most common nondramatic programs, however, a neutral furniture style will prove to be most flexible and will enable you to choose from a wider variety of chairs, desks, tables, and so on.

6 *Have enough identical chairs to accommodate a reasonable number of people.* There is nothing more frustrating for a scenic designer than to have to seat five panel guests in two or three different types of chairs. While extraordinary circumstances requiring more chairs may arise from time to time, every studio should have enough identical chairs for the most common production situations.

SPECIAL MECHANICAL EFFECTS

Special effects are used to create a videospace reality which is either impractical or impossible to actually produce. In Chapter 11, we discussed electronic special effects; here we will cover some basic mechanical special effects which are usually a scenic designer's responsibility.

As a general rule, a special effect is rarely satisfactory by itself. A mechanical effect requires close coordination with talent, the audio engineer, the lighting director, and the program's director to enhance the illusion visually and aurally.

Smoke and Fog

Dry Ice dropped into a bucket of hot water will produce the swirling white mist that is often used for fog, smoke, or a dreamlike fantasy effect. Boiling the water increases the smoke, and a small fan can be used to blow the fog over the studio floor. The smoke produced by dry ice and water hangs low over the studio floor and disappears quickly unless it is continuously replenished. A difficult problem with dry ice is the noise of the bubbling water, which may have to be masked with audio sound effects.

An alternative to dry ice is a special smoke-making machine, which can be rented from film and theatrical supply houses. The machine uses a fog oil to produce the smoke vapor, and produces a continual stream of smoke without the mess involved with dry ice.

Fire

Fire is always a potentially hazardous special effect because the only convincing fire is a real one, and safer substitutes rarely work as well. With the exception of a small, controlled flame, as in a fireplace or campfire scene, avoid attempting special fire effects without the advice of a skilled special effects technician. You should also consult the local fire codes, and have the proper safety equipment on hand whenever you deal with fire in the studio.

For fireplace or campfire effects, most studios use artificial logs made of fireproof asbestos cloth that is molded around a frame of chicken wire. The "logs" will not burn, yet appear real on camera. The fire is produced by burning cans of Sterno, which are positioned under the logs out of camera range. The lighting director can augment the effect by producing a flickering light on the subjects illuminated by the fire.

Breakaway Props and Furniture

Some dramatic or comedy productions require the use of *breakaway props* or furniture, designed to shatter harmlessly on contact. Breakaway props such as glasses, bottles, and windows can be purchased ready-made from theatrical supply houses. Breakaway furniture, such as chairs or tabletops, is usually made to order from balsa wood. The joints should be lightly glued together. Never use nails or screws in breakaway furniture since these items can injure a performer during the action.

Snow and Ice

Falling snow—a typical effect on many annual Christmas specials—can be produced in a number of ways. You can purchase granular plastic snow from theatrical supply dealers and use it year after year. The snow is either dropped by stagehands, who are positioned on ladders out of camera range, or held overhead in a large muslin trough, which is gently agitated to create the snowfall. A simpler and less expensive method is to use the paper confetti which results from hole punching. Of course, only white paper should be used for this purpose.

Icicles are easily made out of polystyrene which is etched away through the application of acetone. Styrofoam is also useful for making blocks of ice or packed snow. Frost on window panes is easily produced with the spray aerosol sold commercially for Christmas decoration.

SETS AND STAGING IN PRODUCTION

The scenic designer—or whoever is responsible for designing the sets and staging—is a key production team member who should be included in the earliest production meetings. Until a set is developed, designed, and approved, the director cannot begin blocking performers or cameras, the lighting director cannot plan the lighting design, and other production activities which depend on the set and floorplan cannot be started.

Preproduction

The largest share of the scenic designer's work occurs during the preproduction process as the set is planned and designed, a floorplan is developed, and set construction is started. As with most other areas of production, set and staging design is a combination of aesthetic judgments and technical factors. Compromise between the two is inevitable, and it takes skill and imagination to produce a set which not only complements the production's creative objectives, but facilitates its technical operation as well.

A set and staging design must be developed by the scenic designer working closely with the program's producer and director. If the show is fully scripted, reading and rereading the script will help the designer begin to conceptualize a design approach. If the show is not scripted or if a full script is unavailable, the designer must rely heavily on the producer and director to establish the production's overall concept, content, and objectives.

As the designer begins to incorporate these elements with his or her own creative ideas, two important design areas must be taken into account: aesthetic considerations and technical requirements.

AESTHETIC CONSIDERATIONS Aesthetic considerations are the artistic or "creative" elements which enter into the set design. Among the most important questions the scenic designer needs answered are these:

1 *What are the program's overall concept and objectives?* Is the show designed to inform, entertain, teach, or instruct? What should the set and staging communicate to the audience about the show? How will the set support the production? Will it serve as an integral part of the show or as an unobtrusive background?

To a certain extent the answers to these and similar questions depend on the script and on the producer's and director's concepts of the show. In the final analysis, however, general ideas and notions must be interpreted by the scenic designer into the show's physical environment.

2 *What atmosphere, mood, or environment should the set convey?* This question involves many individual design elements which must be integrated to produce a unified, workable, and meaningful atmosphere which complements the show's objectives.

3 *Which design style or approach will work best?* Should it be realistic, stark and neutral, stylized, or abstract? There is obviously no right or wrong answer, as each production presents its own unique set of aesthetic circumstances and requirements. While the producer and director may have an idea in mind, the scenic designer must present a broad range of design possibilities and translate

their initial ideas, along with his or her own, into the final design plan.

Many scenic designers keep a file of different period designs, interior decorations, clippings from magazines, books on interior design, and other reference sources for ideas and inspiration. Quite often a final design is a conglomeration of ideas and approaches from many different places.

TECHNICAL REQUIREMENTS Technical requirements are the practical considerations which are concerned primarily with the production process. Among the important technical questions which should be considered are these:

1 *How will the production be produced?* A "live" show or one videotaped without editing requires all sets to be standing since there is no opportunity to strike one set and replace it with another. Productions which are taped in segments to be edited together may permit the scenic designer more flexibility in set design because a set can be struck after shooting is completed and a new one set up overnight for the next day's scenes. Of course, the scenic designer must know the taping schedule since productions are often videotaped out of normal sequence.

2 *What type of set does the director prefer?* Some directors like to work with a three-wall box set, others prefer a two-wall set, and still others like an open, fragmented setting. Although it is not always possible to accommodate a director's personal preference, the scenic designer ought to know if a particular type of set makes any difference to the director.

Some directors also want foreground elements—such as doorways, windows, archways, props, and furniture—arranged in the set to facilitate foreground composition treatment. The scenic designer should also know the director's basic shooting plan: Are there any special angles planned? Do the cameras need extra room for intricate movements? Will cameras need special access to dolly onto the set for extreme close-ups or other special shots?

3 *How will the performers use the set?* A set must be designed as much for the performers who must work in it as for the crew and the viewing audience. Such practical considerations as ease of movement, placement of furniture and size and accessibility of steps, stairs, and levels should all be planned with the performer in mind. On some sets the performer must use practical graphics such as a chart, picture, or drawing board. These must be positioned at a convenient height and location. In general, the set should be designed to facilitate performer actions and to make performers feel as comfortable as possible within their working environment.

4 *Must the set accommodate special equipment or unusual production techniques?* If the production will use such large equipment as a boom microphone, camera crane, or hand-held cameras, the scenic designer may have to design the set to accommodate them. The designer will also need to know if the on-camera set must include large, special equipment such as a rock band's sound gear or unique devices necessary for instructional or demonstration productions.

Similarly, a designer must know whether the director plans to use extraordinary shooting angles. For example, a very low camera angle may require the use of risers for the set and performers and the construction of extra-high scenery flats to prevent the cameras from shooting off the set. A close-up demonstration requires a set which will allow the camera to dolly in close without obstruction.

5 *Will the set be assembled from stock scenery or will it be constructed especially for the production?* Building a custom-designed setting offers the scenic designer great aesthetic

flexiblity, but raises budget, manpower, and construction problems. Using stock scenery eliminates these worries, but forces the designer to settle for whatever already exists in storage. It is not always easy to use the limited range of stock scenery, props, and furniture usually available at most studios. Making this stock material work for a show is a real test of the designer's imagination and ingenuity.

THE FLOORPLAN A floorplan is an overhead view of the set as it is to be positioned on the studio floor. The floorplan includes the size and location of every major item which appears on the set—scenery, curtains, furniture, large props—as well as the general location of major pieces of studio production equipment such as cameras, booms, and video floor monitors. The floorplan is a crucial piece of production paperwork. Not only is it used by the scenic designer, it is also used by the director, the lighting director, the engineers, and by other team members during preproduction planning and setup.

To be most useful, a floorplan must be drawn to *scale*. Most floorplans use a scale of ¼ inch equals 1 foot. The floorplan of the production's set is drawn on a *studio plan* or *studio plot*, which is itself a scale drawing of the entire production studio. The studio plan shows the location of real doorways, cable outlets, the control room, and other important studio areas. Most studio plans are imprinted with an overlay of the lighting grid. This is helpful for two reasons: First, it is easy to refer to the printed grids on the floorplan to locate any actual position in the studio. Secondly, it makes it easier for the set designer to position the set where the lighting battens are located and for the lighting director to use the location of the actual grids in designing the lighting. (See Figure 15-17.)

Since the floorplan is drawn to scale, the scenic designer must know the actual dimensions of all sets, furniture, and large props which will appear on the floorplan. You should have a complete inventory of all stock scenery, risers, and furniture available with their physical dimensions. You will also need to know the approximate sizes of cameras on pedestals, boom microphones, and video monitors which have to be positioned on the

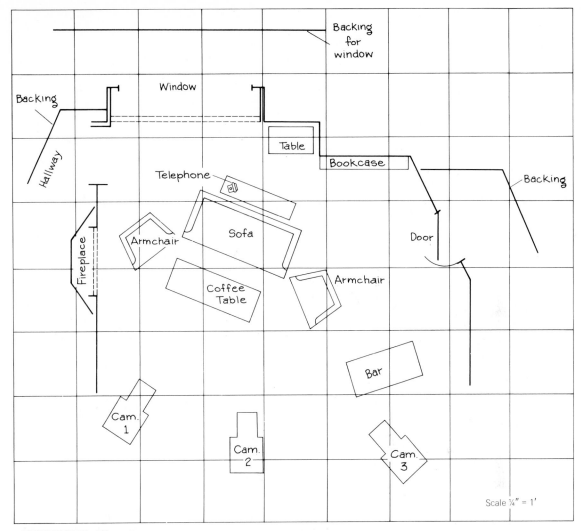

Scale ¼" = 1'

FIGURE 15-17 Floorplan.
The floorplan is drawn to scale and indicates the size and position of major set
elements.

studio floor. If you do not have such an inventory, a day spent backstage with a tape measure will save lots of time and aggravation later on and ensure an accurate floorplan for every production.

Once you know the dimensions of the set and the equipment to be used, you can begin to plot out the floorplan. A drafting ruler, which automatically converts dimensions to scale, eliminates the need for mental calculations and makes scale drawing relatively painless.

A common mistake in working out a floorplan is using studio space which, for any number of reasons, is not actually available for staging. Although a large space may appear on the studio plan, in reality the area may contain a permanent set or may be a portion of the studio that is being used for storage. At facilities with a limited number of lighting instruments, try to position the set where the instruments are already hanging. Another problem is the natural tendency to try and fit large items into smaller areas. You may be able to fit an 8-foot-wide set into a 6-foot space by cheating on the paper floorplan; but once you arrive in the studio, you will find the actual situation less flexible. The floorplan is indispensable in telling you how sets, equipment, and furniture will actually occupy a predetermined space, but only if you design and use it properly.

You will probably have to experiment to find the best location for the set, furniture, and studio equipment. It is typical for the scenic designer to spend a considerable amount of time working out the best combination of set, furniture, and equipment positions before coming up with the final production floorplan. Once the plan is approved by the director, it is copied and distributed to other production team members, where it serves as a foundation for many of the interrelated production activities which take place during preproduction and setup.

SKETCHES AND MODELS A sketch is a perspective drawing which is produced by the scenic designer to enable the producer, director, and other team members to better visualize the proposed setting for a production. Especially on elaborate productions where a set will be designed and constructed specifically for the show, the sketch will give the producer and the director an accurate indication of the scenic designer's intentions before they authorize the set's construction. (See Figure 15-18.)

Some designers like to build a three-dimensional model of the proposed set using cardboard, toothpicks, and construction paper. Of course, the model is built to scale and ideally should reflect the designer's color scheme as well. Models are very helpful because they show the set's true depth, something even a sketch cannot do, and because lighting, shooting angles, and the overall working environment can be carefully studied and planned before actual construction begins. The problem with using models is the strict time limitation most television designers work under. There is often little enough time to design a set and rough out a sketch, much less actually to build a scale model with all the trimmings. However, on elaborate productions where the complexity and costs justify the effort or when the designer has the time, a model can be helpful to everyone in visualizing exactly how the set will appear when constructed. (See Figure 15-19.)

FIGURE 15-18 **Designer's Sketch.**
The designer's sketch is used to give the producer and director an accurate idea of the scenic concept. *(Courtesy: Thomas Fichter)*

FIGURE 15-19 Set Model. The model enables the production team to see how the setting will appear in all dimensions. Camera shots, lighting, and other production variables can be analyzed more accurately in advance by using a scale model. *(Courtesy: WBZ-TV Group W Boston)*

SELECTING COLORS, FABRICS, AND MATERIALS Among the final details which must be determined by the scenic designer are (1) the color of the background; (2) selecting the paint, wallpaper, or paneling for the background; (3) choosing the color and style of furniture; and (4) the set dressing.

The color of the background should be selected according to various aesthetic and technical considerations mentioned earlier. Of crucial importance is avoiding a color and brightness background which equals or exceeds normal skin tone. For best results, the color and brightness of the background should complement but never dominate the foreground subject. Swatchbooks of various paint colors are an easy way to select paint, fabric, and paneling color, texture, and finish. Remember to check all swatches under studio lighting so you can match colors under properly balanced light.

As a general rule, the scenic designer tries to emphasize detail, depth, and texture on a normal background. Unless you are purposely stylizing the set, try to keep detail as realistic as possible since the close-up camera will reveal substitutes on screen which might have easily worked on the theatrical stage where the audience is much farther away. At the same time, you should avoid overly ornate or intricate background detail which looks too busy on screen and can distract the viewer.

Sets should be designed and constructed to appear as solid as possible on camera. Nothing is more unconvincing than to see a supposedly sturdy brick wall sway precariously back and forth each time a door is slammed.

A list of set dressings and hand props should be drawn up by the scenic designer as the set is designed and the director and designer confer about talents' required hand props. Check the list against the stock property list. Special props will have to be obtained in time for rehearsal and production.

Setup and Rehearsal

The scenic designer must be told when the studio will be available for erecting and dressing the set. The setup phase involves a certain amount of coordination between the staging crew and the lighting crew so each can do its respective job without getting in the other's way or delaying the production. The set crew usually erects and braces the basic set pieces before the lighting crew begins because it is difficult, if not impossible, to light a set accurately without the major set elements. As soon as the scenery, risers, and large furniture pieces are in place, the lighting crew can hang and focus lights while the stage crew completes the finishing touches on the set.

As soon as the lighting director has lit the set, the scenic designer should make a last-minute check of the entire set and staging on a studio or control room monitor. The color, tonal values, brightness, and overall atmosphere should appear correctly in the videospace. This is also the time to check for any glare from furniture, props, or background. Glare caused by the reflection of studio lights from a highly specular surface can be eliminated or reduced by the application of dulling spray.

Furniture positions should be marked with a small piece of masking tape, especially if they will be moved during rehearsal and production. During the rehearsal the scenic designer should watch for any floor clutter from props, furniture, or rugs which may be interfering with either the production crew or with the performers.

Production

During the production stage there is rarely much work for the scenic director if all scenery, backgrounds, furniture, and props were correctly set up during rehearsal. In fact, even if a serious problem arose at this point there would be little that the designer could realistically do without completely halting the entire production.

Postproduction

Once the production is completed, the scenic designer is generally responsible for supervising the strike, as the set is torn down and stored away. On some regularly scheduled shows, the basic set may remain permanently standing, although smaller props and furniture pieces—which might be stolen, misplaced, or clutter the studio—are stored away.

Before you strike a set you will not be using again immediately, it is a good idea to take a snapshot for your files. These photos, along with your floorplan and sketches, can serve both as a reference should the same set be needed again

FIGURE 15-20 Set Storage Area.
The storage area, sometimes called a "scene dock," has a series of bins to hold set flats. Each flat is individually numbered and stored in a particular bin. This makes locating specific colors or flat styles more convenient.

and as a possible source of future ideas. If the set is to be struck for another production but will be reset the next day or so, mark the location of the set walls and the furniture on the floor with masking tape so you can easily reposition the background and save the lighting.

SET STORAGE   Flats are usually stored in a large bin outside the studio. The bins should be compartmentalized so that various types of flats can be categorized by their color, type of face, and size. You may find it convenient to label each flat on the outer edge of its frame for ease in identification and access. Risers are stored either stacked on top of each other or standing upright on one end. (See Figure 15-20.)

PROPS   Props should be stored in a large closet, bin, or cage depending on the number of properties you have and the amount of space you need. Props should be carefully indexed and stored so they can be easily found without having to sort through mounds of miscellaneous junk and so everyone knows which props are available in stock. A variation of Murphy's law: The prop you carelessly misplace today is usually the very one you will desperately need for next week's production.

SUMMARY

Sets and staging are the responsibility of the television scenic designer. The four functions of set and staging design are (1) to provide the background and physical environment for the action, (2) to set the time and place and to establish the mood, (3) to give the show a unique style which unifies its visual elements, and (4) to work as an effective production element which complements the show's objectives and the production operation.

The various elements which make up set design consist of (1) style, (2) composition, (3) line and texture, (4) contrast, and (5) color. A set's style can be either neutral, realistic/representative, or stylized/abstract, depending on the script and the designer's approach. All sets should be designed for effective composition, but the designer must also realize that the various camera shots dissect the entire background into smaller segments which must also appear well composed on screen. Line and texture, contrast, and color are the elements which enhance the overall design approach and help to create the atmosphere and tone for the production.

The most frequently used fabric background in television is the versatile cyclorama, which can be used for a wide variety of production situations. Curtains and drops are also used, usually in combination with other sets and furniture pieces.

Flats are individual units which are joined together to make a complete set. Hardwall flats are most commonly used in television owing to their durability and versatility. Twofold and threefold units are simply individual flats which are permanently hinged together to produce backgrounds which can be set up quickly and struck easily. Risers are modular platforms which enable the designer to vary the floor height for a production. Seated performers look better on risers and permit more comfortable camera operation. Risers are also highly versatile because they can be arranged in numerous combinations of steps, heights, and levels. Among the miscellaneous scenery and staging items and techniques commonly used in television are plastic polystyrene, seamless paper, modular set units, and treating the studio floor with paint, tile, or tape.

Props and furniture are used as functional elements for performers and as set dressing to add detail and realism to a set. Set props are mainly used to enhance the background although some are "practical," which means they actually operate. Hand props are properties handled by talent during the production.

Special mechanical effects, such as smoke, fog, fire, snow, and the use of breakaway furniture, are usually assigned to the scenic designer. Some can be produced using materials on hand; others require the rental or purchase of special effects equipment.

The scenic designer's primary role in production occurs during the preproduction stage when the set design is conceived and developed. The floorplan is an overhead view of the studio set, drawn to scale, and used by many production team members to indicate the layout of the sets, furniture, and production equipment. During setup and rehearsal, the set designer's stage crew should coordinate their work with the lighting crew since the two functions are closely related. A final check of the set should be made on the studio monitor once the lighting is completed. Minor changes in sets, furniture, and props can be made once the show begins rehearsal and specific problems arise.

CHAPTER 16

TELEVISION SCRIPTS

Every television production relies on a script of one sort or another. Of course, the enormously wide range of programming requires a variety of scripting approaches. Some scripts are elaborately detailed and include word-for-word dialogue, specific camera shots, and staging instructions. At the other extreme, some scripts are little more than an outline of the production routine. Different as they may appear at first, all television scripts are designed to (1) establish a show's routine or format, (2) indicate the program content, and (3) organize important production information, so it will be most useful during planning, rehearsal, and especially during production.

Although you may not be a writer, every member of a production staff should be sufficiently familiar with script formats either to contribute to a program's script or, at the very least, to be able to translate written material into television script form. Scriptwriting—that is, the development of a situation, story, characters, copy, and dialogue—requires the very special talents and skills of a writer. But the ability to organize written material into a usable format simply takes some practice and an understanding of how a script is used in production. In this chapter we will cover the basic script formats which have been developed over the years and have proved to be among the most useful ways of organizing a program's content for use in production.

WRITING FOR TELEVISION

Writing for television is not the same as writing for print media. Assuming the scriptwriter has the basic skills and talents of a writer, an adaptation to the particular requirements and idiosyncrasies of the television medium is still necessary. A reader can move along at his or her own speed, stop to reread a passage, or quickly skip forward over an uninteresting or irrelevant section. The television viewer cannot do any of these things, which puts an added responsibility on the television scriptwriter. A scriptwriter must understand the medium's capabilities and limitations and be familiar with the basic techniques of television production.

Television Is a Visual Medium

It has almost become a cliché by now, but television *is* a visual medium. The television scriptwriter must be as comfortable working with images as with words. Understanding the best way to combine words and pictures and when to let one or the other carry the program's message is the essence of the scriptwriter's art. Sometimes scriptwriting deals with very little spoken copy; the images may convey the message alone or in combination with appropriate sound effects or music. Since the writer must be as visually oriented as the director, it is no surprise that many television writers are "hyphenates," having combined skills, such as a writer-director, or a writer-producer.

A Script Must Be Written to Be Heard

Although television is a visual medium, all copy or dialogue in a script must be written for the ear, not for the eye. The audience never reads your script; they only listen to its delivery. Copy which might read well will not necessarily sound right when spoken by talent. As a general rule, always read your copy aloud, listening to how it sounds. This, after all, is how the audience will ultimately receive your message.

In general, keep your sentences simple, direct, and to the point. A complex sentence structure is both difficult for talent to deliver effectively and hard for the audience to comprehend. In your writing, strive for a relaxed, informal approach. That is not to say that you should use slang or colloquial expressions for every script, but rather, avoid an overly formal writing style, which sounds stilted and unnatural to the ear and might interfere with the audience's understanding of the message.

The viewing audience must follow the script continuously and cannot review material once it has been presented, so try to establish a clear, logical sequence of presentation. Do not cram too many facts, figures, names, or ideas into a few sentences. If you are about to shift from one concept or story to another, set up a transition so you do not lose part of the audience. Such familiar script approaches as flashforwards, flashbacks, and parallel action can be effective, but they must be handled with care by both the writer and director to avoid totally confusing the audience.

Write for Your Audience

If at all possible, know, and write for, your target audience. Obviously the style, tone, and complexity of a script for the children's show *Sesame Street* will be different from a college instructional series on applied mathematics. The use of language and style also depends on the production situation and the intended audience. What might work in a script for NBC's *Saturday Night* may not be effective on the *Lawrence Welk Show*.

Structure is also an important consideration. For example, it has been found that children do not easily follow parallel action, flashbacks, or flashforwards. A writer must be careful never to get ahead of the audience's sophistication in structuring the show's format and development.

The composition of an audience is another important factor for the writer to consider. For example, a news story about a city government scandal aimed at a local audience could assume that most viewers are familiar with the mayor, the city council, and some of the surrounding cast of characters and events. The very same story would have to be written quite differently for a national network audience since the writer could not assume that everyone watching would have the same familarity with what are essentially local facts. More time would have to be spent introducing national viewers to the background and the significant people involved, in order to put the story into an understandable context.

Consider Television's Aesthetic and Technical Production Capabilities

A script serves as a road map, or point of departure, for a television production. In the end, the viewing audience never sees the script; they see only the program. Scripts should never be written as final, unalterable documents but rather as flexible frameworks around which the production team works to build the videospace. It is not always easy for a writer to accept changes or revisions, especially when they are made for technical reasons. There are times, however, when changes are unavoidable because necessary equipment is unavailable; the budget won't cover the number of sets, locations, or performers indicated in the script; or a show is running long.

The television writer must always work within the production's own range of capabilities and limitations. There is little point in writing a script which requires resources that simply are not available. Many writers are often able to overcome seemingly impossible limitations, however, through the ingenious use of those production techniques at hand.

Time is a limitation which affects every scriptwriter. In the first place, most television scripts are written under serious time pressures. No matter whether you are writing a situation comedy, an episodic drama, a nightly news script, or an instructional presentation, there is rarely enough time to work at a leisurely pace. Time is also an important factor within a program since the television scriptwriter is usually given a very inflexible time frame within which to present an idea or communicate a message. Commercials are an

obvious example where the writer has only ten, twenty, thirty, or at most sixty seconds to deliver the sponsor's message, but many other production situations also require a careful consideration of time. A theatrical play can run as long as the writer and director wish; a television drama must play within the time allotted to the production. A writer who has been asked for voice-over copy to accompany a silent film clip must produce material which runs the length of available film and yet conveys the necessary information to the viewer. It is not uncommon to have twenty-seven seconds worth of script but only eighteen seconds worth of film. Sure, the talent can read the longer material more quickly but will the audience be able to comprehend the message? A stopwatch is as indispensable to a television scriptwriter as a typewriter. Working within these many time limitations and still delivering an effective, high-quality script is the mark of a true professional.

DETAILED SCRIPT FORMAT

A detailed or complete script contains the spoken copy, or dialogue, sound and music information, the major visual elements which should accompany the audio, and important production information, such as timings, video and audio sources, and so on. The basic television script format, which is universally used with minor variations, is illustrated in Figure 16-1.

You will notice that the page is divided, one column labeled "video" and the other labeled "audio." Program content and production information are divided into these two major categories and appear in the appropriate columns.

A script should *always* be typewritten, and all spoken copy should be either double- or triple-spaced. Pages should always be numbered clearly in the upper right-hand corner.

The "audio" side contains the sound elements of the show. Each performer's dialogue or copy is double- or triple-spaced, with the speaker's name (or character if a drama) centered and typed in

PAGE 1

VIDEO	AUDIO
	HOST
HOST SEEN IN CAMEO	Dance has long been one of the most beautiful and interesting of the performing arts. In recent years its popularity has increased tremendously with a wide range of performances from classical ballet to modern dance. Yet for many people dance remains somewhat intimidating. Without an understanding of what it is and where it comes from it's hard to develop an
BACKGROUND LIGHT CUE AS CAMERA DOLLIES PAST HOST INTO REHEARSING DANCER	appreciation for the art. Today with the help of the Modern Dance Company, we'll explore the world of dance in all of its multifaceted DIMENSIONS.
FILM OPENING (:30)	MUSIC: THEME UP AND UNDER
	ANNOUNCER (V.O.)
	DIMENSIONS. . . A weekly series on the performing arts. Today, "The World of Dance" with the Modern Dance Company. Here is the host of DIMENSIONS, Carol Richards.
HOST AT INTERVIEW SET	HOST
	Good afternoon and welcome to DIMENSIONS.

FIGURE 16-1 Detailed Script Format.
The script is divided in half with one side labeled "Video" and the other "Audio." Only important video information is typed onto the script. The rest of the space is kept blank for the director's camera shots and for other production team members to write in their operational cues.

capitals. Sound effects and music cues are also capitalized and underlined to separate them from dialogue. Performers' stage movements are also included in the audio side and are usually capitalized to avoid confusion with spoken copy.

The video side is left almost blank by the scriptwriter to leave room for the director's camera shot notations. However, important visual elements such as titles, graphics, special effects, film or videotape inserts, are usually typed in caps opposite the corresponding audio.

When a page of a script is revised, the new pages should be clearly marked "Revision" with the date of the change. To make certain everyone is using the same corrected version of the script, revision pages can be duplicated on paper of another color than the one used for the original script pages. Scripts which will be used on camera by talent should never be printed on white paper, which reflects light and causes camera and reading problems. Instead, use such pastel papers as blue, pink, or green, which are easier to read and will not upset the camera shading.

Dramatic Scripts

Scripts for dramatic or comedy shows follow the detailed script format with few changes. If the scriptwriter follows a standard margin and types the same number of lines per page, it is easy to estimate the running time of the script by counting pages. For example, you may find that an average page contains dialogue that runs about one minute. If you remain consistent in your typing format, you can count pages to get a rough idea of the program's running time. On the average, forty to forty-five television script pages run about a half hour, but variations in type sizes and formats make it a good idea for you to establish your own guidelines.

Commercials

Commercials also follow the same script format but, since they are designed to convey information in a relatively brief time, the copywriter often indicates the specific visuals which are to accompany the sound track. (See Figure 16-2.)

News Scripts

A news script is probably the most common type of television script since every station prepares them daily. The news script must be written quickly, yet must also be journalistically accurate. The script also serves as the production script, which the director and production team use to produce the show. Since few news programs are actually rehearsed, it is very important for a news script to organize the information in a clear, concise package which can be followed easily.

A news script is usually typed on a multisheet carbon pack so it can be distributed to a number of production team members quickly. If the production establishes a format, including margin size, type style, and the number of lines per page, it is easy to estimate a program's running time by counting pages. If you work out a system where each page holds thirty seconds of spoken copy, it is even easier for the writer, talent, or the director to estimate quickly a show's running time.

Each story should be typed on its own separate page, and each page should be labeled with a heading, or "slug," describing the story. One- or two-word descriptions such as "Arson," "Mayor,"

"Presidential Politics," or "Stock Prices" are usually sufficient to identify each story. With each item on a separate page, it is easy to reorder the sequence of stories or to drop or add a story to vary the show's running time.

FILM OR VIDEOTAPE CUTS The use of film or tape inserts is an important part of the news script. The following examples refer to the script formats in Figure 16-3.

VOICE-OVER A voice-over consists of narration, which is read by talent as silent film or tape footage is run on screen: (1) Indicates the film cut on the reel; (2) is the running time of the film; (3) indicates silent film; (4) notes the talent's name and the fact that it is a voice-over spot.

SOUND ON FILM OR TAPE The use of sound from a film or videotape track requires additional information to be included on the script: (1) Indicates VTR cut six and sound on tape; (2) gives the running time of the tape cut and indicates that the audio should be sound on tape; (3) identifies the final words heard on the tape.

The outcue is especially important for sound on film (SOF) or sound on tape (SOT) scripts since it helps the director, switcher, and audio engineer make a smooth transition from the film or tape cut back to the on-camera newscaster. Unless the outcue words are completely accurate, the cue is of absolutely no value whatever. The outcue does not replace the film or tape running times; the two complement each other, and they serve as a double-check. If the outcue is sound, rather than spoken dialogue, indicate it as follows: *"OUTCUE:" . . . "why I intend to vote for it (APPLAUSE.)"*

B-ROLLING Using a second film or tape cut to insert either picture or sound into an existing story can be organized as follows: (1) identifies film cut number, sound on film, and B-roll; (2) gives the total running time of the entire story; (3) is the start time of the B-roll, which is measured from the top

BDA
TV Script

Client	DELTA AIR LINES	City DTW (MIA, TPA, PBI)	Spct No. 8100

| Date 12/19/ | Job No. D7-9188-9 | Type | Length 30" |

| This Spot effective Jan. 16 | It replaces | Remarks "CHEF" |

KITCHEN SET. CHEF SHOWS TRAY OF 12 BROILED STEAKS, GARNISHED WITH PARSLEY, TO CAMERA

CHEF: This is why every-one's flying Delta to Florida . . .my juicy steaks. . . served on all mealtime nonstops . . .in Tourist, too.

SOMMELIER INTERRUPTS WITH BOTTLE OF CHAMPAGNE

SOMMELIER:

Hah! They fly Delta to Florida for my fine champagne.

CHEF:

In Tourist?

SOMMELIER NODS.

SOMMELIER:

And on all nonstops. . . day-time and Night Coach.

CUT TO ECU STEAK AS KNIFE SLICES IT.

CHEF:

Ah, but you should taste my steaks. . .choice, charbroiled filet mignon.

BDA-9

FIGURE 16-2 Storyboard.
The storyboard uses small sketches to indicate important camera shots or angles. Storyboards are frequently used in writing commercials because a great deal of specific information must be conveyed in a relatively short period of time.
(Courtesy: Delta Air Lines)

436

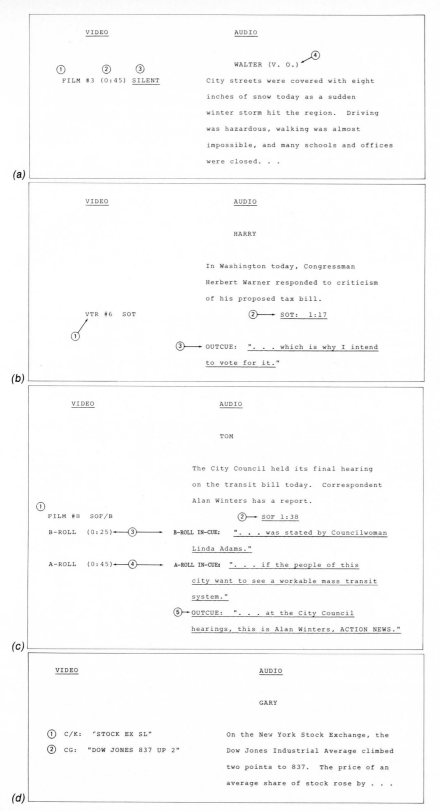

FIGURE 16-3 Script Cues for News. Although these script formats apply to news because of the heavy use of film and videotape inserts, they can be adapted for any production which requires accurate cuing of insert material.

(a) **Voice-over.** (1) Indicates film cut on reel; (2) running time of film; (3) indicates silent film; (4) talent's name and indicates that it is "voice over." The identical format is used for videotape inserts as well.

(b) **Sound on Film/Sound on Tape.** (1) Indicates tape cut and sound on tape; (2) gives running time of cut; (3) final words heard on tape.

(c) **B-Rolling.** (1) Film cut number, sound on film, indicates B-roll; (2) running time of total story; (3) start time of B-roll; (4) A-roll return cue; (5) final audio out-cue.

(d) **Graphics and Chroma Key.** (1) Indicates use of chroma key; (2) "CG" indicates character generator.

of the film story. The cue indicates the last words on the A-roll before the B-roll should be cut in; (4) is the A-roll return cue. Time is, again, measured from the top of the SOF story. The A-roll should be cut in immediately following the outcue lines; (5) is the final audio outcue for the entire film piece.

GRAPHICS AND CHROMA KEY At some stations the scriptwriter is asked to indicate the use of lower-third supers, graphics, and chroma key inserts on the script. Since the script is usually used as a reference in preparing graphics and titles, it is essential that the writer check names, titles, and spelling before submitting the script to the producer: (1) C/K indicates the use of a chroma key insert. The "STOCK EX SL" indicates a stock exchange graphic slide; (2) "CG" is the character generator. The information in between the quotation marks is the word-for-word message which is to be typed out. If your station uses slides instead of a character generator, the notation might read: "SL #7: "DOW JONES 837 UP 2.""

FILM-STYLE SCRIPT FORMAT

The two-column television script was devised for the director who must select shots from multiple cameras covering the action simultaneously from various angles. The popularity of single-camera shooting techniques, which are available using electronic field production equipment, has made the single-camera, film-style production mode a practical option. Some producers and directors now require scripts intended for these productions to be written in a film-style script format.

As you can see from the example in Figure 16-4, a film-style format is quite different from the two-column television script. Notably, the script is not divided into columns but rather into a series of scenes. Each scene is assigned a number and described as to time of day, location, and additional information to help establish the overall atmosphere.

This may seem, at first, to be quite complex, but it actually facilitates single-camera production

better than the two-column television script. Since each scene is described by a different number, the director can easily go through the script selecting those scenes which require the same location, actors, time of day, and so forth. These scenes are all shot at the same time although they may appear in a different sequence on the edited program. In addition, the single-camera technique requires the same scene to be shot a number of times from various angles to give the director and editor a variety of perspectives to use in editing the show. Referring to each scene by its own number is a convenient way to mark the slate for each new take and helps to eliminate confusion in the editing room when many takes, angles, and shots are pieced together into the final, edited show.

PARTIALLY SCRIPTED FORMATS

Not every television production lends itself to a completely detailed script. Obviously, a sports event or news coverage of an unfolding story cannot be scripted in detail. Spontaneous programs such as talk, interview, or game shows also cannot be scripted word-for-word. Instructional programs and demonstration shows often sound dry and stilted when performers—especially nonprofessionals—attempt to follow a detailed script instead of talking in their own words.

This is not to say, however, that these programs are produced entirely without a script. Instead of a detailed format, however, the script provides the skeleton around which the show is organized and produced. Without a carefully written and developed format, the director, talent, and the production team would never be able to know the proper sequence of events and could not put together a tight, integrated, and professional looking production.

Rundown Sheet

A rundown sheet is an outline of the entire show routine from the top of the show until the final fade-out. Rundown sheets can be used on all productions, but they are especially important on semiscripted shows, to establish the basic format routine and the planned sequence of events.

```
FADE IN:

1.  EXT. FRONT OF OFFICE BUILDING IN A DOWNTOWN CITY AREA - DAY      1.

    A taxi pulls up to the curb in front of the office building.
    KATE, an attractive young woman in her early 20's gets out of
    the cab and enters the building.

2.  INT. LOBBY AREA OF OFFICE BUILDING                              2.

    Camera picks up KATE as she walks through the building doors
    and enters the lobby.  She looks around a moment until she
    spots GARY, a business executive in his mid-30's who is standing
    near the elevators.  She walks over to him.

                            GARY
            Well, good morning!  I was afraid you
            weren't going to make it.

                            KATE
            I almost didn't.  You know, I'm
            not happy about this at all.

                            GARY
            Look, it's already been decided.  Now
            let's just do it and get it over with.
            Did you bring the papers?

                            KATE
            They're in my handbag.

                            GARY
            Okay.  Let's go.

    They walk over to an elevator and enter.  The elevator
    doors close behind them.

3.  INT. CONFERENCE ROOM - DAY                                      3.

    A large conference room.  The furnishings are modern, tasteful
    and obviously very expensive.  A secretary is arranging papers
    and pencils at each seat along the long table.  KATE and GARY
    enter.

                          SECRETARY
            Good morning, Mr. Johnson. . . Miss Kelly.

                            GARY
            Good morning, Grace.  I wonder if you
            wouldn't mind doing that later.  Miss
            Kelly and I have some important matters
            to discuss before the meeting begins.

                          SECRETARY
            Of course, Mr. Johnson.
```

FIGURE 16-4 Film-Style Script Format.
Most productions which are shot with a single camera are written in the film-script format. Numbering each scene enables the director to schedule an efficient shooting sequence and to keep accurate track of each scene during planning and production.

As you can see from the sample in Figure 16-5, a well-designed rundown presents all the vital information about a production clearly and concisely. Each segment shows the talent involved, where it takes place (studio, remote, from a VTR or film chain), and its approximate running time. Rundown sheets should be produced and duplicated for every member of the production team.

Outline Scripts

Even though a show may be ad-libbed around an established format it is a good idea to write the opening and the closing of the show to ensure a smooth beginning and ending. In addition, there are sometimes important segments within a production which need to be carefully presented, either because they contain important copy (such as a commercial) or they are used as production cues (such as film or tape roll cues). In these instances, the outline script would contain both a word-for-word, detailed copy of important segments and a more general outline for loosely structured portions of the show.

As an example, take a typical demonstration program. The show is a series on home photography with a professional host and guest photogra-

```
                    THE MORNING SHOW

                  Monday, January 29


   7:00:00    (0:20)   LOGO OPENING                  VTR


   7:00:20    (1:00)   BILL & SUSAN Intros.          HOMEBASE


   7:01:20    (5:00)   NEWS SEGMENT/Curtis           NEWS SET


   7:06:20    (1:00)   WEATHER/Roberts               WEATHER BOARD


   7:07:20    (0:20)   UPCOME INTRO & COMML LEAD-IN  HOMEBASE


   7:07:40    (2:00)   COMMERCIAL #1                 VTR


   7:09:40    (0:15)   SUSAN: ENT. REVIEW INTRO      HOMEBASE


   7:09:55    (2:50)   MOVIE REVIEW                  VTR


   7:12:45    (7:30)   DRUNK DRIVING INTERVIEW/      LIV. ROOM SET

                    Bill and Rachel Warren


   7:20:15    (2:00)   COMMERCIAL #2                 VTR


   7:22:15    (4:00)   COOKING DEMO/Susan & Karen    KITCHEN SET


   7:26:15    (1:45)   SPORTS/Bill                   HOMEBASE


   7:28:00    (0:50)   Bill & Susan Goodbye and CREDITS  HOMEBASE/VTR


   7:28:50    (1:10)   STATION BREAK
```

FIGURE 16-5 Rundown Sheet.
A sample rundown sheet from an early morning news and information show. Although the program is basically unscripted, note how detailed the rundown sheet is, specifying times for each segment and the talent involved, as well as providing an overall running time.

phers. As you can see from the example (Figure 16-6), the opening of the program is fully scripted to ensure a smooth and precise introduction. The middle section, where the demonstration takes place, is outlined loosely. The talent will have no trouble ad-libbing around the format, but the opening, closing, and commercial introductions (as well as important on-air plugs) are carefully written into the script.

The outline script is frequently used in news or sports coverage, where certain information for openings, closings, and breaks must be precisely handled. As you can see in Figure 16-7, this sports script has a specific opening, which is fully scripted. In addition, the roll cue for commercials, which are inserted from the home studio, is written as "With the score, ——— to ——— we'll take a brief time out for these commercial messages." This is an important production cue, which must be delivered accurately in order to ensure smooth transitions from the remote feed to the home studio and back.

Interview Scripts

People are often surprised to learn that most talk or interview shows employ a staff of writers. This does not mean that the interview seen on the air is scripted, but it does suggest that writing and research are necessary to make a difficult interview appear effortless on camera.

Television programs cannot afford rambling, unfocused interviews for two reasons: (1) there is rarely enough time for irrelevant conversation, and (2) the viewing audience probably will not watch for long unless the conversation is interesting and to the point. This is why a talk show must be carefully organized with a series of questions

written to get the interview off to a good start and keep it on track.

A preinterview is often the best way to structure a guest's appearance on a talk show. In a few minutes of chatting with the guest, a staff writer can pinpoint a number of good stories, interesting

```
                                                        PAGE 1



         VIDEO                              AUDIO

FADE IN:

CLOSE-UP PHOTOGRAPHIC PRINT

   AS IT SLOWLY DEVELOPS AND

   THE IMAGE FORMS--PULL OUT

   TO REVEAL HOST                          HOST

                                   There's nothing more satisfying for a photographer

                                   than to watch as the image slowly forms

                                   on a print of a picture which you've taken,

                                   developed, and printed.  It's one of the

                                   most interesting and enjoyable parts of

                                   photography.  By the time our program is

                                   over you'll be able to develop your own

                                   film, enlarge it, and print it in your own

                                   home.  The Photographic Darkroom... our topic

                                   today on "Fun With Photography."

OPENING FILM (:30)                 MUSIC:  THEME UP AND UNDER

                                              ANNOUNCER (V.O.)

                                   Introducing you to the fascinating world

                                   of camera and pictures.  This is Fun With

                                   Photography with your host, Tony Jordan.

                                              HOST

                                   Hello, and welcome to Fun With Photography.

                                   For most folks their role in photography

                                   ends with the exposed roll of film.  They

                                   drop it off at the local drugstore and a few

                                   days later pick up their prints.  Sure it's

                                   easy but if that's how you develop and print
```

FIGURE 16-6 Outline Script. Demonstration shows are an example of a program type which is usually only partially scripted. The opening and the roll cues into a video taped segment are fully scripted, but much of the demonstration is simply outlined, and the talent will ad-lib.

```
                                          PAGE 2


                                       HOST (con't)


                      your pictures you're missing out on some

                      of the most creative aspects of picture taking.

                      But before we go any further, let's take a

                      look at a typical darkroom.

FOLLOW AS TONY POINTS
OUT EQUIPMENT                           (TONY AD LIBS AS HE SHOWS:

                                                -Developing Tanks

                                                -Chemicals

                                                -Sink

                                                -Enlarger

                                                -Print Trays.

                                             HOST

VTR CUT #1 (1:45)          VTR ROLL CUE: Let's take a look and see how

                           to load the developing tank in the darkroom.

                           VTR SOUND TRACK (1:45).

                                             HOST

                           Once your film has been loaded into the

                           tank and you've secured the top cover you

                           can turn on the room lights again.  The

                           next step is to prepare your chemicals.

                           (TONY AD LIBS AS HE PREPARES CHEMICALS ON TABLE)
```

FIGURE 16-6 **Outline Script.
(Continued)**

anecdotes, and other relevant material to form the structure of the host's on-camera questions. Pre-interviews are not always possible, however, especially with many guests who appear on a show to plug a book, movie, television show, or personal appearance and who arrive in town just prior to the production. In this case press releases, magazine and newspaper clippings, and other reference sources are the best way to research possible questions.

```
SATURDAY, JULY 16        BASEBALL GAME OF THE WEEK        OPENING OF TELECAST


                    VIDEO                           AUDIO

1.   VTR      BASEBALL OPENING                        SOT

              AT APPROX :18 FROM TOP ON TITLE FREEZE FRAME--MUSIC CONTINUES BG
                    INSERT SLIDES                 CUE ANNCE FROM MU

2.   INSERT SL:_____VS_____     AD LIB...THIS AFTERNOON......

                                          FROM_____THE_____

                                          VERSUS THE_____

                                          BROUGHT TO YOU BY...

3.   INSERT SL: CHRYSLER/PLYMOUTH         YOUR CHRYSLER PLYMOUTH DEALERS,

                                          WHO INVITE YOU TO BUY OR LEASE

                                          THE NEW CHRYSLER LE BARON...

                                          THE FIRST IN A TOTALLY NEW CLASS

                                          OF AUTOMOBILES.

                                          BY...

4.   INSERT SL: MILLER/LITE               LITE BEER.  EVERYTHING YOU ALWAYS

                                          WANTED IN A BEER...AND LESS.

                                          AND BY...

5.   INSERT SL: GILLETTE/DRY LOOK         GILLETTE...MAKERS OF THE DRY LOOK.

                                          #1 SPRAY FOR MEN.  AVAILABLE IN

                                          AEROSOL AND NEW PUMP SPRAY.
```

FIGURE 16-7   **Program Opening Format.**
An example of the scripted opening for a remote broadcast of a baseball game.
(*Courtesy: NBC Sports*)

SUMMARY

A television script establishes the show's format, includes the program's content, and organizes important production information so it will be most useful in planning, rehearsal, and production.

The television scriptwriter must have the skills and talents of a writer although the material must always be adapted to the characteristics of the television medium. Television is a visual medium, and the writer must deal with images as well as with words. At the same time, copy must be written to sound natural. Since the audience never reads the script, it hears only the performer's delivery. The most effective television writing style is clear, simple, and concise and arranges information logically. If possible, the writer should use a style and approach which best suit the program's target audience, especially if the show is aimed at a specific and easily identifiable group. Finally, the television scriptwriter must be able to deal with the production limitations which exist. These include time problems, budget limitations, and production capability.

The detailed script format includes specific dialogue, visual elements, and such additional audio as music and sound effects. It is often used on dramatic, documentary, news, or commercial productions. The film-style format is an alternative format that is often used for productions which are to be shot with a single camera.

Partially scripted formats are designed to encourage a show's natural spontaneity and still provide the needed structure of an established format. The amount of scripted material varies, but usually openings, closings, and important cue lines within the show are fully scripted, the remainder being merely outlined for routine and sequence.

CHAPTER 17

TELEVISION TALENT

The Performer and the Actor

In television, the term "talent" is used to refer to anyone who appears on camera. The enormous variety of program formats naturally demands a diversity of talent to fit each production requirement. The weathercaster on the local news, the actor in a daytime serial, the emcee of a game show, and the business executive appearing in a corporate training tape are all very different members of the same group. For convenience, we can divide talent into two categories: (1) performers and (2) actors.

Performers appear as themselves, usually in a nonfictional situation, and often speak directly to the audience. Newscasters, announcers, interviewers, and hosts of variety shows are all familiar examples of performers.

Actors, on the other hand, portray fictional roles and attempt to create the illusion of a character.

445

Naturally, actors work in dramatic and comedy productions, but they also appear in such varied productions as musical-variety sketches, commercials, and children's shows.

The basic performance fundamentals for all television talent—actor and performer alike—are essentially similar, and all talent has the same objective: to communicate a message, idea, or emotion to the viewer. But there are some significant differences between the performer and the actor. The performer often works with only a partial script and with a minimum of rehearsal. The ability to ad-lib, to think and act quickly, and to remain as conscious of the program's timing and the director's cues as well as to on-camera performance are all important abilities necessary for a performer. To a great extent, the performer is often as much in control of a semiscripted or unscripted program as the director. Since the actor usually works on a fully scripted show that has been rehearsed in advance, the television actor requires a slightly different set of skills. Rather than deal with unpredictable events, the actor has as his or her primary job developing and sustaining a believable character, often with very little advance time to prepare, and to deliver an effective performance in the midst of the controlled chaos that usually accompanies any television production.

TALENT AND TELEVISION

To perform well in any medium—the stage, film, radio, or television—talent must understand something about how the medium operates and how the actor or performer must interact with it. To do this, we must look at three areas: the characteristics of the medium, the production techniques used, and performance methods.

Characteristics of the Medium: Using the Videospace

Television is an intimate medium. It may sound clichéd by now, but that fact is substantiated by countless studies and observations. Although the audience for a broadcast program may total in the tens of millions, viewers still watch alone or in small groups, on a relatively small screen, often on a regular basis. Television viewing is, in many ways, an experience which is unique for both the viewer and the performer, and this can be used to the television communicator's advantage.

Because of the medium's intimate qualities and the viewing environment, audiences tend to become involved with talent in a one-to-one relationship. For those performers who communicate directly with the audience—a newscaster or program host, for example—this relationship becomes almost "friend-to-friend." For actors who portray a role, the intensity of the identification can result in viewers confusing fantasy with reality. The actor is perceived by the audience as the character he or she plays. A number of soap opera actors, particularly those who play heavies or villains, have reported being accosted on the street by an outraged fan who mistakenly believes the actor is the character they see regularly on screen.

Contributing to the medium's intimate quality is the small screen size, which results in most camera shots being framed and composed rather tight. The predominance of the close-up shot and the ability of the camera and microphone to detect, capture, and magnify the slightest gesture, expression, or vocal inflection mean that a performer must always behave naturally on screen. Television's uncanny ability to uncover the slightest phoniness in a performer is well respected among professional talent.

Finally, the performer must always be concerned with the impact of the videospace and how the cameras and microphones translate image and voice to the audience. There are times when a performer or actor must dress or move slightly differently on camera in order to appear "normal" onscreen. For example, actors and performers who are new to television frequently complain that they must work so physically close together on camera. However, onscreen, the distance between them appears quite normal. If they were to stand at what we would consider to be a normal distance in everyday life, the space between them would appear to be unnaturally large on screen. As with any production element, how a performer or actor looks in person does not matter. It is the appearance in the videospace which counts because that is how the audience will ultimately see and evaluate the performance.

Production and Talent

Talent plays an obviously important role in the entire production team effort and has a specific number of job functions and responsibilities. There is no place in a busy production schedule for the egotistical "star," and most successful performers and actors understand that a good production requires a team effort. Professionals are also aware that the efforts of the crew can make or break their appearance onscreen. Even the best performance is meaningless unless it is produced onscreen with precision and imagination. By cooperating with the crew, talent ensures looking as good as possible on camera.

Talent's first responsibility is to be thoroughly prepared. There is simply no excuse for an actor who has not memorized lines, an interviewer who has not done his or her background homework, or a lecturer or demonstrator who has not prepared and rehearsed the presentation. A full studio crew, forced to stand idly by while talent learns in the studio what should have been prepared earlier, is expensive, wasteful, and infuriating to all concerned.

A good actor or performer knows enough about production operations to help the director and crew during rehearsal and production. This means understanding camera positions and shots, when to hit marks perfectly, how to handle microphones properly, and maintaining an overall sensitivity to production activities and problems.

Talent Performance

The performance you give—how you look and sound on screen—will naturally vary, depending on the program and the production situation. Actors may play very different roles, and performers may have to adapt to a particular type of program format. Regardless of your particular character or role, the key word for all performance is to be "natural." The microphone and camera can pick up and amplify the slightest whisper or expression, so overplaying or exaggerating voice or movements—as you might on the theatrical

stage—is not only unnecessary; it can actually impair your performance.

For performers who must communicate directly to the audience, eye contact is a vital factor in successfully delivering the message to the viewer. Most performers pretend that the camera is simply a person and speak to it as though speaking normally to anyone seated a few feet away in a living room. The more natural, relaxed, and understated you are on television, the more effective your performance is likely to be.

Developing a relaxed and comfortable on-camera style is a job in itself, but a good performer must also be sufficiently flexible to deal with the frenetic pace and complexities of production. Television is often a high-pressured operation with close deadlines, limited budgets, rushed rehearsals, and short production periods. An unexpected technical problem can force sudden changes in the script, revisions in the blocking, even a reshuffling of rehearsal and shooting schedules. Actors used to the leisurely rehearsal pace of theatrical productions are often shocked to discover how much must be accomplished in so little time in television. Experienced television talent is able to cope with these inevitable pressures while still delivering an effective and seemingly effortless performance.

THE TELEVISION PERFORMER

The performer is concerned with a number of production elements including (1) the floor manager, (2) production equipment, (3) the script, (4) prompting devices, (5) clothing and wardrobe, and (6) makeup.

Floor Manager

The *floor manager* (or stage manager) is a central figure in the production team. The floor manager conveys the director's commands and relays cues to talent and other personnel, is responsible for checking on props and costumes, holds cue cards, or follows along with the script, and generally oversees the operation on the studio floor. For talent, the floor manager serves as the basic communication link between the director in the booth and performers on the floor.

FLOOR MANAGER CUES During rehearsal and while the studio mikes are off during production, the floor manager can convey signals and cues verbally to the entire studio floor. Once the mikes are live and the cameras are shooting, the floor manager gives all cues to talent silently.

HAND CUES A series of hand signals has been developed for some of the common direction and time cues. Obviously, these cues are given silently while the show is on the air. Figure 17-1 shows each hand signal, how to give the cue, and what each signal means.

Hand cues must always be given clearly and forcefully. You need not wind up like a baseball pitcher before delivering a cue, but you must not underplay it either. It is the floor manager's responsibility to move to a position where the talent can see the cue without having to move his or her head or eyes noticeably on camera. If the performer is speaking directly to the camera, the cue should be given next to the take lens of the camera that is on the air.

Talent should never overtly acknowledge a cue. Even the slightest nod of the head or wave of the hand will be visible on camera. Usually the floor manager can tell if you have received the cue. If you do wish to acknowledge the cue, however, the least noticeable method is simply to blink your eyes slightly longer than usual.

Although the floor manager must establish eye contact when sending a cue, too much eye contact between the floor manager and performers when there are no signals, is harmful. Some performers are distracted or unnerved, and others may begin talking to the floor manager rather than to the camera.

WRITTEN CUES Some performers, particularly those who are unfamiliar with television, prefer using written cues. These are brief words printed in large type on hand-held cards. The floor manager flashes the card instead of a hand signal to relay the cue. (See Figure 17-2.)

Written cues are often easier to see and understand than a hand cue, and there is less chance for confusion about what a hand signal meant. Such common cues as "commercial," "station break," or a series of time cues (one minute, thirty seconds, and so on) are easily prepared. The only real disadvantage with written cues is that the floor manager must carry them during production. Hand cues, on the other hand, can be given as soon as the command is received from the control room.

COUNTDOWNS A countdown into and out of program segments helps the performer to make a smooth transition and sharpens the production's timing and precision. Whenever microphones are closed and the director is about to cue talent to begin, the floor manager should give a loud "Stand by" and tell everyone on the floor how much time remains until the cue.

As an example, say a program is about to finish a film commercial and return to the studio. The floor manager yells out at the right time, "Stand by. Coming back in fifteen seconds." This alerts talent and crew alike to prepare for the cue. At ten seconds, the floor manager says, "Ready in ten seconds. Coming up on Camera 2." Directing talent to the proper camera will help to avoid confusion and a possible miscue. The floor manager should call out time, in seconds, from ten seconds down. At five seconds, the floor manager must also begin a hand countdown just next to the take lens of Camera 2. The last audible count should be "three" with "two, one, and cue" given silently, in case the audio engineer opens the mikes early. (See Figure 17-3.)

To count down into a film, tape, or other closely timed segment, the floor manager should give the final five-second count with hand signals as he or she hears the time cues from the AD or director in the booth. The hand countdown is always given next to the take lens of the air camera, so talent can watch it while still maintaining eye contact with the audience.

CUE CARDS The floor manager or an assistant

FIGURE 17-1 **Floor Manager's Hand Cues.**

Cue	Meaning	Hand Signal
Stand by	Show or sequence is about to start.	Stand next to the camera which talent should speak to. Hand and arm upraised.

Cue	Start talking or begin action.	Upraised hand is pointed to talent.

Cut	Stop talking or stop action.	Draw hand cross throat in a cutting motion.

Stretch	Slow down. There is too much time remaining.	Pull hands apart as though pulling taffy. Amount of stretching is indicated by relative distance of hands as cue is relayed.

FIGURE 17-1   **Floor Manager's Hand Cues. (Continued)**

Cue	Meaning	Hand Signal
Speed up	Talk faster. Accelerate the action. Not much time is left.	Rotate hand and arm clockwise in a circle above head. Speed of rotation indicates relative urgency and the amount of time remaining.

Cue	Meaning	Hand Signal
OK	Everything is fine. Keep on doing/talking as you are.	Make a circle with thumb and forefinger.

Cue	Meaning	Hand Signal
Thirty seconds to go	Thirty seconds left in show or segment.	Make a T with both hands.

Cue	Meaning	Hand Signal
Fifteen seconds to go/wrap up	Fifteen seconds left in show or segment. Wrap up whatever you are doing.	Grabbing motion into a fist.

FIGURE 17-1   **Floor Manager's Hand Cues. (Continued)**

Cue	Meaning	Hand Signal
Back up	Step back. Move away from microphone.	Pushing motion with both hands.

Come closer	Move up toward camera. Move closer to microphone.	Palms facing floor manager with both arms moved toward FM.

Speak more softly	Tone down. Speak more softly. Your are too loud.	Palm raised to mouth.

Speak up	Speak up. You are talking too softly. We cannot hear you well.	Cup ear with hand.

FIGURE 17-1 **Floor Manager's Hand Cues. (Continued)**

Cue	Meaning	Hand Signal
Speak or look toward this camera 	This is the camera you should be looking at. This is camera that is on the air. 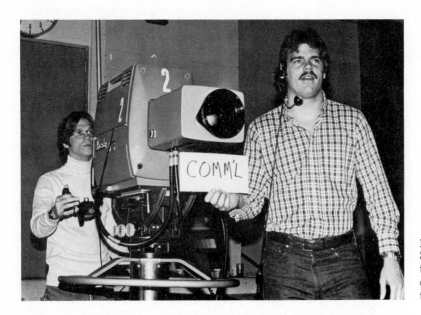	Hand pointing to on-air camera. A waving motion from one camera to the other can be used to wave talent from the camera which is currently on-air to another which the director wants to punch up.

must hold any cue cards to be used on a show. The cards are held directly next to the take lens of the air camera, and the line of dialogue being spoken is always on a line with the lens. As talent speaks, the floor manager moves the cue card upward, always keeping the spoken line right next to the lens. This reduces talent eye movements and eliminates the obvious appearance of reading on camera. The floor manager must be certain to follow talent closely since moving the cards too quickly or too slowly will affect the performer's delivery

PROPS AND COSTUMES   Unless you are working on a large production with separate personnel assigned to handle props and costumes, it is the floor manager's responsibility to keep track of these items. A list of props and costumes should be available to the floor manager prior to rehearsal.

FIGURE 17-2 **Written Cues.** Some floor managers signal talent by holding up small cards with the cues or directions clearly printed on them.

FIGURE 17-3 Floor Manager Countdown Cues.
Countdown cues help talent make a smooth transition into and out of a program segment. They are used at the beginning of a program or segment and whenever talent introduces a film or videotape insert.

Production Equipment

No matter how expert your performance, it must be captured and reproduced properly by the production equipment in order for the audience to see it. A good performer understands the importance of quality production and helps the crew and director whenever possible.

CAMERAS In most circumstances the performer works with two or more cameras which shoot from a variety of angles. In order to work with the cameras, the performer must consider three things: (1) Where are the cameras positioned? (2) What is each camera shooting? (3) Which camera is on the air?

CAMERA POSITIONS You will need to know where the cameras are and their basic area of operation. Keeping this information in mind will enable you to help the director get the best shots and to avoid creating camera problems.

CAMERA SHOTS You ought to know basic camera shots for each camera. This does not mean you must know the shots as well as the director, but you should know the general sequence. Say you are recording a commercial with two cameras. Which camera will shoot you head-on and which camera will shoot the product demonstra-

tion close-up? Knowing the close-up camera enables you to set up the demonstration at the most advantageous shooting angle. Knowing which is your camera will ensure that you work to the right camera at all times.

AIR CAMERA Working with two or more cameras means that you will usually need to know which camera is on the air at any particular time. The tally lights are, of course, one way to tell, but there are times when you will need to know which camera the director will punch up next. Working with three cameras, unless the floor manager tells you in advance which camera will start the show, you will have a one-in-three chance of guessing right.

On some shows where the performer must speak directly to the camera, the program director may wish to change camera shots during the presentation. To do this smoothly and unobtrusively takes coordination between talent, floor manager, and director. First, the director cues the floor manager to ready talent to switch cameras. The floor manager prepares talent for the switch and, on the director's cue, waves talent to the proper camera. It is talent's job to look down momentarily on the switch cue and then look up into the other camera. The director will cut from one shot to the other during the look up, and when the operation is performed correctly, the switch looks perfectly natural on screen.

Should you find yourself talking to the wrong camera at some time during the show, simply look down, find the tally lights out of the corner of your eye, and look up at the air camera's lens. The floor manager will usually help wave you over if he or she can get there before you realize the error. Avoid making comments like "Where are you?" or "Thought you'd gone away for a while!" They not only sound unprofessional, but they also call attention to a mistake which many viewers may not have even noticed.

MOVEMENT ON-CAMERA Before you move on-camera, remember that it takes the camera operator a lot longer to move a camera across the studio than it takes the performer to get up and cross the set. Especially on unrehearsed shows, try to telegraph your moves before actually making them. Saying something like, "Why don't we get up and take a look over here . . ." before rising will help alert the director, camera operator, and other crew members that a move is about to occur. Also, try to rise or sit a bit more slowly than normal. It will make it easier on the crew and yet will appear natural on screen.

CLOSE-UPS ON CAMERA   Since television is a close-up medium, there may be times when you must show an object or graphic in close-up. A close-up shot magnifies both the size of the object and any extraneous movement, so hold all objects as steady as possible by resting them on a table top or by bracing your hand. You can reduce flaring on the shiny side of an object or graphic by tilting it slightly downward. Your natural movement is also greatly magnified in close-up. If you know you are being shot very close, avoid sudden moves which are difficult for the camera operator to follow. Only a matter of inches in where you stand or move can make the difference between being in or out of a close-up shot.

AUDIO   Top quality sound is as important to a production as a good picture. Talent can help the audio engineer achieve this by handling and using microphones properly, giving correct sound levels, and assisting the audio crew whenever possible.

MICROPHONE HANDLING   We have discussed the proper microphone handling techniques in the audio chapters, so we will just review the important points which concern talent here. Most performers use on-camera mikes, usually a lavalier, hand, or desk mike. Lavaliers should be clipped or hung so they do not rub against clothing or jewelry, creating annoying extraneous sounds during the show. Performers should never play with the mike or mike cable, should avoid absentmindedly tapping on table tops equipped

with desk mikes, and should be warned against moving outside the pickup range of a stationary desk or stand mike.

Performers who must use a lavalier or hand mike and move about the set should tuck some slack cable behind them in their pants waistband or belt. This gives you an added safety pad in case you snag the cable while moving across the set. As you move, lead the cable in front of you so you can be certain that the cable is free of obstructions.

Talent should know the microphone pickup pattern of the equipment they use. There is nothing worse for sound than someone speaking into the dead side of a directional microphone or moving outside the pickup range of a stationary mike. If you think the microphone on the set does not permit the flexibility you need, speak with the audio engineer and director during rehearsal.

There are times when a performer must use an off-camera microphone such as the boom. Although the boom microphone relieves talent of any mike-handling chores, in some ways the boom can be the most difficult microphone for a performer to work properly. Booms, like cameras, take time to move large distances, and sudden, unexpected moves about the set should be avoided. Telegraph your intention to move before actually getting up, lest you wind up with a boom in the shot. If you see that the boom operator is having difficulty positioning the boom, just a step or two in the right direction may be all it takes to improve the sound quality.

AUDIO LEVELS   Levels are usually obtained by the audio engineer prior to rehearsal. When you are asked for a level, speak as you will when on camera. There is no point setting levels only to have the performer drastically change volume or delivery once the show begins. The most accurate level is achieved when talent says something from the program's script. This not only eliminates the sudden speechlessness that often accompanies a request for a voice level, but it gives the engineer a better sample of your voice than a "Testing 1, 2, 3." If you are hosting a show with nonprofessional talent, you can help the audio engineer by asking the guest a question or two so the engineer can set levels as the guest replies.

SETS AND LIGHTING The set and lighting together establish the environment in which the performer appears. Usually set and lighting are completed before the performer enters the studio, but there are times when talent is asked to stand on-set for final lighting touch-ups. A person's physical size, hair and skin color, and other individual characteristics sometimes require slight lighting adjustments.

If this is your first time on the set, familarize yourself by taking a brief walk around. Make a note of steps, risers, furniture, and props so you will know your way around and where all necessary items are located.

One of the most distracting and annoying set-related mannerisms is the performer who insists on swiveling, rocking, or tilting in a movable chair. Not only does this make it difficult to frame a shot, but the constant movement makes watching him or her enormously irritating. Many studios use swivel chairs, however, so train yourself to sit still and warn others to avoid such movements on camera.

The director may show you a series of "marks," usually masking tape on the studio floor, which indicate where you are expected to stand for camera shots, lighting, and audio. It is talent's responsibility to hit all your marks accurately without making it look obvious to the audience. On some sets lit for low key, a performer must be in just the right position to be lit properly. Experienced performers develop a sixth sense in finding their light. You can usually feel the location where the lights are strongest on you and when the focus becomes off center. Sometimes only a step or two can make the difference between a poorly lit subject and a properly illuminated one.

Working with a Script

The best advice is to try to work without a full script whenever possible. Reading from a script, even with occasional look-ups into the lens, interferes with eye contact and can reduce the performer's communication impact. For a production situation where it is important to say just the right lines or when there is too much material to memorize, however, a full script may be the only alternative.

Here are tips for using a script on camera:

1 *Type It on Colored Paper.* A script should always be typed, never handwritten. Some studios have typewriters with oversized characters but even a normal typeface can be made more legible by typing everything in CAPITALS and double- or triple-spacing. White paper should not be used because it reflects too much light, can be difficult to read under studio illumination, and may interfere with the proper camera contrast range. Yellow, pink, or blue paper is far less distracting, photographs well, and is usually easier to read under the lights.
2 *Use Notes Whenever Possible.* It is hard to follow a script word-for-word and still maintain eye contact and some degree of spontaneity. If at all possible, use the script as an outline and ad-lib within your typed notes. However, it is a good idea to write out the open, close, and transition cues completely since these must be performed flawlessly and with less time flexibility than most material within the body of the show.
3 *Mark Your Script.* Write important cues, blocking moves, and other information clearly on your script. Some performers also like to mark certain words for emphasis and to make additional notes on delivery in the script. Be careful, however, not to overmark the script to the point that it becomes illegible.
4 *Do Not Staple the Pages.* Never clip or staple the script pages together. Instead, clearly number each page in one corner. During the show you can slip each top page underneath as you finish it.

CARDS Some performers, who use an outline or short copy, prefer to hold index cards rather than sheets of paper. The cards are smaller, easier to handle, and can be tucked away in a pocket when necessary. A 5- × 7- inch index card is usually large enough to hold oversized typed copy. As with paper, use colored cards. Number each card, so you do not inadvertently read them out of the proper order.

Prompting Devices

There are times when a performer cannot use an on-camera script and yet cannot memorize the entire script. In this case, either cue cards or a Teleprompter can be used to prompt the talent while still keeping the copy off-camera.

CUE CARDS A cue card is a large piece of stiff oaktag or poster board with script copy written in large lettering. Cue cards are usually used by talent when speaking directly to the camera. In this case, the copy should be prepared on long, narrow cards. Cutting a standard piece of poster board or oaktag that is 22 × 28 inches in half lengthwise will produce a 14 × 22 inch strip, which is just the right size. The narrow card allows talent to read the copy while minimizing obvious eye movement. The copy must be written in large, legible print with a broad felt-tip pen. Never use a pencil, ballpoint, or narrow felt-tip pen since the words will be illegible. Cue cards are always used

from a distance, although the exact space between talent and the cards will vary, depending on the studio and production circumstances. Usually cue cards must be read from about 8 to 12 feet, so your cards should be legible from at least that distance. Remember, too, that the glare and reflection created by studio lights will make it more difficult for talent to read the cards on the set.

The card is held by the floor manager or an assistant with the line of copy you are speaking right next to the take lens. As you talk, the floor manager must move the card upward, keeping the spoken line always next to the lens. Since a new card must immediately follow a card that has just been finished, never print the second part of the script on the back of a card. The floor manager would have to flip the card over in order to continue, and this would disrupt the performer's delivery. Most floor managers prefer to hold the stack of cards next to the lens and to move only the top card upward as talent speaks. Once the first card has been used, the floor manager hands it to an assistant while the next card is already in position for talent to read. (See Figure 17-4.)

Even on fully scripted shows, where the talent

FIGURE 17-4 Cue Cards. The cue card is held by the floor manager or by a floor assistant directly next to the camera lens. The floor manager should keep the line of dialogue which the performer reads off the card adjacent to the take lens to minimize talent's eye movement.

holds the script on camera, the use of cue cards for the opening, closing, and film or tape transitions is recommended. The cards enable talent to maintain eye contact with the audience while delivering lines exactly as written and with highly accurate timing.

Another way to use cue cards is to outline the sequence of events as a reminder to talent. Johnny Carson's opening monologue for instance, is outlined on a series of cue cards positioned just off-camera. Carson delivers the jokes from memory but uses the cue cards to remind him of the proper sequence. The same idea is useful in demonstration or instructional shows, where the talent knows what to do and what to say, but may not always remember the proper order or sequence of events. Cue cards used as outlines need not be written on narrow strips or held next to the camera lens since talent does not read them line by line. In fact, the cards can be positioned anywhere off-camera as long as talent can easily glance at them from time to time.

TELEPROMPTER The Teleprompter is a device which enables talent to read continuous copy while maintaining maximum eye contact with the audience. A close-circuit vidicon camera is used to scan the written copy, and a television monitor and mirror system mounted on the studio camera are used to display the copy to talent. The script appears on the face of a one-way mirror, which is positioned directly in front of the camera lens. Although the copy is invisible to the viewer and does not interfere with camera operation, the performer can read the lines and still look directly into the camera lens. The speed of the moving copy is variable and controlled by the prompter operator. A single vidicon camera/copy stand can feed any number of individual prompters, each mounted on a separate studio camera. (See Figure 17-5.)

While the Teleprompter is much more expensive than cue cards, it offers some obvious advantages. Cue cards must be hand lettered from the original script. It would take literally hundreds of cards and hours of full-time work to prepare cue cards for a half-hour show. You can imagine the time and expense involved if this were to be done every day. Since the Teleprompter needs no

special type of script preparation, any script page that has been normally typed can be run immediately on the prompter. On most daily newscasts, for example, the script is written on a multisheet carbon pack. One page is given to talent, another to the program's director and a third is used for the Teleprompter. Last-minute script changes present no problem since any page can be easily replaced by a new one. For production situations involving a great deal of copy, where the performer needs to maintain maximum eye contact and still follow a script—such as on newscasts, commercials, lectures, and speeches—the Teleprompter is a valuable production tool.

Clothing and Wardrobe

Performers generally wear normal street clothing for work, but your on-camera wardrobe should be selected carefully. Your understanding of videospace should tell you that there are some clothing styles, colors, and fabrics which may look fine in person but can appear unflattering onscreen. A performer's clothing must be selected as much for the effects of videospace as for style and appearance.

IMAGE The clothing you wear can say a great deal about you even before you have said a single word. We tend to associate types of clothing with certain roles and images, and we become uncomfortable when clothing does not match role. You would probably be suspicious if a doctor wore overalls while examining you or if you spoke with a banker who wore a suit dotted with sequins and rhinestones. If you want to be taken seriously on television, do not show up wearing a tennis shirt and jeans.

Your clothing selection will depend on your program, your audience, and the image you want to convey. This does not necessary mean you must always dress formally for a television appearance; rather the clothing you select should complement the program and your communica-

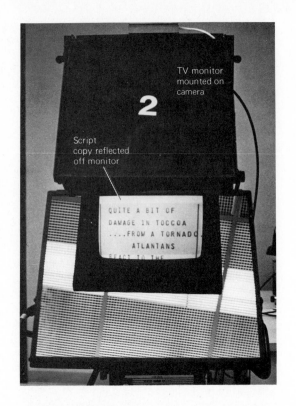

TV monitor
mounted on
camera

Script
copy reflected
off monitor

QUITE A BIT OF
DAMAGE IN TOCCOA
....FROM A TORNADO
ATLANTANS
REACT TO THE

tion objectives. A performer wearing a tie and jacket to conduct an interview on Malibu Beach would look as ridiculous as Dan Rather delivering the network news in a floral print body shirt.

COLOR The same color considerations already discussed about lighting, sets, and graphics also apply to your television wardrobe. The choice of color depends a great deal on your own taste and style, but stay away from colors that are too bold, bright, or deeply saturated. These colors tend to reproduce poorly, do not enhance your looks on screen, and can actually impair your communication effectiveness. Most performers look best on television in muted, lightly saturated colors.

The set in which you will appear is naturally an important consideration when selecting your clothing. You will want to achieve a good color

FIGURE 17-5 Television Prompters.
The camera prompter uses a mirror device which superimposes the script copy over the camera lens. Talent can read the prompter and maintain direct eye contact with the viewing audience. The script is run along a treadmill and is photographed by a closed-circuit television camera and fed to the monitor on top of a studio cameras. The speed of the script movement past the prompter camera can be varied by the operator to keep pace with the talent's delivery.

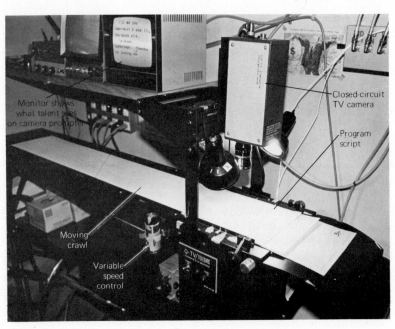

Monitor shows
what talent sees
on camera prompter

Closed-circuit
TV camera

Program
script

Moving
crawl

Variable
speed
control

balance between the set and your clothing, and the proper amount of contrast for both color and black and white reproduction. Clothing colors with too much contrast—a white shirt worn against a black limbo background, for example—can interfere with the camera's operation and affect your appearance. On the other hand, insufficient contrast will not separate you from the background, and it creates an uninteresting and ineffective impression. One of the costliest clothing mistakes in television history occurred during the 1960 presidential debates. Richard Nixon showed up for the broadcast wearing a light gray suit which was almost identical in color and brightness value to the light gray set. The insufficient contrast between Nixon in the foreground and the gray background set contributed to his overall poor appearance, which, in turn, adversely affected viewer perceptions and cost him the debate.

The proper contrast between a performer's own articles of clothing—shirt and pants, for example—as well as the contrast between clothing and skin tone are equally important. Avoid wearing pure white or deep blacks. A white shirt with a black suit, for instance, may upset the picture's normal contrast ratio and cause camera shading problems, which will hurt your appearance. Shirts in pastel blues, yellows, and grays are preferable to white, and usually photograph better. Dark-colored clothing tends to make a pale subject appear even lighter than reality. On the other hand, light-colored clothing will make dark-skinned or heavily tanned performers look even darker on screen.

An extremely important color consideration concerns the use of chroma key. If the color of your clothing too closely matches the chroma key window on a set, you may find yourself doing an unintentional imitation of the invisible man as your clothing keys out along with the background. If there is any question in your mind about a particular color or article of clothing, check it on the set to make certain it will not interfere with the chroma key.

PATTERNS AND FABRICS The correct choice of clothing patterns and fabrics are two additional wardrobe considerations. Avoid closely lined patterns such as herringbones, checks, and small plaids, which may produce a "moiré effect." This is the annoying flash of multicolored light produced when the clothing's pattern interferes with the camera's scanning operation.

Avoid patterns or fabrics with very small or highly intricate detail. The resolution limitations of the camera may completely wash out the detail on screen. Stay away from very busy patterns which will distract the viewer's attention from the most important part of the show: you and your message. Simple patterns and solids tend to look best, and variation can be added through such accessories as scarfs, ties, and jewelry.

Material that is highly reflective, such as a starched shirt, or material which absorbs a great deal of light, such as velvet or velour, should not be worn on camera. The highly reflective fabric may glare or wash out completely, while velour and velvet are so light-absorbent that all detail is lost as the various dark tones merge together.

Fabrics or patterns with vertical lines tend to give a thinner and taller appearance. Horizontal lines usually exaggerate size and weight. Since television tends to add a few pounds to most people anyway, avoid fabrics or patterns which emphasize your size and weight in an unflattering way.

LINE The way clothing is cut often has an impact on a performer's appearance. Baggy, or loose-fitting, clothing will always make you look worse, especially because television gives the impression of added weight to most talent. Clothing that is tailored and shaped is generally best because it produces a clean and tapered line on screen. Of course, you should not try to wear tight-fitting, tapered clothes if you are not built for it. In general, clothing should fit you well without appearing overly tight or form-fitting. Since most productions use multiple cameras, which shoot from a variety of angles, select clothing which will flatter your appearance from all sides.

ACCESSORIES Ties, scarfs, and jewelry can help to vary your wardrobe and enhance your on-camera appearance, but they should never distract or overwhelm the viewer. A three-carat diamond drop dangling from your neck may look elegant but it is also taking the viewer's eyes and attention away from you. Unless you have a good reason for doing so, stay away from overly shiny, flashy, or garish jewelry, which can cause glare and specular reflections. Sequins, rhinestones, and glitter should also be worn sparingly, and only for effect.

Makeup

Makeup is one of the easiest and most effective ways of improving virtually anyone's appearance on camera. Yet at many production facilities makeup is rarely, if ever, used. There is a mistaken notion that makeup belongs in "show business" and has no place outside major entertainment studios. There is also the inaccurate impression that today's modern color cameras are so sensitive that people can look acceptable on screen without any makeup at all. While performers may look "acceptable," they certainly will not look their best, because the television camera plays tricks on many people's facial features and skin tones. Cameras have a tendency to exaggerate minor skin blemishes which may go unnoticed by the naked eye. Men's beards are usually much more prominent onscreen, and some men appear with a "five o'clock shadow" even though they may have shaved minutes before air time. Color cameras also emphasize the natural red, yellow, and green tones found in skin pigmentation and can produce unflattering and unnatural results onscreen.

When we talk about makeup, we refer to *straight makeup,* which is a basic application of makeup designed to make talent look natural in the videospace. We are not referring to either "corrective" or "character" makeup, two far more elaborate techniques, which are used to alter talent's appearance and which requires the skills of a trained makeup artist. Straight television makeup is easily learned, can be applied quickly

with a minimum of supplies, and makes an enormous difference on camera. The idea behind straight makeup is not to transform a plain person into a ravishing beauty but simply to make talent appear natural and looking their best. If you have any reservation about the difference makeup can provide in a performer's appearance, try a simple before-and-after test in your own studio. The results are certain to convince you that makeup should be considered a standard part of television talent preparation for all on-camera work.

MAKEUP SUPPLIES Television makeup is available from a number of manufacturers, each with their own color and shade classification system. As a general rule it is not a good idea to use commercial street makeup for television use. This is because one important reason for using makeup on color television is to counteract the tendency of the camera to exaggerate reds, yellows, and greens. Most commercially available makeup emphasizes yellows and oranges, the very hues we try to tone down on-camera. However, if a subject is allergic to conventional makeup, the hypoallergenic brands can be used as long as they do not contain much yellow or reddish tones and if they register naturally on camera.

Makeup designed specifically for television is available from a number of manufacturers, including Max Factor and Bob Kelly. All of the makeup and supplies you will need can be ordered by mail at a reasonable cost.

BASE MAKEUP Base, or foundation, makeup is used to completely cover the face and to create the desired skin tone and color. Most base makeup is supplied in either pancake or stick form. *Stick makeup* is grease-based and must be applied with a sponge and then powdered to reduce shine. *Pancake makeup* is a powder and dries with a matte finish. Grease-based makeup covers more easily and is most commonly used. A makeup kit must include a number of different color shades, so the base makeup can be closely matched to the subject's natural skin tone. (See Figure 17-6.)

ROUGE Rouge is available in both liquid and brush-on formats for men and for women. It is

Alcohol

Tissues

Cream sticks   Lipsticks

Makeup
brushes

Velour   Foam-rubber   Rouge   Makeup   Eye shadow   Sable hair
powder puff   sponge   pencils   brush

FIGURE 17-6 **Makeup
Supplies.**

used to accentuate the cheekbones in women and to break up the flatness of the base makeup in both men and women.

POWDER Translucent powder is a light, color-less powder, which is applied over base makeup to set the makeup and reduce shine.

EYE MAKEUP Eyeliner, mascara, and eyeshadow are used for both male and female subjects. These can be purchased from any cosmetic store, but, as a general rule, avoid blue, green, and purple shades. While these colors may look fine for street makeup, they reproduce poorly in color. Instead, emphasize soft brown tones, which register much more naturally on screen.

PENCILS Makeup pencils are used for eyebrows, eyeliners, and to provide detail and definition. Pencils are available from any cosmetic store and should include black and a variety of brown shades.

APPLICATORS A foam rubber sponge is recommended for applying grease-based foundation makeup. A sable-hair brush is useful for applying rouge as are velour powderpuffs for applying translucent powder.

MISCELLANEOUS ITEMS Cleansers—including

a mild soap, cold cream, and baby oil—are necessary for makeup removal. A well-supplied makeup kit should also have tissues, cotton balls and swabs, combs and brushes, hair spray, razor blades, scissors, and an apron to protect the subject's clothing during makeup application.

MAKEUP STEPS FOR MEN

CLEAN THE FACE Have the subject wash his face thoroughly with soap and water to remove all dirt, oil, and perspiration. If the subject has an exceptionally heavy beard, he may have to shave prior to makeup. Tuck in tissues around the talent's shirt collar to protect it from smudges, and drape a towel or a makeup apron around his neck and shoulders.

APPLY THE BASE MAKEUP Select a correct shade of creamstick, which is closest to the subject's natural skin tone. If you cannot find the particular shade you need, use two or more creamsticks, and blend the different colors to achieve the proper shade. The grease makeup is applied with a dry, foam rubber sponge. Lee Baygan, director of makeup for NBC, suggests applying the creamstick with one side of the sponge and then using the other side to wipe off the excess makeup. The makeup which remains is all that is usually necessary.

Apply the base makeup over the face, neck, and ears. Spread it evenly, leaving no clumps or smears. Bald men or those with receeding hairlines should have their heads covered with foundation as well.

If there are discolorations under the eyes, extend the base color to cover the discoloration or cover it with a touch of "highlight". A highlight is a base shade two or three shades lighter than the base color. Be sure to blend the highlight smoothly with the base makeup until there is no obvious demarcation between the two shades.

APPLY POWDER Powder the entire face, neck, and ears with translucent powder. The powder is applied with a puff and lightly pressed onto the skin, not rubbed or "powdered."

APPLY MALE ROUGE Apply a touch of male rouge over the cheekbones, over the forehead, nose and chin. This is to break the smooth, flat look of the base makeup and to bring out the natural coloring in the subject's face. Don't overdo it. Using the rouge brush, add just enough to make a slight difference but not enough to call attention to the makeup.

EYES AND HAIR Use mascara on lashes and makeup pencil on eyebrows only if they are so light that they will wash out under the studio lights. If you do apply mascara, eyeliner, and pencil, never add so much that the makeup looks obvious.

If the subject has a receding hairline, makeup pencil will bring out the definition. Also around the temples, pencil or mascara can provide a slight darkening to prevent blonde or white hair from completely washing out. Do not use pencil or mascara to the point that it appears obvious on camers.

EVALUATE THE MAKEUP The overall result should be a healthy, normal face without the appearance of makeup or obvious corrections. (See Figure 17-7.)

MAKEUP STEPS FOR WOMEN

CLEAN THE FACE All street makeup should be removed with soap and water. Protect the talent's blouse with tissues which are tucked in around the neck and a makeup apron which covers her shoulders.

APPLY BASE MAKEUP Select a shade of base makeup nearest to the subject's normal skin tone. If necessary, use two or three different shades and blend them together to produce the shade you need. As with male makeup, the base should be applied with a foam rubber sponge and spread evenly across the entire face and neck as far down as will appear on camera. Remember that only a very light application is necessary, and this will avoid the unnatural, caked-on look, which too much makeup can produce.

Discolorations under the eyes should be covered with a highlight two or three shades lighter than the base and smoothly blended to the base color.

Subjects with a very broad nose may benefit from the application of "shadow" along both sides of the nose. Shadow is a base makeup two or three shades darker than the overall base. The shadow should be applied sparingly, and smoothly blended with the basic base shade.

APPLY EYE MAKEUP The eyes are an especially important feature for women, and the makeup is usually more extensive than we use for male subjects. However, eye makeup should still be applied sparingly, so it will appear natural on-camera.

A brown eyeshadow should be applied to the brow bone to shape the eye. Always apply eye shadow with the subject's eyes open so you can see the effect of the makeup as it will appear on-camera. Avoid using eyeshadows of green, blue, or purple, which do not register well on-screen.

Apply a thin line of eyeliner in a medium brown shade. The liner can be applied with a pencil, or a liquid eyeliner and brush can be used. For women with thin or narrow eyes, a very thin white line drawn on the bottom ledge of the eye can make the eyes appear larger and wider.

Mascara in a brown shade is applied to both upper and lower lashes. If you will use artificial

(a)

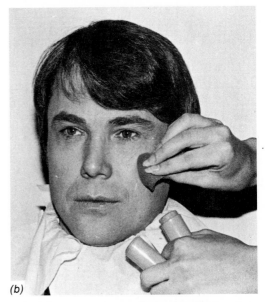

(b)

(c)

FIGURE 17-7 Applying Television Makeup.
The subject appears in the first picture *(a)* without makeup. The makeup is applied with a foam rubber sponge. Two shades of creamstick are mixed to achieve the proper skin tone closest to the subject's natural face color *(b)*. The completed makeup job *(c)* should not appear drastically different. The beard has been covered, and the eyes have been made up and accentuated so that they appear natural on-camera.

APPLY POWDER Using the powderpuff, gently press on translucent powder to set the makeup and reduce shine. You want to achieve a "finished look" that does not appear artificial.

APPLY ROUGE The application of rouge under the cheekbones helps to contour the face and provides a three-dimensional look. The rouge is applied with a brush directly under the cheekbone and lightly brushed up and over the bone. Use the same brush and rouge to add a slight touch of color to the forehead and chin and under the eyebrows. This will break up the flatness of the base makeup and add color to the face.

APPLY LIPSTICK A woman's lips are an important and prominent feature and under normal circumstances should have makeup applied. Avoid bright reds or oranges which do not repro-

eyelashes, work them in a bit to eliminate the stiffness. The lashes should be cut so they are no longer than the normal lashes.

Using an eyebrow pencil, shape the eyebrows, and fill in the outer ends, which may wash out under the lights. Use discretion in penciling the eyebrows since too much will look unnatural on screen.

duce well and can upset the color balance of the face. Select lip colors in brown tones with a subtle touch of pink or orange. Lip gloss can be applied last if you want to give the lips a moist look.

EVALUATE THE MAKEUP The end result should be a healthy, natural-looking subject without an obviously "madeup," or artificial, appearance.

MAKEUP FOR BLACKS AND ORIENTALS The very same techniques just described apply to all subjects regardless of skin color. It is not recommended to attempt to change a black subject's natural skin color by choosing lighter or darker base colors. There are a number of makeup shades produced specifically for the black performer, and the makeup artist should select the shade which appears closest to the natural skin tone or should use two or three different shades and blend them to produce the shade closest to normal.

HAND AND BODY MAKEUP Unless the hands or body appear prominently in the show and look too pale, too dark, or otherwise do not reproduce well, no additional makeup is necessary. However, some older performers may have age spots on their hands, which can be covered with base makeup if their hands are going to be shot in close-up during the show.

EVALUATING THE MAKEUP Makeup, like lighting, can be properly evaluated only on camera. If at all possible, check the performer's makeup while he or she is on the production set. If the performer will be working under hot lights for an extended period of time, the makeup may become blotchy as sweat begins to shine through. A light application of colorless powder will repair minor problems; powder should always be kept handy on the studio floor.

REMOVING MAKEUP Performers who use makeup regularly must be sure to take exceptionally good care of their skin. All makeup should be removed as soon as the production is over. Use cold cream and tissues to remove facial makeup; avoid harsh rubbing, which will only irritate the skin. Eye makeup can be removed using a little baby oil on a tissue or cotton ball. Once all makeup has been removed, wash the skin with mild soap and water. If you wish, a mild alcohol astringent can be applied last to close the pores.

THE TELEVISION ACTOR

A good television actor must be, before anything else, a good actor. No amount of technical expertise can compensate for the ability to develop and sustain a believable performance, whether on film, on the stage, or on television. With that not insignificant talent taken for granted, the next requirement is that the actor have some familiarity with the techniques of television production. With only a few exceptions, all of the points we have made concerning the television performer apply equally well to the television actor. In this section we will briefly discuss some of the major differences.

The Actor and the Floor Manager

The floor manager is the actor's communication link with the control room. However, since actors usually work on fully scripted shows with some prior rehearsal, the floor manager's role is slightly different when working with actors than when working with performers. Actors rarely need the continual stream of time and directional cues which are necessary on unscripted and unrehearsed programs. In fact, the floor manager should avoid all eye contact with an actor unless a cue is being delivered. That is because establishing eye contact can break an actor's concentration and spoil the performance. Of course, when the actor must receive a cue, it is the floor manager's responsibility to move to a position where the actor can see the cue while still remaining in character.

Working with Production Equipment

The ability to deliver a believable and apparently spontaneous performance while still paying close attention to the many technical aspects of a production are among the most difficult skills for a beginning television actor to learn. Unlike the stage, where minor improvisations will not upset anyone, the television actor who misses his mark or who intentionally or unintentionally changes a blocking move can seriously upset the camera operator's shots and the director's entire shooting pattern. Of course, a good director and crew will try to make the actor feel as comfortable as possible, but that does not relieve the actor of his or her own responsibilities during the perform- ance. If you see that an over-the-shoulder shot is being blocked, for example, a step to one side will help the camera operator get the shot faster. If you know you are being photographed in a tight close-up, keep your movements slow and to a minimum. If you are involved in a boom split where the mike is farther away from you, a bit more projection may help the audio engineer with the sound.

During rehearsals, the actor should keep the camera positions in mind and make all move- ments as consistently as possible. It does make a difference which hand you use to open a door, how you face in a two-shot, how far you cross when you walk over the set. Marks are especially important in complex foreground-background shots where talent and camera must be precisely aligned for the shot to work.

Working with Prompters

There is no arguing the fact that the best perfor- mances are delivered by actors who have com- pletely memorized their lines. An actor who is not confident with his or her part may give an inse- cure and hesitant performance that is made even more obvious by the cameras and microphones. There are times, however, when the rapid produc- tion pace of television makes the use of some prompting device necessary.

Cue cards are difficult to use for plays because of the enormous number of cards necessary to cover the show's script. However, if you do use cards, remember to position them in the direction the performer will naturally look toward. This is not necessarily next to the air camera. If more than one performer will be using cards, color coding each actor's lines can help to reduce confusion.

The Teleprompter is usually a more effective way to prompt a play, but the monitor device

mounted on cameras does not work well since the actor rarely looks directly into the camera lens. Instead, some productions use the older type of prompter whereby the script is typed in large print on a long continuous roll of paper. A small motor runs the paper, and the speed is variable to keep pace with the actors. Rather than mounting the prompters on cameras, floor assistants hold them just out of camera range, where talent can easily see them.

If you need a line which you have forgotten during a performance, the worst thing you can do is to jerk your head around abruptly to read the cue card or prompter. Instead, take a pause, look down and glance at the prompter, look back and deliver your line. If you have done the move properly, the pause often seems natural in the context of the play, and your grasping for the next line of dialogue may go completely unnoticed by the viewer.

Regardless of how good an actor is at reading prompters or cue cards, sooner or later the televi- sion medium usually reveals the use of these devices. Unless absolutely necessary, memorize your lines, and come to rehearsal and production fully prepared.

Costumes and Props

Most costumes are rented especially for a pro- duction since few facilities use them often enough to justify the expense of making and storing them. The only problem with renting costumes is that most are designed for use onstage, where the distance between the performer and audience is far enough to mask imperfections or the lack of realistic detail. The same holds true for props. A wooden gun may look fine from the balcony of a theater, but on television, in close-up, it looks absurd. Check all props and costumes far enough in advance so changes or additions can be made to correct their appearance for the camera.

Character Makeup

Character makeup is designed to alter an actor's

appearance to suit the role or character being portrayed. Although modern makeup artists can create almost any illusion, given the time and money, few production studios can afford to keep a full-time makeup artist on staff. Doing your own character makeup is risky because the close-up camera will reveal the makeup unless it has been expertly applied. If your production will require extensive character makeup, it is a good idea to hire an outside professional to design and apply the makeup on a consultant basis.

Actor and Director

Television can be a very difficult medium for the actor because, just as the character begins to develop and full rehearsals start, the director is up in the control booth inaccessible to the actors. To make matters worse, during early studio rehearsals the actor is often asked to play what is essentially a prop or stand-in, while camera shots are learned and other production elements are rehearsed. It is not that the director is uninterested in the actor at the moment, only that camera shots and other technical problems must take precedence once studio rehearsals begin.

This means that actors must have a firm idea about character and performance before the studio rehearsals begin. During dry rehearsal outside the studio, the director can give you his or her undivided attention, and this is when problems or questions concerning a role should be raised. Once you enter the studio, stopping rehearsal to discuss a line or a character is expensive because a studio crew and full facilities must wait. Television actors in all performance situations—commercials, situation comedies, daytime serials, and so on—must be able to produce an effective and believable performance with a minimum of actor-director introspection.

There is a story about a young actress fresh out of school who was hired for a minor part on a long-running daytime serial. As the actors and director assembled for the first read-through, the actress kept interrupting the rehearsal by asking

the director a continual stream of questions about her character. At one point she said to the director, "I don't really understand my motivation for the cross to the couch in this part." By this time the director had had enough. "Your motivation," he replied, his voice dripping with sarcasm, "is the paycheck you'll receive on Friday."

Of course, the director must be sensitive to an actor's problems and questions, but talent must also realize that time pressure on the director make rehearsal time a precious commodity. The better prepared you are, the easier it will be for everyone to accomplish the most possible during rehearsals. A successful television actor can develop a character or role quickly and still deliver a real, believable, and multidimensional performance on screen.

TALENT IN PRODUCTION

In this section we will cover the responsibilities and functions of talent during each of the four production stages.

Preproduction

Talent's role in preproduction depends, in large part, on the program format. Actors usually are not very busy during the earliest stage of preproduction except to read through their scripts and begin forming the character. Performers, on the other hand, may need to do some background research or writing in preparation for their work. Often interviewers, reporters, and other similar types of performers must write their own material and help develop the program's outline and rundown in conjunction with the director and the producer.

Interviews are a common production format, which require considerable preproduction work on the part of the interviewer. Unless you are familiar with the people you will interview, know and understand the issues involved, and have prepared your line of questioning, the interview will not be effective. On those programs with a full staff, some guests may be preinterviewed, and a list of possible questions then supplied to talent. These questions should only be used as a guideline, however, since an interview can move in any direction once it begins, and the host must be able to guide the interview as it progresses on the air.

Performers who must work on a demonstration, presentation, or a speech should practice it thoroughly until it can be delivered flawlessly and without hesitation. Once you get into the studio, you are apt to be somewhat nervous, so the more familiar you are with the script, demonstration, or presentation, the better the performance you will give.

Setup and Rehearsal

Talent is rarely involved during the studio setup unless you are using special props, set pieces, or other production elements which require your supervision.

Instead, talent is usually involved with dry rehearsals or preproduction meetings at this time. For actors, the first stage of rehearsal is the dry rehearsal, where the actors and director meet to read through the script, develop the characters and the approach, and work out actor movements and blocking. Dry rehearsal is the time for the actor to raise any problems, suggestions, or modifications. Once the production moves into the studio for facilities rehearsals, performer variables should be fairly well settled, so the director can begin work with the cameras and technical crew.

CAMERA BLOCKING The first part of the studio rehearsal involves the director and camera operators working out individual camera shots. Camera blocking can be a tedious time for talent, as you are asked to repeat a move or deliver a line over and over until the director is satisfied with the camera shots. The performer or actor should simply understand the reason for this (it is not your fault; it is for the cameras), expend a minimum of energy by walking through your blocking, and stay as relaxed and patient as you can. Unless the director specifically asks you to perform during camera blocking, do not go all-out in your performance only to find that you have exhausted your energy before the actual production. However, it is very important for you to walk through your blocking and movements as consistently as possible since this is when the camera

operators and director are planning their shots and camera moves. Their work depends on talent's ability to repeat actions without significant variation.

RUN-THROUGHS As soon as the director has completed the camera blocking he or she will begin a series of run-throughs, in which all the various production elements are integrated into a unified show. Early run-throughs are often a series of stops and starts as problems arise and must be resolved before the production moves on to the next part of the show. Most of the problems at this point are technical, but talent may be asked to change some aspect of performance in order to improve camera shots, audio, or other elements of the production. As the run-throughs progress, the stops and starts become less frequent, and the show should begin to run at its normal pacing and speed. This is when the director can start to evaluate your performance and how well it plays when combined with the camera shots, sound, lighting, and other production elements.

DRESS REHEARSAL The last rehearsal before production is a *dress* when all the elements are rehearsed as though it were the final "air" show. By this time you should be in wardrobe and makeup so that the director, lighting director, and video engineers can see how everything looks together and can make changes where necessary. Since the dress is the final performance before air, some directors will videotape it and edit together the best performances of the dress and air versions into the final production which the audience sees.

NOTES After each rehearsal the director will give out notes to performers. Notes are important because they pertain not only to your own performance but to problems which relate to camera shots, audio, lighting, and other production operations.

Production

Talent's role in production varies, depending on the type of program and how it is produced. A show that has been fully scripted and rehearsed should present few surprises since all major problems ought to have been ironed out during rehearsals. On taped programs, major mistakes can usually be corrected through retakes and subsequent editing. However, talent should never stop performing until the director or floor manager calls out, "Cut." What may seem like a major error to you may not be considered by the director serious enough to stop tape. It is even possible that the mistake did not appear on the air. In any event, performers and actors should never stop unless told by the director or floor manager. The same goes for ending a show or segment. Always remain in character until you receive the "all clear." You may think a mike is closed or a camera off only to learn they were not. If you treat all mikes as "live" and all cameras as "hot," you will avoid potential embarrassment and costly mistakes during production.

Performers who work on shows with a minimum of rehearsal naturally have to deal with both their performance and the many elements of production, such as timing and pacing. You are controlling the show as much as the director, and you must stay alert and flexible at all times.

One of the hardest lessons for some talent to learn is to *listen* as well as to speak. Actors who actually "listen" while others deliver their lines are considered by their peers to be among the best in the trade. Not only do they tend to react more naturally, but they also help other actors to motivate their performance and delivery. Careful listening is especially important for such ad-lib performers as interviewers, hosts, and emcees. You may be thinking ahead to the next question or cue but do not neglect what is being said at the time.

To illustrate the consequences of not listening carefully, consider this story of an unfortunate game-show host who was introducing a new contestant:

"What do you do?" the emcee asked the lady contestant.

"I'm a housewife," she replied.

"Wonderful!" said the host. "How many children do you have?"

"Three," she answered.

"Wonderful!" exclaimed the emcee. "And what does your husband do?"

"My husband is dead," she said. "I'm a widow."

"Wonderful!" said the host without a moment's pause. "Now let's play our game . . ."

LIVE PRODUCTION No other production situation rivals a live broadcast for the excitement and tension it produces in even the most experienced television talent. A live show demands precision, flexibility, and the ability to do it right the first time since there is no chance for retakes. Talent must be able to react to cues quickly and still make their performance appear natural and effortless on screen.

On some live shows where the event itself is unplanned—such as a sports event or special news coverage, for instance—the director may wish to talk directly with talent from time to time. To do this, talent wears an unobtrusive earpiece connected to an *interrupted feedback* (IFB) system. By pushing a button in the control booth, the director can talk to talent even while they are on camera. It is not easy to listen to one voice while trying to speak coherently, but most performers develop the knack after some practice. A good director will keep IFB commands to a minimum and will try not to interrupt while talent is in the middle of a sentence. (See Figure 17-8.)

TAPED PRODUCTION Videotape recording and editing offer obvious production advantages, which we have discussed in earlier chapters, but they can present some difficulties for talent as well. A common problem is shooting out of sequence. For convenience and production efficiency, it may be necessary for an actor to tape the opening scene of a play and then immediately afterward tape the dramatic closing scene, which occurs in the same set or location. Although the two scenes will appear in their proper context and sequence on the final, edited version, the performer is expected to reach the right level of intensity and delivery in production without benefit of the intervening parts. Out-of-sequence pro-

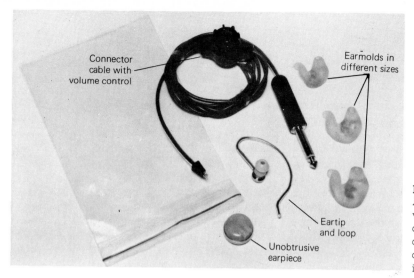

FIGURE 17-8 IFB Earpiece. An unobtrusive earpiece which is worn by talent on-camera. Director sends cues and production information over the IFB while the talent is on the air.

duction requires actors to match themselves perfectly without the natural momentum possible with sequential shooting.

Another problem associated with videotape is the tendency to lose some spontaneity and the natural "edge" which comes from a live or one-time-only production. Performers (and the crew) may adopt the attitude that mistakes can always be reshot, and this can result in a dull and lackluster performance. After a long day in the studio you will be tired, maybe even exhausted, but the audience who watched your show neither knows nor cares how much effort went into the production. All that matters to them is your performance as it appears in the videospace, so be sure to discipline yourself to give the best performance you can on each and every take.

Postproduction

Talent's primary job is usually finished when the production is completed. Obviously, on a live show there is no postproduction at all. On some taped programs, however, the director may ask you to remain for *pickups,* which are small sections of action, dialogue, or both which are taped for later use in editing. A pickup can be a close-up shot which was not possible during the actual production, a reaction shot, a move, or the delivery of some dialogue. The most important part in doing pickups is to match yourself as closely as possible. If you are unsure of the right movements or your delivery, ask the director to play back the part of the show which occurs just prior to the pickup point.

On some interview shows the director may ask the host to reread the questions for reverse-angle shots. These are not usually necessary when taping with multiple cameras but may be required on single-camera productions using ENG equipment. It is important for you to repeat the question exactly as it was first asked and with a natural inflection. Replaying either the videotape or an audio cassette recording of the interview can help you with the precise wording and the proper delivery.

SUMMARY

In television, talent refers to anyone who appears on camera. There are two talent categories: performers and actors. Performers appear as themselves, usually in a nonfictional situation, while actors portray a character different from themselves in real life.

All television talent must consider three factors when preparing for their performance. The first—characteristics of the medium—refers to the talent's use of the videospace. Television is an intimate, close-up medium, which often involves the viewer in a one-to-one relationship with the performer or actor. Talent must always recognize the impact of the videospace on their appearance and their actions because this is how they will be judged by the audience.

The second factor—production role—refers to talent responsibilities as a member of the production team. Aside from their own performance, actors and performers are expected to understand the basics of television production operations and to cooperate with the director and crew during rehearsals and performance.

The third factor—performance methods—is closely related to the preceding two elements. The camera and the microphone are capable of capturing and magnifying the slightest gesture, expression, or vocal inflection. Talent must work to appear natural, relaxed, and completely normal in both voice and movement. Exaggerating gestures and projecting your voice as you might on stage are not only unnecessary but they can actually impair your communication effectiveness.

The floor manager is the talent's primary communication link with the director in the control room. During rehearsal and production, the floor manager relays time and directional cues from the booth to talent on the studio floor, usually with silent hand signals. In addition, the floor manager is responsible for cue cards, props and costumes, and most floor activities during production.

The performer may work with either an on-camera script or with such off-camera prompting devices as cue cards or the Teleprompter. Actors should memorize their lines for the best performance and rely on cue cards or prompters only in an emergency.

Clothing and wardrobe must always be selected with their on-camera appearance as the most important consideration. The clothing's color, fabric, contrast, and line should be determined by the type of program and the image talent wishes to convey, the set and lighting, and how well the clothing complements talent's appearance on screen.

Makeup is necessary to make talent appear normal and look their best on screen. Variations in skin color or imperfections in complexion can be corrected for the camera using simple straight makeup and minor corrective makeup. Character makeup is used to alter an actor's appearance for a role, and is more complex, usually requiring a skilled makeup artist for best results.

The television actor must follow the same basic guidelines used by the performer but with some major variations. The performer on a semiscripted or unscripted show is often controlling the program's pacing and flow as it progresses. The actor works on a fully scripted and rehearsed production but must develop a believable character in a short period of rehearsal time and is expected to deliver an effective performance amidst the distractions of a television production set.

During dry rehearsal the actor and director work on characterization, dialogue, and blocking outside the studio. These elements should be settled once the production begins in-studio rehearsals, since camera shots and other technical factors revolve around the actor's delivery and movement. Consistency in performance is a crucial element for any television actor.

Videotaped productions require talent to maintain the appearance of spontaneity even after multiple retakes. In addition, actors may be asked to perform out of sequence and are expected to build a character in bits and pieces, all of which must flow naturally together once the program is edited. Live productions create their own natural spontaneity and a one-chance-only pressure. They require talent to think quickly, remain flexible, and still deliver a top-quality performance in what can be a stressful production situation.

CHAPTER 18

PRODUCING

It has been said that the theater is an actor's medium, motion pictures a director's medium, and television a producer's medium. While these broad generalizations may not apply universally, in each case they do identify the individual who is traditionally the focal point of creative activity for each medium.

There is no question that television is a team operation and, like any team, it needs a leader: someone with the authority, responsibility, and talent to plan the operation and make the difficult decisions. In television, that is the role of the producer. The producer develops the program idea, supervises the entire production from the first preproduction meeting to the last videotape edit, and is the one person with the ultimate responsibility for every element—both technical and creative—that goes into the production.

One reason the producer has all this authority and responsibility is the unique nature of the television business. Television requires a constant source of program material, and, while many people are continually involved in producing television programs, most team members join a show when their specific talents are needed, perform their jobs, and then leave the production once their contribution is over. Only the producer and perhaps a few of the producer's staff actually stay with the show from its earliest inception until its ultimate broadcast or distribution. The producer must give a sense of continuity and unity to the production since he or she is the one person who knows how all the different parts of the show will ultimately fit together.

Regardless of a program's size or budget, a producer's job is pretty much the same. Sure, there is a difference in scale between network coverage of the Super Bowl and a local station's broadcast of high school football. But when you come right down to it, the producer on both shows is doing essentially the same job, with the same responsibilities, and with the same overall objectives.

There are some superficial distinctions among different types of producers. Most producers are known as *staff producers*. These are regular employees of a network, station, or production organization and are responsible for developing and supervising program production. Staff producers at the three major networks are usually assigned to the news and sports divisions. At local stations and at production facilities, staff producers generally work on a wide variety of shows. The production staff working along with the producer also consists of regular employees. The show is invariably produced at the station or production organization's studios. While special equipment or personnel are hired from the outside when necessary, for the most part, these shows are primarily in-house productions.

A slightly different breed of producer is the *independent producer,* an entrepreneur with a production company that sells programming to networks and stations. In fact, independent producers are responsible for almost all network and syndicated entertainment shows. The independent producer assembles a creative "package,"

which he or she sells to a program buyer. The package consists of the program idea, the script, the director, the performers, and the production team. The show is produced at any facility, not necessarily the studios of the network or station which buys the show. Once the project is completed, the free-lance production team is dissolved and the members go their own ways.

Some very successful independent producers, such as Aaron Spelling *(Fantasy Island, The Love Boat)* or Dick Clark *(American Bandstand* and many musical and dramatic specials), often supervise a number of programs simultaneously. In this case, they serve as *executive producers,* who primarily develop and sell the initial idea. The day-to-day supervision on the studio set is handled by one of their subordinates, who is called a *line producer.* In addition to entertainment programs, many commercials and even industrial productions are produced by independents who bid for the jobs and work closely with advertising agencies or corporate clients.

THE PRODUCER'S ROLE

Television's answer to the Renaissance man is the television producer. That is because a producer has to know a little about a lot. Since the producer is responsible for every element in a show, he or she must be sufficiently knowledgeable in all areas of production. Even though the producer may never dolly a camera, light a set, or edit a videotape, he or she must know enough about these and other production aspects to make intelligent decisions. At the very least you as producer must know the capabilities and limitation of every production element so you will know whether or not you have stretched the production team beyond reasonable limits. You must also know enough about every team member's job to understand and evaluate the contribution that each makes toward the entire production effort

A producer must also be a creative individual: someone with a broad and varied background who is conscious of the world around him or her, sensitive to events, and able to undertake different responsibilities and varied program assignments. As the producer you must create a vision of the show—how it should look, how it should sound, and how it should communicate its message. Perhaps it is best summed up by Paul Rauch, executive producer for NBC's daytime serial *Another World,* who says simply, "A producer must have good taste." In other words, producing involves the ability to know from both instinct and experience what will "work" in a show and how to use your imagination and ingenuity to achieve it.

But creativity alone is not enough for a successful producer. Television is a complicated and technical medium, which demands an efficient organizer and an effective businessman or woman to coordinate hundreds of different details. According to George Heinemann, an Emmy-winning producer for NBC, "Producing is 60 percent organization and 40 percent creativity." The point is that without the organizational ability, there's little chance you will be able to transform a creative concept into a successful television program.

Finally, the producer must be an effective communicator. You are dealing with people after all, and you must be able to convey ideas and enthusiasm to the production team. You must motivate the unit, inspire, guide, and lead it, but most of all, keep a group of diverse and sometimes temperamental people heading in the same direction for a common goal. As one producer commented, "Producing is as much group psychology as it is television production." Remember, it is the production team that must ultimately turn your vision into a television reality. How well they do their jobs will affect the entire show and ultimately affect you as the producer.

DEVELOPING A PROGRAM IDEA

Every program begins in a slightly different way. Staff producers are usually assigned a production

as a normal part of their jobs. For example, a local station may have sold a department store the idea of sponsoring a series of one-hour musical programs featuring local talent. The task of developing the series concept and supervising its production is assigned to a staff producer. Or a station's management may wish to produce a public-affairs documentary program for a particular time slot and assigns the job to a staff producer. In corporate communications, a company may have a new product line, and management asks the television production facility to prepare a program introducing it to regional sales representatives.

On other occasions, a producer may develop a show completely from scratch. Independent producers must do this continually by first trying to discern programming trends and then by developing shows which will meet network and station program needs. Also, a staff producer may simply come up with a good idea for a show or series that he or she tries to sell to management as a program possibility.

No matter how a program idea actually originates, the producer's job is to develop the initial concept, which usually begins as little more than a vague notion, into a viable and effective television production. To do this, the producer must move through a series of program development steps. Although these steps may vary somewhat from program to program, for the most part all shows travel along the same basic road from the initial idea to the completed production.

Among the important program development steps are (1) analyzing the audience, (2) researching the program idea, (3) deciding on the production mode, (4) developing the treatment, and (5) preparing the program budget.

Audience Analysis

The fundamental reason for producing any show is, of course, to show it eventually to an audience. The audience's composition, program needs, and program preferences are important factors for the producer to consider at the very start of program development. Just as the public speaker selects his or her material and speaking approach for the audience to be addressed, so too must the producer develop a program with a "target audience" in mind. A target audience can be a rather small and specific group of people who are easily identified, or it can be a great mass of diverse individuals. Take the first instance, in which we can identify the audience. An industrial program, for instance, might be specifically produced to show factory welders new safety procedures to use on the job. In this case, the producer knows the size and composition of the audience and can tailor the content and presentation of the program to be most effective for them.

Audience composition is not as easily defined for most broadcast programs, but, even here, we can begin to narrow it down somewhat. Are we producing for a national audience or for a local audience? Is the target audience mainly men, women, or children? What age? Obviously, there is a great difference in the producer's approach between the *Lawrence Welk Show* and *American Bandstand.* Although they are both musical programs, the first is aimed primarily at mature adults, the latter, at teenagers and young adults. Fast cutting and loose, hand-held camera shots are fine for *American Bandstand,* but simply will not do for the older audience watching *Lawrence Welk.* As producer, therefore, you must always consider the composition of your audience and develop a show that you feel will best complement their viewing needs and tastes.

AUDIENCE RESEARCH A good producer relies heavily on his or her own instincts and experience in matching a program idea and approach to its audience. Still, "gut reactions" are not always reliable. Sometimes it is helpful to have more scientifically obtained data to use as an additional tool in analyzing audience program interests and attitudes.

All broadcast stations are required by the Federal Communications Commission to conduct *ascertainment surveys* on a regular basis to learn of the particular broadcasting needs within the community they serve. The results of the ascertainment surveys are supposed to be used by the station to develop programming that will better serve its audience.

Other research techniques can be used to test specific program concepts or production approaches. The popular and highly successful children's series, *Sesame Street,* is the product of

years of elaborate audience research on children to learn the most effective ways to present material to young viewers. *Good Morning America,* a daily magazine show on ABC-TV, is another example of a show which was developed through extensive audience research. The results obtained from audience surveys and interviews played a significant role in the introduction of various program segments, the selection of regular talent for the show, and the kinds of program material that viewers indicated they most wanted to see. Many local news formats are designed in cooperation with program research consultants, who survey the audience in the market area and use the resulting data to develop a program style and approach that will best appeal to the station's target audience.

There is a lot of controversy over the value of this sort of· research and testing of program concepts and audience attitudes. Some program executives make extensive use of the research data; others dismiss it as worthless. Of course, simply following a research-developed "recipe" will not guarantee you a successful show, but audience research that has been conducted, analyzed, and interpreted properly can be one of a number of useful factors which enter into program development decisions.

Researching the Idea

Once the producer has the basic idea for a show, the next step is to begin background research. If you are producing an informational program—a news, documentary, educational, or instructional show—the first step is a trip to the library to learn as much about the subject as you possibly can. At the same time, you may wish to contact some content experts to help you focus on the important issues, or to suggest additional sources for research. Local universities, public officials, professional associations, and consumer interest groups are all good resources for content specialists.

If you are producing a purely entertainment

show, your research needs are somewhat different. This type of program depends largely on two important factors: (1) the script or "property" and (2) the talent. Often one, or both, of these key ingredients gives a program idea its audience appeal. Since the show hinges on your ability to come up with a group of acceptable performers and a worthwhile script or property, preliminary negotiations with the artists or their representatives is your first step. You will need to know immediately whether or not the talent and performance rights are available and, if so, whether you can afford to purchase them.

As you research the program concept, you must constantly ask yourself a key question: "Is this idea suitable for television and for my production situation?" Television is a visual medium and needs interesting and exciting pictures to make it most effective. Talking heads or relatively static scenes rarely play well on the screen. There are, however, some notable exceptions to this rule. Nothing could seem as nonvisual as the televised congressional Watergate hearings, yet the intense drama and impact of the hearings made them compelling television. We usually think in terms of a large cast and as elaborate and realistic a set as possible for most dramatic productions. Yet such one-man shows as Hal Holbrook's *Mark Twain Tonight* or James Whitmore's *Give 'Em Hell Harry* were both highly effective television programs, even with a limited set and simple close-up shots. This does not mean you can stand a performer on an empty set with only a blackboard and expect to produce a dazzling half hour on molecular biology. In general, look for a program idea that will lend itself to television's visual capabilities, but use your judgment in deciding when to break the rules for an exceptional idea which has the strength to maintain interest without the usual visual accompaniment.

Another important question is the extent of your production facility's resources. Some producers try to produce a show that requires the resources of Twentieth-Century Fox in a tiny studio with limited facilities. You must match your ideas with your available resources. That is not to say you must always think small, just realistically. Given the talent, the facilities, and means available, is your idea too ambitious to be produced well? On the other hand, what can you do to develop the show to take best advantage of television's unique qualities: its timeliness, its intimacy, and its ability to closely involve an audience in a program?

As the once vague program idea slowly begins to take shape, you must consider if the program topic is too broad or too narrow. A show that tries to cover too much ground can wind up sketchy and incomplete. At the same time, an idea that is too narrow or specific may turn dull and uninteresting because you have exhausted all that there is to do or say about it. Also, ask yourself if you can present your program material successfully within the time allotted. Half an hour may be insufficient for some shows and much too long for others.

An obvious, but often overlooked, rule is to try to work from your strengths. Some producers are more comfortable working primarily with informational programming, such as news and documentaries whereas others may be better at dramatic or musical productions. Consider also your station's facilities and resources. Maybe your station is a leader in sports coverage and has the facilities and expertise to produce a regular sports series. Perhaps a local university has a number of renowned experts who could lend credibility to a documentary you are producing. A local station's strongpoint is often the fact that it can deal with *local* issues which are produced specifically to attract and interest an audience which simply can not be served in the same way by the national networks regardless of their large production budgets. The point is to recognize your strengths and try to use them whenever possible.

Every producer will tell you of the countless ideas that seemed so promising when they first came up but, after some background research, turned out to be totally uninteresting or unworkable. Never try to bend an obviously infeasible idea into a program. Rely on your judgment, research, and experience to tell you which ideas are suitable and which are not. The secret is to learn when to cut your losses and move on to another, perhaps more promising, project or approach. The sooner you can make this important decision, the less time and expense you will waste on a program concept that simply will not work. (See Figure 18-1.)

Deciding on the Production Mode

Once you have gathered sufficient background information to determine that your program idea is suitable, the next step is to decide how you will produce the show. The two major considerations are these: (1) Should you produce the show inside the studio or on location? and (2) What specific production mode should you use: live, videotape, or film?

STUDIO VERSUS LOCATION   Quite often the show itself determines whether it is produced in the studio or outside on location. A baseball game, for instance, must be covered as a remote. But many other production situations are not so clear-cut, and they offer the producer a number of shooting options. To use the studio or not to use the studio depends on the type of show you are producing and the facilities, time, and money you have available for the production.

Studio work offers you maximum control over most production variables. Inside the studio you will not encounter cramped floor space, poor working conditions, weather problems, and the countless hassles that inevitably accompany location shooting. For these reasons a studio show is usually less expensive than the same program shot on location although there are exceptions to the rule.

Some shows, however, obviously benefit from outside production, especially those which require sets, props, or locations which cannot be easily duplicated in the studio. The new generation of portable production equipment has released the producer from the confines of the studio and made location shooting a realistic option. Certainly there is no way you could adequately re-create inside the studio the atmosphere of a romantic beachfront, a bustling factory, or a historical monument. But location shooting generally increases both production costs and production time. As producer you must decide if

the trade-off in time, money, and inconvenience is worth the unique atmosphere and environment you can achieve with location shooting. (In Chapter 21, on remote and field productions, we will discuss some of the specific considerations which apply to on-location production.)

PRODUCING LIVE   Going live is a unique combination of the very best and the very worst that television production has to offer. On the plus side is the immediacy that is available only on live television. Not only can this heighten audience interest, but the one-time-only aspect of a live show often inspires both cast and crew to superior performances since they know retakes are impossible. Because there is no postproduction on a live show, videotape recording and editing expenses are unnecessary and can help to keep down production costs.

But a live show leaves no margin for error. Technical problems and production mistakes cannot be fixed by editing, and the producer generally has the least control over a live production than over any other production mode. Of course, sometimes content automatically dictates whether or not a show is to be produced live. Events coverage, such as news or sports programs, are almost always produced live. Complex entertainment programs, which do not require the immediacy of a live program, are almost always recorded. If you have the option of producing live or on tape, consider the trade-offs involved. What is gained in immediacy and impact is lost in the amount of control you can realistically expect to have over a live production.

PRODUCING ON VIDEOTAPE   Videotape is by far the most commonly used production mode and offers the producer four different recording approaches: (1) live on tape, (2) recording in segments, (3) single camera/single VTR, and (4) multiple camera/multiple VTR. Since we have already discussed the advantages and disadvantages of each recording technique in Chapter 10, we will not repeat them here. Your recording

PROGRAM DEVELOPMENT CHART

1. Developing the Program Idea

    A. Need Assessment

        1. Producer is assigned the program project
        2. Producer develops idea based on stations programming needs
        3. Producer develops idea independently and proposes it to station/network

    B. Target Audience Analysis

        1. Who is target audience?
           a. General Audience (all ages)
           b. Children (6–11; teenagers)
           c. Young Adults
           d. Older Audience
           e. Specific audience for program content (sports, drama, music, etc.)

    C. Proposed Broadcast Schedule

        1. Special one-time program
        2. Monthly or irregularly scheduled series
        3. Once-a-week program
        4. Daily "strip-show" in regular time slot

    D. Will Program Be Sponsored or Sustaining?

        1. If sponsored, will additional budget be available?
        2. If sponsored, will there be any conflicts between sponsor and program content?

    E. Potential Program Distribution

        1. Will program be aired only by local station?
           If so, program content can be tailored to local market.
        2. Will program be offered for syndication meaning that content must be more general to apply to a variety of potential markets?
        3. Will program be broadcast over regional or national networks?

    F. Initial Idea Research

        1. Does program content appear to be suitable for the intended amount of time available? If too broad, can it be narrowed down; if too specific, can additional content be added?
        2. Does idea seem suitable for presentation on television? Is opportunity for visual presentation available? Is idea too complex for accurate presentation via television?
        3. Will a content specialist be necessary during script preparation and be available for consultation during program production to check content accuracy and detail?

    G. Acquisition of Program Rights and Talent

        1. Are program rights available for purchase? (eg., books, script, title, etc.)
        2. Can you receive commitments from various performers, writers, directors, etc. who you would like to work on the project?

Figure 18-1 **Program Development Checklist.**

PROGRAM DEVELOPMENT CHART

2. Developing the Treatment

  A. Initial Research

    1. Have you contacted content specialists if necessary to add their input into program idea?

  B. Determining the Production Mode

    1. Will program be produced and broadcast "live" from studio?
    2. Will program be produced and broadcast "live" from remote location?
    3. Will program be taped in studio (live-on-tape) for later broadcast?
    4. Will program be taped in studio in segments with heavy postproduction editing later?
    5. Will program be taped on location in either segments or live-on-tape with subsequent postproduction editing?
    6. Will program incorporate segments on film or tape for later editing?
    7. Will program be produced entirely on film or will it use film segments within the taped or live show?

  C. Prepare Preliminary Program Budget

    1. Above-line costs—writer/script, producer, director, performers
    2. Below-line costs—studio facilities, technical personnel, sets and costumes, transportation costs, videotape and film-stock costs, postproduction editing and facilities

  D. Presentation of the Treatment

    1. Written treatment
    2. Proposed budget
    3. Additional supporting material such as target audience, station programming needs, desirability of producing the show, potential sponsors, etc.

---

approach should be based on the show's production requirements, the facilities available for production and editing, and the time and money you have to spend.

PRODUCING ON FILM With the exception of network prime-time entertainment shows, few television programs are produced entirely on film these days. Most television film is used for news and documentary coverage, for commercial production, and for short segments and program inserts, which are used within a larger show. Modern videotape equipment offers such increased production speed, versatility, and lower costs that many operations are replacing their film equipment entirely with electronic gear.

But film can be a viable alternative to videotape, particularly when you must shoot under difficult production conditions, where equipment portability is crucial and where lighting or working conditions are unsuitable for electronic equipment. For example, say you are producing a show about a mountain climbing club. You could conceivably shoot the rugged climb with electronic cameras and VTRs, but 16-mm film equipment is so much more portable and convenient for such a shooting situation that it would probably be the producer's best choice.

Writing the Treatment

A *treatment* is a brief outline of the proposed

program (or series) that is used by program executives in deciding whether or not to authorize you to begin production. (See Figure 18-2.)

The treatment should contain a description of the basic show idea, a justification of its worth and audience appeal, the production mode you have chosen, and any other relevant information which you think will help to sell the idea. The last item is often the most important because it can set your idea apart from similar ideas proposed by others. There just are not that many unique and original ideas, and your project may be approved because you found a fresh, new angle to a standard format.

For example, WXIA-TV was assigned to produce the Atlanta Emmy award presentation. Producer Neil Kuvin decided to differentiate the show from many other awards programs by producing it on location at a local hotel complex with a spectacular indoor ice-skating rink. The use of well-known skaters performing entertaining musical numbers on the ice was a fresh angle that was clearly emphasized in the program treatment and subsequent presentation. Other examples might be a unique location, a popular celebrity as host, or a topic of widespread and intense audience interest. But beware of going too far in your treatment. Although a certain amount of "hype" is inevitable, program decision-makers are intelligent and experienced executives who have heard on countless occasions why a proposed idea will make the greatest television show since *I Love Lucy.*

Treatments are usually necessary in broadcast television, where the program idea is "sold" to a station or network buyer. In corporate or educational operations, where you do not always have to sell a program, writing up a treatment is still a good idea. Forcing yourself to sit down and develop a treatment will help you to better organize the program, to be more specific about the production's goals and objectives, and to communicate your concept to other production team members more effectively. If you cannot write out a clear and concise treatment, you may not have adequately thought out the show.

Program Budgets

A program budget is a detailed estimate of the costs involved in producing your program. A program budget, like a household budget, is only as useful as the time and care you spend to develop it and how strictly you follow it. There are two important figures to keep in mind when you devise a budget: (1) How much money do you estimate the show will cost to produce? and (2) How much money do you expect to have available to produce it? Of course, the two figures should be fairly close. There is no point in attempting to produce a show unless you expect to have enough money on hand to do the job right.

Most program budgets are divided into two sections: (1) *above-the-line costs,* which include all creative personnel, such as producer, director, writer, and talent; and (2) *below-the-line costs,* which include the production expenses, personnel and technical facilities necessary to execute the production.

The sample budget in Figure 18-3 will give you an idea of how a production budget is broken down into specific categories. Sometimes a category such as "producer" includes not only an individual's salary but also staff and office expenses. The proportion of the above-line to below-line costs usually is about 45 percent of the total budget for above-the-line costs, 55 percent for below-the-line costs. This will vary, however, depending on the particular show and its requirements. A program with expensive performers, writers, and directors will make the budget top-heavy. Shows that feature exotic locations or expensive production variables, such as sets or costumes, will inflate the below-line figure.

You will need to research your costs carefully to find out exactly what your expenses will be. All production studios have a *rate card,* which lists facilities and their costs on a per-hour and per-day basis. Various suppliers will quote you prices on such items as videotape and film stock, set construction materials, costumes, and special production equipment. If you are shooting on location, also figure in the costs of transportation, meals, and, if necessary, lodging. The trick in developing a budget is both determining *what* you need and for *how long* you will need it. You must accurately estimate the amount of time that will be spent in preproduction, rehearsal, produc-

```
                    Program Treatment
                          for
                    S E G M E N T S


     SEGMENTS is a weekly, half-hour series designed to run in the 7:30 PM

to 8:00 PM time slot.  The program will be produced in cooperation with

Journalism students at the University of Georgia with its target audience being

students, faculty, and staff at the University as well as the Athens community.

The diversity of people, activities, and interests both at the University as

well as in and around Athens will provide a wide variety of program material.

The "home town" approach lends itself to immediate audience interest and appeal.


     Each SEGMENTS show will consist of three to five separate stories or

"segments."  Each story is introduced by the regular series host who provides the

necessary transition between segments.  SEGMENTS will be produced entirely on

location with video equipment and edited on 3/4" video cassette.  The use of EFP

gear will keep production costs to a minimum while maintaining viewer interest

through location shooting and providing creative production flexibility.


     The SEGMENTS format offers a number of significant advantages.  Each individual

story can be planned, produced, and edited independently.  Completed stories can

be stored in the can and used when necessary.  Those stories which are "timeless"

in nature can be produced during slack production periods when equipment and

personnel are readily available.  Stories requiring current production will be

produced during the week of broadcast, but since all stories in a SEGMENTS show

need not be produced during the same week, deadline pressures can be kept to

a minimum.  The story length is flexible depending on subject matter and treatment.

When special circumstances or events warrant, an entire SEGMENTS show can be

devoted to a single story.


     SEGMENTS is unique since no other show in Athens is designed to address

local issues, events, and personalities either on campus or within the city.

The 20,000 students plus 5,000 faculty and staff members comprise a natural

audience along with the many local residents who point to the University and

its activities with justifiable pride.  Including various town events will

only increase the potential audience for the show.  The on-location production

will give the show an exciting, interesting look, and the student viewpoint

in production promises to offer a fresh and original perspective.

     Some sample stories are included on the next page:
```

Figure 18-2 **Program Treatment.**

tion, and postproduction before you can figure your costs.

Sometimes a budget can be stretched considerably through trade-offs. Television exposure is a valuable commodity, and you may be able to obtain goods or services in exchange for an on-air credit or for commercial air time. For example, suppose you are doing a series of instructional shows on tennis. A tennis equipment manufacturer may offer you free tennis gear in exchange

PROGRAM BUDGET

Program _____   Date Prepared _____

Producer _____

ABOVE-THE-LINE COSTS Item	Estimated Cost	Actual Cost	Under/Over
Producer			
Director			
Writer/Script			
Talent (a)			
(b)			
(c)			
(d)			
Music			
Office Overhead			
Miscellaneous (a)			
(b)			
(c)			
Pension & Welfare			
ABOVE LINE SUBTOTALS			
BELOW-THE-LINE COSTS			
Studio w/technicians			
Assistant Director			
Lighting Director			
Scenic Designer			
Set Construction			
(a) Supplies			
(b) Personnel			
Special Equipment (a)			
(b)			
(c)			
(d)			
Film Stock			
Videotape Stock			
Wardrobe			
Props			

Figure 18-3 Program Budget.
The sample budget form includes budget items most commonly used on an average production. The first column, "estimated cost" is used to develop the budget. "Actual cost" indicates how much the item ultimately cost the production. The "under/over" column is used to figure cost overruns and items which came in under budget, and is very helpful as a reference in planning future budgets.

Location Expenses			
(a) Food			
(b) Lodging			
(c) Transportation			
Graphics			
Videotape Editing & Postproduction			
Miscellaneous (a)			
(b)			
(c)			
(d)			
Contingency			
Union Benefits			
BELOW-THE-LINE SUBTOTALS			
ABOVE-THE-LINE SUBTOTALS			
PRODUCTION TOTALS			

for an on-air credit. A local tennis club might permit you to tape the shows on their courts in exchange for another credit. If another show will require some large, elaborate sets and you lack the necessary set-construction facilities, perhaps the local dinner theater will be willing to construct your sets free of charge in exchange for commercial air time. Be sure to check with station management for their approval before making any final trade-off arrangements.

The most essential part of developing a budget is to remember to include all the necessary items. Such additional expenses as "pension and welfare" payments for union employees, insurance, and transportation costs must always be included in your budget. Once you have worked up all of the items you think you will need, add another 10 or 15 percent for "contingencies." Remember Murphy's law: If something can go wrong, it will. The contingency pad will come in handy for emergencies or if you encounter some unexpected cost overruns.

It is difficult to overemphasize how important a role the budget plays in determining the final program. Although the budget may appear only a series of numbers on a sheet of paper, in reality the budget determines—long before you ever enter the studio—exactly what you can or cannot do from a creative standpoint. Given a certain amount of money, it is up to the producer to allocate it intelligently. You can devote more money to providing increased production time in the studio, but the trade-off might be less editing time during postproduction. There is only one hard-and-fast rule about budgets: They are never large enough. One of the producer's most difficult and important jobs is knowing where to spend—and where *not* to spend—so that the final product is as good as can possibly be accomplished given the limitations of the working budget.

At many smaller studios and stations, the producer need not develop an extensive budget since staff, crew, and production facilities are always available. In this case, the budget consists largely of expendable supplies and outside costs, which are above and beyond normal salaries and overhead expenses. Some small operations still require the producer to prepare a budget though since they assume that the staff and production facilities assigned to one project are unavailable

483

for other work assignments. Although such a budget consists mainly of "paper money," it does give the producer and station management a good idea of how much of the station's resources are expended on a particular program or project.

It is a good idea to develop an *estimated* budget before the production and an *actual* budget afterward to reflect the real costs of production. By comparing the two figures, you can see where your estimates were significantly over or under the show's actual expenses. This information is helpful in preparing a more accurate budget the next time around.

PRESENTATIONS

Once you have written the treatment and developed an estimated budget, you may be asked to make a formal presentation to station management or other executive decision-makers, who must give their approval before production can begin. The presentation is an opportunity for you to expand on the information in the treatment and to convey personally your enthusiasm for the project. It also gives executives an opportunity to hear about the production directly from the producer and to raise any questions or objections. Quite often you, as a producer, are as much on display as your program idea. If you can show others that you have done your homework and are thoroughly prepared, you greatly improve the chances for approval.

It is quite possible that during the presentation a number of changes or revisions in the original idea will be requested. Your program budget may be too high; part of the show may not be completely acceptable; a production facility you expected to use may not be available. These problems are usually solved through compromise, but the one area on which to remain firm is the production budget. You may have added some extra expenses which are not essential and can be traded away during negotiations, but you should also have in your mind the rock-bottom figure that you think necessary to bring in a program of sufficient quality. If you agree to a severe budget slash, you may have to totally revise your production concept or risk running out of money before the show is completed. Even if you can finish the show within the reduced budget, the shortcuts and compromises you may be forced to make can seriously damage production values. If you believe your budget is accurate and fair, stick with it. You may be better off not doing the show at all rather than risking an inferior production with an insufficient budget.

MUSIC CLEARANCE

The performance rights to copyrighted musical material are protected under the law. In order to use the material legally, the producer must arrange for permission and payment to the music composer and publisher.

ASCAP and BMI

Two major organizations—ASCAP (American Society of Composers, Authors, and Publishers) and BMI (Broadcast Music Incorporated)—serve as clearinghouses for music performance rights. Virtually every tune which is written and performed is registered with one of these organizations, which, in turn, licenses all radio and television stations to use the music. The stations pay a yearly fee which is determined by the size and income of the station. The blanket fee permits the station to use any, or all, or the organization's musical catalogue during the year.

Production Music

Music which is written and recorded especially for production use is also covered under copyright law. The purchase price of the record may not cover performance rights. It is best to check with the production music company to determine exact details about performance rights and fees.

At most operations the producer is required to submit a music performance form after each show, indicating all the ASCAP, BMI, and other music organizations' tunes used. These include not only major performances but also theme, background, and incidental music. You can learn the organization which holds the rights by checking the record label.

UNIONS AND GUILDS

At many stations and production facilities across the country, some of the staff and free-lance positions are represented by unions or guilds, which negotiate wage and working agreements with management. Since these contracts have a direct impact on both your program budget and production operation, you ought to become familiar with some of the largest and most representative unions.

Talent Unions

American Federation of Radio and Television Artists (AFTRA) Represents radio and television talent (announcers, newscasters, reporters, sports commentators, singers, dancers, comedians) who appear on live or videotaped programs and commercials.

Screen Actors Guild (SAG) Represents performers in all filmed television programs and commercials and on some videotaped productions as well. SAG agreements are in effect primarily in the major film production centers of Los Angeles and New York.

Both unions have a minimum wage for employees called "scale." Of course, better known or principal performers are commonly paid much more than scale. A major distinction in AFTRA salaries is between "under five" and "over five" performers. "Under five" performers are those who have less than five lines of dialogue, and their scale is lower than that for artists with more dialogue.

American Federation of Musicians (AFM) Represents all musicians working in television production. The AFM is a national organization with various local chapters in the largest cities.

Directors' and Writers' Unions

Directors' Guild of America (DGA) Represents program directors, assistant directors, floor managers, and, at some stations, production assistants in both film and television. In addition to

stipulating scale wages for staff and free-lance directors, the DGA contract spells out working and travel conditions, fringe benefits, and the director's "creative rights" on a production.

Writers' Guild of America (WGA) Represents television script writers for live, tape, and television film production. In addition to basic wage agreements, the complex WGA contract stipulates residual payment fees for reruns and specifies the creative rights and screen credits to which a writer is entitled.

Technical Unions

National Association of Broadcast Employees and Technicians (NABET); International Brotherhood of Electrical Workers (IBEW); International Association of Theatrical and Stage Employees (IATSE) Try not to let the different names confuse you. These three technical craft unions all represent employees who work on studio crews in both film and television. The first, NABET, currently represents all NBC and ABC network technical personnel; IBEW represents all CBS network technical personnel. Both NABET and IBEW also represent technicians at various stations throughout the country. IATSE represents stage hands, carpenters, lighting grips, and some film crews at all three networks, as well as some television crew members at various stations.

These powerful unions negotiate highly intricate agreements that specify each unions's jurisdiction over equipment operation and working conditions. In addition to establishing basic wage and salary scales, the contracts stipulate overtime for extended work hours, the way in which employees can be scheduled, "penalty" fees for violating union work agreements, the timing of meal and rest breaks, and a host of other details.

THE PRODUCER IN PRODUCTION

The producer is one of the few members of the production team who is intimately involved in each of the four production stages, lends continuity to the entire show, and keeps it moving toward the objectives which were outlined during the preproduction planning stages. The producer is naturally concerned with several important production decisions simultaneously, but in order to present these logically, we will have to deal with each one separately. Keep in mind, though, that in a real-life situation, events are rarely as well organized and as clear-cut as we have made them seem here. We will refer to a number of production team members as individuals; however, at some stations or studios, one person may actually undertake a number of different production responsibilities. The set designer may also be responsible for lighting, or the director may double as the technical director and operate the program switcher. Whether you work with a full production team or with a smaller utility staff, the producer's job is the same, so you should not find it difficult to adapt the information below to your specific production situation.

Preproduction Planning

For the producer, preproduction planning is the busiest and most important part of the entire production. This is when the program concept is refined and focused, when the show's overall objectives are established and communicated to the production staff, and when the entire production effort is planned and coordinated. The better the job of planning and organizing now, the smoother the production will proceed later on.

During preproduction, the producer is concerned with countless details. Some are crucial; some, less important. At this point you must begin to delegate authority to your staff. You must naturally be fully involved in the important decisions about the script, the direction, the casting, and the budget, but let someone else order the coffee and doughnuts. As Bob Shanks, a network producer and program executive, has written, "There is a high risk of bruising necessary objectivity by getting too close to production. A produc-

er must learn to protect his senses. The first and best advice is to let other people do their jobs. If a producer is writing the jokes, mouthing the dialogue, painting the set, adjusting the lights, moving the props, and seating the audience he is likely to develop callouses on his sensitivity. . . . "[1] As a producer, your most important function is to develop the program concept and then hire people you trust to do their jobs. There is simply too much to do for any one person to attempt it all. NBC's George Heinemann suggests that in the long run, the producer must be concerned primarily with those elements which will appear onscreen and affect what the audience sees. Select your production team carefully, explain what you want, and trust them to do their jobs for you. You will find that if you communicate your concept to others and give them the opportunity and freedom to contribute their ideas as well, you will wind up with a happier and more productive staff. As producer, you will be able to direct your attention to those production elements which most need your time during preproduction—the script, the director, casting, and preparing the production notebook.

The Program Script

The foundation of every show is the program script. Of course, sports or events coverage has no detailed script, but you will still need someone to outline the production and to provide background material for the director and talent to use in formulating the coverage and on-camera commentary. Even on such ad-libbed interview programs as the *Tonight Show* or *The Merv Griffin Show,* a staff of writers is assigned to research and preinterview guests and then to prepare a series of questions for the host to use on the air.

Because the script is so fundamental to the production, some producers prefer to write the show themselves. These "hyphenates" (so called because of their title: "writer-producer") are often found on comedy, dramatic, or documentary programs. If the producer does not write the script, then either a staff writer must be assigned to the production or a free-lance writer must be hired from the outside. If you are working on a program

[1]Bob Shanks, *The Cool Fire,* Norton, New York, 1976, p.10.

with historical, technical, or scientific subject matter, you may also wish to hire an outside content specialist who can serve as a consultant to the writer and producer. The job of the content specialist is to help develop the program's focus and to verify facts and information in the script.

A producer of an ongoing series which involves a number of different writers will sometimes develop a "writer's bible." This is a detailed explanation of the show's premise and objectives and a description of the major characters in the show. The "bible" serves as the writer's basic reference and eliminates the need for the producer to sit down with every new writer to explain the show's style and approach.

Almost every television script goes through a series of revisions and rewrites before it is produced. Even the best-written script may have problems with characterization, exposition, sequence, or style. Sometimes the writer's imagination may wind up costing more money than the budget will allow, and the producer will ask for certain scenes to be eliminated, for a number of different characters to be combined into a single role, or for an exotic location to be changed to one that is more accessible and less costly. At some stations the final script must also be approved by "program practices," which checks for "offensive" or inappropriate material and for possible legal problems.

Once the writer has responded to the producer's suggestions and has submitted a second draft, the producer may wish to hold a script conference with the writer, the director, and key production members. The director may offer some suggestions or point out potential production problems. The set and lighting designers can begin to use the script to formulate rough ideas on how to approach scenery and lighting. At this point be sure you have read through the script for running time. If it is too short, material must be added. If too long, the writer must make some cuts.

After all the script revisions and "polishes" are completed, the final approved version becomes the shooting script. It is typed, duplicated, and distributed to the members of the production team. This does not mean, however, that the script is not revised any further. Almost every show makes script changes throughout rehearsal

and even into production, and the writer is often asked to remain available for quick revisions. Some even set up a typewriter in the studio so they can more easily work on the necessary rewrites during rehearsal and production.

Director

Except for the producer, the program's director is the most important member of the production team. While the producer's job is to develop the idea and to supervise the production, the director must interpret and execute the producer's concept into the actual show. As the producer you make an extremely important decision when you select a director for a production. Producer Norman Lear has said that the people he has selected to direct his various programs have each made a distinctive mark on the overall style and approach of every show.

The person you select to direct your show ought to be one with whom you can work closely and whose talents and abilities you trust and respect. The latter is particularly important because there will be times when the producer and director come into direct conflict over how to approach or interpret elements in the show. A good producer should always be open-minded about any suggestion, especially when it comes from the director. After all, the director is a creative and imaginative artist who might bring a fresher and more interesting perspective to the show than the one you already had in mind.

As soon as the director joins the production, he or she becomes involved in almost all preproduction meetings, especially those dealing with the script, casting, and production. The director's first job is usually to read and carefully analyze the script, consider any potential production problems, and suggest possible solutions or script improvements.

Casting

Selecting talent for a production is always an

important job, and this applies to nondramatic as well as to dramatic roles. Casting even as deceptively simple a part as the narrator or on-camera "teacher" in an instructional production must be done with care. The talent on screen will have to carry much of the show and contributes enormously to a program's success or failure. Good television talent is not always easy to find. The performer must be comfortable in the role, be able to work on a studio set with all its distractions, and still relate well to the viewing audience.

Many variables enter into casting decisions: the availability of talent, the cost of their salaries, and whether they look and sound "right" for the show. The director will usually assist the producer during casting since he or she must ultimately work closely with the talent during rehearsal and production.

There are a number of ways to go about casting. If you can pay for performers, contacting talent agents is usually the best approach, since they handle a number of artists who might be suitable for your needs. Advertising open auditions will usually produce a large number of eager applicants, but you must be prepared to spend time wading through countless try-outs before you will find performers who are right for your program. If your budget is limited and you are unable to pay for talent, try some of the community theater groups and university drama departments which often have a large pool of interested volunteers willing to trade their talent for television exposure.

Production Book

With the program script either finished or in final revision, the director assigned to the show, and work begun on various production elements, the producer must now coordinate the many complex operations that go into a television program. There is no way you can keep all this information in your head (even if you wanted to), so the producer must begin to work out production scheduling and organization on paper.

One reason producing may have appealed to you is the ability to work in a creative environment without the paper-pushing tedium associated with many desk jobs. But the amount of information a producer must deal with and the communication and coordination between so many different people is so enormous that organizing on paper is the only practical way to get the job done. Paperwork is rarely fun, but, if done carefully and accurately, it will save you much time and aggravation later on during rehearsal, production, and postproduction.

The *production book* is the producer's bible, reference, and security blanket all rolled into a large loose-leaf binder. Most producers prefer to use a loose-leaf notebook so they can add or remove pages easily. Your production book should contain just about every piece of information you are likely to need about the show. What to include and how to arrange your production book is a matter of individual taste and habit, but most producers include the shooting script, complete budget information, and copies of all the various production forms, memos, and schedules which are written and distributed. The following schedules and production forms are examples of the kinds of material most often included in a production book.

MASTER PRODUCTION SCHEDULE   The master production schedule is designed to give you an overview of the entire production. Every producer has a different method of outlining the master schedule. Some even prefer to mount it on a large bulletin board on the office wall and tack color-coded index cards on it for various activities. Having at least a copy of the master schedule in your production book is a great convenience, however, when you need to refer to it outside the production office.

The master schedule includes every production operation which must be completed: what is to be done, who is to do it, when its completion is expected, how it is to be integrated into the show. A television production involves many separate activities which proceed independently, but must all come together precisely on schedule and in the proper sequence. Unless the set is constructed, lighting cannot be hung or focused. Pretaped insert segments must be produced and edited

SCRIPT	*OUTLINE* 8/31	*1st DRAFT* 9/15	*FINAL* 9/21	*DUPLICATED+DISTRIB* 9/23	*CUE CARDED* 9/30
TALENT	*CASTING COMPL. BY 9/25 DRY REH 9/30, 10/1 STUDIO REH 10/2, 10/3 PROD 10/4, 10/5*				
SETS	*MEETING 9/22 FINAL DESIGN 9/25 CONSTR 9/26, 9/27 STUDIO SET-UP 10/1*				
LIGHTING	*MEETING 9/22 FINAL DESIGN 9/25 LIGHT 10/1*				
AUDIO	*MEETING 9/22 MUSIC+SFX BY 10/1 / AUDIO SETUP 10/1 / AUDIO POST PROD 10/7-10/9*				
GRAPHICS	*DUE 9/30*				
FILM/VTR SEGMENTS	*REMOTE VTR PROD 9/28*				
WARDROBE	*WARDROBE FITTING 9/28 CAST IN STUDIO 10/2 - 10/5*				
PROPS	*ASSEMBLE PROPS 9/30, 10/1 IN STUDIO 10/2 - 10/5*				
EDITING	*REMOTE VTR SEGMENT 9/29 - SHOW POST PROD 10/7, 10/8, 10/9*				
FACILITIES SCHED	*REMOTE VTR CREW - 9/28 / EDIT 9/29 / STUDIO SETUP +LIGHT 10/1 / STUDIO FAX REH 10/2, 10/3 / PROD 10/4, 10/5 / EDIT 10/7-10/9*				

IMPORTANT DUE DATES: Final Script By 9/21
Casting By 9/25
Set/Light Approval 9/25
Studio Set-Up Lights/Set: 10/1
Studio FAX Reh: 10/2, 10/3
Production: 10/4, 10/5
Postproduction: 10/7-10/9
Final Master Completed: 10/9

Air Date: 10/15

Figure 18-4 Master Production Schedule.
The master production schedule is designed to show the status of all important production elements at a glance. The sample schedule illustrated here is only one of many ways the information can be organized.

before they can be used in the actual production. Music, graphics, costumes, and props must all be ready by rehearsal time. The master schedule will help you to keep track of a large number of operations and to check on the show's progress and possible trouble spots before they become serious problems. (See Figure 18-4.)

CAST/CREW LISTING   As its name suggests, the cast/crew sheet is a complete listing of the names, phone numbers, and addresses of everyone involved with the show. In the case of remote

productions, it includes everyone's hotel address and phone number as well.

REHEARSAL SCHEDULE   The rehearsal schedule lists the day, time, and location for all production rehearsals. Usually, dry rehearsals—preliminary rehearsals without any cameras or costumes— are held outside the production studio in a rehearsal hall, an office, or some other rented space. That is why it is important to be specific about the location of each rehearsal session.

REHEARSAL SCHEDULE

PRODUCTION TITLE: STARRY NIGHT

DIRECTOR: Robin Tellman

PRODUCER: William Rogers

Monday, Sept. 30th	9:00–12:00	Entire Cast Readthrough	REHEARSAL HALL #1
	12:00–1:00	LUNCH BREAK	
	1:00–2:30	BLOCK ACT I–Bill, Robert, Eileen, Paul	
	2:30–5:00	BLOCK ACT II–Bill, Eileen, Sandy, Karen	
			REHEARSAL HALL #1
Tuesday, Oct. 1	9:00–12:00	RUN THROUGH ACT I–Bill, Robert, Eileen, Paul	
	12:00–1:00	LUNCH BREAK	
	1:00–4:00	RUN THROUGH ACT II–Bill, Eileen, Sandy, Karen	
	4:00–5:00	RUN THROUGH COMPLETE SHOW–Entire Cast	
Wednesday, Oct. 2		STUDIO REHEARSAL-PRODUCTION STUDIO "C"	
	9:00–12:00	CAMERA BLOCK ACT I–Bill, Robert, Eileen, Paul	
	12:00–1:00	LUNCH BREAK	
	1:00–4:00	CAMERA BLOCK ACT II–Bill, Eileen, Sandy, Karen	
	4:00–5:00	NOTES–Entire Cast	
Thursday, Oct. 3		STUDIO REHEARSAL-PRODUCTION STUDIO "C"	
	9:00–12:00	RUN THROUGH ACT I–Bill, Robert, Eileen, Paul	
	12:00–1:00	LUNCH BREAK	
	1:00–3:00	RUN THROUGH ACT II–Bill, Eileen, Sandy, Karen	
	3:00–4:00	NOTES AND COSTUMES AND MAKEUP — Entire Cast	
	4:00–5:00	DRESS REHEARSAL	
	5:00–6:00	NOTES	

Figure 18-5 **Rehearsal Schedule.**

The schedule should also indicate which parts of the show will be covered during each rehearsal session and those cast/crew members expected to attend. You may not need everyone present if you intend to rehearse only a portion of the show during a rehearsal period. On complicated shows, the director may wish such key production team members as the assistant director, floor manager, and audio engineer to attend a final dry rehearsal. (See Figure 18-5.)

FAX SHEET The "facilities," or FAX, sheet is a detailed description of all the production facilities required for your show. The producer and director run down the FAX sheet and check off the equipment and technical personnel they wish to order. Of course someone must pay for the facilities you request, so act reasonably. There is no need to ask for five cameras, a studio crane, and a multitrack audio tape recorder to do a simple two-person interview show. (See Figure 18-6.)

The information on the FAX sheet is needed by a number of different studio departments (production, engineering, traffic and scheduling, accounting, set and scenery, lighting, and so on), so the FAX sheet is often a multipage form. Each duplicate page is usually color-coded and sent to the appropriate department.

OPERATIONAL SCHEDULE   The operational schedule, or "ops sheet," as it is commonly called, is a rundown of the studio technical equipment and personnel assigned to the production. The ops sheet contains a description of every studio activity (setup, camera, blocking, taping), its scheduled time, the technical personnel required, the equipment that has been ordered on the FAX sheet, and any additional personnel, such as wardrobe, costumers, and makeup artists. The sample in Figure 18-7 will give you an idea of an ops sheet content and layout.

CALL SHEET   The call sheet is generally reserved for cast members and indicates when and where they are expected both for rehearsal and for production. Be sure to consider makeup, costume, and transportation time when you work out your call sheet. For instance, if you want your performers in makeup and costume by 9 A.M., you may need to call them for 8:15 A.M., so they have time to put on their makeup and costumes and still make the nine o'clock call.

SHOOTING SCHEDULE   If your show is produced either live or live on tape, the shooting schedule is simply the program's rundown sheet, since everything occurs in normal sequence. However, if you are taping the program out of sequence, you must devise a shooting schedule to indicate the specific scenes which are to be shot, the day, and the times for which they are scheduled. The shooting schedule should be arranged carefully to make the most efficient use of your cast, crew, and production facilities. You may be able to save both time and money on

costly talent and on the rental of special production equipment by scheduling all the scenes in which they are involved on the same day. (See Figure 18-8.) (See Chapter 10 for a detailed discussion on how to develop a shooting schedule.)

RELEASE FORMS AND PERMITS   Everyone who appears on your television program should be asked to sign a standard release form. By signing the form, the individual grants permission to be photographed on television. This is especially important for nonpaid talent who appear as guests. If you are paying for on-camera talent, a signed performance contract eliminates the need for a release form.

You should also secure written permission whenever you utilize a copyrighted work such as still photos, clips from films or television shows, illustrations, and so on. Even productions designed only for closed-circuit distribution are subject to certain limitations on the use of copyright material. Before you plan to make a slide from an illustration in a book or use a clip from a film which is available, make certain the proper permission has been obtained, and always file a release form to protect yourself should any question arise later on.

Most cities require a production unit to obtain a police permit before shooting on remote locations. This does not apply to news crews (who already have a police "press pass"), but to more elaborate production crews that will be working in a public area for an extended period of time. The permit authorizes you to shoot on public property, provides a parking area for your production vehicles, and sometimes includes police assistance for traffic and crowd control. Check local regulations for the permit requirements in your area.

Production Meetings

Throughout the entire preproduction phase, the producer is involved in a countless series of production meetings. Often these include only the producer, director, and a key production team

PRODUCTION FACILITIES REQUEST FORM

Date: _____

Production: _____ Prod. Number _____

Producer: _____ Director: _____

Day/Date Requested: _____ Time From _____ to _____

FACILITIES ( ) STUDIO SETUP/LIGHT ONLY ( ) CONTROL ROOM ONLY ( ) STUDIO & CONTROL

CAMERAS   Mounting   Lenses   Cable Length (if other than normal)

1. _____
2. _____
3. _____
4. _____

AUDIO

Microphones: (1) _____ (3) _____ (5) _____
(Type &
Number)   (2) _____ (4) _____ (6) _____

Boom: _____   Studio Flr Speaker _____   Other _____

TELECINE   (1) ( ) 35 mm Slide   ( ) 16 mm film   ( ) Super 8 mm

         (2) ( ) 35 mm Slide   ( ) 16 mm film   ( ) Super 8 mm

         ( ) Interlock

VIDEOTAPE   Time From _____ to _____ RECORD ( ) Quad ( ) 1 inch ( ) ¾ cassette

           Time From _____ to _____ PLYBK ( ) Quad ( ) 1 inch ( ) ¾ cassette

                                         ( ) TBC

MISCELLANEOUS

     ( ) Char. Gen

     ( ) Video Monitors (number and location) _____

     ( ) Other _____

                     _____

LOGGED   _____ Engineering   _____ Production   _____ Traffic

---

Figure 18-6 FAX Sheet.
The FAX sheet indicates all equipment required for the production. At most facilities, the FAX sheet is a multipage form, duplicate pages being routed to various departments.

THURSDAY, MARCH 18	STUDIO #1
LOAD-IN, SETUP, LIGHT	7:00 AM — 10:00 AM
ENGINEERING SETUP	10:00 AM — 10:00 AM
CAMERA BLOCK	10:30 AM — 12:30 PM
LUNCH BREAK	12:30 PM — 1:30 PM
CAMERA BLOCK	1:30 PM — 3:00 PM
NOTES & RESET, MAKEUP	3:00 PM — 3:30 PM
DRESS REHEARSAL	3:45 PM — 4:15 PM
NOTES & RESET, MAKEUP	4:15 PM — 5:00 PM
TAPE	5:00 PM — 5:30 PM
PICKUPS	5:30 PM — 6:00 PM
STRIKE	6:00 PM — 7:00 PM

ENGINEERING PERSONNEL   (7:00 AM — 7:00 PM)

1 TD	3 CAMERA OPERATORS
1 LD	2 CAMERA ASST/ASST FM
1 AUDIO	1 BOOM OPERATOR
1 VIDEO	

ENGINEERING EQUIPMENT   (STUDIO AT 7:00 AM)

3 CAMERAS ON PEDS

1 MOLE BOOM

(2) QUAD VTRs (FAX ORDERED FROM 4:30 PM — 6:00 PM)

PRODUCTION PERSONNEL   (9:00 AM — 6:00 PM)

1 ASST DIRECTOR

1 FLOOR MANAGER

1 MAKEUP/WARDROBE/PROPS

Figure 18-7 **OPS Sheet.**

member to discuss a specific area of the production, such as set design, the script, or casting.

A number of preproduction meetings, however, must be larger and should include all the key production team members. This is when the various facets of a production must be integrated so all subsequent planning and execution will be based on the overall program objectives which are understood by the entire production team.

Of course, some regularly scheduled programs require fewer production meetings because everyone's role has been defined over a period of time. A show which is produced completely from scratch would naturally require more meetings since everyone is essentially working on the project for the first time.

Regardless of how many or how few production meetings are scheduled, some meeting of the

TAPING SCHEDULE

TITLE "The Dancers"

PROD. NO' 5.32-11

PAGE 2 OF 8

DIRECTOR Robinson P.A. A.D U.M.

DAY, DATE AND TIME	SCENE, SETUPS AND DESCRIPTION	D/N INT/EXT	LOCATION AND CAST	SET PIECES AND EQUIPMENT
MONDAY, JAN 29				
9:00 AM	SCENE #18—Dance Rehearsal	D/I	DANCE SCHOOL Robert Jill Martha All Dancers	LIGHTING MUST BE SETUP AND READY
9:45	SCENE #32—Robert & Jill Fight	D/I	DANCE SCHOOL Robert Jill	
10:30 A	SCENE #33—Jill & Martha Find Out	D/I	Jill Martha George	
12:00-1:00	LUNCH BREAK			
1:00 P	SCENE #35 Robert & Jill Together	D/E	OUTSIDE SCHOOL Robert Jill	REFLECTORS, BOOSTER LIGHT, TAXI & SUITCASES
2:30	SCENE #1—Jill Arrives at Dance School	D/E	OUTSIDE Jill Martha	TAXI — Same Lighting
3:00 P	SCENE #1A—Dancers outside School	D/E	Dancers-	Same Lighting

Figure 18-8 Taping Schedule.
The taping or "shooting" schedule is used to indicate the specific scenes which are to be shot and at what dates and times. The producer and director work out the most efficient shooting schedule, taking into account performer and location availabilities, necessary sets, and the requirements of the shooting script.

minds among the key production team members is absolutely essential to ensure a coordinated and cohesive effort once the project moves into the setup and rehearsal and production stages.

SETUP AND REHEARSAL

As the production moves into the setup and rehearsal stage, the producer's emphasis shifts from attention to specific details toward the broader task of integrating all the production elements. By this time, specific production details are covered mainly by the production team members working in their respective areas: the director rehearses the cast and prepares the camera shots; the audio engineer designs the audio pickup for the show; the scenic designer supervises the construction of the set; and so on.

The actual rehearsal is a crucial time when the show should begin to take shape. Before long it will be too late to make any major changes in the production, so the producer must watch each run-through with a critical eye. This is when he or she must become the "surrogate audience," watching and reacting to the program as the viewer would. As rehearsals continue, the producer watches the line monitor and makes copious notes on all the production aspects that need attention. These include the direction and camera shots, performances, sets and lighting, costumes and makeup, audio, music, and sound effects. Between run-throughs, the producer gives notes to the appropriate team members. Use discretion, however, in when and how to give notes. The first rough camera run-through is hardly the time to expect perfection, and the entire production team is better left alone until everyone settles down a bit.

While the producer watches the program monitor, he or she must also keep an eye on the studio clock. It is not uncommon for a director and cast to become so engrossed in a particular problem that they spend far too much time on what is basically a minor part of a much larger show. If your director lingers at one point in the production for too long, discreetly remind him or her to keep moving and to return to the problem later as time permits. You ought to have some idea of how much the director should accomplish —so many script pages, a number of scenes, and so on—during each rehearsal session so you do not wind up with a show that has been only half-rehearsed.

By the end of the rehearsal stage, the program should begin to look and sound the way the producer had envisioned it. During the final run-through before dress rehearsal, look carefully and critically at the entire production. To give you an idea of the kinds of things a producer looks for during a final run-through, here is how Paul Rauch ran a notes session with his key production personnel from *Another World*.

The program is a one-hour weekday drama produced in segments on tape. Since it is a continuing series, the entire production team works together closely every day, so everyone involved is already familiar with the show and with the program's basic objectives.

As producer, Rauch busies himself with a variety of details and problems during the morning, turning his attention specifically to the day's show at run-through time. He watches from his office on a conventional television receiver. By not watching from the control booth, thus avoiding its distractions, he is able to gauge more realistically the program's impact on a home audience.

As the run-through proceeds, Rauch jots down many notes and suggestions on virtually every aspect of the show. His eye is critical, picking up subtle problems or inconsistencies which would probably have gone unnoticed by most viewers, yet taken as a whole, his comments and suggestions improve the entire production.

At one point, Rauch noticed that the lighting in a scene appeared too intense on an actor, and he asked the lighting director to correct it. The background area of another set appeared too dark, and this was also noted. Rauch wanted one character to use a hand prop for some added business during a scene. Since the character was a wealthy businessman, it was decided to use a copy of the *Wall Street Journal*. A boom shadow appeared during an actor's cross to a chair, and this was noted for both audio and lighting. A variety of notes concerning performances were given to the program's director. Some covered minor nuances in voice or inflection; others were more substantial changes, such as a better way to establish a relationship between two characters. Costume problems were noted, and Rauch ordered an actress playing a scene in a garden to wear gardening gloves as she trimmed flowers. The dialogue in one scene read unconvincingly, and Rauch asked for a rewrite. Even a musical bridge between scenes did not escape his critical ear. Rauch opted for a lighter and less orchestrated version than the one which had been played on the runthrough.

Some of the production team who were not directly involved in the run-through viewed the show with the producer in his office and received their notes immediately. Others, such as the director and the audio engineer, were given notes after the run-through was completed. The key personnel then went off to meet with their respective crew members while the director conveyed both Rauch's notes and his own suggestions to the performers.

Needless to say, *Another World* is a very special sort of show: a one-hour drama produced at a network studio five times a week. The notes from a producer will be different on a musical program or a news show, a sports event or an instructional production, but the intent is the same: to keep on top of everything that appears on screen and to serve as a substitute audience by evaluating the production as the viewer would.

The dress rehearsal is crucial for any prerehearsed program since it is an exact replica of the air show that the audience will see. Any major changes you want to make in the show must be done before dress; otherwise, there will be no time to try them out before the actual production. Most production changes create a domino effect. A change in one actor's blocking can affect the camera shot, the lighting, the position of the boom microphone, the audio, and so on, down the line. Naturally, a serious problem must be corrected regardless of when you see it, but a problem of this magnitude should have been spotted much earlier and corrected long before the dress rehearsal.

Some producers like to videotape the dress rehearsal, believing that it keeps the production crew and the performers more finely tuned and results in sharper performances. The extra tape footage is also helpful during editing, particularly if someone has turned in a one-time-only performance that you are lucky enough to have recorded during the dress rehearsal.

PRODUCTION

According to George Heinemann, "during the production a good producer is the loneliest guy in the world. That's because if you've done your job properly, you get the feeling you're in the way since all of the major decisions were made long before now."

Of course, there are some notable exceptions to this rule, particularly in live news, special events, and sports programming, where the producer must make split-second decisions continuously the whole time the program is on the air. But for most preplanned shows, by production time the producer's job is essentially over; the show is now in the hands of the director and the production crew. The producer is still there, of course, but more to give encouragement than for anything else. By this time the show you see on the line monitor should be the realization of what you saw in your mind's eye so much earlier, before all the meetings, the paperwork, the negotiations, the deadlines: a concept that has come to life in the videospace on screen.

POSTPRODUCTION

Postproduction often involves the producer in more than just the completion of the actual program. Of course, a show must be edited, but the producer is also concerned with such activities as promotion and publicity to ensure the largest possible audience for the show and with analyzing the audience response to the program to determine how well it was received and how close it came to satisfying the producer's initial production objectives.

Editing

Many producers prefer to leave the preliminary editing decisions to the director. This is generally a good idea, since the director probably shot the show with an editing strategy specifically in mind. The producer should always view the rough-cut, however, and make any comments or suggestions before the program is edited into the final version. Needless to say, there are some producers who insist on playing a more direct role in editing and others who leave the job entirely in the director's hands. This decision depends on your own judgment, how much "creative" editing is involved, and the amount of time you can spend in the edit room.

Promotion and Publicity

Even the best produced program is not worth much unless there is an audience to see it. Closed-circuit producers of educational or industrial shows usually do not have to worry about promotion since they have a "captive" audience for which the program was originally produced, but broadcast programs are an entirely different story. The producer, along with the station or network, must try to publicize and promote the show so as many viewers as possible are made aware of its air date and time.

Larger stations and the national networks all have promotion and publicity departments responsible for promoting upcoming programs. Usually the producer is asked to consult with the promotion department to suggest possible publicity angles and to help in the development of the promotional campaign. At smaller operations with no promotion department, you may have to take on the promotion and publicity chores yourself.

PRESS RELEASE One way to publicize a show is through a press release sent to all radio and television newspaper and magazine editors within the broadcast area. Editors need a constant supply of copy material and a well-written and interesting release about your show stands a good chance of being included as a feature or news item in a column.

The release should contain all the essential information about your show: its title, the air date and time, the channel, a brief summary of the

show's content, the names of principal performers and guests. It is also a good idea to work up some interesting production sidelights which make good copy. Anything different or unique about the show ought to be included. For instance, a press release about an upcoming sports remote might discuss the extraordinary preparation that went into the show, any special new production techniques that will be employed, the number of cameras, miles of cables used, and so on.

Plan on sending out your press releases as early as possible. Daily newspaper deadlines are pretty flexible, but most magazines and newspaper TV supplements have deadlines that may be weeks in advance of your air date. If you are unsure about specific deadlines, call and find out. A press release is a great promotional opportunity since it costs nothing for the print space and can help to generate audience interest and curiosity about your show.

PRINT AND BROADCAST ADVERTISING A television program can be advertised and marketed just as any other product. Print advertising space can be purchased in newspapers, magazines, and even on billboards and posters. You can also produce radio and television commercials and purchase broadcast advertising time on local stations. Buying advertising space and time is expensive, though, and many smaller stations do not have the money necessary to run extensive advertising campaigns. Sometimes you can work out a trade-off where advertising space or time is exchanged for other services. Another possibility is to tie-in a program's promotion with its sponsor's regular advertising efforts.

PROMOTIONAL SPOTS An on-air promotional spot or "promo," as it is commonly called, is a commercial that you produce for your own station. The spot is run in the place of a paid commercial, and it promotes your program. The advantage of a promo spot is that it costs you nothing. However, at most stations a "sustaining" promo will be replaced with a paying commercial if the sales department can sell the air time to a sponsor.

Variations on the promo are voice-over announcements which are run over the ending credits of other shows, and shared-ID graphics which are used during station breaks to identify the station and to promote the show.

Audience Evaluation

In the final analysis, it is the audience that determines whether a program is a success or a failure. The criteria you use in evaluating the audience's response to your show can vary, however, depending on your specific production situation or the program's objectives. The size of the audience is often crucial in broadcast television, but noncommerical public stations are generally less concerned with audience size. Closed-circuit instructional or informational programs usually use entirely different program evaluation criteria from those which broadcast television uses. These programs are produced with a specific audience and with particular objectives in mind. How effective the production was in teaching a process or in conveying information becomes the means for evaluating this type of program's success or failure.

You can learn about an audience's reaction to your show in a number of ways. Some are more formal and objective measures, such as program ratings and postviewing tests or interviews. Other types of audience feedback are more informal such as letters, phone calls, advertiser response, and word of mouth.

PROGRAM RATINGS   Program ratings provide you with an estimate of the size of your viewing audience. Because they are only estimates (it is impossible to hook up every television set to a computer), they are subject to variation and error, particularly when the rating figures for two competing shows are very close.

Ratings can tell you only the size and the composition of your audience. They cannot tell you how much the audience enjoyed your show, whether or not they understood and appreciated what you were trying to communicate, whether they paid much attention, or even if they watched your show at all! The rating simply means that a certain number of television sets were tuned to a particular show at a particular time. (In fact, during a congressional hearing on ratings some years ago, it was learned that a woman with a ratings meter on her set kept the television on all day to entertain her dog while she was at work. The ratings company assumed that whatever shows the set was tuned to were also being watched by some 50,000 other television homes!) With all their faults, however, ratings have proved to be fairly reliable estimates of audience size, and, until a better system comes along, they will remain the industry standard for audience measurement.

Most local and national ratings are conducted and reported by two major organizations: the A.C. Nielsen Company, and the ARBITRON Company. They send regular ratings reports to their paying subscribers, who include local television stations, national networks, advertisers and ad agencies, and independent program producers.

The ratings book reports two important figures for every show: (1) the program rating and (2) the program's share.

RATING   The program rating is a percentage of all television households tuned to a program from among all the television households who potentially could tune in. As an example, say we have a sample of 100 television homes included in our rating survey and out of the 100, some 20 households are tuned to your program. The rating would be 20/100 = 20 percent, or a rating of "20" (the percentage sign is always dropped when reporting a ratings figure).

It may have occurred to you that the size of the rating figure depends not only on the popularity of the show but also on the number of people who are actually watching television at the time. Naturally, you would expect more people to watch television at nine o'clock at night than at nine o'clock in the morning. Regardless of a show's popularity, the rating figure of a mediocre evening program will invariably be higher than the rating figure of a highly successful daytime program. In order to account for this difference among the number of available viewers and to permit us to make comparisons between programs that are broadcast at different times of day, we must use the second important program rating figure, the *share*.

SHARE   Share is short for "share of audience," which is exactly what the second ratings figure refers to. Unlike the program rating, which is a percentage of homes watching from among all those with television sets, the share is calculated as the percentage of those households watching a show from among all homes that are *using television at the time*. (See Figure 18-9.)

Now look at our sample again. There are still the same 20 households tuned to our show, but assume that only 60 percent or "60" households are actually using television (this figure is often referred to as the "HUT"—households using television). To calculate the share, we divide the total HUT by the number of households tuned to a show, or 20/60 = 33 percent or a "33 share" (remember, drop the percentage point). If your show was competing directly against two other programs, the 33 share would be a respectable portion of the total audience. Since all the program "shares" added together for each time period always equal 100, we can use the share to provide an indication of a show's relative popular-

ity against direct competition in its particular time slot.

Overall program ratings and shares are important but they tell only a part of the story. A ratings report also breaks down the audience for a show into its demographics—various age, sex, and income groupings—which are helpful in determining a show's audience appeal, planning its future development and buying advertising time. (See Figure 18-10.)

POSTVIEWING EVALUATION TESTS   It is rarely difficult to learn the size and composition of a closed-circuit educational or industrial program audience since all you need do is count heads to find out who and how many people are watching. What is frequently more important in this situation is to discover how effective a program has been in communicating a skill or in conveying information to the audience. Even some broadcast programs with similar instructional objectives—such as *Sesame Street* or the *Electric Company*—will not find these answers in the program ratings. They must be discovered through postviewing evaluation tests.

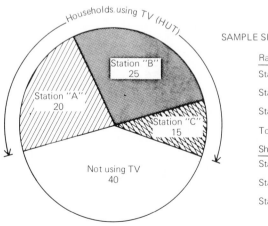

SAMPLE SIZE = 100 Television Homes

Rating

Station "A" $\frac{20}{100}$ = 20

Station "B" $\frac{25}{100}$ = 25

Station "C" $\frac{15}{100}$ = 15

Total HUT =   60

Share

Station "A" $\frac{20}{60}$ = 33

Station "B" $\frac{25}{60}$ = 42

Station "C" $\frac{15}{60}$ = 25

100 (Note: Total shares always equal 100%)

Figure 18-9 Television Ratings.

EVE. SUN.   OCT. 24

W E E K 2

ABC TV

RIPLEY'S BELIEVE IT—NOT	MATT HOUSTON (OP)	ABC SUNDAY NIGHT MOVIE — THE BIG RED ONE (9:00–11:18PM)(OP)

TOTAL AUDIENCE (Households (000) & %)
- RIPLEY'S BELIEVE IT—NOT: 12,660 / 15.2
- MATT HOUSTON: 14,660 / 17.6
- ABC SUNDAY NIGHT MOVIE — THE BIG RED ONE: 22,990 / 27.6

AVERAGE AUDIENCE — rows: Households (000) & % ; SHARE OF AUDIENCE % ; AVG. AUD. BY ¼ HR. %

Program / ¼-hr	HH (000) & %	Share %	Avg ¼ hr %
RIPLEY'S	8,660 / 10.4	16	9.2
	11.3*	19*	11.8
	9.6*	17*	10.0
			10.7
MATT HOUSTON	11,080 / 13.3	21	12.1
	12.6*	20*	13.1
			13.6
	14.1*	22*	14.5
BIG RED ONE	13,910 / 16.7	27	17.2
	16.7	27	16.9
			16.3
	17.1*	25*	17.0
	17.1*	25*	17.1
	17.1*	25*	17.3
	16.0*	27*	15.7

CBS TV

60 MINUTES	ARCHIE BUNKER'S PLACE	GLORIA (OP)	JEFFERSONS	ONE DAY AT A TIME	TRAPPER JOHN, M.D.

TOTAL AUDIENCE (Households (000) & %)
- 60 MINUTES: 25,490 / 30.6
- ARCHIE BUNKER'S PLACE: 18,330 / 22.0
- GLORIA: 16,740 / 20.1
- JEFFERSONS: 20,240 / 24.3
- ONE DAY AT A TIME: 20,240 / 24.3
- TRAPPER JOHN, M.D.: 18,660 / 22.4

AVERAGE AUDIENCE

Program / ¼-hr	HH (000) & %	Share %	Avg ¼ hr %
60 MINUTES	20,580 / 24.7	42	22.8
	24.0*	42*	25.2
	25.3*	42*	24.9
			25.7
ARCHIE BUNKER'S	16,330 / 19.6	31	19.4
			19.8
GLORIA	15,580 / 18.7	29	18.6
			18.8
JEFFERSONS	18,080 / 21.7	32	20.6
	21.7	33	21.9
			22.8
	22.1	33	22.3
TRAPPER JOHN	18,410 / 22.1	33	21.9
	18.7	31	18.9
	18.7*	30*	18.6
	18.7*	32*	18.5

NBC TV

VOYAGERS	CHIPS (OP)	NBC SUNDAY NIGHT MOVIE — LITTLE GLORIA:HAPPY AT LAST, PART 1 (OP)

TOTAL AUDIENCE (Households (000) & %)
- VOYAGERS: 11,250 / 13.5
- CHIPS: 16,910 / 20.3
- NBC SUNDAY NIGHT MOVIE — LITTLE GLORIA:HAPPY AT LAST, PART 1: 23,740 / 28.5

AVERAGE AUDIENCE

Program / ¼-hr	HH (000) & %	Share %	Avg ¼ hr %
VOYAGERS	7,750 / 9.3	16	7.4
	7.8*	14*	8.2
	10.8*	18*	11.6
			10.1
CHIPS	13,740 / 16.5	26	14.8
	15.5*	24*	16.1
			17.5
	17.6*	27*	17.6
NBC MOVIE	16,410 / 19.7	31	18.9
	19.7	29	20.0
			20.6
	20.0*	30*	20.2
	20.4*	33*	20.6
	19.3*	29*	19.6
	19.2*	33*	18.6
			19.8

TV HOUSEHOLDS USING TV (See Def. 1)

	WK. 1	WK. 2
	60.5	55.2
	61.3	58.2
	63.4	59.7
	63.9	60.9
	63.5	62.5
	64.7	64.2
	65.9	65.1
	67.5	65.9
	67.7	67.1
	69.3	68.0
	68.1	67.7
	68.0	67.3
	69.3	68.0
	67.7	67.1
	67.5	65.9
	64.4	63.4
	62.6	61.4
	60.5	59.6
	58.7	57.9

For explanation of symbols, See page A.

U.S. TV Households: 83,300,000

Figure 18-10 **National and Local Television Ratings.**
The national ratings (above) show the ratings and shares for the three commercial networks. Demographic data is not shown on the page illustrated. Local market ratings (opposite page) show ratings and shares for a four-week "sweep" period, with viewers broken out by demographics. The term "DMA" refers to "Designated Market Area," which is the area where the station's signals are received by viewers. (Courtesy: A. C. Nielsen Co.)

Time and program · Ratings · Previous ratings · Demographic breakouts

WK1 2/04–2/10 WK2 2/11–2/17 WK3 2/18–2/24 WK4 2/25–3/03

DMA HH

Time	Station	Program	WK1	WK2	WK3	WK4	RTG	SHR	NOV '81	MAY '81	FEB '81
		R.S.E. THRESHOLDS 25+%	2	2	2	2	1				
		(1 S.E.) 4 WK AVG 50+%	1	1	1	1	LT				
4.30PM	WABC	PEOPL COURT MF	8	6	9	7	7	20	19	26	20
	WCBS	BARNABY JONES1	4	6	6	4	5	13	14	16	19
	WNBC	MARY T MOORE	5	5	5	5	5	13	15X	16	14
	WNET	SESAME STREET	4	3	2	3	3	8	6X	2	4
	WNEW	WHATS HAPPNING	6	6	5	5	5	15	10	15	15
	WNJU	AVG. ALL WKS	<<				<<				
		ROMANCE					<<				
		SHW DE MENUDO		<<	1	<<	<<				
	WOR	4 OCLOCK MOV	6	4	5	6	5	14	15X	10	10
	WPIX	AVG. ALL WKS					4	12	16	10	16
		LITTLE HOUSE	3				3	7			
		GOOD TIMES		6	4	5	5	13			
	WXTV	JUEGOS-DESTINO	1	<<	<<	<<	<<			3	2
		HUT/PUT/TOTALS *	36	37	37	39	37		35	32	39
5.00PM	WABC	5.00 EYEWT NWS	7	6	7	6	7	16	15	21	18
	WCBS	CH 2 NWS-5	5	6	5	5	5	13	13	16	14
	WNBC	LIVE AT 5	6	8	5	6	6	15	19X	21	19
	WNET	MISTER ROGERS	4	3	3	2	3	8	7X	3	5
	WNEW	BIONIC WOMAN	6	7	5	5	6	14	12	11	17
	WNJU	FIESTA	<<	<<	1	1	<<				
	WOR	4 OCLOCK MOV	6	3	5	7	5	13	13X	10	10
	WPIX	AVG. ALL WKS					6	14	14	15	17
		LITTLE HOUSE	5				5	13			
		LITL HOUSE 5PM		6	6	7	6	15			
	WXTV	CRBNA-AMBROSIO	<<	<<	<<	<<	<<		X	2	
		HUT/PUT/TOTALS *	42	43	39	42	41		39	32	41
5.15PM	WABC	5.00 EYEWT NWS	7	5	8	5	6	14	15	20	18
	WCBS	CH 2 NWS-5	4	7	5	5	5	13	13	15	13
	WNBC	LIVE AT 5	7	8	6	7	7	16	19X	22	17
	WNET	MISTER ROGERS	4	3	4	2	3	7	7X	2	4
	WNEW	BIONIC WOMAN	7	8	5	7	7	16	11		9	16
	WNJU	FIESTA	<<	<<	1	1	1				
	WOR	4 OCLOCK MOV	6	3	6	7	5	13	12X	10	9
	WPIX	AVG. ALL WKS					6	14	18	15	19
		LITTLE HOUSE	5				5	12			
		LITL HOUSE 5PM		6	5	7	6	14			
	WXTV	CRBNA-AMBROSIO	<<	<<	<<	<<	<<		X	2	
		HUT/PUT/TOTALS *	45	42	41	44	43		42	34	44
5.30PM	WABC	5.00 EYEWT NWS	6	4	7	5	6	12	14	19	18
	WCBS	CH 2 NWS-5	5	7	6	5	6	13	14	15	14
	WNBC	LIVE AT 5	7	8	7	8	7	16	20X	23	18
	WNET	ELECTRC COMPNY	2	2	3	2	2	5	5X	2	3
	WNEW	BIONIC WOMAN	7	9	6	6	7	16	13	10	17
	WNJU	FIESTA	<<	<<	1	1	1				
	WOR	4 OCLOCK MOV	7	4	6	7	6	13	11X	10	9
	WPIX	AVG. ALL WKS					6	14	16	16	17
		H DAYS AGN 1	5				5	11			
		LITL HOUSE 5PM		7	5	8	7	15			
	WXTV	QUIRME SIEMPRE	<<	<<	<<	<<	<<		X	2	1
		HUT/PUT/TOTALS *	44	46	42	45	44		44	35	46

DMA RATINGS

Program	P 2+	P 18+	W 12-34	W 18+	W 18-34	W 18-49	W 25-49	W 25-54	W WKG	FEM 12-24	PER 12-24	M 18+	M 18-34	M 18-49	M 25-49	M 25-54	TNS 12-17	CH 2-11	CH 6-11
THRESHOLDS 25+%	1	1	1	1	3	2	2	2		2	3	1	3	2	2	2	4	5	6
(1 S.E.) 50+%	LT	LT	LT	LT	1	LT	1	LT	1	1	1	LT	1	LT	1	LT	1	1	1
PEOPL COURT MF	5	5	6	6	7	7	6	6	4	10	7	3	2	2	2	1	9	3	4
BARNABY JONES1	2	3	1	3	1	2	1	1	1		3	2	1	1	1	1	1	1	
MARY T MOORE	3	3	2	4	3	3	3	2	4	2	2	2	2	2	2	2		1	1
SESAME STREET	2			1	1	1	1											10	3
WHATS HAPPNING	3	1	4	2	3	3	2	1		6	6	1	1	1	1	1	10	10	13
4 OCLOCK MOV	3	3	1	3	1	1	1	2	2		1	4	2	2	2	3	1	1	1
WPIX AVG. ALL WKS	2	3	2	3	2	2	2	2		5	5	1	2	2	1	1	6	10	10
LITTLE HOUSE	2	1	2	1	3	2	2			1	2	1					2	7	
GOOD TIMES	4	2	4	2	3	2	2	3	1		6	2	3	2	1	1	8	12	11
HUT/PUT/TOTALS *	22	18	20	21	20	20	18	18	12	30	26	14	11	11	9	10	33	38	35
5.00 EYEWT NWS	5	5	5	7	6	6	5	6	4	7	6	4	3	2	2	2	4	1	1
CH 2 NWS-5	3	3	1	4	1	2	2	2	1	1	1	3	1	1	1	1	1		
LIVE AT 5	4	4	2	5	2	3	3	4	3	2	2	3	2	2	3	2	2		1
MISTER ROGERS	2			1	1	1	1	1										11	4
BIONIC WOMAN	4	2	4	2	3	3	3	2		6	6	2	2	3	2	2	9	8	10
4 OCLOCK MOV	3	3	2	3	1	2	2	2	1		2	4	3	2	3	3	2		
WPIX AVG. ALL WKS	4	2	4	3	6	5	4	4	2	7	6	1	2	2	1	1	6	12	11
LITTLE HOUSE	3	2	4	3	5	5	4				5						5	13	
LITL HOUSE 5PM	4	2	4	3	6	5	4	3	3	6	4	1	2	1	1	1	6	13	11
HUT/PUT/TOTALS *	25	23	19	26	21	23	22	22	16	26	23	19	12	13	13	14	27	36	31
5.00 EYEWT NWS	4	5	4	6	5	6	6	6	4	5	5	4	3	2	2	2	4	1	1
CH 2 NWS-5	3	3	1	4	1	1	1	1		1	1	3	1	1	1	1	1		
LIVE AT 5	4	5	2	6	2	3	4	4	3	2	2	4	2	2	3	3	2		
MISTER ROGERS	2			1	1	1	1	1										11	4
BIONIC WOMAN	4	3	5	3	3	3	3	3	3	7	6	2	3	3	3	3	11	9	11
FIESTA						1	1	1	1										
4 OCLOCK MOV	3	3	2	3	1	2	2	2	1		2	4	3	3	3	3	2		
WPIX AVG. ALL WKS	4	2	4	3	4	4	4	4	2	7	5	1	2	1	1	1	5	12	11
LITTLE HOUSE	3	2	4	3	8	5	5	4		6	4	1					5	10	
LITL HOUSE 5PM	4	2	4	3	6	5	4	3	3	8	6	1	2	2	2	2	6	13	12
HUT/PUT/TOTALS *	26	23	20	27	21	23	23	23	17	26	24	19	13	14	13	14	29	37	32
5.00 EYEWT NWS	5	4	4	6	5	5	5	6	4	4	4	3	2	2	2		4	1	1
CH 2 NWS-5	3	4	1	4	1	2	2	2	1			3	1	1	1	1	1		
LIVE AT 5	4	5	2	6	2	3	4	4	3	2	2	4	2	2	3	3	2		1
ELECTRC COMPNY	1			1	1	1	1											9	5
BIONIC WOMAN	5	3	5	3	3	3	3	3	3	7	7	3	2	3	3	3	11	9	10
FIESTA						1	1	1	1										1
4 OCLOCK MOV	3	4	2	3	1	2	2	2	1		2	4	3	2	3	3	3		
WPIX AVG. ALL WKS	4	2	5	3	6	4	3	3		8	6	1	2	2	2	1	7	12	12
H DAYS AGN 1	3	2	4	2	5	4	3	3		6	4	1	1	1	1	1	6	8	
LITL HOUSE 5PM	5	3	5	3	6	5	4	3	3	8	6	2	2	2	2	2	7	14	13
HUT/PUT/TOTALS *	26	23	20	26	20	22	22	22	18	25	24	20	13	14	14	15	31	36	33

You will probably need the help of an instructional development specialist in the design and administration of postviewing tests or questionnaires. You must first outline the basic objectives —in other words, what it is you want the audience to learn or take away from the show they have just seen. The specialist will develop a test instrument or questionnaire designed to tap these concepts and to provide you with data that can be used to evaluate how effective your show was in meeting its objectives.

Sometimes the best way to evaluate the effectiveness of a training production is to analyze subsequent audience performance. Do sales figures show a rising trend after the sales staff has seen a motivational television production? Do pre-med college students' test scores show a significant improvement after viewing a series of supplemental programs on biology? Has the number of work-related accidents decreased after employees saw a tape on safety procedures? A careful study of performance data and test scores can not only help you find out the usefulness of the program as a whole, but also identify specific segments which may have proved more or less effective than others.

INFORMAL AUDIENCE FEEDBACK   Phone calls, letters, and the audience response to advertisers' commercials constitute more informal audience feedback. All stations receive both letters and phone calls in which viewers register their approval or their dislike for particular programs. Of course, the trend of letters and calls can sometimes be biased by a highly vocal and visible minority, but many writers and callers are sincere both in their criticism and in their praise. Paying attention to various calls and letters and noting the overall trend of such messages can give you some interesting perspectives on viewer reactions to your show.

Another informal way to gauge the impact of a program on an audience is through audience response to advertisers' messages. The number of sales or the response to offers that were advertised on commercials can be carefully followed. Audience response to an advertiser's message often has nothing to do with the program itself, but a sizable reaction or a successful sales campaign suggests a substantial and attentive audience.

SUMMARY

The producer is the individual who is ultimately responsible for every element—both creative and technical—that goes into a television production. The producer must be familiar with all aspects of television production in order to supervise the production team and to make the maximum use of every production capability. A successful producer combines the creativity of an artist with the organizational efficiency of a business executive. The organizational ability is essential in order to realize the creative idea in the highly complex television medium.

The producer is responsible for the development of the program idea. Whether the concept was originally assigned to the producer or developed to fill an existing programming need, the producer must take the initial concept through a series of program development steps which include (1) analyzing audience needs and tastes: (2) researching the program idea for content and production feasibility; (3) determining the production mode— whether the show is produced in studio or on location, live, on tape, or on film; (4) developing the program outline or treatment; and (5) preparing the program budget.

The producer is one of the few production team members who is directly involved in each of the four production stages. The preproduction stage is the busiest and most important for the producer since this is when the program concept is developed and the show's approach and production are planned and organized. The script serves as the foundation for every show, and it is usually one of the producer's earliest concerns. The script can be written either by the producer or by a writer working closely with the producer.

After all necessary changes and revisions are made, the final draft becomes the shooting script. Another crucial preproduction concern for the producer is hiring the director, who must ultimately interpret and execute the producer's concept. Both the producer and director work closely on script revisions, casting, and production approaches. To help the producer organize the various elements which must go into the show, the production notebook is put together. The notebook is a collection of all necessary production information, including (1) the shooting script, (2) master production schedule, (3) cast/crew listing, (4) rehearsal schedules, (5) FAX sheets, (6) operational schedules, (7) call sheets, and (8) shooting schedules or rundown sheets.

During the rehearsals, the producer's most important function is to serve as a substitute audience, watching and critically analyzing the show as it appears on the program monitor. During rehearsals, the producer confers with the key production team members and hands out notes concerning all production elements that need attention. Changes must be made early enough in rehearsal so they do not completely upset the performers or production crew. The dress rehearsal—a replica of the show as it will appear on the air—should always be run before actual production.

Once into the production stage, the producer turns the show over to the director and production crew. If the producer has done his or her homework during the earlier stages of production, there really is not much for him or her to do since the production should look and sound as the producer envisioned it early in preproduction.

The producer's postproduction activities include approving the final, edited, videotape version of the program; working on promotion and publicity to attract the largest audience possible; and evaluating the audience's response to the show. For broadcast programs, a widely used measure of audience reaction is the television ratings. Other objective measures of audience response are postviewing evaluation tests and observation of audience performance on a job or task which the show addressed. More informal, but useful, audience feedback can include phone calls, letters, and response to ads which appeared on the show.

CHAPTER 19

DIRECTING

Creating the Videospace

Television directing is a unique art requiring a multitalented individual who combines technical agility with artistic ability. Unlike a painter who works directly with his or her tools, using brush and paint to transform his or her ideas into art, the television director's thoughts and visions must first be interpreted through electronic equipment that is operated by a production crew before they can appear in finished form. The way in which a television director must work is like asking Picasso to create a painting without ever touching a paintbrush. He could only instruct a number of assistants where to apply paint to canvas, which brush to use, how thick or thin to make a line or a curve, where a shape begins, and where it should end. The nature of television requires the director to create a program by supervising a crew and the operation of complex equipment while never losing sight of the fact that all this hardware and personnel exist only as a means to an end: the communication of ideas and messages to a viewing audience.

It is the director's job to transform the producer's program concept into a show—in effect, by creating both the videospace and the audiospace. This is done through the director's development and selection of shots, the combination of shots into sequences, sequences into a show, and the use of sound to complement and enhance the visuals. All this first takes place only in the director's mind. It must then be translated to the television screen, using the array of equipment and production techniques we have discussed throughout the book.

In this chapter, we will consider some of the important creative and aesthic elements facing the director. In the next chapter, we will see how the director supervises the production team to actually create the look and sound of the show.

CREATING THE VIDEOSPACE

For all intents and purposes, a television show begins with a blank screen. How you fill the screen—create the videospace—determines how the viewer perceives your program and the message you are trying to convey. Somerset Maugham, the author, once told an admirer. "Madame, all the words I use in my stories can be found in that dictionary. It's just a matter of arranging them into the right sentences."

The television director faces a similar situation. Just as every writer begins with the same set of words, all television directors start with essentially the same tools and techniques. There are only so many possible variations of camera shots or angles, only a limited number of lighting methods, just a few standard ways to change camera shots, only so many microphones and audio sources available, and so on. What does separate one director from another is how each uses the production elements which are available—how they are selected, combined, and integrated into the show. That is why you can give the identical program to two different directors and you are likely to get two very different shows.

Directing is a highly subjective art, which makes giving rules and guidelines difficult. But there are three important concepts which are helpful in understanding how the director goes about creating the videospace: (1) choice, (2) viewpoint, and (3) visualization.

The Director and Production Choices

You might think of the director as someone faced with an almost unlimited number of choices. A good director is able to make the right production choices consistently, in the right order, at the right time. In effect, a television program is the sum total of the director's choices. Of course, just being able to recognize the choices open to you and knowing how to take advantage of them are two fundamental aspects of the art of directing.

There is a story about Mike Nichols—the highly acclaimed stage and film director—that illustrates the idea of a director faced with unlimited choices. Nichols had had a very successful stage career when he was assigned to direct his first film, *Who's Afraid of Virginia Woolf*? When he arrived at the studio for the first day of shooting, the first scene involved a simple matter of an actor coming in a door. As the story goes, the cinematographer asked Nichols how he wanted the performer to enter so the camera could be positioned. "Just have him come through the door," said Nichols. "No, no—you don't understand what I mean," said the cameraman. "Do you want to see the actor from a high angle, a low angle, in a wide shot, in a close-up? Should we pan with him or let him walk out of the frame? Does the camera dolly in or track along. . . ." Coming directly from the stage, where things were a lot simpler, Mike Nichols was surprised to confront the many choices and decisions facing the film (or television) director.

Of course stage directors also face many important decisions, but the television director's choices are particularly important because they determine what the viewer actually sees and hears. Needless to say, as you become more proficient as a director, the choices become more automatic based on your experience, skill, and a sixth sense which tells you which choices will work and which will not. But as you begin directing, try to be very conscious of all the choices which confront you and the decisions that you make. Unless you are aware of the possibilities, you may forfeit some opportunities to make the program even more effective.

Viewpoint

The many production decisions made by the director are obviously not determined by chance. The decisions are based on an overall approach, style, and concept which we define as the director's *viewpoint*. The director's viewpoint is shaped by the various creative and emotional responses the director has to the program's material. As long as the director views himself or herself simply as a transmitter of events, there is no established viewpoint to the show. The process of directing must involve a transfer of the director's emotional, intellectual, and creative values to the production. Without a viewpoint, which is established and firmly fixed in your mind at the outset, it will be difficult for you to produce a unified program which realizes the production's overall objectives. A television show can be viewed as a series of many decisions which are strung together to create the program. It is easy to stray from the show's overall objectives unless you have a definite viewpoint, which is constantly used as a reference, or guide, throughout planning, rehearsal, and production.

For example, ABC Sports claims its coverage of sporting events is designed to be "close-up and personal." It is a conscious effort by the producers and directors of ABC to show an event from a more intimate perspective than simply to present detached reporting of the action on the playing field. Naturally, the game is of primary importance, but the personal point of view, as conveyed by various camera shots of losers, as well as winners, and the many activities which surround the basic action, gives their productions a special look and feel, which conveys the network's established viewpoint.

Do not think that viewpoint applies only to entertainment programs. An instructional show, for example, offers a variety of potential viewpoints, and it is up to the director to determine

which approach to use. For instance, does the director intend to cover the demonstration from an objective point of view, or is the camera used to put the viewer directly into the action by employing subjective camera techniques? When directing an interview program, should the director shoot the subject of the interview dispassionately and objectively, or should he or she focus in tight on subtle mannerisms and nervous facial tics or hand gestures? There are no rules or guidelines to follow in selecting a program's viewpoint. Obviously, the director's viewpoint must complement and enhance the program's objectives, but precisely how to do this involves the director's own taste, style, and creativity.

Visualization

The television director works, in large part, with pictures and images. Not only does the director reproduce reality as it exists, he or she also manipulates the reproduction of "reality" to create new images through the use of camera angles, lenses, and special effects. Before you can do this, however, you must be able to visualize the program in your mind's eye. That is not just an overall "image" of the show; it is every scene, every camera shot, every shot transition. In effect, you run the television show in your mind before you ever walk into the studio to produce it.

Learning to visualize takes some practice and concentration but, fortunately, no special equipment or supplies. Just concentrate on seeing (and hearing) the show that you are planning to direct, and try to picture it in your mind's eye. Before long you will find it becomes easier and easier to actually see the show in your head. Until you can visualize the show in your imagination, you cannot begin to plan specific camera shots and performer blocking on paper.

CAMERA SHOTS

One of the director's primary concerns is to develop the camera shots which convey the essence of what is happening to the viewing audience. Remember the audience's only frame of reference is what they see in the videospace and hear through the television speaker. What you show and how you show it are crucial factors in communicating to the audience.

Think about a stage play for a moment. While there are lots of tricks that stage directors use to focus your attention on what they feel is the most important aspect of the show at any particular time, there is nothing to stop your eyes from wandering from the actor who is speaking, for instance, to the actor who is listening or to look across the background players or perhaps to concentrate on the stage furniture and props.

This freedom is not possible in television. Unlike the theatrical audience, television viewers can see only what the director wants them to see. You decide what the audience can look at and, equally important, what the audience cannot look at. A good way to think about this is to show the audience (1) what they *want* to see, (2) what they *need* to see, (3) when they *want and need to see it.*

The audience knows—if only unconsciously—what it *wants* to see, but it is the director's job to decide what the audience needs to see and when they should see it. Take a scene with a young woman as she enters her apartment late at night. She walks into the door, flips on the switch, and her eyes widen in fear and astonishment. At that point the audience wants to see what she sees and wants to see it right away. But the director might decide that the audience need not see it immediately. Perhaps the show's suspense would be heightened if there were a pause between showing the girl's face and showing the cause of her astonishment. In a detective show, the element of mystery is crucial to our enjoyment of the story. There are other possible reasons why the director might not wish to show the audience what it wants to see at that particular time: to emphasize the girl, to fool the audience, and so on.

Quite often, of course, what the audience wants to see and needs to see are identical. Yet the director must also choose how to show it to the viewer. Should he or she use a wide shot or a close-up, a subjective angle, or a dolly shot? There are no rules here, because every situation presents its own set of possibilities or choices. You should remember how important it is to consider carefully what the audience sees, how they see it, and when they see it onscreen as you develop your camera shots.

One way to become more sensitive to camera shots is to start watching television with the sound turned off. Without the audio, you can concentrate on the visual presentation and watch how the director uses camera shots, lenses, angles, and movement to create the videospace and communicate to the audience.

The Shot

The shot is the basic visual element in a television production. A shot is a single, continuous image taken by one camera. It can be as short as one-thirtieth of a second (the length of a single video frame) or as long as the length of the entire program. We almost never use a single shot for an entire program. Rather, we assemble a number of different shots together into a sequence to show the viewer the action from the best possible angle, distance, and viewpoint.

Developing Camera Shots

In every television program there are certain elements which are more important than others and which require the viewer's fullest attention in order to understand and enjoy the program. It is the director's job to develop camera shots in such a way that the audience sees not only what they want and need to see, but also to emphasize and punctuate those elements in the videospace which require audience attention.

In order to do this, the director works with three related visual elements: (1) the size and content of the shot, (2) the camera angle, and (3) the movement of the camera or the subject in the shot.

SIZE AND CONTENT What you decide to include in the shot and what you decide to exclude are fundamental decisions which every director makes throughout a program. Visual emphasis is directly related to the size of the subject in the shot, since a close-up, by definition, excludes all other picture elements except the principal sub-

Figure 19-1
The subject seen in a wide shot must share the viewer's attention with the surrounding background area. The subject's dominance is greatly increased when shot in close-up since she is not only closer, but very little of the background competes for the viewer's attention.

ject from the viewer's attention. As the shot widens out to include less of the principal subject and more of the background area, the principal subject becomes physically smaller and shares the videospace with other elements in the shot. The viewer's attention can wander to a number of objects within the shot. (See Figure 19-1.)

CAMERA ANGLE You already know that certain camera angles convey specific impressions to the viewer. A low-angle shot usually suggests a powerful and dominant subject whereas a high-angle shot makes a subject appear less powerful and physically smaller. A subjective shot is a much different view of a scene from what we are normally used to seeing and tends to involve the viewer. Think of the difference between showing an auto race with the cameras positioned in the grandstand (objective shot) and strapping the camera onto a racing car's fender (subjective shot) to show the viewer what the drivers see and feel as they speed around the track.

CAMERA MOVEMENT Camera movement can help to create dominance or to establish visual emphasis in a shot. Dollying or zooming in to a subject is an obvious way to focus audience attention. The opposite effect is achieved when

you dolly or zoom out. Other camera moves, such as trucks, arcs, or on-air pedestaling are additional ways to focus audience attention where the director wants it. The speed of the movement is yet another method to vary audience attention—faster speeds create more attention than slower, more restrained movement.

SUBJECT MOVEMENT We tend to direct our attention toward a moving object rather than one standing still. Veteran actors are well aware of this and sometimes use such distracting movements to steal a scene by drawing the audience attention toward them and away from others on stage. The director can use the effects of subject movement to advantage, however, to focus audience attention.

ADDITIONAL SHOT ELEMENTS The use of lenses and optical effects, as well as the use of "plastics"—the varied design elements such as set, lighting, costumes, and makeup—are additional ways to build a shot and direct viewer attention.

LENSES AND OPTICAL EFFECTS You already know that lenses of varying focal lengths produce different perceptual effects. Subject size and depth perspective are exaggerated with a wide-angle lens and compressed by a telephoto or long focal length lens. Distortion and depth manipulation are some effective ways to attract audience attention. Focus also plays an important part in directing viewer attention in the videospace. If you have two subjects, each on a different depth plane, a shallow depth of field allows you to place one subject in focus, the other out of focus. Viewer attention is naturally aimed toward the in-focus portion of the screen. Similarly, throwing the background out of focus by using a longer focal length lens will make foreground subjects appear more prominent in the frame.

PLASTICS   Set design, lighting, costumes, make-up, graphics, and other similar art design considerations—often called "plastics"—are additional elements to use in creating the videospace and directing audience attention. A performer in a white suit will stand out prominently, particularly if positioned against a dark background set. Set areas which are brightly illuminated will draw viewer attention away from darker areas. Using no set—as in a cameo or a limbo approach—highlights the foreground subject even more since nothing in the background competes for viewer attention. A very busy and complicated set, on the other hand, may distract the viewer's attention away from foreground subjects.

EDITING THE SHOW

After developing the camera shots, the director must decide how to join individual shots to make a program. This assembly is called "editing" or "cutting." The term "cutting" comes originally from motion picture film, where the film editor physically cut and spliced together individual pieces of film with different shots. In television, editing is accomplished in a number of ways. The most common editing method still requires the director to watch a number of camera monitors simultaneously and to edit "on the fly," making instantaneous decisions as to which camera shot to put on the air at which time. Electronic videotape editing is another way to edit a show. Instead of making split-second decisions, the director using tape editing has more latitude to experiment

with the most effective juxtaposition of shots. Regardless of the editing approach used, however, the aesthetic considerations are identical.

Editing involves working with three related elements: (1) the juxtaposition of shots, (2) the timing of shots, and (3) the transitional device used to go from shot to shot.

Juxtaposition of Shots

In the early 1920s, two Russian filmmakers and theorists, V. I. Pudovkin and Lev Kuleshov, ran an interesting experiment. They filmed a close-up shot of a male actor with a blank expression on his face. Next, they shot three different scenes: (1) a bowl of soup, (2) a small child playing, and (3) an old lady lying in a coffin. Each of the three images was edited next to a shot of the man's close-up, so that they had the following three segments: (1) close-up—soup—close-up; (2) close-up—child—close-up; and (3) close-up—coffin—close-up.

Different groups of people were asked to watch one of the three scenes and then were questioned about what they had seen. Those watching the first version with the soup thought the actor looked absolutely famished. The audience who saw the second version commented on the obvious love the man had for his daughter. The group that saw the third series of shots remarked how grief-stricken the man appeared.

What the Russian filmmakers were trying to demonstrate was how the juxtaposition of shots—the order and sequence into which they are assembled—affects an audience's perception of what it sees. In every case, the audience psychologically related the two completely different shots together, creating an association or relationship between them. This is a fundamental concept in editing since the shots you select and the sequence, or order, in which you use them has an impact on how the viewers perceive your message.

Juxtaposition of shots is often used to make the audience think they saw something they really did not. The terrifying stabbing sequence in the shower from Hitchcock's classic film, *Psycho* is a perfect example. Although audiences swear it is one of the most violent and bloodiest scenes they have ever seen, a careful inspection of the sequence in a film viewer reveals that there is not a single frame showing the knife actually penetrating the murder victim. But Hitchcock shot and edited the sequence so cleverly that the audience thought it had seen something absolutely horrible.

Used properly, the juxtaposition of shots plays a major creative role in a show's direction. Given the same raw footage, three different directors or editors will create three different sequences, with variations possible on the emphasis of the scene, the characterization, the dominant elements or events, even where the audience's sympathy or emotions lie. But juxtaposition plays an important role in nonfictional programs as well. News and documentary programs are particularly vulnerable to manipulation through the sequencing of shots. Events can be intentionally or unintentionally distorted by the addition or elimination of something as seemingly innocent as a reaction shot at a strategic point in the show.

Timing the Shots

The director must determine not only what to show and in what sequence, but must also consider the timing of each shot and how quickly or slowly to switch from one shot to the next. The timing, or rhythm, of the shots is an important element in the program's overall "pacing." *Pacing* is the audience's perception of the speed of the show; a psychological and emotional impression. Of course editing alone is only a part of a program's pacing, but the timing of the shots, along with the performer's delivery, determines the audience's perception of screen time.

Naturally the shooting rhythm you establish ought to reflect the program's content and the director's objectives. Watching a singer perform a moody, romantic number, we would expect the timing of the shots to be fairly slow to match the overall mood established by the song and the performance. On the other hand, a documentary about a busy factory might call for a rapid cutting tempo from shot to shot to convey the feeling of great activity and movement.

The timing of the shots depends on a number of interrelated factors, including the content of each shot, the complexity of the shot's image, and the context of the shot within the overall sequence. A shot containing essential content or a shot including a very complicated image should remain on screen longer than a simpler shot or one without important content. The audience's familiarity with a shot's content is also significant since, if we are already familiar with what is included in a particular shot because we have seen it before, the director can usually cut more rapidly without disorienting the viewer. A reaction shot of an interviewer listening to a guest's remarks can usually be made quite rapidly since we are familiar with the host, and this shot it not terribly important at the time. On the other hand, a close-up of a complex demonstration in an instructional program might well require a much longer time on screen before the director cuts to the next shot.

The timing of the shots is a highly relative concept since the rhythm of cutting and the transitions used in prior sequences can play as important a role as the absolute speed of the cuts. A show with relatively little cutting can be speeded up appreciably with a moderately paced cutting rhythm. A fast-action chase sequence, however, which involves many short cuts, would require much faster cutting and shorter shots in order to convey a similar increase in cutting tempo. The director should always be concerned with the cutting rhythm and with pacing, even in such simple programs as an interview or panel show, demonstration program, or instructional presentation. Try to avoid settling into a predictable shot routine which can diminish the viewer's interest.

Transitional Devices

The particular device we use to switch from one camera to another depends on the production circumstances and the particular impression that the director wishes to convey to the audience.

There are four basic transitions: (1) the cut, (2) the fade, (3) the dissolve, and (4) the wipe.

CUT   A cut is an instantaneous change from one shot to another. You might think of a cut as the shortest distance between two shots. A cut from one shot to another is similar to what our eyes do as they rapidly focus on various parts of our surroundings. But the cut can also radically change time and place in the videospace and, if used improperly, can disorient the viewer.

Used correctly, the cut is the least obvious transition because it occurs so quickly and appears so natural. The speed of a cut is fixed and constant and, unlike other transitional devices, it cannot be varied.

MOTIVATING THE CUT   A cut should be motivated by some element in the show. The motivation might be action, the beat of the music, dialogue, or some other apparent reason to change camera shots.

CUTTING ON ACTION   Except in the rarest of circumstances a cut should be as unobtrusive as possible. Cutting on the action is one way to accomplish this since the movement will usually cover the transition from one shot to another. If, for instance, you had a wide shot on Camera 1 of someone entering a room and a close-up on Camera 2 of the performer sitting down, the cut from one camera to another should occur during the action as the subject makes the move. Cutting before the action occurs or after the action is completed is not as fluid and usually does not look "right" on screen. (See Figure 19-2.)

CUTTING ON MUSIC   Music is frequently used to establish the timing of the cuts. Cutting on the beat has been found to be an effective and

Figure 19-2   Cutting on Action
A cut is least noticeable when the transition between shots is made *during* the action. In the illustration, the subject reaches for the telephone in a wide shot, and the director cuts to the close-up shot on another camera during her movement.

aesthetically pleasing way of timing the cuts from one camera shot to another.

CUTTING ON DIALOGUE A very common motivation for cutting is to use spoken dialogue as a cue for shot transitions. It is best to cut from one speaker to another at the end of one person's statement or thought and just before another person begins speaking. This applies equally to a spontaneous interview show as to a prepared dramatic scene or on-camera speech. There are exceptions, though, particularly when you want to cut to a reaction shot to show another individual's response to what is being said. In this case, the reaction itself is the major motivation for making the cut. You should try to establish a cutting pattern which complements the on-camera dialogue. Cutting too frequently may create a choppy impression, which inhibits the viewer's ability to follow the discussion. At the other extreme, remaining on one shot for too long a time can be visually boring.

ADDITIONAL USES FOR CUTTING There are some special situations where cutting is used to achieve a special visual effect. One is a *montage* —a rapid succession of shots in sequence. The dramatic impact of a montage is derived from the total effect of all the shots as a whole and not from any single shot or cut. Because a montage is traditionally a series of short, rapid-fire cuts, it is usually produced on film or videotape, so the many individual shots can be edited together. Using a montage you can race through the 200 years of our nation's history in a few minutes or quickly establish a setting, emotion, or background for a program. Montages are frequently used for opening and closing program titles and to establish a situation or mood in a very short period of time.

Cutting between two or more shots can also provide the illusion of movement or *animation*. You can prove this to yourself by setting up two camera shots of the same subject. Have Camera 1 show the subject in a long shot positioned on the left side of the screen and Camera 2 show the same subject on the right side of the screen. By cutting rapidly between the two cameras you create the illusion of movement as the subject jumps back and forth across the screen.

FADE In a fade the picture either turns from black to an image (a *fade-in*) or from an image to black (a *fade-out*). Unlike the cut, the fade is an obvious transitional device, which punctuates a program segment much as a period ends a sentence. The fade is often thought of as being similar to the curtain in the theater that is used to separate each act. The fade can also be used to separate different program elements such as the show material from the commercials. Fades are often used in drama to indicate a major change in time or space.

The speed of the fade is variable—it can be almost as rapid as a cut or as slow and deliberate as you wish. One way to use a slow fade is to give the audience time to reflect on what they have just seen. For example, if you just showed a particularly tragic news story, a slow fade to black gives the audience a chance to think about what has been shown. The same effect can be quite dramatic when used appropriately within a play. In a particularly emotional episode of *All in the Family*, Archie Bunker and his family watch from their front door as an acquaintance's car is exploded by a terrorist bomb. The impact of the scene was made even more devastating through the director's painfully slow fade to black.

DISSOLVE A dissolve is a simultaneous fade-in on one shot and fade-out on another. The effect on screen is a gradual blending of one shot into the other. If the dissolve is stopped midway, the effect produced on screen is a superimposition, or *super*. Since a dissolve is always a super during part of the transition, it is important to compose both shots so they will complement each other.

The speed of a dissolve is variable and can be produced either rapidly or slowly. The dissolve is frequently used as a less abrupt video transition than a cut. If you were directing a slow ballet, for example, cutting from camera to camera during the fluid movements of the dancer would appear too harsh and disturbing. Dissolves, on the other hand, permit the transitions to flow smoothly, suiting the mood more appropriately than a series of cuts. In a dramatic show dissolves are sometimes used to suggest a minor change in time or place or simultaneous actions occuring at two different locàtions.

Because a dissolve can be made effortlessly with the video switcher and because it produces such an interesting visual effect, it is often subject to overuse or abuse. Remember, a dissolve is a much longer route between two shots than a cut, and it should be used sparingly to be effective. If your intention is to show the viewer what is happening on screen in the most direct way, the director should travel as quickly as possible from camera shot to camera shot. In the case of an interview show, for example, a dissolve between panelists would be totally inappropriate since the cut is a more direct and straightforward transition which better complements the particular production situation.

MATCHED DISSOLVE An interesting variation on the dissolve is a *matched dissolve*. This is accomplished by matching two different camera shots so the effect produced is of one shot blending subtly into another. For example, shooting identical shots of two clocks, one at nine o'clock and the other at five o'clock, and then producing a matched dissolve between them conveys the passage of time of to the audience. Another example of the matched dissolve is the opening of a rock concert in which the director locked a camera on a pedestal on a wide shot of the empty arena. As the audience began to fill the seats, the director recorded a few minutes of videotape at regular intervals until the seats were completely full. During postproduction, the director set up "A" and "B" VTR rolls for a co-ord and by dissolving between various taped segments produced the effect of the hall filling up with an audience in only a matter of seconds.

The trick to a matched dissolve is to set up both camera shots properly. Whether you will be producing the dissolve between two or more cameras in "real time" or will be using videotape and producing the dissolve in postproduction as with the rock concert example we've just mentioned, it is essential for the different shots to match as closely as possible. If your camera viewfinders

can show the mix/effects output, it makes it easy for the camera operators to align their shots precisely by using each other as a reference until a perfect match is achieved.

PAN/DISSOLVE A variation on the dissolve is to pan, or tilt, both cameras simultaneously while dissolving between their shots. The effect that is produced is a very smooth and fluid movement in the videospace that is often highly effective in musical or dance numbers. The usual way to set up this shot is to have one camera start on the subject while another camera begins by shooting off the subject, usually into a dark or neutral area of the set. On cue, the camera that is shooting the subject pans or tilts so the subject moves out of frame while the other camera pans or tilts to include the subject in the frame. At the appropriate time the director dissolves between cameras, creating the effect on screen. The transition must be carefully rehearsed since its success depends on the combination of cameras moving at the right speed and the precise timing of the dissolve. The shot looks best against a limbo or cameo background, which eliminates distractions in the shot and provides the illusion of the subject moving across infinite videospace.

FOCUS/DEFOCUS DISSOLVE As its name suggests, the third variation on the dissolve involves one camera going out of focus while the second camera comes into focus. As the cameras simultaneously focus/defocus, the director dissolves between the two. The diffused and dreamy quality of the effect can be used to show a change in time or place; to suggest altered perception, hallucinations or dreams; or simply as an aesthetically pleasing transition in a musical or dance production.

WIPE In a wipe transition, a new picture literally wipes across the screen replacing the existing image with a new shot. The pattern and direction of the wipe are variable, and most modern production switches offer a tremendous assortment of wipes. A wipe is the most artificial transitional device and, consequently, the most obvious to the viewer. It is useful, however, when you especially want to call attention to the transition. For example, in sports coverage the transition from live action to a videotape replay is frequently made with a wipe to show the viewer clearly the change from live action to reply.

A variation on the coventional hard wipe is a *soft wipe*, which features diffused, graduated borders. This lessens the impact of the transition and makes the soft wipe appear more like a dissolve than a hard wipe.

Switchers which are equipped with a digital video effects unit can produce a number of special wipe transitions such as push-offs, flips, image compression, and so forth. Although these effects appear quite artificial and call attention to themselves, they are very effective transitions in nondramatic applications such as news, sports, commercials, or training productions.

SPLIT-SCREEN   The speed of a wipe is variable, and when a wipe is stopped midway, it creates a *split-screen* effect. The split-screen is very useful when you must show two or more events simultaneously or when you wish to show a single event from various perspectives.

In the first situation, split-screens are frequently used in baseball games to show the batter at the plate and the runner leading off first base. In the second case, a director of a cooking show might use a split-screen to show both the on-camera talent at the top of the screen and a close-up of her food preparation on the bottom half of the screen.

As many as four split-screen images are possible with many production switchers (even more if you use special effects equipment and editing), but beware of excess. The size of the television screen is limited, and the viewer can process only so much visual information at one time. The greater the number of split-screen images you use at one time, the smaller the size of each. In addition, the amount of information the viewer receives may reach the "overload" stage, where it is either ignored or becomes confusing and distracting.

When to Change Shots

Knowing when to change a camera shot is one of the many important choices facing the director. To a great extent the decision you make must be based on the specific situation at the time, but there are a number of guidelines to help you in cutting from shot to shot. Incidentally, the term "cutting" is often used to refer to all video transitions although cuts are the most frequently used transitional devices.

In general you should cut to another shot when you have exhausted all the information contained within the existing shot, when you want to show the viewer something entirely different, when you want to show the same subject from a different perspective, or when you want to vary the emphasis and direct the audience's attention to other elements in the videospace.

SHOW SOMETHING NEW   The most obvious time to cut is to show the viewer something new. If we start an interview show with a close-up of the host, we will want to see the guest when introduced. Similarly, if a performer in a full shot picks up an object for us to examine more closely, the director will cut to a close-up to show the new object that has come to our attention.

SHOW A DIFFERENT ANGLE   Usually a camera is positioned for a particular kind of shot. The same camera position may not show a close-up as well as it shows a long shot. Since the viewer often needs to see the same scene from a different and more advantageous angle, the director should use the opportunity to cut to the better camera shot. This is done constantly in sports for news coverage to give the viewer "the best seat in the house." We also do it on dramatic programs when, for example, we cut from a two-shot to an over-the-shoulder angle or in an interview show when we cut from a wide, establishing shot of the panelists to a tighter close-up of the speaker.

WHEN THE SHOT IS MOTIVATED ONSCREEN
Imagine a medium shot of a man working at a

desk when we suddenly hear a loud crash off-camera. The man turns to see what has happened, and this movement, accompanied by the offstage sound, motivates a cut to a new shot to show us what caused the commotion. Motivation for a cut need not be so obvious, though. A glance, a turned head, a pointed finger, a line of dialogue can all be used as motivating factors to cut to another shot.

REACTION SHOTS As its name suggests, a reaction shot shows the viewer the reaction of a subject or a group to onscreen events. Although the reaction shot may not carry the program's principal message, reactions are useful in establishing character involvement or in showing the audience a more complete picture of a situation or event. A shot of wildly cheering fans during a ball game is a common example of a reaction shot. Although the ball game is of primary interest, screaming fans and banners waving after a big play help to show another side of the same event and establish the relationship between the game on the field and the fans in the stands.

Interview or talk shows also benefit from the director's use of reaction shots. Of course, frequently in a drama it may be more important to see a character's reaction than to see the performer who is actually speaking.

When Not to Cut

One of the most difficult things for a beginning director to learn is when *not* to cut. Too much of any good thing is bad, and overcutting a show will not necessarily make it more interesting or more effective. In fact, overcutting can actually hurt your program. That is because every time we cut, we risk pulling the visual rug out from under the viewers, breaking their concentration, and disorienting their perspective of the events onscreen. As the program's director you have probably seen and heard the show so often during planning and rehearsal that you are apt to become bored with a shot long before the viewer. Remember that the audience is seeing your show—and each shot—for the first time. In fact, you may have to restrain yourself from cutting too often, especially on instructional or demonstration programs, because the viewer who is trying to learn a

technique or understand the presentation needs extra time to process all the visual information contained in any shot.

As a rule of thumb, ask yourself before changing shots: "Am I showing something new or different by cutting to this shot?" If yes, then make the cut. If you find you are cutting only for visual variety, you may risk overcutting the show. From a purely production point of view, overcutting can tie up some of your cameras with unneccessary shots, making it more difficult for you to use them later for some really important ones.

ALTERNATIVES TO CUTTING Although cutting from shot to shot is one way to show new information or give the viewer a different perspective, there are a number of alternatives to cutting that you ought to consider. Among these are camera movement, performer movement, use of focus/depth of field, and the use of plastics.

CAMERA MOVEMENT Simply moving the camera—pan, tilt, dolly, arc—can often sufficiently change the audience's point of view and eliminate the need to cut to another camera shot. Since a zoom often resembles camera movement (although it lacks the depth perspective of a dolly), we can consider it in the same category. While overzooming is definitely to be avoided, combining a zoom with performer movement can disguise the zoom move and still vary the point of view of the shot. For example, say you wish to show an actress enter a room in a medium shot and then show a wide shot as she crosses from the door to the middle of the set. You could accomplish this with two cameras: Camera 1 shoots a waist shot at the door, and Camera 2 shoots a wide shot of the set. As the actress leaves the frame on Camera 1, the director would cut—on the action—to the wide shot. An alternative would be to use only one camera to pick her up in a waist shot as she enters the door and looks around. As she moves forward, the camera zooms out with her to produce the wide shot.

Remember that cameras are mounted on wheels and can move on the air as well as off. You might want to combine the zoom we mentioned above with a slight arc, so the camera's wide shot is more centered and provides an even better perspective for the audience.

Moving cameras also enhance depth in the videospace, particularly when they move past foreground objects or through doors or arches. Pedestaling up or down can also eliminate the need to cut. By pedestaling down as a performer walks into a room and sits in a chair, you maintain the proper camera angle without cutting to another camera shot.

Camera movement offers you a number of important advantages. First, it is a more fluid way of changing the viewer's point of view than resorting to cut after cut. Secondly, economy in cutting means that your cameras have more time to prepare for their next shot, and both you and the camera operator can spend more time on framing and composition. Of course, moving the camera presents its own set of difficulties, but practice and coordination between the camera operator and a floor assistant can make even the most difficult moves possible.

PERFORMER MOVEMENT Another way to change a shot onscreen is to move your on-camera performers to frame the shot for you. For example, let us start with two actors photographed in a medium two-shot. As the scene develops, one moves forward a few steps forming a new shot—a foreground/background shot. If the foreground performer turns around and faces the background subject we now have a third shot—an over-the-shoulder shot. Finally, if the actor facing the camera walks off and the other person turns to look at him, we have a fourth shot—a close-up. Notice that we did not move the camera at all except to reframe the picture slightly as the performers moved, yet we produced four different shots for the price of one. (See Figure 19-3.)

Performer movement requires precision blocking of both talent and cameras. You will have to rely on talent to hit their marks consistently, and it is a good idea to have the floor manager put down masking tape to indicate each position. It may take some time and effort, but varying the shot with talent blocking can be especially valuable in maintaining the flow of a scene and increasing the number of different shots you can produce with a limited number of cameras.

USE OF FOCUS/DEPTH OF FIELD We naturally turn our attention to the areas of the videospace which are in sharp focus, and we look away from those areas which are out of focus. Consequently, the director can use focus and depth of field as another alternative to cutting. Using a shallow depth of field and throwing focus from subject to subject permit you to vary the audience's attention within the same shot. (See Figure 19-4.)

PLASTICS The use of lighting and sets can also change your camera shots without cutting to another camera. A commonly used lighting technique is to illuminate a foreground subject in "cameo," with the background completely dark. By assigning background lights to their own separate dimmers, we can light the background on cue, creating a new "camera shot". Other lighting variations are changing from silhouette to normal lighting or varying the color of a cyclorama.

Sets can be cleverly brought into the shot or removed either on wheels or "flown" above the studio on counterweight battens. The content or composition of the shot changes without cutting to another camera.

Screen Direction

Because television subjects move about the videospace, we must always be conscious of their screen direction. Violating some of the basic rules of screen direction can disorient the viewer and create some unintended confusion. The audience is used to having screen direction established through cutting and on-camera movement. The borders of the television frame are constant and provide a reference point for the audience. That is why in a cops and robbers chase we must show both the pursuers and the pursued traveling in the same direction. It would look awfully odd if the robbers were moving left to right while the pursuing cops moved right to left.

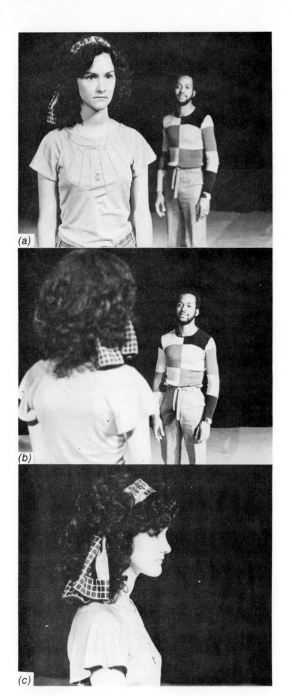

(a)

(b)

(c)

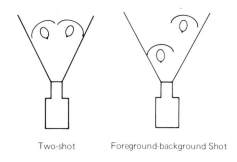

Two-shot   Foreground-background Shot

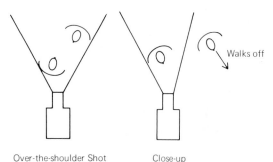

Over-the-shoulder Shot   Close-up

**Figure 19-3 Using Performer Blocking
to Vary the Shot**
By carefully coordinating performer movement and
camera shots, the director can produce a number of
visually different shots without cutting to another
camera. In this scene, the director started off with
both subjects in a two-shot. As the woman walks
toward the camera (a), a new shot—foreground-
background—is created. When she turned to answer
the man, a different shot (b) is produced—an over-
the-shoulder shot. When the man walks out of the
frame, the woman watches him leave (c), creating
another shot—a close-up.

runs left to right through city streets, the park,
down a bank of stairs while we intercut to show
the woman running right to left through hallways,
into the street, and finally into her lover's arms. By
establishing the basic screen direction for each,
we establish a sense of direction in the viewer's
mind that should not be unintentionally violated.

Not all screen direction problems are as romantic. Take the case of a football player, for instance.
We must always see the football players run
across the field in the same direction from one
shot to the next. The way to maintain consistent
screen direction is to use an imaginary line run-

On the other hand, opposite screen direction
implies that two forces are about to meet. Think of
the typical running-lover scene where the man

ning through every subject that is called the "action axis." As long as all your camera shots are made from the same side of the action axis, the direction onscreen remains consistent. If you cross the action axis, you will reverse the screen direction. That is why football games are covered with all cameras on the same side of the field. Otherwise, one or more of the shots would cross the action axis and literally flip the runner into the opposite direction on screen. This electronic version of Wrong-Way Corrigan would obviously disorient and confuse the viewer. Naturally you can change screen direction, but you must clearly establish the change for the audience. The easiest and most direct method is to show a subject change direction in a shot. Another way is to shoot the subject head-on, which is a neutral direction, and then cut to a camera shot from the other side of the action axis. (See Figure 19-5.)

Jump Cuts

A jump cut is a radical, or startling, transition from one camera shot to another. Generally, jump cuts are to be avoided because they call attention to the transition and can easily disorient the viewer. Of course, you can use a jump cut when this dramatic or startling impact is the very reaction you would like the audience to feel. Still, jump cuts must be used carefully because too many in the wrong places will irritate the audience rather than involve them in your show.

Here are some of the classic jump cut problems that crop up in every director's programs from time to time and some ways to avoid them.

CHANGE IN SUBJECT SIZE Cutting from an extremely wide shot to a tight close-up or vice versa produces an obvious jump cut. Again, you can use this for a dramatic effect—for example, when someone receives startling or unexpected news. To avoid the jump cut, cut to a shot or subject size midway between the two extremes to help smooth the transition.

Figure 19-4 Using Depth of Field to Vary Audience Attention.
The director can change the audience's focus of attention through the use of depth of field. In this shot, the foreground subject is in focus, and our primary attention is naturally focused on her. If the director wished to emphasize the subject in the background, she could have the camera rack focus to throw the foreground subject out of focus and bring the background subject into focus.

CHANGE IN CAMERA ANGLE OR POINT OF VIEW A radical change in the camera angle or the audience's point of view of a scene produces a jump cut. Cutting from a high-angle to a low-angle shot or from a straight-on shot to a stark profile are some examples of a radical change in camera angle. Similarly, changing the audience's point of view without sufficient notice should also be avoided. For instance, avoid cutting from an objective shot to a subjective shot without first preparing the audience for the change.

CHANGE IN SCREEN DIRECTION As mentioned earlier, a radical change in screen direction without reestablishing the viewer's perspective is a jump cut that can be avoided by keeping all cameras on the same side of the action axis.

CHANGE IN SCREEN POSITION Inadvertently changing screen position is a common jump cut

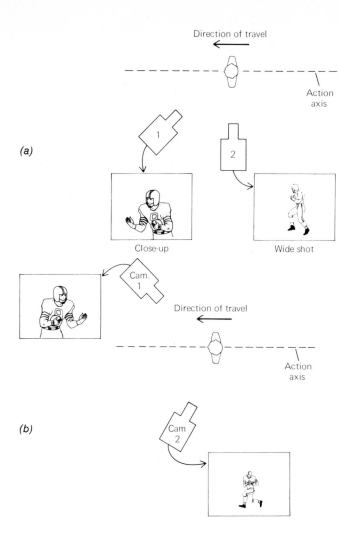

(a)

Direction of travel

Action axis

Close-up

Wide shot

Cam. 1

Direction of travel

Action axis

(b)

Cam 2

Figure 19-5 Action Axis.
In figure *(a)*, both cameras are located on the same side of the action axis. Although their shots are different, they maintain the proper screen direction. In figure *(b)*, the cameras are positioned on opposite sides of the action axis. If the director cut from Camera 1 to Camera 2 in this case, the subject would reverse screen direction in the videospace and disorient the viewer.

that is usually associated with two different shots containing two or more subjects. For example, say three performers are on the set. Camera 1 may have subjects A and B; and Camera 2, subjects B and C. Cutting between these shots will make subjects B appear to leap across the screen. You can eliminate the jump by cutting to a wider or tighter shot first and then zooming either in or out or simply panning across to the other subject. (See Figure 19-6.)

CUTTING FROM A MOVING CAMERA TO A STATIONARY CAMERA

Cutting from a moving camera to a stationary camera is the visual coun-

terpart of slamming on the brakes of a fast-moving car. The effect can be distracting although on some live sports or news programs it may be difficult to avoid all the time. Try to get the moving camera to stop—if only for a moment—before you cut to the next shot.

DISCONTINUITY IN THE ACTION

A major change in continuity—such as a subject seated in one shot and standing in the next—is an obviously jarring cut for the viewer. Discontinuity is rarely a problem when you are shooting with multiple cameras since you are cutting the show in real time. It can be a problem, however, when you edit

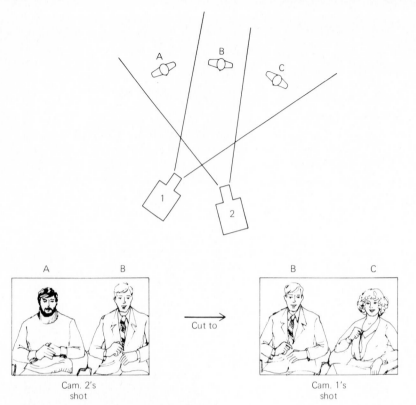

Figure 19-6 **Change in Screen Position.**
The director must be careful not to create a jump cut inadvertently by magically moving a subject across the screen when cutting from one camera shot to another. In this illustration, the three subjects are being shot by two cameras in a series of two-shots. However, when the director cuts from Camera 2 to Camera 1, subject B jumps from the right side of the screen to the left side of the screen. The jump cut can be eliminated by using a zoom or dolly to vary shot size, panning across subjects, or shooting each subject alone.

tape or film, since the shots were taken at different times. You can cover such a discontinuity in the action by inserting a cutaway, such as a reaction shot, to cover the jump cut.

CREATING THE AUDIOSPACE

You have probably heard the adage time and time again that "television is a visual medium." Though true, that does not mean the director should ignore the program's sound. On many productions the audio carries a significant part of the program message and deserves the director's

attention as much as the visual portion of the show. Although the director leaves the actual audio production primarily in the hands of the audio engineer, it is up to the director to establish how sound will be used in the show. This includes considering both the technical problems of audio pickup and reproduction and the creative uses of sound.

Sound Pickup

The audio engineer must understand your basic program objectives before he or she can begin to plan how to pick up and reproduce the sound technically. If your program will require off-camera microphones, the placement of the boom and the blocking of the performers are as crucial to the audio engineer as they are to the director and camera operators. Sometimes the director must modify certain shots or some actor blocking to accommodate the audio better. If you are working on a remote production, audio considerations can become even more complex because of the difficult sound characteristics of your shooting location and the usual problems when shooting outside a soundproofed and controlled environment.

Even a relatively simple in-studio show, which uses such on-camera microphones as lavalier or hand mikes, must consider audio. The audio engineer needs to know if the talent will remain stationary or if they will move across the set. Also, must other individuals be miked, such as audience participants? On a demonstration show, can the talent move freely without audio problems?

The director who takes the time to discuss the program with the audio engineer in advance can save an enormous amount of time and frustration later during rehearsal and production.

Sound Perspective

One of the important aesthetic audio considerations the director must keep in mind is the sound perspective. This refers to the quality of the sound and its relationship to the picture. We would expect the sound perspective to suggest a more intimate sound presence when we shoot in close-up and to reflect less presence when shooting in a wide shot. On some programs, the director may wish to alter sound perspective by manipulating the audio through special electronic effects to create the sound of a telephone voice, someone inside a cavernous, echo-producing environment, and so on.

Music and Sound Effects

Music and sound effects contribute enormously to the overall production. Selecting just the right piece of theme music, background music, or a musical bridge is not easy and often takes a considerable amount of time. But the right selection, used at the proper time, can result in a

synergistic effect where the audio and the video produce an impact greater than the sum of their parts.

Remember that music (like most sound) can be used to either *parallel* the video or provide a distinct *counterpoint* to the images on screen. The possibilities are endless, but the effective marriage of sound and picture requires planning and a considerable amount of attention to audio elements.

Sound effects are also indispensable aids in many productions. For example, your lighting director may produce some fabulous lightning effects, but the entire illusion is incomplete without the sound of thunder. The use of sound effects under a cocktail party scene to convey the sounds of clinking glasses and party chatter, the sound of an auto engine as you show characters driving, the noise of an airplane as two performers converse in their seats are all subtle touches which can make the difference between a barely credible program and a really effective show. Remember that some sounds must be captured naturally, but the extra effort is more than worthwhile. What is baseball coverage without hearing the crack of the bat, a basketball game without the sound of the ball being dribbled downcourt, a parade without the sound of marching feet and spectator cheers?

The Director and the Audio Mix

Be sure to listen to the audio mix, not only prior to the production, but also after you have begun to integrate sound and picture. You may find that a well-balanced audio track sounds fine when you listen to it separately, but it just does not work when combined with the visuals. Once you can see and hear the effect of both sound and picture, you may have to remix or rebalance the sound to compensate for the combined impact that the video and audio create together.

If you have the obvious luxury of mixing audio later in postproduction, you need not divide your concentration between directing the visual elements and listening critically to the audio during rehearsal and production. Still, most programs are not produced that way, which means the director must give audio some attention during the hectic rehearsal and production sessions. Remember, though, a program that has been beautifully shot and edited can be ruined by a poor audio track. While poor sound pickup or unintentional reverberations are usually obvious, the director must also listen for the more subtle audio elements such as balance and perspective, which are essential in creating the audiospace.

Ideally, sound and picture should complement each other to the extent that the resulting whole program which is seen and heard by the audience is greater than the sum of its component video and audio parts.

SUMMARY

Creating the videospace is one of the director's most important jobs. This involves three concepts: (1) directorial choice—the program director faces a wide range of choices, and the finished program is the sum total of the choices the director has made; (2) viewpoint—the director should approach every program with a particular perspective which provides unity and focus for the production choices; and (3) visualization—the director must be able to conceive the visual elements of the show in his or her mind's eye before translating them onto the television screen.

The shot is the basic visual element in a television show. The director develops camera shots by selecting the size and content of the shot, determining the appropriate camera angle, and deciding on camera or subject movement within the shot.

Individual shots must be edited, or joined together, into a meaningful sequence. The director edits a show by working with three factors: (1) the juxtaposition of shots—the sequence and order of individual shots have a definite impact on the viewer; (2) the timing of the shots and the transitions; and (3) transitional devices, or the way in which we change camera shots. There are four basic transitional devices: the cut, the fade, the dissolve, and the wipe.

The timing of the transition and knowing when or when not to change shots are important directorial decisions. In general, we change shots when we want to show the viewer something new or a different or more effective angle or view of the scene, or when there is obvious motivation for the cut. There are a number of alternatives to changing shots which allow the director to vary the viewer's perspective without actually switching to another camera. These include: camera movement, performer movement, use of focus and depth of field, and plastics.

Since subjects often move within the television frame, the director must be conscious of the rules of screen direction. Using the action axis—an imaginary line which runs through a subject—can help keep track of screen direction. If all cameras photographing the subject remain on the same side of the action axis, unintentional changes in screen direction will be avoided.

Jump cuts are abrupt or radical changes from one shot to another and usually startle or disorient the viewer. Among the most common jump cuts are radical change in shot or subject size, severe change in camera angle or point of view, jump in screen direction, discontinuity, and cutting from a moving camera to a stationary camera shot.

Although the director is concerned with creating the videospace, audio must never be ignored. Sound carries a great deal of important information, and an inferior audio track, because of poor technical pickup or a bad mix, can ruin what is otherwise a good show. The director must consider audio when planning the production and should listen carefully to the sound during rehearsal and production. Sound perspective—the relationship of sound to picture—should match the visuals and enhance the program. Music and sound effects are indispensable audio aids in establishing a mood and reinforcing visual elements. Ideally, the sound and picture should complement each other to the extent that the whole program produces an effect greater than the sum of its component video and audio parts.

CHAPTER 20

DIRECTING

The Director in Production

The television director must function on two levels. The first level—the aesthetics—has already been discussed in the preceding chapter. Visualizing the program and making the right series of choices at the right time, however, are only a part of the director's responsibilities. In this chapter we will discuss the second aspect of the director's job—the transformation of ideas and concepts into actual sounds and pictures.

A good director is involved in virtually every facet of a production. Every chapter in this book has mentioned the director because it is impossible to discuss any particular aspect of production without including the individual who must integrate the many diverse production elements to create a program. The specifics of every production area and the director's role in each have already been covered in earlier sections. In this chapter we will see how the director plans, organizes, and supervises the production team and uses production techniques to create a television program.

THE DIRECTOR AND THE PRODUCTION TEAM

It is sometimes easy to lose sight of the fact that all the equipment used in a production must be operated by *people*, and it is the crew that plays a large part in any program's success or failure.

"You're only as good, only as effective, as the people you have with you," says Dwight Hemion, one of the top directors in television. "It's a big team effort and you've got to respect the crew, get to know them, give them the feeling that they are contributing a great deal to the production—because they are." Among successful directors, the importance of the team is a vital ingredient. "The director may work with actors and plan out the shots," says Bob LaHendro, a director of such programs as *All in the Family*, "but when you come down to it you need that crew behind you working as a team. To get that kind of working attitude where everyone is pulling for the show simply demands that you, as director, respect everyone else's contribution. They like you, you like them, and there is mutual trust and respect. That way when you ask the crew to go above and beyond they'll do it willingly."

Even the most experienced directors make mistakes from time to time, but a production unit that works closely together can "save" each other through close teamwork and cooperation. If you and your crew have a mutual respect, you will avoid the experience of an arrogant young director who arrived on the set of a variety show one day. "I just wanted to let you know," said the director as he met the crew, "that I'm in charge here and I expect all of you to do exactly as you are told—no more, no less." Later, when the show went on the air—live in those days—the director told one of the cameras to dolly in on a girl singer. As the camera began the move, the director's attention was diverted from the air monitor as he set up the next shot. When he glanced back, he was horrified to see the camera continue to dolly past the singer, through the curtain, and all the

way to the rear of the set. "What do you think you're doing!" he shrieked over the PL system. "I followed your directions exactly," replied the camera operator. "You said, 'Dolly in,' you didn't say to stop."

Another point to remember when dealing with the crew is to know and respect each person's capabilities and limitations. You are never going to find the "perfect" crew, so learn to make the best use of the one you have. Put your best operator on the camera with the toughest job. Do not ask an inexperienced camera operator to accomplish intricate moves and manuevers which you know he or she cannot do.

Because the director is so closely involved in a production, it is easy to forget that the crew may not be as familiar with the program. Even the best crew member cannot do a great job unless he or she knows what is expected. A short meeting with the crew just prior to rehearsal can help to familiarize them with the show and give everyone a better idea of what you expect them to do. A crew that knows what is to come and what is expected of them will feel more comfortable, work more confidently, and do a better job than a crew that is unfamiliar with the project and uncertain of what the director expects from them.

DIRECTOR COMMANDS

A visitor to a busy control room would probably wonder about the strange series of commands issued by the program's director: "Ready one on the cross. Take one and zoom in. Insert the graphic and ready to wipe to a split of one and three. Ready to roll VTR. Dissolve to the effect, roll VTR. . . ." To the uninitiated, the words and phrases may seem strange and almost unintelligible as the director spews them out in a steady stream of chatter. In reality, though, each of these commands refers to a specific production operation and is followed by the production crew during rehearsal and production. Because everything moves so quickly during production, the director must be able to communicate to the crew accurately, quickly, and efficiently.

In each chapter of this book we have given the basic terminology associated with the equipment and production techniques covered. These terms —with a few variations—have become standard within the television industry, and they enable the director and the production team to communicate quickly and accurately. Although the director can communicate with anyone on the crew who is hooked up to the PL system, during rehearsal and production the director speaks primarily to the camera operators, to the switcher, and to the floor manager.

Camera Commands

The director will probably speak to the camera operators more than to any other production team members in order to set up shots, perfect the framing and composition, and direct camera movement while the shot is on the air.

Here are the guidelines in giving camera commands.

1 *Give a "ready" whenever possible.* Before you punch up a camera on the air or ask the camera to execute a move, try to give a "standby," or "ready," cue. The camera operator will know to hold the shot still as soon as he or she hears the "ready." It is also a good idea to ready a camera move. For example, "Camera 1—ready to pan right with the host." This is particularly important when the on-camera subject is about to make a major move, such as getting up from a chair, sitting down, or crossing the set. Although it is not always possible to give a ready on a spontaneous and unscripted show, try to give them whenever you can.

2 *Call each camera by number.* You ought to know each camera operator's name, but ignore names while you are calling shots. What if you have two camera operators named Susan? Also, other team members listening

over the PL may not know the camera operators by name but will be able to easily identify the cameras by number.

3 *Preface the command with the camera number.* Say: "Camera 2, pan right to the door," not "Pan right to the door, Camera 2." Prefacing the command with the camera number will avoid confusing the camera operators and prevent them from making unintentional moves. It will also help the switcher, audio engineer, and floor manager since camera operations can affect their jobs.

4 *Remember the difference in pan and tilt commands and the resulting movement on screen.* When you "pan right," you move the image onscreen to the left. When you tilt up, you move the image onscreen down. Be sure to consider the onscreen effect of the pan or tilt before you give the command.

5 *Be specific in your commands.* While you should use as few words as possible, do not eliminate specific references which will help the camera operators to get their shots. For instance on a three-person interview show, the command "Camera 1, zoom in" is too vague. Instead, say: "Camera 1, zoom in to a close-up of the host." The same applies to pans, tilts, dollies, and so on. Rather than say, "Pan right," say "Pan right to a two-shot." The more specific your commands, the faster and more accurately the camera operator can get their shots for you.

Switcher Commands

At some studios, the director operates the switcher during production. At other facilities, switching is performed by someone else, who operates the switching console on the program director's command. Even if you will operate the switcher yourself, get into the habit of giving verbal switching commands. If you do not, the production crew members listening to the PL for directions will not know about a video transition until it has already happened onscreen.

Here are guidelines for giving switcher commands:

1 *Give a "ready" before each transition.* A warn-

ing before each switching cue is helpful to the switcher, as well as to the camera operators. Try to give the ready cue just before the actual transition. A ready that comes too soon loses its impact, and a ready that comes too late gives the switcher too little time to prepare. It is particularly important to give the switcher advance warning on such complex transitions as dissolves, wipes, or inserts, which require a preset before the transition can be performed.

2 *Be specific.* The more specific your commands, the less chance for error. If, for example, you want to insert one camera source over another picture, say: "Ready to insert Camera 1 over Camera 3. Fade-in insert." Also specify how you want the insert to appear since an effect can usually be "popped-in," "faded-in," or "wiped-in."

3 *Indicate the speed of a fade, dissolve, or wipe.* Use your voice or a hand motion to indicate how fast or slow you want to see the transition on screen. A verbal countdown is even more precise and can help the audio engineer, who may have to accompany the visual transition with an audio fade or segue.

4 *Give the switcher advance warning on complicated effects.* The commands "fade," "dissolve," "super," "insert," "key," or "wipe" should always remind you that the switcher must preset a number of video sources before the transition or effect can be produced. Especially on chroma keys or inserts, where clipping levels must be established, try to give as much advance notice as possible so the effect can be set up and checked on the preview monitor.

5 *Economize on command words.* "Rather than saying, "Ready to cut to Camera 1 . . . cut to Camera 1," say: "Ready 1, take 1." Fewer words take less time and reduce the chance of confusion. Since cuts are the most often used transition, you might arrange with the switcher and camera operators to consider a "ready" without further qualification to be a straight cut. For example: "Ready 2 . . . Take 2" (a cut), but "Ready dissolve to 2 . . . Dissolve to 2" (a dissolve).

6 *In a sequence of commands, cue the switcher last.* Since the switcher is positioned next to the director in the booth, he or she usually needs less time to react to a command than other production team members, who must receive the cue through the PL or the floor manager. So, if you must give a sequence of commands such as "Open mikes, cue talent, fade in on Camera 3," give the cues in the following order: (1) to audio, (2) to talent, and (3) to the switcher. This will also reduce the chance of fading-up on a performer who might have missed the floor manager's cue.

Other Director Commands

Most of the other commands given by the director during the show involve the floor manager; the talent; and such areas as audio, lighting, telecine, and videotape. Since we have covered many of these commands in their respective chapters, we will not repeat them here. However, we will discuss some of the floor manager and talent commands since they are given frequently by the director during rehearsal and production.

TALENT AND FLOOR MANAGER COMMANDS
The cue to start talking or to begin some other action is the most common command given by the director, usually through the floor manager. The same rules mentioned above apply equally to cuing talent:

1 *Be precise.* Say: "Cue Tom," *not* "Cue him."

2 *Be economical.* Say: "Cue Anne," *not* "Tell Anne to start talking."

3 *Be specific.* Say: "Show the host five minutes remaining," *not* "Show him the time."

Avoid giving talent a "cold" cue without any preparation. Talent readys are always important, especially when you are coming out of a prerecorded tape, commercial, station break, or other out-of-studio segment. Have the floor manager inform talent how much time is left until the show comes back to the studio. A silent hand countdown from the FM is helpful in smoothing the transition since it gives talent a more precise idea of time remaining. The countdown is particularly important when the performer must lead into a film or videotape, which requires a precise preroll cue.

Often the floor manager will convey time signals to help talent gauge how much time is remaining in a segment or in the overall show. Be sure that the floor manager and the talent have coordinated all time signals before production. There is a big difference between "five minutes left in the program" and "five minutes left before you begin to wrap up the show." Confusion here can result in a program ending early or running late. The director should also ask talent the kinds of time signals and other cues they need and then be sure to deliver them during the production.

On some live programs, such as sports or news coverage, the director may use an *interrupted feedback device* (IFB), which is an unobtrusive earpiece worn by the performer. Pushing a button in the control room permits the director to talk directly to the talent, even while the performer is on-camera. Be cautious in how you use the IFB, though, since it is hard for the performer to talk while listening to another voice giving instructions.

ASSISTANT DIRECTOR

By now it should be apparent that a director is awfully busy during rehearsal and production. The director must set up and call camera shots; check for framing and composition; cue talent; follow the script; listen to the audio; watch all line, preview, and camera monitors; cue other production elements, such as audio, VTR, and film; and keep track of the program's time. In addition to all this, the director must constantly evaluate the performances and the interaction of sound and picture to be certain of accomplishing the production's preset goals and objectives.

There is no doubt that the director can use some help in doing all this, and that is where the assistant director (AD) comes in. An AD is not always used at all production facilities, but if you have the opportunity to work with one, you will find the job of directing to be a lot easier. The director is still responsible for the entire production; but having someone to help keep track of time, ready cues, and even preset camera shots enables the director to concentrate more fully on what is happening now without having to worry too much about what is to come.

Exactly what the AD does during production depends pretty much on the program's director. Some directors give the AD enormous responsibility; others use the AD sparingly. Ideally, the AD and director should form a closely coordinated team where each complements the other. Usually the AD is responsible for (1) timing the show and individual segments, (2) readying upcoming cues, and (3) presetting camera shots.

Timing the Show

Time is one of the most important and difficult variables to work with during a production. Even if a show is to be videotaped in segments and edited later, accurate timing during production will avoid bringing in a show that is far too long or much too short for the time allotted. Accurate timing during production will also ensure a tightly produced show that looks better on screen.

CALCULATING TIME   It is not hard to calculate time, but you must start thinking in terms of minutes and seconds. Adding or subtracting time requires that you use a base of six (sixty seconds = one minute) instead of the usual ten. For example, say we are given the following segment times to add together:

1	Intro segment	0:35
2	Opening tape	1:00
3	Interview	4:15
4	Demonstration	2:40

To calculate the total time, first add all the seconds and divide the total by 60, keeping the remainder in seconds. Then add all the minutes, and carry over the total number of minutes from the seconds column. The result is the total time (8:30) in minutes and seconds. For those who feel uncomfortable even with these simple calculations, you can buy an inexpensive pocket calculator which is programmed to figure time without the need for any conversion into seconds and minutes.

BACKTIMING Often the director must know how much time remains in a program. In order to "backtime" the show, you will need to know both the present time and the program's end time. Then subtract the present time from the end time to find out how much time remains. For example, say we are 23:30 into a show that is scheduled to end at 28:50. To find remaining time:

```
  28:50 (show's end time)
− 23:30 (present time)
   5:20 (time remaining to end of show)
```

You can check your calculations quickly, because the time remaining plus the present time should equal the show's end time.

Backtiming calculations also allow you to determine the dead-pot roll time for a record, audiotape, videotape, or film. Say you have recorded a musical theme that runs 3:45 and your show ends at 58:30. To find the time the music must roll in dead-pot, subtract the theme's running time from the end time.

```
  58:30
−  3:45
  54:45 (dead-pot time)
```

The tape should be rolled in dead-pot at 54:45 into the show. It will end precisely at 58:30. By the way, notice that in order to subtract the time we had to "borrow" sixty seconds from the minute side to subtract the seconds. Again, adding the calculated dead-pot time plus the theme running time should always equal the show's total running time.

ESTIMATING RUNNING TIME Knowing the running time as the show progresses is always helpful and avoids unpleasant surprises at the end of the show when you find there is too much (or too little) time remaining. Although segments

should be timed during rehearsals and each segment's running time indicated on the program's rundown sheet, these times will vary once the performers get into the studio and the director starts to call shots. Using a timing sheet like the one in Figure 20-1 can help you keep track of each segment's time. In the left-hand column is the estimated time; the first column is used to record the actual running time during the first rehearsal; the next column is used for the second rehearsal, and so on. Next to each time is a space where the AD can mark how much over or under the segment is running. It is easy to tell the director how much the actual running time deviates from the estimated time so that minor timing corrections can be made throughout the show rather than waiting to make a major correction at the end of the show.

TIMING AND COUNTDOWNS Accurate timing is important in almost every production, so get into the habit of starting your stopwatch when the segment begins onscreen or as soon as the film or VTR material appears. Often a director or AD must use a number of stopwatches simultaneously. One might indicate total program running time while another is used to time the individual segments.

Some directors prefer to have the AD give the verbal countdown when going in and out of prerecorded material. This way the director can concentrate on his cues while the production team hears the AD counting down over the PL system.

VIDEOTAPE TAKE SHEETS On productions which are videotaped for postproduction editing, the AD may be asked to keep track of each take as well as timing the segments. The *take sheet* includes the scene and take number of every take to permit accurate identification during editing. Some ADs find it convenient to combine the take sheet information and timing information onto one piece of paper.

If the AD is responsible for tracking each take, he or she must be certain that the floor manager has the proper scene and take number on the slate before every recording.

Readying Cues

The AD can help the director by readying cues for audio, lighting, film, VTR rolls, announcer, and talent. Usually, the AD "readies" the operation, but it is still the director who cues the music, effect, film roll, and so on at the precise time it is to occur.

Presetting Camera Shots

On a fully scripted or semiscripted show, where a specific sequence of shots is planned, the AD can be used to set up the next camera shot for the director. To do this the AD must copy the director's shooting script, shot by shot. During rehearsal and production, the AD stays one shot ahead of the director and readies each upcoming shot. The director will usually touch up the framing and composition before punching-up the shot on the air. Using the AD to preset shots allows the director to concentrate completely on the on-air camera and the timing of the video transitions rather than on the mechanical operation of alerting cameras for their upcoming shots.

FIGURE 20-1 **Program Timing Sheet.**
The program timing sheet enables the director or the assistant director to keep track of a program's timing and to identify those segments of a show which are running too long or too short.

PROGRAM TIMING SHEET

PROGRAM SEGMENT	ESTIMATED	RUN THRU	DRESS	AIR/TAPE
1 OPENING	:30	:30	:30	:30
2 INTRO	1:30	2:15 /+:45	1:45 /+:15	1:40 +:10
3 INTERVIEW w/SALLY	5:00	4:45 /-:15	5:00 /	5:00
4 DEMONSTRATION #1	6:30	7:00 /+:30	6:45 /+:15	6:45 +:15
5 INTERVIEW w/BILL	4:15	4:00 /-:15	4:15 /	4:00 -:15
6 DEMONSTRATION #2	8:30	1:15 /+1:45	9:30 /+1:00	9:00 +:30
7 CLOSE + CREDITS	2:15	2:00 /-:15	1:30 /-:45	1:35

Labels (annotations pointing to columns):
- Segment times as originally planned → ESTIMATED
- Segment time → RUN THRU
- Times from first rehearsal → RUN THRU
- Amount over/under from estimate → DRESS
- Segment times from dress → DRESS
- Actual times during production → AIR/TAPE

TOTAL 28:30   OVER by 2:15   OVER :15

Having the AD preset shots is a great convenience as long as the AD and director do not work at cross-purposes. Too much talk over the PL will make everyone's directions impossible to understand, so the AD must learn to fit in his or her commands between those of the director. The shot or action on the air always takes precedence over anything that is to come.

THE DIRECTOR IN PREPRODUCTION

Production is certainly the most exciting time for the director, but without the proper preproduction planning, the production stage can be a disaster. Because crew and facilities time is too valuable to waste, the director must know what is needed and how to accomplish it long before rehearsal and production begin.

As Bob LaHendro says, "I've found that if you do your homework and know what you want before you go into the studio, everything moves along much more quickly. The crew is happier because they know what you expect from them, it gives the performers more confidence, and if you've planned everything out, it just makes for a much smoother and more productive day."

Production which takes place outside the studio at a remote location demands the same amount of preproduction planning by the director, if not even more so. Inside the studio you will find a perfect working environment and all the equipment you need; on location, if you have not planned what to bring along and how it will be used, there is no way for you to get the job done properly. Regardless of where your production takes place—inside the studio or on location—preproduction planning is an essential part of the director's job. It is impossible to overstate the importance of the preproduction stage to an effective and efficient production.

The Director and the Producer

Sometimes the director also doubles as the program's producer; at other times the director works with a producer in developing the concept, planning the production, and interpreting the script into a show. In Chapter 18 we discussed the

many details which the director and producer must handle during the preproduction. Rather than repeat them here, we will just emphasize that the most important aspect of preproduction is for the director and producer to agree on the director's basic interpretation and approach to the show. Once this is done, most other problems can be solved. If the producer and director do not agree on interpretation at this point in the show, you may find yourselves in conflict over fundamental concepts during rehearsal and production, which result in frustration, damaged egos, a confused cast and crew, and ultimately a bad show.

During the initial preproduction period, the director will meet with the scriptwriter to iron out final script details and with the set and lighting designers to work out the setting and lighting. This should be done as early as possible because the director's most essential tools during preproduction planning are (1) the floorplan and (2) the script.

Floorplan

The floorplan is an overhead view of the set, which should always be drawn to scale. Using the floorplan, the director can plot camera shots, performer movements, and the positions of production equipment.

Take a careful look at the studio layout as it is represented on the floorplan. You are interested as much in the off-set production areas as in the on-camera set because the cameras, additional equipment, and crew must have enough room to move about during the production.

Begin by roughing out where you intend to position major pieces of equipment, such as cameras, boom microphones, video monitors, and so on. You will need the approximate dimensions of each piece of studio equipment in order to determine the amount of space needed. You are fooling yourself if you plot the equipment on paper as being smaller than it really is. The reality of the situation will become painfully obvious once you arrive in the studio and find insufficient room for your equipment to operate.

If it looks as though you will need more production room than you have available, consider reworking the location of the set in the studio to give you more flexibility. For example, on a show where you need a very wide cover shot, position the set so your cameras can dolly back as far as possible. On those productions where lateral camera movement is of primary importance, position the set to give you maximum horizontal working space. Remember that it is easy to experiment with the best set and equipment positions on paper as long as everything has been drawn to scale. Once the set is constructed in position in the studio, there is not much you can do to correct a serious problem, and even minor changes are far more difficult to accomplish.

REMOTE FLOORPLANS If you will be producing all or part of your production at a remote location, a floorplan of the various areas where you will be shooting will help you to plan your camera positions and to work out shots before you arrive for production. Of course, few locations have ready-made floorplans, but you should take the time to visit the location, measure the space dimensions of the shooting area, and draw a floorplan as close to scale as possible. For more information about the director's job on remote productions, see Chapter 21.

CALCULATING HORIZONTAL CAMERA ANGLES Under most circumstances you will not have any trouble getting close-up shots with your cameras. The combination of powerful zoom lenses and the ability to dolly in for a close-up will usually permit you to shoot as tight as you wish. What is often more critical, however, is the maximum width you can expect from your widest cover shot. Once you have zoomed all the way out to your shortest focal length and dollied back as far as you can physically move, unless the camera can photograph everything you must include in the wide shot, you will encounter some serious shooting problems. Knowing what your widest possible shot will be in advance is a great help in planning the positioning of sets and cameras, and helps to avoid unexpected surprises when you begin work in the studio.

In our chapter on lenses we said that you can determine the horizontal angle for any 16-mm format lens by using the following formula:

$$\frac{676}{\text{Focal length in millimeters}} = \text{the horizontal angle}$$

Using this formula and the studio floorplan, we can easily determine whether or not our widest shots will encompass all the sets and subjects we need.

As an example, consider a panel discussion with a large number of participants. The middle camera must be able to shoot a cover shot of the entire set, and the director knows that his or her widest zoom lens is 12mm. Using the formula, he or she finds that 676/12 = 56°. In other words, the wide-angle lens will produce a 56° horizontal angle. Using an inexpensive protractor, the director measures the 56° angle on the floorplan from the farthest possible camera position. As you can see in Figure 20-2, the horizontal angle safely includes the entire set, so the director can plan shots with the assurance that the widest shot will be sufficient.

The Script

The program's script is the foundation for your show. Although some programs are produced with a partially detailed script and some are produced with virtually no script at all, for now let us assume that you are working with a complete script since the director's approach is essentially the same regardless of the program format or the amount of script detail. As the director reads through the script and begins to work with it, he or she must deal with four related stages: (1) read-through, (2) performer blocking, (3) camera shots, and (4) marking the script.

READ-THROUGH As soon as you receive a copy of the script, read it thoroughly, not just once but a number of times. Each time you read the script you will likely find new ideas and possibilities which you should jot down on a pad or in the margin of the script. As you continue to read, try

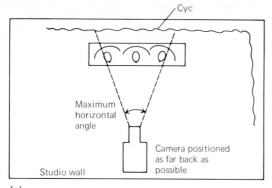

(a)

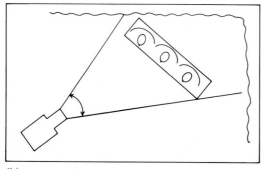

(b)

FIGURE 20-2 **Using Horizontal Angle of Lenses to Plan Shots.**
Using the horizontal angle formula for lenses and a copy of the studio floorplan enables the director to learn if the camera shots, set positions, and shooting requirements of a production will work before entering the studio. In the first illustration *(a)* the horizontal angle of the widest available lens is not wide enough to include the entire set as it has been positioned in the studio. By changing the location of the set and taking advantage of the studio's length, the director can see that the revised floorplan will work with the space and the camera lenses which are available *(b)*.

to visualize the show as you would like it to appear on screen. At the same time listen to the words. Does the dialogue sound natural or stilted? Do the lines make sense, or are they confusing? Are there any serious production problems in the script which need to be corrected? Do some scenes contain production difficulties, such as the need to cover many actors, an action sequence, or some very tight action in close-up? As you continue to read the script, your visualization will become more defined and focused as the

overall program and point of view begin to crystallize in your mind.

PERFORMER BLOCKING Once you have become familiar with the script and have developed your point of view, you can begin to work out some tentative performer blocking. Where do we first see the talent? How and when do they move during the show? These are considerations that apply not only to dramatic programs but to any production where talent appears on-camera. Of course, the blocking you develop now is tentative. Once you begin talent rehearsals, you will undoubtedly revise or completely change some of the blocking to accommodate the artists, as well as camera shots and other production problems.

As you block out performer moves, use your floorplan to help you to visualize the set and the relationship of the actor to cameras, microphones, monitors, and so on. Do not worry too much about the exact camera shot at this point, but try to keep the approximate camera position in mind so you do not trap yourself by moving performers into positions where they cannot be easily photographed by the cameras.

DEVELOPING CAMERA SHOTS Once you have established the basic performer blocking, you can go back to your script and floorplan to work out camera shots. The idea is to visualize the overall traffic pattern of moving talent, cameras, and equipment, in addition to considering the camera shots themselves. As a director, you must plan your performer and camera blocking so the talent meets the camera at the right place at the right time to produce the shot you want. Of course these shots are bound to change during the course of rehearsal and production, but they do provide you with a starting point. All of this preplanning paperwork is demanding and requires considerable time and concentration, but the time spent with paper and pencil on preproduction will pay off later during rehearsal and production.

In the preceding chapter we discussed the various aesthetic considerations which determine the director's choice of camera shots. Now we will concentrate on how to get those shots that you need on the air. As you begin to work out camera shots, try to establish the basic shot and camera position. Do not worry too much at this point exactly what size the shot will be. Whether a camera shoots a subject in a close-up, a bust shot, or a medium-wide shot does not matter much since the camera operator need only zoom in or out from the camera position to vary the shot size. What is important is the camera-to-subject positioning and the overall sequence of camera shots and angles. Once these have been established, it is relatively easy to change a particular shot's size and visual emphasis.

As you develop your camera shots, there are some basic points to keep in mind. These include adherence to the action axis, cross-shooting when possible, and planning sufficient time to break a camera from one area of the studio to another.

ACTION AXIS   Remember to consider the action axis when planning your performer and camera blocking, so you do not inadvertently change screen direction or create a disturbing jump cut.

CROSSING SHOTS   Usually the most favorable and effective shot of a performer is one taken from a position that is relatively straight-on. Profile shots are rarely flattering, and, because we see little of the talent's face, a great deal of expression and communication can be lost. In order to shoot straight-on, try to cross your cameras as indicated in Figure 20-3.

BREAKING A CAMERA   Often a camera must "break" from one area of the set to another, and the director must realistically estimate the amount of time necessary for the camera operator to break and set up the next shot. Underestimating the time may result in your calling for the shot before it is ready. Overestimating the amount of necessary time results in your losing a possible camera shot when you need it. As you plan camera breaks and set design, remember the

concept of videospace. All the audience sees is what you show them onscreen, so you can easily position a camera card stand or the credit crawl just off the set, where the camera will have a short traveling distance in order to set up the next shot. The closer the camera's new position, the faster the camera operator can get you the shot, and the sooner it will be available for you to use.

MARKING THE SCRIPT   As you work out camera shots, they must be indicated on your copy of the script. Each director usually develops an individual method for marking a script, but, regardless of how you indicate shots and cues, the objective is clarity, legibility, and economy. You will need to follow your markings during the fast-paced activities in rehearsal and production, so make all script notations as easy to read as possible. A sloppy or illegible script is hard to follow and increases the chance of losing your place at a critical point in the show.

Figure 20-4 lists some of the common director shorthand symbols and their meanings. It is a good idea to write directions in pencil since you will be able to erase and make changes more easily than if you write in ink. If you have time between rehearsal and production, you should completely copy your cues and shots onto a fresh program script.

CAMERA SHOTS   Camera shots are the most common script notation and are written on the blank "video" side of the script. Indicate the camera number, the type of shot, and the subject being photographed. If the camera must move, zoom, or rack focus during the shot, include these directions as well.

PERFORMER MOVEMENT   Remind yourself of major performer moves so you can alert the camera operators and other crew members before the move occurs.

UPCOMING READIES AND CUES   Note important readies and cues, such as a camera break, a film or VTR roll, a talent cue, or dead-pot time in your script. Bracketing these directions in a box will help to differentiate between a ready and an on-air cue or command.

STORYBOARD NOTES   Some directors like to

make small drawings to remind them of a particular visual composition. These notes are conveniently made on the blank page opposite the script page.

PREPARING SHOT SHEETS   On a fully scripted show, you may wish to prepare *shot sheets* for each camera. A shot sheet is a description of every camera's shots listed in order from beginning to end. You can use the shot sheets in a number of ways. If you never intend to vary the preset shot sequence, each camera operator can simply go to the next shot once the camera's tally lights go off. This makes shot setup virtually automatic, but it does cut down on the director's flexibility to cut between cameras spontaneously during a scene. Once the director has cut from a camera, the operator will automatically move on to the next shot.

If you are working with an AD, the assistant

director can use the shot sheets to ready each camera shot by number alone. This eliminates extraneous chatter over the PL system since it is a lot faster to say "Camera 3, shot 17" than "Camera 3, close-up of the host and follow to the door."

Probably the most flexible way to use shot sheets is to tell cameras to remain on each shot until they are released by the director or AD. Then, if your cutting pattern between cameras is slightly different from what you had planned, the cameras will still be holding their shots. Once the director releases the cameras, however, they can set up their next shot immediately since it is indicated on the shot sheet.

Obviously shot sheets are not much use on an

FIGURE 20-3   Cross-Shooting.
Shooting subjects straight-on results in profile shots *(a)* which are usually unflattering and prevent the viewer from seeing the subject's face clearly. Cross-shooting *(b)* solves the problem, and results in a full-face image of the subjects.

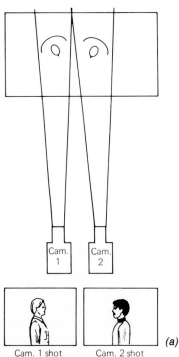

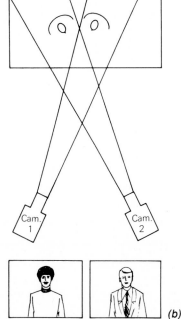

(a)

Cam. 1 shot   Cam. 2 shot

(b)

Cam. 1 shot   Cam. 2 shot

SYMBOL	MEANING
C1—————————	Camera number. Written alone symbol indicates a take to the camera. The line is used to indicate precisely on what word or action the cut should occur.
WS	Wide shot
MS	Medium shot
BS	Bust shot
CU	Close-up
ECU	Extreme close-up
2-SHT	Two-shot
3-SHT	Three-shot
OS	Over-the-shoulder shot
FG-BG John/Mary	Foreground/Background shot (Names used to specify subjects)
ROLL Q	Roll cue (for film or VTR)
Q	Cue
ANNC	Announcer
SOF/SOT	Sound on film/Sound on tape
SFX	Special effect
C/K	Chroma key
NSRT	Insert
✕ or DIS	Dissolve
X to	Actor cross (movement to new position while on camera)

FIGURE 20-4 Director Script Markings.
All directors develop their own method of indicating camera shots and other production cues on the program script. The symbols and script markings illustrated here and on the following two pages should be adapted to your own style.

unscripted show, where the director and camera operators must wing it, but when you are able to plan all shots in advance, shot sheets can be an enormous help to a busy director.

Production Meetings

Throughout the preproduction stage the director is involved in a series of production meetings.

Some of these are often long and intensive sessions, in which each program element is carefully planned with the key production personnel. Other meetings are less formal but still enable the director to make certain that everyone involved with the show understands the program's objectives and their individual contribution to the production effort.

One crucial production meeting takes place in

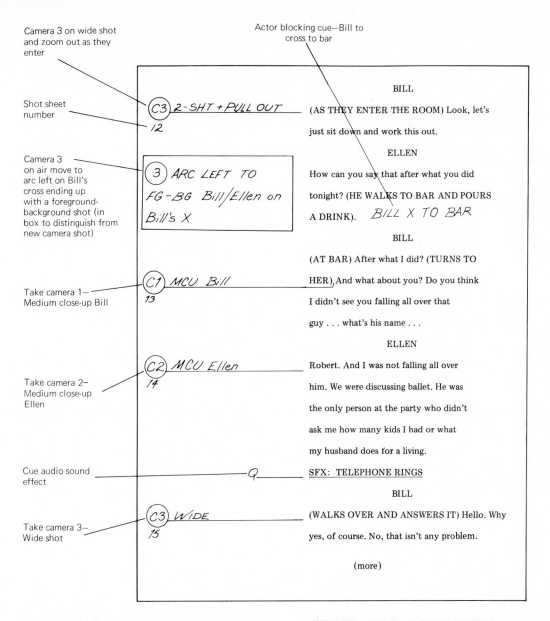

Camera 3 on wide shot and zoom out as they enter

Actor blocking cue—Bill to cross to bar

Shot sheet number

Camera 3 on air move to arc left on Bill's cross ending up with a foreground-background shot (in box to distinguish from new camera shot)

Take camera 1—Medium close-up Bill

Take camera 2—Medium close-up Ellen

Cue audio sound effect

Take camera 3—Wide shot

(C3) 2-SHT + PULL OUT
12

(3) ARC LEFT TO
FG-BG Bill/Ellen on
Bill's X

(C1) MCU Bill
13

(C2) MCU Ellen
14

Q

(C3) WIDE
15

BILL

(AS THEY ENTER THE ROOM) Look, let's

just sit down and work this out.

ELLEN

How can you say that after what you did

tonight? (HE WALKS TO BAR AND POURS

A DRINK). BILL X TO BAR

BILL

(AT BAR) After what I did? (TURNS TO

HER) And what about you? Do you think

I didn't see you falling all over that

guy . . . what's his name . . .

ELLEN

Robert. And I was not falling all over

him. We were discussing ballet. He was

the only person at the party who didn't

ask me how many kids I had or what

my husband does for a living.

SFX: TELEPHONE RINGS

BILL

(WALKS OVER AND ANSWERS IT) Hello. Why

yes, of course. No, that isn't any problem.

(more)

the studio just prior to rehearsal and production. This is when the director assembles the crew for a five- or ten-minute orientation. It is unlikely the crew knows much about your production, so this brief introduction can answer a number of questions at one time. A few minutes of explanation to everyone at this point can avoid countless later frustrations for the director. Because the crew will understand what is to happen, and why, it is in a better position to make a more meaningful contribution to the production.

SETUP AND REHEARSAL

The director is usually not closely involved with the actual studio setup since these functions can be handled by the production team following the director's prepared instructions. Nevertheless, the director should always take the time to check on the progress of the setup and approve the final set construction, lighting, and equipment.

The rehearsal stage is a demanding time for the director because that is when the various produc-

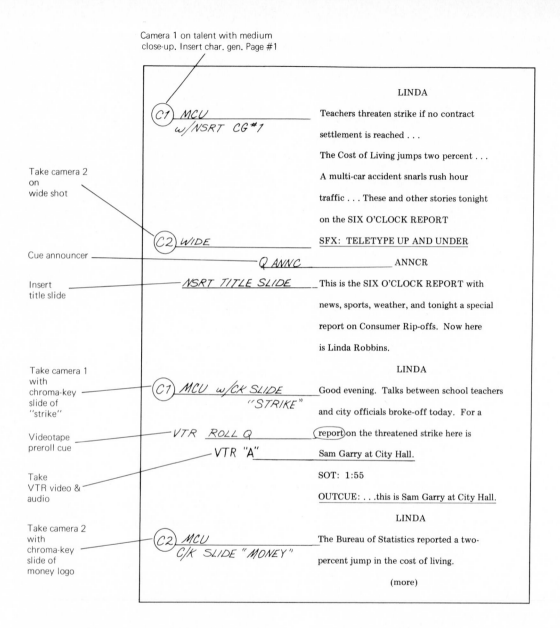

Camera 1 on talent with medium close-up. Insert char. gen. Page #1

Take camera 2 on wide shot

Cue announcer

Insert title slide

Take camera 1 with chroma-key slide of "strike"

Videotape preroll cue

Take VTR video & audio

Take camera 2 with chroma-key slide of money logo

C1 MCU
w/NSRT CG #1

C2 WIDE

Q ANNC

NSRT TITLE SLIDE

C1 MCU w/CK SLIDE "STRIKE"

VTR ROLL Q

VTR "A"

C2 MCU
C/K SLIDE "MONEY"

LINDA

Teachers threaten strike if no contract

settlement is reached . . .

The Cost of Living jumps two percent . . .

A multi-car accident snarls rush hour

traffic . . . These and other stories tonight

on the SIX O'CLOCK REPORT

SFX: TELETYPE UP AND UNDER

ANNCR

This is the SIX O'CLOCK REPORT with

news, sports, weather, and tonight a special

report on Consumer Rip-offs. Now here

is Linda Robbins.

LINDA

Good evening. Talks between school teachers

and city officials broke-off today. For a

report on the threatened strike here is

Sam Garry at City Hall.

SOT: 1:55

OUTCUE: . . .this is Sam Garry at City Hall.

LINDA

The Bureau of Statistics reported a two-

percent jump in the cost of living.

(more)

tion elements must be coordinated and integrated into a television program. Exactly how the director approaches rehearsals will depend, in large part, on the particular show and the production mode being used. You already know that a program can be produced in a variety of ways, ranging from a live broadcast, where the director must work with the performers and crew in real time, to the single-camera-VTR, film-style approach, where every shot is individually set up, lit, rehearsed, and videotaped before moving on to the next. The particular production mode will obviously influence the director's rehearsal approach. A director who uses a segmented or a shot-by-shot approach can concentrate on small portions of the show, perfecting and taping each one before going to the next. A show that is to be produced live or live-on-tape must be rehearsed in its entirety since that is how it must ultimately be produced.

Dry Rehearsal

The earliest rehearsal sessions take place outside the studio in a rehearsal hall. This is when the performers and director first assemble to read through the script and to discuss the production. Since no production facilities are used, these sessions are called *dry rehearsals*.

A dry rehearsal period is always used in dramatic, musical, or comedy shows, but it is often necessary in nonentertainment productions as well. An instructional or demonstration program needs rehearsal as much as any other type of program. A series of dry rehearsals in an office or large room will help the in-studio rehearsals and production of any show proceed more smoothly.

Once the initial read-throughs are completed and the necessary script changes are made, the director begins to work out performer blocking. Although you may be working in a room without sets or props, you can use tables, chairs, and other available furniture to suggest the actual set pieces and their relative positions. Some directors mark the exact dimentions of the set with masking tape to give the performers a more accurate idea of their working area. However you simulate the studio set, try to keep your talent moving in an area that closely approximates the size of the actual setting. (See Figure 20-5).

The director should give the performers some idea of where each camera will be positioned, how far the boom will extend to cover performers, when they must hit their marks exactly, and so on. As you begin to take the performers through the preplanned blocking, you are likely to find some situations in which your early ideas are not working and others when someone comes up with improvements.

Dry rehearsal should be a time of innovation and experimentation. Although you want to keep the rehearsal proceeding on schedule, this is also when you can afford to be most flexible in trying out new ideas and alternative approaches. By the end of the dry rehearsal, both director and talent should be satisfied with the basic blocking. If you expect talent to have their lines memorized, give them an exact date when you want them to be "off-book." If you will be using cue cards or other prompting devices, have them available so talent can practice working with them. This is particularly important for nonprofessional talent who are unfamiliar with using cue cards or prompters.

Studio Rehearsal

Studio rehearsal is a crucial time for the director.

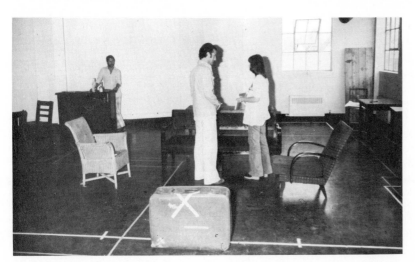

FIGURE 20-5 Rehearsal Hall.
During dry rehearsal, the director and performers work out characterization, line delivery, and movement. Various pieces of furniture and tape on the floor are used to indicate major set pieces and furniture so the performers can become accustomed to working in a space similar in size to the studio set. *(Courtesy: Thames Television Ltd.)*

In too little time with too much to accomplish, the director must establish the shooting sequence, integrate all production elements, and, at the same time, maintain an objective eye and ear for the creative aspects of the program. Juggling both the technical matters and aesthetic judgments is not easy, and the time pressures, which all directors encounter, compound the difficulty. The secret is to use your rehearsal time as efficiently as possible: have a plan worked out in advance, establish a rough rehearsal schedule of what must be accomplished by what time, and try not to get blocked by a minor problem which may take too much time away from the overall production.

The studio rehearsal generally proceeds through three phases: (1) camera blocking, (2) run-throughs, and (3) dress rehearsal.

CAMERA BLOCKING The director has already established the performer blocking during dry rehearsal. Now is the time to determine the shots, angles, and positions for each camera. Camera blocking can be tedious and repetitious, especially for talent. The director should warn performers of what to expect. Advise them to avoid expending all their energy in blocking and to save it for the dress rehearsal and actual production. If you are working with performers who are inexperienced with television, explain what is happening so they will understand why they are asked to repeat a move over and over.

Some directors like to block cameras from the control booth; others prefer to work directly on the studio floor. The floor approach has the advantage in that the director can communicate easily to everyone involved and can see any problems instantly as they occur. If the director wears a PL headset connected to the switcher in the control room, any of the cameras can be punched up on the studio monitor as the blocking proceeds.

If you are working with a tightly scripted show, your shots are already established for every camera. If you are working on a semiscripted or unscripted production, give each camera an idea of what it will be responsible for shooting. For example, in an interview situation, Camera 2, which is positioned in the center, might be assigned to maintain a fairly wide shot and to cover the host, while Cameras 1 and 3 might be told to photograph whoever on the panel is talking.

Do not spend a lot of time during camera blocking perfecting the framing and composition of every shot; this will come later. For now, simply establish the camera's basic shot to be sure that, first, the camera operator understands what is to be photographed and, secondly, that the camera is in position to shoot the subject.

CAMERA RUN-THROUGHS Once the camera shots, angles, and positions are established, the director can begin to run-through the actual show. Understandably, the earliest run-throughs proceed much more slowly than normal, as production problems are encountered and resolved. The director should try to speed up the pacing of the run-through as soon as possible, though, until the production is moving along at close to normal speed. Only then will problems in pacing or with blocking moves of cameras and talent become apparent. It is a good idea to direct most run-throughs from the control booth since the director can watch all camera monitors simultaneously and can also begin to work with the rest of the production team on their cues and operations.

The run-through usually flows in a start-and-stop manner in the beginning, with the director stopping to correct problems as they develop, backing up a little, and moving on until the next problem occurs. Once you have run through the entire production at least once and corrected what appear to be the major problems, successive run-throughs should become increasingly smoother as everyone starts to coordinate their operations.

The director is naturally concentrating on technical matters at this point but should not forget to keep an eye on the performances as well. Remind talent of the necessity to remain as consistent as possible in their movements and to be careful in handling graphics or other objects that are to be shot in close-up.

GIVING NOTES After each rehearsal period is

over, the director should give out suggestions, criticisms, and changes in the form of notes. If you have a production assistant available to jot down notes as they occur to you during rehearsal, you will not have to stop to write them down yourself. Otherwise, make some indication in your script so you will not forget your comments once the rehearsal is over. A good director hands out notes discreetly. Constant criticism of cast or crew will not necessarily result in superior performances. The note session should be a combination of the director's thoughts and the production team's feedback. Remember, the source of a good suggestion is unimportant; if the suggestion improves the show, it will make the director look better.

DRESS REHEARSAL   It is impossible to overestimate the importance of a dress rehearsal. The dress is the final chance for everyone to do the show before it is actually produced. If you never have a dress rehearsal—a complete show from start to finish—your air show will actually be the dress. Whatever problems might have been solved earlier will be a part of the production. Naturally, not all shows have a dress, but even a spontaneous and unscripted program can rehearse the opening, the closing, and some of the segments within the show.

The dress should include every element that will appear on the final production, including credits, music, sound effects, lighting cues, film, VTR inserts, and so on. Some directors like to tape the dress rehearsal since there are times when the best performances of both cast and crew occur in dress, not in the actual air show. If you will be doing much videotape editing, you might combine the best performances from the dress and air shows into a superior production.

PRODUCTION

Strange as it may seem, in most artistic endeavors the actual production effort is almost anticlimatic. All the real work and planning are poured into the preproduction and rehearsal periods. If the preparation was done properly, the production itself will flow smoothly. A symphony conductor spends hours carefully analyzing the musical score to develop a point of view and a style. Then

he enters into hours of intensive rehearsal as he molds the members of the orchestra into an ensemble unit, shaping, coloring, and building the sound. By the time of the actual performance, replicating what has already been achieved during planning and practice is paramount.

It is much the same for the television director. After so much time in preproduction and rehearsal, the actual production usually leaves little left to be discovered. Of course, a live program poses its greatest challenges during the actual broadcast. Even so, the director should have already decided on the basic approach, the camera coverage, and how to use all the production equipment and techniques at his disposal.

The Psychology of Directing

Directing—like sex—is one of those activities that must be experienced to be fully appreciated. We have only scratched the surface here, but regardless of how much reading you do about television directing, there is no substitute for the actual experience.

Before you sit in the director's chair, however, there are some points you ought to consider, which we will call the "psychology of directing." First, remember that the director is the leader of the production team. Consciously or unconsciously, the team relies on the director for guidance and for support. A director who appears to be in command will convey that feeling to the cast and crew. A director who appears tentative and uncertain will communicate this insecurity, which can result in a nervous and unsure performance. A good director is open with the crew, solicits suggestions and advice, but is also capable of making a firm decision when necessary.

When you sit in the director's chair, you must be the *director*, not the *directed*. Beginning directors are often overwhelmed by the barrage of information thrust upon them—watching the camera monitors, following the script, listening to the audio, tracking the time, cuing shots and performers, watching the show on the line monitor, and trying to maintain some sort of objective perspective of the entire production. If you are not careful, these many elements can begin to run out of your control, forcing you to respond to events that have already occurred rather than to make things happen at your command. It is almost like Frankenstein's monster, who turns on its creator. Instead of the director controlling the show, the show controls the director. All the director can do in that situation is to hold on and try to stay with the rapidly increasing momentum of a production gone out of control.

How do you avoid this? It is easier said than done, and a lot of it will come with practice and experience. One way to start is to develop the idea that everything which happens should occur because you, as director, want it to happen. The minute you feel control slipping away (it happens to every director at one time or another), realize what is happening and try to recover by slowing the pace by going to a wide shot and staying there until you find your place and by mentally composing yourself. Much like Hemingway's

characters, a good director exhibits "grace under pressure."

When Things Go Wrong

Even on the most carefully planned show, things are bound to go wrong sooner or later. A camera suddenly goes dead on a live broadcast, the film projector jams in the middle of a crucial insert, an actor blows lines, upsetting your carefully prepared shot sequence. You cannot avoid these situations, but you can learn to develop the right attitude to deal with them.

"One thing I've learned," says John Litvack, a director of CBS-TV's *Guiding Light*, "is that nobody out there watching the show has a copy of your shooting script. There's not one viewer that knows you should have been on Camera 2 when you were on Camera 3." Quite often a mistake that seems so obvious to you goes completely unnoticed by the average viewer. And even when mistakes are obvious, getting excited or upset about them will not correct the problem. A calm director who thinks ahead quickly can respond to a problem with a minimum of damage to the show.

If you are videotaping a production, you might take a ten-minute break after a disaster has forced you to stop shooting. Sometimes all it takes

FIGURE 20-6 Director in Production.
At many facilities, the program director must also function as assistant director and switcher. To function in these multiple roles and still direct the production effectively requires complete familiarity with the operation of the equipment and excellent preplanning. *(Courtesy: 3M/ Datavision Video Products)*

is a break in the tension to help everyone (including the director) relax and recharge. Rather than push everyone to the limit, call a short break; it can be the most effective medicine for a hyper-tense production team.

Consider the Team

One final point is to consider the effort of the production team. It is easy to forget the enormous concentration and physical energy expended by everyone working on a show. Be sure to give rest breaks from time to time even if you are under time pressure. A refreshed crew and cast will invariably turn in a better performance. Finally, remember to thank everyone once the production is over. Regardless of how well or poorly the show ran, a "thank you" is a small courtesy that is always largely appreciated.

POSTPRODUCTION

The director's primary role in postproduction is to supervise the editing of videotaped shows. There are times when the editing is a simple assembly of various segments into a completed production. Other editing sessions may involve the elaborate editing of individual shots to build each sequence and then the assembly of each sequence into the final show.

Viewing the Footage

The director's first job is to sit down and watch all the recorded footage that was marked "good" or "maybe" on the production take sheets. If you can make a copy of the footage into a helical cassette, you will be able to view your tape off-line under less expensive and less distracting conditions. If you are editing with time code, you can begin to write down possible edit points as you view.

Screen the footage critically but with an open mind. First impressions in the control room can be deceiving. You may think, for instance, that there is nothing worthwhile in the dress rehearsal footage while you are in the booth only to find, when viewing the recording later, that there is a lot of strong material that can be used. Although you need not spend time viewing takes which were

labeled "no good," do not throw them away yet. You may still be able to use a shot or a short segment from a bad take even though the entire sequence was unsatisfactory.

It is a good idea to use a fresh copy of the script as an editing script. Make notes about the type of shots on both sides of a potential edit point to avoid producing a jump cut or a poor visual match. It is to be hoped that you will have recorded some protection, or cover shots, reactions, and pickups which can be used to smooth out difficult video transitions during editing.

Do not concentrate only on the picture, but listen carefully to the audio as well. Poor audio can be as good a reason to reject a take as a poorly framed shot. If you are working with multi-track audio recorders and postproduction audio mixing, start to think how you will be mixing the audio with the picture. The final audio mix is usually done when the video has been edited together, but considering both audio and video simultaneously will give you more control over both production elements.

Finally, remember that, as you edit the show, you may inadvertently alter the pacing of the shots. After you have completed editing a series of shots or segments, go back and view the entire sequence straight through with an eye on the pacing to make certain it remains consistent with the rest of the show.

COMMON DIRECTING SITUATIONS

The term "television production" applies to a wide variety of programming ranging from commercial broadcasting to educational and industrial closed-circuit communications. Yet, from the director's point of view, it is possible to group all programs into three major production categories: (1) fully scripted programs, (2) semiscripted programs, and (3) unscripted programs. In this section we will discuss some of the most common programs within these three categories and show some of the directorial considerations for programming within each of the three. Our purpose is not to give you a detailed "recipe" to follow since that is technically impossible and aesthetically undesirable, but rather to give you some idea of the director's role in working with various types of programs and to point out similarities and differences among the most common directing situations.

Fully Scripted Programs

As the name suggests, for shows in this category, the director uses a detailed script with specific dialogue and action that is rehearsed completely before production. The most common types of fully scripted shows are newscasts, dramatic or comedy show, and musical programs.

NEWSCASTS The newscast has become a staple of broadcast and cable television. Although the newscast is fully scripted—insofar as the newscasters read from a prepared script—owing to time limitations the program is often produced without a complete rehearsal. The standard operation, such as the use of chroma key, lower-third graphic inserts, and film and VTR rolls, are done so often that once the pattern has been established, there is little need for daily rehearsals with an experienced production team.

The important factor in a newscast is to establish the basic operating procedures, which are used day after day. It doesn't matter which graphic is actually used or which visual is chroma keyed onto the screen. As long as the crew is able to perform the operation smoothly, the program will have a polished and precision on-air look. Any change in program format or a major change in the talent or production crew is a good time to schedule some rehearsals and run-throughs until the director is satisfied that the team is well coordinated again. News can occur so quickly at times that director and crew must work largely on instinct and habit when deadlines run close to air time. As long as the production has an estab-

lished and workable production format, the director can handle the fastest breaking news with little difficulty.

Here are some of the important points a newscast director should consider:

1 Make certain everyone knows and understands the preroll cue time that is used for film and VTR. Confusion here can result in disastrous introductions and transitions.

2 Assign someone on the production team to check all graphics (on slide, camera card, or electronic still frame) before the broadcast. If no one is available, leave yourself enough time to check them for accuracy and correct sequencing.

3 Use the format sheet or daily rundown sheet to preview all film and videotape clips. Check for the proper sequence and to make sure that the running time, indicated on both the rundown sheet and the script, is accurate.

4 If you will be A- or B-rolling either film or videotape, be sure you know the right cues and times for switching sound or video from one projector or VTR to the other.

5 Look ahead to the next series of major events. Have some alternative plans in mind in the event of such unexpected problems as the failure of the live remote feed or a film clip which has jammed or broken in the film chain.

6 Timing is critical on a newscast because of the number of short film and VTR segments, the A- and B-rolling operation, which requires precision coordination of sound and picture; and the unexpected flow of news events, which can cause instant changes in a previously planned newscast. Whether you work with an AD or keep time yourself, you must be able to time all program segments accurately and to calculate running time, program time, and dead-pot time, quickly and efficiently.

DRAMA AND COMEDY Although drama and comedy may appear to be quite different types of programs, from the director's perspective the two can be approached in much the same way. The director assigned to either must read the program's script until he or she is sufficiently familiar with it to develop the approach and point of view that will be used to interpret the writer's script into picture and sound.

Once the director is familiar with the script he or she can begin to work out camera and performer blocking. The particular shots are determined by the director's point of view and interpretation of the script and the characters.

In any dramatic or comedy show the actors are of paramount importance. Although you are using television technology to convey the performance to a viewing audience, the play comes from the actors. The director must do whatever is necessary to make the actors feel comfortable, so they can deliver the best possible performance. During the early dry rehearsals, the director should encourage the give-and-take of ideas, suggestions, and changes from the cast. This will help to make the actors feel more at ease with their dialogue and blocking and can result in a better overall show.

Timing—both in terms of actor performance as well as cutting from shot to shot—is an essential ingredient in both comedy and drama. Unless the director is able to meet the actor at the right place, at the right time, with the right shot, the dramatic or comic moment that is being so carefully developed can be completely lost. The choice of shot, angle, and other aesthetic factors must never be casually decided in comedy or drama, but should be carefully selected to complement the script and the performances. Remember, too, that there are times when a simple reaction shot at the right moment or a nonverbal expression shot in close-up can be more powerful and revealing than pages of dialogue.

MUSICAL PROGRAMS Musical production situations vary in complexity from a relatively simple tune sung within a daily variety program to an elaborate opera performed by a full cast and symphonic orchestra. Musical programs can be difficult at times because they do not always have a written script with lines and action to follow. If you will be directing the performance of a singer or a group, you can transcribe the lyrics to use as the basic shooting script. For musical breaks between lyrics, indicate the solo instrument and the number of bars played between the lyrics.

If you are to direct such classical works as

symphonic or chamber music, you might use an actual musical score as the script. If you cannot read music notation, have a musician help follow the score for you. Simply indicate your shots on the score itself, and follow the notation as you would follow conventionally written dialogue.

How much you can do with a musical group in terms of shots depends in large part on how the performers are arranged in the studio set. Before deciding on any setting, consult the group to discuss its performance. Some musicians may have to be physically close together or at least maintain a line of sight so they can follow each other as they play. Work out a staging arrangement which will offer you some interesting pictorial composition with strong foreground and background elements. Avoid a setting in which important members of a group are in positions where the cameras cannot shoot them easily during the show.

Once you know the specific selections to be performed, listen to the music repeatedly until you begin to visualize what you want to show the audience. Make certain that the arrangement you are using to plan your shots is identical to the one which will be performed. A different arrangement might require a radical alteration of your planned shot sequence.

Avoid the common pitfall of scheduling too many shots for any musical number. Music that is being performed can go much faster than you might imagine while listening to a tape in your office. It is better to underestimate the number of shots you can comfortably use and to add more later than to have overplanned shots only to find that your shooting sequence is in a shambles when the cameras cannot get the shots in the time they have available. A final point to remember when planning musical programs is to check the volume on the PL communication system. Music is often loud when performed—this applies to all music and groups short of a classical guitarist—and the volume can overpower a weak PL system, making director commands unintelligible to camera operators.

Semiscripted Programs

A semiscripted show follows a definite program format, or rundown, but does not use a script containing word-by-word dialogue. Semiscripted programs are frequently used in instructional or demonstration-type programs, where the sequence of events is planned prior to production, but the actual dialogue is spontaneous within the preset format. Although semiscripted programs do not use a complete script, much of the show can still be thoroughly planned and rehearsed prior to production.

DEMONSTRATION AND INSTRUCTIONAL SHOWS The program format category of demonstration and instructional shows spans a wide range from the familiar cooking and gardening shows to instructional, closed-circuit programming used in many schools, universities, and industries.

Generally the demonstration and instructional show is produced in one of two ways. In the first approach, the talent speaks directly to the viewer. In the second approach, the talent is accompanied by a host or interviewer who acts as a substitute audience, asking questions, clarifying points, and providing emphasis where necessary. The advantage of the latter approach is that an experienced host can help nonprofessional talent by handling all the technical chores, such as timing, receiving cues, and remembering which of the cameras is on the air at a particular time.

Most demonstration and instructional programs use a fully scripted opening and close, but only a detailed rundown sheet for the middle sections of the show. Even though talent may not need a complete set of cue cards, it is a good idea to put the outline of the show on a cue card so the talent will not alter the planned sequence of events once you are into production. Experts may know what they want to say, but they can easily confuse the order of events which can ruin your camera blocking and shot sequence.

A demonstration can usually be covered by assigning at least one camera to shoot talent while another camera shoots a close-up of the action. This way you can vary the size of the talent shot from wide to close-up while still following the demonstration easily on the other camera.

Shooting effective close-ups takes some planning and coordination between director and talent. Make certain talent knows which camera will shoot the close-ups, and remain consistent with your shots. Instruct talent on the best angle to hold the objects for the most effective close-up shot. A studio monitor placed nearby, within sight of talent, can help performers to coordinate their moves with the camera operators (See Figure 20-7).

TALK AND VARIETY PROGRAMS The talk and variety show is a potpourri of interviews, music, and features. Such shows are either partially or completely unscripted but rely on a very precise rundown sheet, which carefully outlines and times each program segment. The director uses the rundown sheet as a guide and tries to preset camera shots beforehand. Usually musical segments and complex demonstrations are rehearsed before production.

For interview or talk segments, most directors develop a standard format that is followed on every show. On a three-camera production, for example, the middle camera might be used to cover a wide shot of the panel. One of the side cameras could cross-shoot whoever is talking at any given time, and the other side camera could be used to cross-shoot the opposite side of the panel. Once the basic shooting pattern is established, the director relies heavily on the camera operators to get their shots as automatically as possible.

Unscripted Programs

The term is somewhat misleading because even an "unscripted" show contains certain segments which are preplanned prior to production. In addition, most unscripted programs follow some sort of rundown sheet, which indicates the basic structure of the show and the way openings, closings, and transitions are to be produced. Since this is the case, the open, close, and some

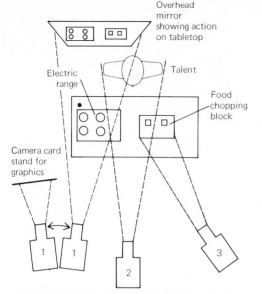

Overhead mirror showing action on tabletop

Electric range

Talent

Food chopping block

Camera card stand for graphics

1 1

2

3

FIGURE 20-7 Demonstration Program.
For the cooking demonstration show illustrated in this figure, the director has assigned each camera its own area of shooting responsibility. Camera 2 primarily shoots the performer and varies the shot size, depending on the director's commands. Talent is instructed to speak to Camera 2. Camera 3 shoots primarily close-ups of the demonstration on the tabletop. Performer is instructed to "cheat" objects and graphics toward Camera 3 whenever they must be covered in close-up. Camera 1 shoots the demonstration as it appears in the overhead mirror, which is suspended directly above the demonstration table. When necessary, Camera 1 can break to cover camera card graphics on nearby stand.

of the internal sequences can be practiced and rehearsed prior to production.

SPECIAL NEWS COVERAGE Special news events, such as election returns, extraordinary legislative sessions, parades, and rallies, are examples of unscripted coverage. However, certain aspects of these programs can be planned and rehearsed to establish a production pattern for the director, crew, and talent. Again, the idea is to work out the use of cameras and equipment in the production, not to worry about specific content. Naturally, openings, closings, and transitions can all be rehearsed prior to air time.

Since the director is busy directing what is

usually a live broadcast, the producer must take an active role during the production by alerting the director to the most important events and coordinating the various facets of the production. For example, on election returns the producer should watch the different stories unfold and tell the director which story (and, therefore, which cameras, locations, talent, and so) to cut to next. The same is true for any special news coverage. On many large, preplanned events, the organizer of the event will usually hold an advance press conference to give details concerning the sequence of the event, anticipated running times, and other pertinent facts, which can be helpful to the director. If at all possible, both the director and the producer should attend these conferences and then hold their own preproduction briefing with the production team prior to broadcast.

SPORTS COVERAGE The popularity of football, baseball, basketball, hockey, and other sports events has made sports coverage a programming standard for local stations and cable systems, as well as for regional and national networks. In Chapter 21, we will discuss the specific details for preparing and setting up these broadcasts. For now, we will cover in a more general way, the director's planning and preparation for directing sports programming.

As with any production, the director of a sports event must be thoroughly familiar with the rules and play of the game, as well as with the particular teams or participants involved. The director must know the participants' style of play, what they tend to do in various situations, who the outstanding players are, and the possible color stories which surround the main event. Unless you know these things you cannot do an effective job of covering a fast-moving game, regardless of the sport.

If you have been assigned to direct a game with which you are not completely familiar, the first thing to do is to watch as many events as possible prior to production. Also, do some reading about the sport and the strategies and tactics employed. Generally the publicity offices of the teams will be able to provide you with team rosters, background information, and, sometimes, helpful tips or hints on what may happen during the game. Not only is this information helpful to the director, but it should also be passed along to the play-by-play commentator.

The director relies more on the production crew during a sports event than in any other directing situation. There simply is not time for the director to call for every shot, graphic, replay, or audio cue. The entire team—not just the camera operators—must be completely involved in the production, all doing their jobs with a minimum of directorial commands.

To illustrate, say you are directing a football game and have just covered a successfully completed pass from the quarterback to a receiver. As soon as the live play is over, the videotape replay operator immediately recues to the start of the play. Simultaneously, the character generator operator punches-in the receiver's jersey number to retrieve the appropriate lower-third graphic. The switcher has the lower-third graphic supered over the freeze-frame and ready on preview. As soon as the director asks for the effect on the air, the VTR operator begins the playback, varying the speed to emphasize the crucial elements of the play in slow motion and replaying less critical parts at normal speed. Except for the director's command to punch up the freeze-frame super and to start the replay playback, everything else was done automatically by a crew working closely together.

Ted Nathanson, a sports director for NBC, tells a story that illustrates how alert a crew should be. "One day we were covering a game when all of a sudden a dog ran out on the field, temporarily halting play. Before I said a word all five cameras had the same close-up shot of the dog. Unfortunately at that very moment the dog was responding to nature's call in front of millions of viewers. Since all my cameras had the same shot there wasn't anything I could cut to. It was a little embarrassing, but I had to give my guys credit for getting the important shot before I ever had a chance to call for it."

Ted Nathanson's preprogram routine is a good example of how a director approaches any sports show. Nathanson spends the week prior to the game studying both teams, their players, and their tactics. He also speaks with the production's play-by-play announcer and the color analyst to coordinate isolated shots and possible color coverage from the bench, sidelines, or dugout. Just prior to the game he meets with the production team for a pregame briefing. He outlines the two teams, explains their overall strategies—what might be expected of each team in various situations, which players to watch carefully—and any other relevant details. Each camera operator is assigned a particular area of responsibility, depending upon the game situation. The camera operators will be expected to get their shots with a minimum of direction since not every shot can be called and set during the game. Nathanson gives similar instructions to the VTR operators, the audio engineer, and the character generator and still-frame unit operators. The AD is used primarily for timing and to coordinate communications between the remote unit and the home studio. The producer calls all isolated and replay instructions and helps to alert the director to important events during the game. You will note that Nathanson tries to prepare the crew to react properly under different circumstances and situations. After the orientation, he expects them to use their own skills and judgments since it would be impossible for him to call out every cue for every production operation during the frenetic activity of any live sports broadcast.

Finally, the director should remember that many viewers expect more than just the game coverage itself. Although the game is the center of attention, stories from the sidelines, the crowd in the stands, cheerleaders, and other color events enhance the viewer's enjoyment of the entire sports scene. But do not become so engrossed in using your cameras for color stories that you neglect the main action on the playing field.

Remote Production

Any production which is shot outside of the studio is considered to be a "remote" production. Of course, all the usual director considerations for a studio production apply to remotes as well, but the nature of location shooting adds a number of additional complexities for the director to deal with during the various production stages.

ENG PRODUCTION News coverage is generally accomplished on a regular basis without a program director per se. Usually, the reporter serves as a "producer-director" and works closely with the camera crew in setting up shots and camera angles. The fast-breaking nature of news events generally does not lend itself to much preproduction planning, and most day-to-day news situations are covered "on the run."

EFP PRODUCTION Electronic field production (EFP) is the term which is generally applied to single-camera shooting/videotaping at a remote location. EFP differs from ENG because the former is usually planned thoroughly, requires much higher overall production values, and generally includes a director who supervises the production.

EFP productions vary in size and complexity from a relatively simple shoot with only a camera operator and another crew member serving as "producer-director" to an elaborate production involving a large production crew and a full-fledged "director." Common EFP situations are entertainment programs, magazine-type feature shows, commercials, and training/industrial productions. Since the EFP technique utilizes only one camera, the director can concentrate on each camera setup without having to cut between cameras as in a control room situation. However, the EFP production must be carefully preplanned so that all camera angles necessary for postproduction editing are shot and to make certain that continuity requirements are kept in mind. Another

crucial consideration for the director working on an EFP production is to determine the crew members and equipment which must be brought to the location. Remember that you will not have the luxury of a fully equipped studio, so lighting, electrical power, props, and even such minor items as gaffer tape and extension cords must either be brought along or provision made to have them available at the location site.

MULTIPLE-CAMERA REMOTES A multiple-camera remote involves the use of multiple cameras and a portable control room which is usually contained inside a van or truck. In the large scale remote (LSR) situation, the director works from a control room setting and functions exactly as he or she would working in a studio control room by cutting between cameras during production. Although the LSR situation functions much like the conventional studio operation, it is still a remote; and cameras, microphone, lighting equipment, and so on must be set up specifically for the production.

In Chapter 21, "Remote Production," we will cover each of these production situations in much greater detail. The important point for you to remember now is that remote shooting requires all the director's usual skills and abilities but makes added demands on the director's capability to work with people and equipment efficiently and effectively and often under conditions which are less than ideal. Careful preproduction planning—while important to the success of any production—is an absolute necessity when you are directing a remote.

SUMMARY

The director's primary responsibility is to supervise the production team in order to transform an idea, concept, or script into the program's actual sounds and pictures. In order to work effectively with the production team, the director must establish a feeling of cooperation, teamwork, and mutual respect among the crew and production staff.

During production, the director must communicate with the entire crew, but especially with the camera operators, switcher, and floor manager. A series of operational commands have been established and should be used consistently. Commands must be precise, accurate, and economical to be most effective.

The many crucial operations performed by the program's director during rehearsal and production can be shared by an assistant director. The AD generally assumes three functions: (1) timing the show and individual segments, (2) readying upcoming cues, and (3) presetting camera shots on fully scripted shows.

During the preproduction stage, the director should meet as early as possible with the producer to agree on the interpretation, point of view, and approach to the show. Once this has been accomplished, the director can meet with key production personnel to assign specific areas of responsibility. Working with the program script and floorplan, the director (1) blocks camera and performer movements, (2) sets up individual camera shots, and (3) develops the overall shooting sequence.

While the director is not usually needed on stage during studio setup, all sets, lighting, and equipment should be checked and approved prior to rehearsal, so changes and corrections can be made in time. The director's approach in rehearsal varies depending on the show's production mode. Programs which are taped in small segments are usually rehearsed differently from live shows or programs taped in longer sections. Most rehearsal periods are divided into two parts: *dry rehearsals,* which are held outside the studio, and *studio rehearsals,* which are held in-studio with full production facilities. During dry rehearsal the director and cast read through the script and work out performer blocking. Studio rehearsal generally consists of three operations: (1) *camera blocking,* in which shots are established; (2) *run-throughs,* when the director begins to integrate the various production elements into a unified show; and (3) *dress rehearsal,* which should be a replica of the final air show.

One of the director's most difficult jobs is to provide leadership and direction to the production team. In addition, the director must always be in control of the program—as the *director,* not the *directed.* As soon as events occur which are not determined by the program's director, the show can run out of control, and it may become impossible to meet the program objective determined in preproduction. A good director must also remain flexible and have the presence of mind to deal with unexpected emergencies or mistakes which can happen at any time.

Postproduction consists of viewing taped footage and editing the show together. Regardless of how simple or complex the editing job, the director must maintain an objective perspective, even while dealing with relatively small elements of a much larger show.

Although there is a great variation among program formats, from the director's viewpoint all productions can be divided into four categories: (1) fully scripted programs, (2) semiscripted programs, (3) unscripted programs, and (4) remote production.

CHAPTER 21

REMOTE PRODUCTION

A *remote* is any television production which takes place outside the studio. In Chapter 12, we discussed a special type of remote production, electronic news gathering (ENG). In this chapter, we will discuss the other two major categories of remote production: (1) *electronic field production* and (2) *multiple-camera remote production*. What distinguishes electronic field production and multiple-camera remote production from ENG is the ability to preplan the first two and to give more careful attention to the various production elements than is usually possible when covering most news events with ENG.

Every remote production—no matter how simple or complex—uses all the television production techniques, which we have discussed throughout the book. But shooting on remote also involves a number of additional considerations which must be properly handled for an effective production. In this chapter, we will discuss the various types of remote productions and how they are approached by the production team. Since remotes always involve basic television production techniques which have been dealt with in detail in other chapters, our intention here is not to repeat each of these but to discuss them within the context of the unique requirements of a remote production.

WHY SHOOT ON LOCATION?

The answer to this question is self-evident for some remotes. A baseball game, a congressional hearing, a Thanksgiving Day parade all take place at a specific location, which is where television must go to cover them. The decision in these cases is not whether to shoot on location but whether to cover the event at all.

There are many other production situations, however, where the producer and director have a choice of shooting either inside the studio or out in the field. Commercials, entertainment programs, and industrial or instructional productions offer a wide choice of production approaches and locations. Why should a producer choose to produce a show either on location in the field or inside the studio? As with most other production decisions, the producer must weigh the advantages and disadvantages from both a creative as well as a technical perspective.

ADVANTAGES OF REMOTE PRODUCTION A remote location offers realism, detail, and an atmosphere which is often impractical or impossible to recreate inside the studio. The producer of an industrial training tape on steel mill safety procedures could never duplicate the blast furnaces of a real mill inside the largest production

studio. The producer of a commercial, written to be shot on a busy city street, could never hope to replicate New York's Fifth Avenue inside the studio. A musical program taped with the Rocky Mountains for a background offers a setting which not even the best scenic designer could ever build inside the studio.

Another advantage of remote production is that set design and construction are usually unnecessary. While the art director might dress a location setting or modify it slightly to suit the production's specific requirements, the use of the existing background is one of the reasons for shooting on location in the first place. It has also been found that many performers—especially nonprofessionals—feel more comfortable working within their natural surroundings, so shooting on location can often result in better performances.

Finally, television viewers have come to expect a higher degree of realism and authenticity in virtually every aspect of television, from news and documentaries to entertainment and sports programming. Shooting on location is one way to enhance a production and add interest and excitement although even the most beautiful location will not compensate for a weak script or a poorly planned production.

DISADVANTAGES OF REMOTE PRODUCTION
Before you contemplate shooting every production in the field, let us look at the other side of the coin. A television studio is a safe, comfortable, and well-equipped place to do a television show. It offers maximum production control over many different aspects of a production. It is soundproof, weatherproof, and lighttight and includes the equipment, electrical power, and physical space necessary to produce a television program. Studios also offer heating and air conditioning, bathrooms, telephones, and so many other items we take for granted, which are not always available at a remote location.

A remote production requires carefully detailed preplanning and coordination to work. A minor mistake or miscalculation on a studio production can turn into a major disaster if it occurs in the field. Even when the production is meticulously planned, however, it is still subject to such uncontrollable problems as weather, noise, and the usual set of difficulties we group under Murphy's law.

ELECTRONIC FIELD PRODUCTION

Electronic field production (EFP) is a term which refers to a wide range of production techniques which all have one thing in common: the use of a portable camera and videocassette or videotape recorder employing the single camera/single VTR film-style production approach. At first glance, EFP may seem quite similar to ENG, which is not surprising because EFP grew out of the technology and experience which was gained with ENG. However, the two are very different production approaches which have evolved in separate directions for different production applications. ENG is designed to cover fast-breaking news events, and this must be accomplished with little preparation and a minimal ability to control events. EFP is used when immediacy is not crucial and when higher production values and better overall technical quality are important considerations. Unlike ENG, EFP offers you the ability to preplan the shoot carefully and to attend to all the aspects of the production with the same care as if it were produced inside the studio. So when we talk about EFP, we are referring not only to a production approach but also to a production "attitude" which involves overall control of the production in ways which are impossible when dealing with news events.

Over the past few years, EFP has become the fastest-growing area of television production because of its ability to combine flexibility and efficiency with the special flavor which only a remote can bring to a production. EFP techniques are used in the production of entertainment shows, industrial and educational programs, commercials, and for the magazine-format feature programs which appear on many local stations and cable channels. The ability to apply EFP methods to a variety of production situations at every level of sophistication is one of the technique's greatest strengths. An example of a relatively simple EFP production would be an interview with the chief executive of a corporation which is taped inside his or her office by a two-person crew and edited on a basic two-VCR editing system. At the opposite end of the EFP range are commercials, training programs, or dramatic presentations which are shot in dozens of locations employing many performers, a crew of a dozen people or more, and a great deal of sophisticated video and audio postproduction.

One of the most important decisions you have to make with EFP is deciding on the level of production sophistication to use for each shoot. The ability to match precisely the amount of equipment and personnel and the level of production sophistication which are necessary to produce a particular remote will enable you to produce a program both effectively and efficiently. In this section we will discuss the equipment and techniques which are used in EFP and how to apply them in each of the production stages.

EFP Equipment and Operation

The wide range of production situations which employ the EFP approach result in a very wide selection of EFP equipment. For very simple EFP productions, the crew may use equipment which is virtually identical to equipment used by an ENG unit, although usually with some modifications. On the other hand, complex EFP productions may utilize cameras, audio equipment, lighting gear, and videotape recorders which are as sophisticated as those found inside the studio.

EFP Mobile Unit

A very simple EFP shoot may require nothing more than a conventional automobile to carry the basic equipment and crew to the production location. However, most production facilities have modified a station wagon or van to serve as the EFP mobile unit. A dedicated EFP van offers a number of important advantages. First, the van—which is usually of the same size as an ENG van—has adequate space to hold a great deal of equipment and a fairly large crew. Since microwave equipment is unnecessary in EFP, the space saved can be used to store extra production equipment. Also, the roof of the van can be modified to make a convenient camera position for high-angle shots. (See Figure 21-1.)

A well-designed EFP van should have a number of built-in racks for equipment storage. The built-ins not only keep the equipment from being damaged during travel, but they also keep everything neatly arranged and enable the crew to locate the equipment quickly. Also, compartmentalized storage space is highly efficient and enables the van to carry much more equipment inside a relatively small space.

ELECTRICAL POWER Most EFP equipment can be operated from either battery power or conventional ac. An EFP mobile unit which can hook up to an ac outlet offers increased production capability. An internal electrical generator which is powered off the van's engine increases its operating flexibility, by permitting totally independent power operation for extended periods of time.

COMMUNICATIONS The EFP unit's primary communication needs are internal. Walkie-talkie units are the most useful means of establishing a communication link between crew and production team members, who may be widely separated throughout the location site.

EFP Cameras

EFP cameras are usually selected from among the high-quality, shoulder-mounted cameras or, for even better picture reproduction, the line of convertible cameras which deliver the same picture quality as their larger studio versions (see Chapter 2). Since EFP units rarely need the lightweight portability necessary for news coverage, they can afford to compromise slightly on size and weight to obtain better picture quality.

To provide for greater production flexibility, EFP camera systems are usually equipped with such accessories as special lenses, filters, and a collapsible dolly in addition to the usual tripod and shoulder mounts.

Audio

EFP audio requirements can vary from a relatively simple and straightforward hand or lavalier micro-

FIGURE 21-1 EFP Van.
The van contains all of the production equipment necessary for EFP production. The top of the van serves as a convenient camera platform for high-angle shots. *(Courtesy: RCA Corp.)*

phone to the use of boom mike and wireless microphones for more elaborate productions. A basic audio setup should include a number of hand-held and lavalier microphones for situations where on-camera mikes are acceptable. Shotgun mikes and a fishpole boom are necessary to pick up audio when microphones cannot appear in the picture.

Since EFP often requires the use of multiple microphones, the unit shoult include a battery-powered audio mixer to permit the audio engineer to balance and mix a number of microphones simultaneously. For production situations where some audio playback will be necessary (such as for musical lip sync), you will have to bring an audio tape recorder, speakers, and amplifiers.

Wireless microphones are becoming increasingly common on EFP shoots because of their versatility. They produce excellent sound quality when hidden on performers, yet avoid the problems associated with trailing cable or the need for off-camera boom microphones.

The EFP audio section should include a great deal of additional microphone cable; gaffer tape for securing cable to floors and walls; headphones for monitoring the audio pickup, and spare batteries, fuses, and so on for microphones, mixers, and recorders. Since you will often shoot outdoors, be certain to carry windscreens for all microphones.

Lighting

The higher production values expected with EFP require more than the basic lighting kit used in news productions. A standard EFP lighting package should contain lensless spotlights, floodlights, mounting devices (stands, gaffer-grips, and wall units), barndoors and screens, gels, and sufficient amounts of electric power cable. Naturally, the amount of lighting equipment you will need varies from shoot to shoot, but a standard package ought to enable the lighting director or camera operator to control the illumination effectively and to create whatever lighting effects are required.

While shooting outdoors with EFP, you may have to use booster illumination to lighten dark shadow areas on the subject. Without that additional fill light, the contrast ratio of the picture would exceed the camera's reproduction capability and result in a poor picture on screen. Reflector boards are the cheapest and most convenient way to produce fill light outdoors. If you intend to use electrical instruments outside, be sure you use dichroic filters to convert their 3200°K indoor light temperature to the higher 5600°K color temperature necessary to match outdoor illumination. (See Figure 21-2.)

Windows in an interior shot can cause lighting problems, especially when you must match the brightness and color quality of outdoor light with an artificially illuminated indoor scene. The large sheets of window gel, which both color correct outdoor light for indoors and reduce the brightness level to better match interior illumination, are very useful for these situations and should be included in the EFP lighting kit.

Videotape Recorders

All EFP productions are videotaped for editing later during postproduction. Probably the most commonly used EFP recorder is the 3/4-inch videocassette machine which is battery-powered, highly portable, and very reliable. For those EFP productions where higher quality is required, 1-inch VTR machines have become a popular choice owing to their small size and flexible production capability.

Battery-powered VTRs and VCRs enable you to shoot your production at locations where conventional ac power is unavailable. But batteries have

FIGURE 21-2 Reflector Unit.
The reflector unit is a flexible and economic way to add booster illumination to lighten shadow areas when shooting outdoors.

limited power life and take a relatively long period of time to recharge. It is important to have a number of fresh battery packs available so you won't have to stop shooting because of rundown batteries.

If you will be shooting any or all of your EFP production where ac power is available, by all means use it to power the VTRs (and other equipment) whenever possible. Ac power is reliable and will not run out at crucial times in production, so save the batteries for those situations where absolutely no ac power is available to you.

It goes without saying that the EFP unit should carry sufficient videotape or videocassettes for the day's shooting requirements. Always bring more stock than you actually think you will need. You can always return the unused tape, but you can't replace the time wasted if you run out of tape before you finish shooting at the location site. The videotape area on the EFP van should also have a slate for identifying each take and pre-gummed labels, which should be affixed to every cassette or tape reel and to every tape box to identify the tape and its content accurately.

MULTIPLE-CAMERA REMOTE PRODUCTION

As the name suggests, multiple-camera remote (MCR) production utilizes more than one camera, enabling the director to cut between two or more cameras while the production is either recorded on tape or broadcast live. Unlike either ENG or EFP production, which use a single camera/single VTR approach, an MCR production functions exactly like a conventional studio production; only it takes place in the field. The program director sits inside a miniature control room situated inside the mobile van and switches between cameras as the event is covered in real time. Some MCR productions are complete programs which are broadcast "live." Other MCR productions utilize multiple cameras to record the production in segments which are later edited into the completed show during postproduction. Although the MCR production team functions much as they would inside a studio, the fact that the production takes place in the field requires a number of additional considerations.

As with EFP production, the MCR approach offers a wide range of production capabilities. At the simplest end is a small van equipped with portable cameras (often the same as those used in an EFP unit) which are modified to feed a production switcher. Inside the minivan is a small production switcher which enables the director to cut, wipe, dissolve, and sometimes create a number of basic electronic effects. At the opposite end of the MCR range are the mammoth production vans which are literally the size of a Mack truck and contain control room and production equipment which is identical to that found inside a state-of-the-art television studio. The key element that all MCR productions have in common is the ability to cut between cameras in real time while the production is either recorded or broadcast live.

MCR Mobile Unit

The MCR mobile unit must be large enough to transport all the necessary production equipment to and from the location and, once at the site, serve as a control center during the actual production. Small-scale MCR units are often vans, minibuses, or recreational-type vehicles which have been specially modified to handle the increased weight load of the equipment and operating personnel. Large-scale MCR vans are generally trailer trucks which can run up to 45 feet in length and contain a completely equipped control room, videotape, and camera control areas. (See Figure 21-3 and 21-4.)

POWER The vans are powered with utility-supplied current, which must be available at the remote site. Smaller vans operate from conventional 120 volt, 20 amp circuits, but the larger units require special power ranging from 220/240 volts at 50 to 200 amps, depending upon the size of the unit and the amount of production equipment which must be powered.

COMMUNICATIONS All the necessary PL communications between the control room and the production crew which are used inside the studio are duplicated in the MCR mobile van. Camera cables carry the intercom communication between the camera operators and the director. Additional intercom systems are available for the

FIGURE 21-3 Small Scale MCR Truck. The specially adapted van is compact, easy to drive, and convenient to park at the production site. The interior control area contains all of the control room equipment necessary to produce with multiple cameras. *(Courtesy: Midwest Video Corp.)*

floor manager and other production personnel on large vans. On smaller vans which do not have secondary intercoms, the FM can plug into a nearby camera in order to communicate with the control room.

Walkie-talkies are useful for any MCR remote but indispensable on those productions where the activity which is covered is some distance from the location where the production vans are parked.

Business telephones are an important communications link at the MCR site, especially if the unit is feeding the program back to the studio for live broadcast of an event. Most large MCR vans

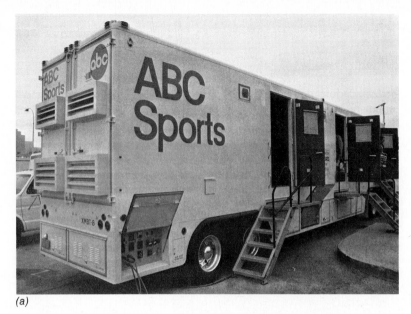

(a)

FIGURE 21-4 **Large Scale MCR Van.**
The exterior (a) shows a typical van-sized trailer. The interior (b) is divided into several production sections.

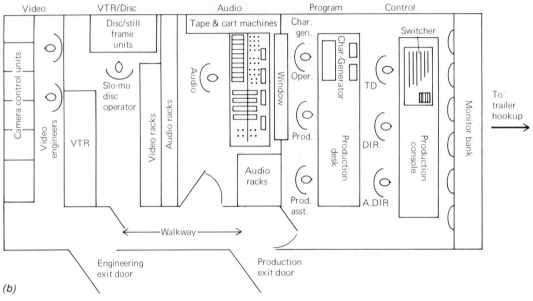

(b)

contain a number of built-in telephone instruments which can be connected to an outside phone line which is made available by the telephone company at the remote site. However, these phone lines must be specially ordered in advance so that the phone company can have them ready when the production van arrives.

CONTROL ROOM The control room area of the mobile van is a miniature replica of a conventional studio control room. There is a bank of monitors which display all camera and other video sources, and also a production switcher enabling the director to cut between them. The size of the van determines how large a production crew the control room can accommodate and the additional equipment which can be installed. Large-scale vans have enough room for a director, switcher, and AD to sit at the production console and for a producer, character generator, and still-store unit operator seated directly behind them. Smaller units generally can hold only three or four people inside the control area.

Usually the audio control console is located in an adjoining area which is separated from the control area by a glass window, enabling the audio engineer to monitor the program sound and communicate with the audio team without interfering with the control room operation. We will cover the audio equipment in greater detail later.

Behind the control area is another area for technical operations. This is where the videotape recorder operators and their machines are located. Often the camera control units are also located in this area of the truck. Since a multiple-camera remote uses a number of cameras operating simultaneously, an engineer must shade each camera as he or she would in the studio. (See Figure 21-5.)

MCR Cameras

The cameras utilized on an MCR production are generally the same studio-type cameras which are found inside a production studio, or they are

convertible cameras which are equipped for MCR with a large electronic viewfinder and a powerful zoom lens. In either case, the output of the camera is fed first to the CCU controls where the picture is shaded and then on to the production switcher located inside the control room area of the mobile van.

Newer mobile units generally use cameras which utilize triaxial cable which is one-fifth the size and weight of conventional multiconductor cable. The triaxial cable permits much longer cable runs between the camera head and the mobile unit and is easier than multicore to handle and store; thus, the mobile truck can carry much more triaxial cable in a smaller space.

CAMERA OPERATION Basically, the operation of a camera on an MCR production is the same as if the camera were being operated inside the studio. The camera operator listens to the program director over the PL headset, and a tally

FIGURE 21-5 **Large Scale MCR Van Control Room.** *(Courtesy: NBC-TV)*

light indicates whenever the camera has been punched up on the air.

Of course, working a camera outside the studio adds a number of special complications to the camera operator's already difficult job. The weather can be controlled inside the studio, but is highly unpredictable on a remote which takes place outdoors. Covering a football game in Minnesota in December, or a stock car race from Florida in July are two examples of extreme weather conditions which can drastically affect the camera operator's performance. Also, many MCR productions involve sports or special events where a full rehearsal is impossible and the camera operator must be able to react quickly to the director's commands as events unfold. Finally, camera positions on a remote are often dictated by the remote location, and you may find yourself working a camera in a variety of uncomfortable, or even hazardous, situations. It is not uncommon to suspend a camera from a crane high above a musical concert in a park or to squeeze camera and operator into a tiny corner at the back of a theater with instructions not to disturb the audience while you photograph the action on stage. These are only a few of the reasons why really good remote camera operators are not easy to find.

CAMERA MOUNTING EQUIPMENT For most productions, the camera is mounted atop a portable tripod which is unfolded and positioned at the camera site. Unless the tripod is equipped with wheels, you will not be able to move the camera once it is positioned. While this is not always a problem, be sure you have enough tripods with wheels for those cameras which must be moved during the production.

There are special situations which may require a pedestal or a dolly unit such as a crab dolly. While these mounting devices increase your production capability, they are heavy and take up a great deal of space on the truck. They also require additional production personnel to operate them.

FIGURE 21-6 **Camera Scaffold.**

CAMERA LENSES The most important question when deciding on camera lenses is the distance between the camera's position and the subject. For some remotes which take place at a sports stadium or inside an auditorium or theater, your cameras will be positioned a considerable distance from the subjects and you will need zoom lenses with a powerful telephoto capability. Range extenders are useful for increasing the telephoto power of the lens, but they also require a higher-than-normal light level which may not always be available from existing illumination.

PORTABLE CAMERAS Hand-held cameras are often an excellent complement to a number of tripod-mounted cameras and permit the director increased camera movement. Of course, the portable camera must be cable-connected so that its picture feed runs through the production switcher inside the production truck, or linked to the production van via portable microwave.

Audio

An MCR production may require as many as a dozen or more individual microphone feeds. This means the audio section of the production van must have an audio control console large enough to handle multiple audio inputs and, ideally, the ability to equalize each incoming feed to correct for acoustic problems which may occur at the location site. Most audio areas also contain the patch panels, reel-to-reel and cartridge tape machines, and audio monitoring equipment which are found inside a conventional studio audio control area.

The usual complement of microphones including lavalier, hand, and directional shotgun mikes are standard, depending upon the audio requirements. For sports or special events coverage, many audio engineers and talent prefer headset mikes for better sound quality and no-hands operation. Wireless microphones are also commonly used both as hand mikes and to hide lavaliers on talent.

Videotape

The type of VTRs used on an MCR remote depends primarily on the size of the truck and the level of production. Smaller vans frequently use 3/4-inch cassettes although a number of facilities have installed high-quality 1-inch VTRs. Large vans generally use only 1-inch VTRs which are equipped with a slow-motion control unit enabling the director to use the slow-motion, replay capability for sports and for special effects.

Additional Equipment

On large-scale MCR vans which are used to produce an entire program, the van is equipped with a number of additional equipment items. Most vans carry a character generator to permit the director to add graphics during the production. Still-store units are also commonly used,

especially in the production of sports and special event programming.

Lighting

The lighting requirements of the MCR production depend on the particular production situation. Sometimes the unit must provide all lighting, and this can involve transporting a great deal of portable lighting equipment to the remote site. Other times, the unit may be covering an existing event such as a sports event and can use the available natural or artificial illumination. For coverage of a concert or play which is being produced in front of a live audience, you may have to cover the event with existing illumination. This requires close coordination between the television lighting director and the stage lighting director to ensure that sufficient illumination is available for the technical operation of the camera.

REMOTE PRODUCTION

Since both EFP and MCR production utilize the same four production stages, we can discuss how the equipment and techniques are applied in each of the stages for both remote techniques at the same time.

Preproduction Planning

Preproduction planning, though important for every show, is indispensable when producing a remote. Once you are on location you cannot run down to the supply room for some gaffer tape or pick up an extra audio cable you forgot to order. Everything your production will need must be planned in advance and brought with you to the location. Detailing the production team's needs and predicting what additional items may be necessary takes skill, experience, and a great deal of preparation.

WHAT KIND OF REMOTE IS IT? As with any production, your first job is to define the particular characteristics of the remote you plan to cover. Remotes can be classified into two categories: (1) events which are *covered*, and (2) events which are *staged*. A covered event occurs whether or not television is present, and the production unit has little, if any, control over the event itself. A staged event, on the other hand, offers maximum production control because it is produced specifically for television. Although a live audience might be present (as in a theater or concert hall), the spectators are themselves a part of the event, which is designed and produced primarily for television.

The nature of the event will also determine the type of remote production approach you can use. A single-camera EFP approach is cheaper and more flexible than a multiple-camera remote with its large mobile unit and dozens of technicians. Yet not all events can be covered with a single camera, especially those which require multiple cameras photographing the action in real time.

Finally, the producer and director must identify the program's objectives and indicate which elements are most important to the show. Until this information is available, none of the production and engineering planning and preparation for audio, lighting, cameras, and so forth can begin.

Remote Survey

Where should the cameras be positioned? How much light is available at the location? Where do we park the mobile unit? Is there sufficient electric power for the equipment? Before you can produce at a remote location, you must run a *remote survey,* which is designed to answer these and countless other important production and engineering questions. The purpose of a remote survey is threefold: (1) to determine the location for the production, (2) to determine where all production equipment and personnel will be positioned, and (3) to determine whether all the production's needs and requirements can be handled at the remote site.

For many productions the answer to the first question, "Where is a suitable production location?" is determined by the event itself. Wherever a sports event, parade, or conference is being held *is* the production location. But many other productions—commercials, entertainment shows, instructional programs—do not necessarily specify locations as much as an overall mood, atmosphere, or background. In this case it is up to the producer and director to find a suitable location which fits the production's aesthetic needs for atmosphere and detail, and meets the production's engineering and logistical requirements.

As a general rule, make your survey visit at the same time of day the event is scheduled to take place. This is especially important for outdoor remotes since the cameras should ideally be positioned with the sun behind them. Surveying a location in the early morning, with the sun in the east, does little good if the production is scheduled to begin at 4 P.M., when the late afternoon sun is in the west. In addition to accurately estimating the sun's position, the location of shadows, the contrast between light and dark areas, and other time-related factors make it important to carefully review the site as close to the time of production as possible.

Interior locations may be immune to these problems, but they generally offer enough of their own. For example, a visit to a downtown hotel on a Saturday morning to survey the ballroom for an upcoming remote would likely reveal a deserted area with plenty of parking space and access for equipment. The same location during a weekday afternoon, when the remote is scheduled, can look quite different, with the parking lot full to capacity and the street clogged with pedestrians and auto traffic which block easy access to the building.

Finally, if you have never done a program at the remote site, take nothing for granted. Are all the electric outlets you see actually working? Does the window through which you will run cable actually open? Do you need a special key to operate the service elevator? Better to have oversurveyed a site than to overlook an important item which can prove to be disastrous later on.

Although every remote is slightly different, here is a rundown of some of the most important production and engineering points to be considered by the survey team during their visit to the location. (See Figure 21-7.)

REMOTE SURVEY FORM

Survey Date: _____

PROGRAM: _____

PRODUCTION DATE: _____   Live/Air Time: _____ VTR _____

LOCATION SITE/ADDRESS: _____

Director: _____   Producer: _____   TD: _____

LOCATION CONTACTS

Primary Contact _____   Phone: _____

Secondary Contact _____   Phone: _____

Permits _____   Phone: _____

Parking Location _____

Credentials _____

CAMERAS   (add sketch of cam positions)

	Position/Location	Lens	Cable Run
CAM #1			
CAM #2			
CAM #3			
CAM #4			

AUDIO

	Mike Type	Location/Cable		Mike Type	Location/Cable
1			7		
2			8		
3			9		
4			10		
5			11		
6			12		

LIGHTING   (add sketch of light plot if needed)

Available Light in fc: _____

Lighting Instr: _____

_____

FIGURE 21-7   **Remote Survey Form.**

PROGRAM: _____

POWER

Location Electrician Contact: _____

Power Requirements: _____

_____

AC Outlets (location/voltage/connector type) _____

_____

_____

Special Power Instructions: _____

_____

COMMUNICATIONS

PL: # Cam Headsets _____   Special PL drops: __(1)_____

__(2)_____   (3)_____   (4)_____

TELEPHONE:

Location

Private Line: _____   _____

_____   _____

Location

Business Phone: _____   _____

_____   _____

CONSTRUCTION:

SPECIAL INSTRUCTIONS:

LOCAL CONTACTS AND CLEARANCES Always establish a local contact who can provide access, information, and help with various details. Obtain the individual's name and phone number and those of any assistants who might be helpful. Among important secondary contacts are electricians, plumbers, and maintenance people, who usually know about everything you will need for the production.

ACCESS Establish where you need access to the site, when you intend to arrive with the equipment, and who will be there to meet the crew. Setup often occurs at odd hours, and, unless someone has been notified of your arrival, you may find the gates locked and no one around when your truck shows up at 4 A.M. to begin an early setup.

PERMITS Most cities require a production unit to take out a police permit either to park vehicles on the street or to shoot on public property. Be sure you have the necessary permits well before the production is scheduled, and arrange with the police for crowd or traffic control if necessary. (See Figure 21-8.)

PERMISSIONS Always obtain written permission to cover an event or to produce the show on private property. Sometimes the production unit must provide special insurance or other legal considerations, and these should be clearly outlined in the permission agreement.

PARKING Arrange for parking space for your production vehicles and for any private cars used by the cast or crew. If you must park on city streets, your police permit should indicate where and how you can position your vehicles. A parking lot next to the remote is a real convenience, but make certain that your space has been reserved by the location contact in advance.

Where to position the mobile unit is an important production and engineering consideration. There are times when you will have little choice in the matter, but if you do have an option, find a location which requires the shortest average cable run to cameras and equipment. You will also need easy access from the mobile unit to the location site since both equipment and produc-

tion personnel require unrestricted movement between the truck and the actual production area.

CAMERAS

POSITIONS The location of your cameras is a crucial production decision which must be made during the survey. The director should already know how many cameras are available and approximately where they are to be located to cover the event. If it is to be an EFP single-camera shoot, the director must know which camera setups will be necessary and approximately where the camera will be positioned for each one.

For multicamera productions, select the best combination of positions to cover the action from a variety of perspectives. Some cameras must be located fairly high so they can provide a wide cover shot of the action. Other cameras must be located at lower levels for closer shots, as well as to photograph the various activities which surround the primary event.

Make certain you have considered screen direction by keeping all the cameras on the same side of the action axis. Otherwise, you risk inadvertently changing screen direction as you cut from one camera to another. Cameras must be positioned where they are free of shooting and movement obstructions. It does not take much skill to avoid planting a camera behind a pole or column, but beware of potential obstructions, which might not be as obvious. For example, a camera positioned behind a seated theater audience might be blocked completely if someone gets up or if the audience rises to its feet during the production.

Whenever possible try to position your cameras in easily accessible locations. This is not always practical, however, and there are times when you may have to construct a special scaffold or use special equipment (such as renting a cherry-picker crane) to provide a unique camera vantage point.

Portable cameras offer much greater production flexibility than stationary cameras and can

MOTION PICTURE – TELEVISION PERMIT   CITY OF NEW YORK   PERMIT NO.
MAYOR'S OFFICE FOR FILM, THEATRE AND BROADCASTING
110 WEST 57TH STREET NEW YORK N.Y. 10019

This permit is issued to the applicant to film or televise on streets or property subject to the jurisdiction of the City of New York at the times and locations designated below. The permit must be in the possession of the applicant at all times while on location. For additional assistance call the Permit Division: 489-6714. Police Unit: 592-6226.

APPLICATION NOT ACCEPTED UNLESS TYPED.

Date _____

1. Company: _____ Production Contact: _____

2. Address: _____ Tel. No. _____

3. Locations: (If more than 2 use Schedule "A") _____

4. Dates of filming: _____ Approx. times: _____

5. Scene to be filmed must be described accurately: _____

6. Animals, firearms, special effects or unusual scenes: _____

7. List production equipment: _____ # in cast & crew _____

   No. of Trucks & plate #s _____

   No. of Autos & plate #s _____

   Other vehicles & plate #s _____

8. Feature Film: ☐ TV Movie or Special: ☐ TV Series: ☐ Other: ☐ (Give title, producer, director and identify celebrities) _____

   Asst. Director: _____ Prod. Mgr. _____

9. If TV commercial name product: _____

10. Public Liability Insurance Company, Policy # and Agent: _____

    Amount: _____ Expiration date: _____

The applicant agrees to indemnify The City of New York and to be solely and absolutely liable upon any and all claims, suits and judgments against the City and/or the applicant for personal injuries and property damages arising out of or occurring during the activities of the applicant, his (its) employees or otherwise. The applicant further agrees to comply with all pertinent provisions of New York laws, rules and regulations. This permit may be revoked at any time.

VEHICLES LISTED ABOVE ARE PERMITTED TO PARK IN ANY AVAILABLE PARKING SPACES IN THE IMMEDIATE VICINITY OF THE ABOVE LISTED FILM LOCATIONS. EXCEPT: FIRE HYDRANTS.

Date _____   Signature of Representative _____   Title _____

DO NOT WRITE BELOW THIS LINE
The Mayor's Office Seal must be embossed on original copy.

Dated _____ 19 _____   Film Coordinator, Mayor's Office of Film, Theatre and Broadcasting

FIGURE 21-8 Police Permit for Remote Production.
A permit is often necessary to park the mobile unit on city streets and to shoot in public areas. *(Courtesy: Mayor's Office for Film, Theatre, and Broadcast, New York City)*

provide some interesting shots which literally take the viewer into the action. Their use must be carefully planned and coordinated, however. A portable camera with a cable must be positioned where the cable will not interfere with the action but will provide enough unrestricted movement. A portable microwave link completely frees the camera from cable limitations but requires additional crew members to handle the unit.

As you plan camera positions, consider what each camera is responsible for photographing during the production. For example, events which use an announce booth must have one camera positioned to cover the announcers on-camera, as well as the action itself. Other cameras may have to be located so they can easily shoot the scoreboard, the audience, backstage, graphics, and so on.

LENSES Depending on the camera-to-subject distance and the type of shots you have planned, you will have to decide which camera lenses to use. You may have to request special lenses or range extenders to provide better shooting possibilities.

CABLE RUNS How will you run cable from camera positions to the mobile truck? On EFP shoots, the VTR is often directly near the camera and long cable runs are unnecessary. For multiple-camera productions, however, every camera must be linked to the truck by either cable or microwave. Cable runs of over 2,000 feet, using multiconductor cable, are not advisable because of possible video problems. Plan your cable runs so the cable does not interfere with the event itself, with other equipment, or with spectators, or run too close to electrical power lines, which can cause serious technical problems.

AUDIO Producing high-quality television audio in the studio is never easy, but the many acoustical problems, combined with the extraneous noise frequently encountered at a remote site, make the audio engineer's job that much more difficult.

Among the basic points you will need to know before surveying the site are: (1) Who and what must be miked? (2) Can the microphones appear in the camera shot? (3) What potential audio problems exist (cavernous sports arena, widely spaced participants at a meeting)? (4) Must any special audio arrangements be made to coordinate television coverage with the public address audio system being used at the remote site?

MIKING TALENT Announcers, commentators, and play-by-play sportscasters can usually be miked with a lavalier or hand-held microphone. A headset mike is also a good choice because it offers talent hands-free operation, reduces extraneous noise, and maintains a constant subject-to-mike distance, even as the announcer follows the action.

How much movement flexibility does talent require? If they will remain in one location, any type of microphone or cable combination will work well. If they must be able to roam around the site, a wireless microphone provides unrestricted movement.

If the microphone must remain hidden off-camera, will a fishpole boom or a shotgun work? Sometimes hiding a lavalier under a performer's clothing will produce the best audio quality.

ATMOSPHERE SOUND The natural sounds of an event are the subtle touches which make a

production more realistic and enjoyable for the viewer. The crack of a well-hit baseball, a down-court dribble, and the swish of a golf swing are examples of atmosphere sounds, which are picked up with either a parabolic or a directional shotgun microphone. Crowds should be recorded with a microphone positioned for a very general pickup pattern. Avoid placing the microphone where it will pick up a few specific people instead of general crowd noise or too near a house PA speaker. Referee and umpire calls can be covered using either special desk mikes or shotguns, or by equipping the referee or umpire with a wireless RF mike. (See Figure 21-9.)

PUBLIC ADDRESS FEEDS Sometimes the television production uses a direct feed from the house PA system. This is often the case in govern-

FIGURE 21-9 **Miking Natural Sound.**
The audio engineer must plan for the pickup of natural sound at a remote event. Often a parabolic reflector or a shotgun microphone is held by an audio assistant outside camera range. Another approach, illustrated here, is to mount a shotgun mike atop a camera to pick up the sound.

mental or committee hearings, concerts, or various indoor events, where a PA is required for the house audience to hear the speakers. The use of a PA feed does not eliminate the need for additional television mikes, since announcers, crowd noise, and other audio sources are not covered by a normal PA system.

If you choose to take a PA feed, remember that the television audio engineer does not control the individual microphones used by the PA system, but simply receives a single, premixed audio feed. This can be a convenience, but you also run the risk of relying on an important audio source which may not be well balanced.

CABLE RUNS Estimate the number and lengths of audio cables necessary for every microphone position. Audio cable should be run around congested areas and away from power lines.

RECORDING ROOM NOISE If you are recording EFP or MCR for later editing, remember to record some room noise which is simply the natural sound of the location without anyone speaking into the microphone. The room noise will be indispensable during postproduction in helping to smooth out audio edits.

POWER Estimate the production's total power requirements, and check to see that the necessary power will be available at the remote site. In addition to noting the voltage and amperage, make certain that the wiring configuration is compatible with your equipment.

MOBILE UNIT POWER If you will use the mobile unit to supply power to all your equipment, learn where the power source at the remote site is located and whether it is compatible in voltage, amperage, and wiring with your van. If you intend to use an internal power generator, will heat and noise present any serious problems?

POWERING INDIVIDUAL EQUIPMENT EFP equipment is frequently powered individually. The simplest approach is battery power, but batteries limit the amount of operating time. If you will operate the equipment from power mains, make certain the correct ac outlets are accessible where you plan to shoot. Check all outlets in advance. Never assume that the presence of outlets where you intend to shoot automatically means that all the power outlets are actually working.

Check with an electrician at the remote site as to the location of the fuse or circuit breaker box. If possible, obtain the wiring schematics for the area. This way you can spread out your electrical load evenly by plugging into a number of outlets wired to different circuits. This reduces your chances of blowing a fuse or breaker and lessens the possibility of voltage fluctuations which can interfere with equipment operation.

It is often possible to tap major cable lines to provide additional power for special production requirements. However, this requires a professional electrician, and you should never attempt to rewire or tamper with electrical circuits unless you are fully trained to do so.

CABLE Power cable runs must be planned for cameras, VTRs, lighting instruments, and other production equipment. Never use ordinary home extension cords, which are too thin to carry the necessary current and are susceptible to fraying or breaking, which can prove highly dangerous. Use heavy-duty, insulated electric cable, which provides the durability, safety, and power load for a remote situation. Determine the number and the length of the cables you will need.

When you plan electric cable runs, avoid leading the cable near video or audio lines since there is a possibility of interfering with the program signals when they are located too close to power cables.

COMMUNICATION All the communication systems normally used for a production inside the studio must be brought to a remote location. This includes both the internal communication system for the production unit and external communication links between the remote unit in the field and the home studio.

PRODUCTION COMMUNICATION If you are

using an MCR van with multiple cameras, the normal intercom PL system, from cameras to control room, will operate once the cameras are connected to the control truck.

Additional PL lines, or "drops," for the floor manager, floor assistants, spotters, and additional production personnel must be planned during the remote survey and arrangements made to provide them. Determine how many headset units are needed and the length of cable necessary from control truck to each PL drop.

Commentators and announcers are usually supplied with an *interrupted feedback* (IFB) system, which permits the director and producer in the truck to talk with talent in the booth while they are on the air. An IFB can be either wired into a dual earpiece headset (one side contains program audio; the other, IFB) or fed to a small Telex earpiece which is worn by talent.

VIDEO AND AUDIO FEEDS During the survey you must determine all the locations which require video or audio program feeds. Obviously, the announce booth needs a video monitor displaying the on-air picture and a feed of program audio. Other locations for video or audio feeds might include the floor manager's location—backstage, on the sidelines, or by the spotter's booth. Needless to say, separate cables must be run from the control truck to every video or audio feed.

TELEPHONES Telephones are almost always necessary, especially on live broadcasts which require a great deal of communication between the remote production site and the home studio. Internal PL telephones between areas of the remote site must be requested from the telephone company (Telco) in advance. Specify how many you will need and their exact locations. Since noise is always a problem, you may want to order PL phones which light rather than ring.

Business telephones are necessary for the remote unit to communicate externally. Again, the survey team must determine how many business telephones are needed and must specify their exact locations. Most LSR vans already have telephone units installed, and Telco simply has to connect their instruments to the proper wires at the location site.

PROGRAM TRANSMISSION Program video and audio for a live broadcast must be fed from the remote site back to the home station. If the remote occurs within range of the station's microwave system, this is an inexpensive way to link the remote unit to the broadcast studio. Otherwise, you must arrange for the telephone company to transmit the program over land lines or microwave links, which they provide especially for the broadcast.

SECURITY Will you require any special security arrangements? This includes security for your equipment, which may stand in place all day or perhaps overnight, depending on the production. Crowd control may also be a problem. Does the remote location have its own security force, or will you have to make special arrangements?

Credentials and *passes* are often necessary to permit production personnel access to all locations at an event. Arrange for these through the location contact in advance of the production. You will need to specify how many people will require passes. This includes production staff, technical crew, and talent.

MISCELLANEOUS Every remote production has its own unique set of requirements. Some additional survey items you may have to take note of are these:

FOOD AND LODGING Are you responsible for providing food for the unit? If not, are there commercial restaurants nearby where food is accessible quickly? It may be faster and easier to bring food to the location rather than to give everyone a lunch break and hope they all return in time.

Even if you do not have to provide a full meal, coffee and cold drinks are greatly appreciated. The cost is small, and the benefits in improved morale are worth the additional effort.

If your unit is staying overnight, you must arrange for lodging. The hotel or motel should be located as near the remote as possible to eliminate travel time.

TRANSPORTATION Will you have to provide transportation to and from the location site? Depending on the size of your unit, this can be a relatively simple job or a major logistical operation.

CONSTRUCTION Will any special construction be necessary? Camera platforms, the announce booth, and lighting units all may require the erection of temporary scaffolding. These should be planned for, if possible, set-up before the remote unit arrives with the equipment.

GRAPHICS If you are not using a character generator, you must plan for a graphics area to handle opening and closing credits, as well as any other graphics necessary within the show. Sometimes a graphics area is set up adjacent to the production truck and uses a small black and white camera to photograph the lettering which is used for credits and for lower-third inserts. Another approach is to use one of the conventional cameras to double on graphics. A portable card stand and light are set up where the camera can quickly pan over to photograph the visual when the director needs it.

Production Meeting

Once the survey has been completed, the key team members must meet for the usual preproduction conference. The producer or director runs down the survey detailing special requirements for each of the key production areas. If the team has never done a remote at the location before, taking a few snapshots or slides of the site can help everyone who was not at the survey to visualize the overall area and specific locations.

During the meeting, the producer must plan the remote production schedule. First the necessary arrival times for the mobile unit and technicians should be determined. If you will have to construct special platforms or lighting grids, plan to have these crews arrive early so that their jobs will be finished by the time the cameras and lights arrive at the location.

Setup will proceed much more efficiently if the entire production unit is divided into a number of smaller groups, each handling its own area of responsibility. This way one crew can unpack and mount the cameras while another group runs audio cable and connects microphones.

Setup and Rehearsal

The activities during the setup and rehearsal of every remote depend on the production approach and on the show itself. A large-scale remote requires an enormous setup compared to a single-camera EFP. In addition, the EFP unit usually sets up a single shot, rehearses it completely, and then tapes it before moving on to the next. A multiple-camera remote must completely set up the entire unit before any rehearsal or production is possible. Events which are staged for television can be carefully rehearsed, shot-by-shot if necessary. Covered events cannot be rehearsed at all even though some of the basic format can, and should, be planned in advance.

With these differences in mind, let us look at setup and rehearsal procedures, focusing primarily on large-scale remote operations, which offer the most difficult challenges.

SETUP Unlike a studio production, where certain elements exist from the moment you enter the studio, every remote situation must be set up from scratch. As soon as the mobile unit arrives, it should be parked and power supplied to the vehicle. The various crews which were organized in preproduction can now begin their individual assignments.

Cameras must be unpacked, carried to their previously planned locations, and mounted atop tripods or pedestals. Any special towers or stands for the cameras should already be in place. As the cameras are set up, another crew should be running cable from the control truck to each camera position. Once the cameras are mounted and cable runs completed, connect the cameras and switch on the power to give the camera's electronics a chance to warm up and stablize before the engineers begin setup and alignment procedures.

The audio crew will already be unpacking their equipment, running cable to each location, and setting up the proper microphones (hand, lavalier, boom, and so on) at their correct positions. If necessary, label each microphone with the performer's name to avoid any confusion later on. The audio crew is usually responsible for setting up all communication systems (except telephones, which are Telco's responsibility), including IFBs, video, and audio feeds.

As each microphone is positioned, run a brief sound check to make certain it has been connected, identified, and patched in properly. Equalizing and balancing microphones should wait until the audio setup is finished. Once you have checked out a particular microphone, make certain it is secured in a safe location where it will not be misplaced or damaged.

All necessary lighting instruments should be hung and focused during the setup period. Since the lights must be in position and working before the production rehearsal can begin, the lighting setup should be planned to prevent holding up the entire unit. For elaborate lighting situations, it may be advisable to have the lighting crew arrive at the location early enough to ensure enough time to finish the job.

Once all the equipment is set up and connected, the crew must go through a complete engineering checkout. This is not a production rehearsal but an engineering operation to ensure that the equipment has survived the trip in working condition and is properly set up. After the cameras are aligned, audio feeds are equalized and balanced, and the other production equipment has been checked, the unit is ready for production rehearsal.

REHEARSAL The remote's rehearsal proceeds much as any normal studio production. EFP operations usually rehearse with the director positioned next to the camera. Each individual camera shot is set up, rehearsed, and taped until the director is satisfied and moves on to the next cycle of setup, rehearsal, and taping.

Staged productions can be carefully rehearsed by the director in exactly the same way a show would be rehearsed inside the studio. For those spontaneous events, where no detailed rehearsal is possible, the director can still work out each team area's individual responsibilities, based on the production plan and format.

NBC Sports director Ted Nathanson always holds a brief meeting with key production crew members prior to every broadcast. Nathanson explains what he expects each camera to cover and specifically details what shots they should have for a variety of expected situations. The same instructions are given to audio, to the slow-motion and VTR machine operators, and to graphics. Nathanson's approach is easily applied to any remote which cannot be completely rehearsed but follows along a preplanned sequence of events.

According to Ted Nathanson, the top remote camera operators are the ones who constantly practice their shots before air time. Just as the players on the field practice before the game, the camera operators warm up on their camera moves. It takes some time to become accustomed to a strange remote site or an awkward camera location. Practicing how far you need to zoom in or out to get certain shots, what it takes to pan quickly over to the scoreboard or to a graphic, or how much freedom of movement a portable camera operator really has can all be rehearsed prior to production and will result in smoother camera work during the actual show.

Production

The production stage for a remote is usually no different from that for in-studio shows. Of course, there are always problems encountered working in the field which cannot occur inside a studio that is specifically equipped for television production. For example, taping on location with a number of lighting instruments inside a small room may produce excessive heat which forces the crew to cease production periodically, turn off the lights, and allow the room temperature to cool. Otherwise, the high heat could interfere with the camera or VTR operation, not to mention the discomfort caused to both talent and crew.

Naturally, a live broadcast from an MCR van offers a special kind of excitement regardless of how much planning and rehearsal went into the show. During a live broadcast the director must concentrate totally on the pictures and sound which are to go out over the air. He or she must

FIGURE 21-10 **MCR Production.**
On many small-scale MCR productions the director also serves as the production's switcher.

rely on the assistant director and the producer to help with time cues, coordination, graphics, and videotape inserts or replays.

For example, on a live remote, the AD keeps an open telephone line between the production in the field and the home studio. This is necessary to cue the remote unit when to begin the show and to smoothly coordinate filmed or taped commercials and inserts, which are rolled-in from the studio. An off-air monitor, tuned to the local station's programming, is used to show everyone inside the truck when the station has punched up the remote feed on the air.

The routine for a live remote broadcast is this:

1   Prior to air time the AD establishes contact with the studio and synchronizes the remote's clock with the master clock at the station.

2   All time cues and coordination operations are double-checked before air time.

3   About one minute to air time, the AD readies the control room. The director tells the TD to punch up the opening shot on the line. This allows the station to confirm that the feed is reaching them and is ready to go out over the air.

4   As air time approaches, the AD simultaneously watches the clock and the off-air monitor

and relays the home studio AD's countdown to the production unit.

5   The AD should hear the "go" cue over the telephone just as the air monitor switches from the station's feed to the remote's line picture. The remote production is on the air.

6   During the broadcast, the AD coordinates all commercial breaks, VTR or film rolls emanating from the home studio, and the remote's final end time.

Postproduction

Once the production is completed the crew must strike all equipment, pack it safely, and return to the studio. A remote strike requires the same planning and coordination as the setup to do the job efficiently and effectively. After the work and effort in producing a remote, the strike is likely to be anticlimatic, but do not let the crew get careless now. It takes as much concentration and attention to detail to pack equipment as it does to set it up.

All equipment should be carefully and neatly stored away. This makes it easy to count the number of cables, headsets, microphones, and other small items, which can be easily left behind. Storage space is often at a premium, and neatly coiled cable will fit better than a jumbled tangle of wire. Assign someone to make an inventory check of the equipment as it is brought back to the truck so that you can be sure you are leaving the remote site with as much equipment as you brought.

Before leaving the remote location, the producer, director, or engineering supervisor should check with the location contact to clear up any remaining business details. It may be obvious, but a production unit which leaves the location as clean as they found it and is courteous and cooperative with the location staff will find it easier the next time they must do a production at the same site.

When you return to the studio, file away the remote survey and additional paperwork for future reference. Programming which was videotaped will begin the editing process. Live programming should have been recorded off-air as an air check. It is a good idea for the producer, director, and key team members to review the tape later and hold a postmortem evaluation session. From these meetings come improvements in production and engineering operations which will benefit the next remote production.

All EFP and some MCR productions which were shot in segments will require a postproduction editing stage. In either case, the editing approach is no different from that for a studio-produced production. For details on videotape editing and editing single-camera/single-VTR productions, see Chapters 9 and 10.

COMMON REMOTE SITUATIONS

Any television show can be done as a remote, but there are some formats which are frequently produced in the field. In this section we will cover some of the special production requirements for some typical remote situations.

Sports Remotes

Whether a high school football championship covered by a local station or the Super Bowl covered by a national network, sports has become a regular part of television programming. Of course, not every station has the equipment and manpower resources which are available to the

networks, but that does not mean locally produced sports must be second-rate. It is not difficult to cover a football game with fifteen cameras and three slow motion VTRs. The real challenge is to cover the game with only three or cameras and still produce an exciting and entertaining show.

Quite often ingenuity and creativity can overcome some equipment limitations. In addition, recent technological advances have drastically reduced the price of slow-motion and still-frame storage units, making these important sports production tools increasingly accessible. When you plan your camera positions, select strategic locations which will permit each camera to be used to the fullest extent possible. Over the years a number of basic camera positions for various sports have been developed, and they are illustrated in Figure 21-11. Although you will have to adapt them for your own particular needs, they can provide a departure point in planning sports coverage.

SPORTS DIRECTOR The director of any sports event must be thoroughly familiar with the rules and the play for every event he or she is assigned to direct; not only how the game is played, but some of the basic strategies involved. Unless you know this information you cannot begin to determine camera positions and plan your production coverage.

The director must also do his homework prior to each production to learn how each team or player approaches the game and what they are likely to do in typical situations. Once these are established, the director can work out each camera's shot responsibilities based on a knowledge of the game and what is likely to occur during the particular game or match.

A good sports director never loses sight of the primary event in selecting camera shots and directing the game coverage. Yet, at the same time, realizing that there are a variety of perspectives surrounding a sports event, he or she tries to show these to the viewer as well. There is a

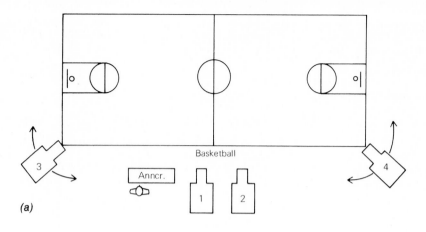

(a)

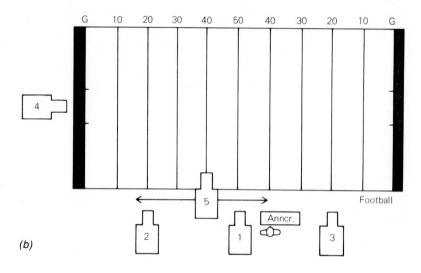

(b)

FIGURE 21-11 Camera Positions for Sports Events.
Every sports event offers its own set of problems and possibilities, but the camera positions illustrated can be used as a general guide. They must be modified, depending on the number of cameras you have available and the production situation. *(a) Basketball.* Cameras 1 and 2 are positioned above court level and follow the action with a zoom lens. Cameras 3 and 4 are optional, but can be used for shot variety, close-ups of foul shots, and to cover the announcer's table. *(b) Football.* Cameras 1, 2, and 3 are positioned above the playing field, one on the 50-yard line and the others on the 20-yard lines. Camera 4 can be used for kicking attempts and as an isolated camera. Camera 1 doubles as the announce booth camera for opening and closing segments. Camera 5 is mounted on a truck or dolly to provide a field-level view. *(c) Boxing and Wrestling.* Cameras 2

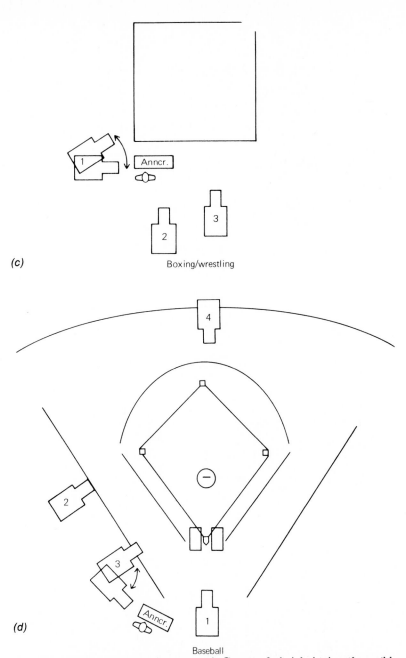

(c)

Boxing/wrestling

(d)

Baseball

and 3 are positioned above the ring, with Camera 2 slightly higher if possible. Camera 1 doubles on the announcer table or for fan reaction shots. *(d) Baseball.* Camera 1 is high above home plate; Cameras 2 and 3 are along the foul lines above the playing field. Camera 3 can shoot the announce booth when necessary. If possible, Camera 4 is positioned on a scaffold behind the centerfield fence and is equipped with a powerful telephoto zoom. This camera can cover the pitcher, catcher, batter, and umpire from an interesting viewpoint.

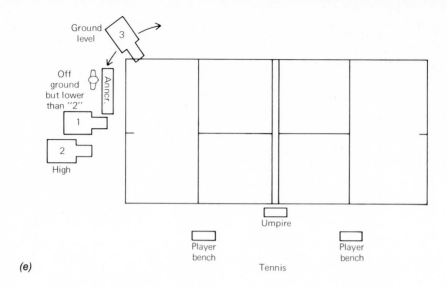

(e)

Tennis

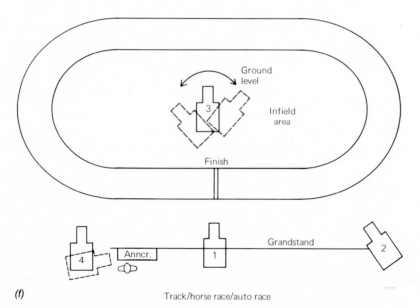

(f)

Track/horse race/auto race

(e) Tennis. Cameras 1 and 2 are positioned at one end of the court, with one camera higher than the other. Depending on the position of the announce booth, either camera can shoot the announcer when necessary. Camera 3 is at ground level and shoots close-ups of either player, as well as the umpire and player benches. *(f) Track events.* Events run on an oval track—such as horse racing, auto racing, or track events—can be approached in similar ways. Camera 1 is positioned high in the stands across from the finish line. Cameras 2 and 4 are at the ends of the stands and follow the racers within their fields of view. Camera 3, positioned at ground level inside the infield, can be used for closer shots as racers move along the far end of the track.

human side to competition which takes place along the sidelines, in the dugout, on the bench, or in the stands. These can be used to increase the viewer's enjoyment of the game, as long as the atmosphere shots do not distract attention from the primary event on the playing field.

PRODUCER The sports producer plays a much more active role during a sports broadcast than on most scripted or preplanned productions. Usually, the producer is involved in establishing the overall game plan the director will follow, helping the director select elements to show the audience, and supervising the isolated camera and the instant-replay operation.

Mike Weisman, a sports producer for NBC, recommends that the producer establish a definite point of view or production game plan prior to the event. This is based on your research and experience with the players, coaches, and their past strategies. While the game plan must be flexible so that it can change on a moment's notice, depending on what happens on the field, it does provide a framework around which the basic coverage can be organized.

One of the producer's most important responsibilities is to coordinate the isolated camera and the instant replay. Effective use of the isolated system requires a combination of thorough planning, good judgment, and plain luck. Based on your research and experience, you should have some idea where the key action is likely to occur for particular situations. This information is the basis for setting up the isolated-camera feed to the slow motion disc or VTR. A producer with only one slow motion unit must decide whether to play it safe and feed program video or risk missing the primary action by isolating on another camera. The safe approach virtually guarantees a replay, but from the same perspective the viewer just saw. The more daring approach produces some unique viewpoints, but increases the chances that you will guess wrong and wind up without the replay of a key event. Sometimes a producer who misses the primary action on iso can salvage a replay anyway and use it to show "why the play worked" by isolating on a particular player's individual contribution. Producers communicate with the slow motion operator via an open PL mike and with on-air talent via the IFB.

SPORTS ANNOUNCERS Most sportscasts utilize two announcers: one for play by play and the other as an analyst or color commentator to provide insight into the game. Both announcers must keep one eye on the playing field and the other on the air monitor in the booth, which shows the game as the viewer sees it. Most announce booths are equipped with additional monitors which show the video being recorded and replayed over the isolated camera and the slow-motion unit.

Veteran ABC sportscaster Keith Jackson says the role of a television sports announcer is to "amplify, clarify, and punctuate" the action on the field. Talent must be able to watch the activity on the field, keep an eye on the air monitor, listen to the director or producer over the IFB, and still make sense to the viewing audience.

Sportscasters are expected to prepare for each broadcast by studying the teams and players, their strategies, and whatever additional information can be obtained from team publicity offices, contacts, and press clippings. It is also important for sportscasters to be sufficiently familiar with each player's name and number so they can quickly identify them during the game. (See Figure 21-12.)

A *spotter* is often used to help the announcers and the director identify various players and to point out significant developments. Sometimes a production will use a number of different spotters: one assigned to the announce booth to locate events for the director and another assigned to help keep statistics for the character generator to use in updating lower-third graphics.

Indoor Theatrical Events

Rock concerts, awards banquets, symphony orchestras, and stage plays are some of the common indoor theatrical events which are frequently covered as remote productions. The production approach is usually dictated by the characteristics of the event. A staged event—like a rock concert or an awards program—offers maximum

Slow-motion
replay
monitors

Line monitor

Headset
volume controls

Dual-channel
headsets

FIGURE 21-12 **Sports Announce Booth.**

production control because the event is produced primarily for television. The appearance of cameras, equipment which blocks the audience's view, or the production's control over the timing and sequence of the event are not problems with staged remotes. A covered event, however, presents added challenges, because the production equipment must remain unobtrusive and must not interfere with the house audience. The television unit is invited as a guest, but it is expected to adapt to the situation and circumstances as they exist.

In either case, lighting is an important production consideration. Stage lighting is rarely sufficient for television cameras, and the lighting approach itself must be modified to produce the best television pictures. The addition of color media to the instruments, a widely used theatrical lighting approach, must be carefully planned when television will cover the event. Varying the color quality of the light can affect proper color reproduction.

In positioning cameras for an indoor theatrical remote, you may have to rope off a section of seats to provide an unrestricted view of the action

area. Be careful when running cable to avoid heavily traveled aisles and doorways. Check all camera locations to make certain the camera's view will not be blocked if someone in the audience gets up during the performance.

Audio quality can be a real problem, especially when you must hide the microphones from the cameras. A number of directional mikes located behind the footlights will pick up the performers, but the sound quality will not be very good. Wireless microphones are becoming a standard theatrical technique for sound reinforcement and are easily adapted for television by either patching into the theater's house PA system or arranging to pick up the microphone transmitter's feed with a separate receiver.

If you are covering a symphony, opera, or ballet inside a concert hall, you may be able to record excellent audio with only one or two microphones. The mikes are located to pick up the overall sound, preserving the natural acoustical ambience of the concert hall. You will need to experiment for the best microphone position, but once the location is determined, little, if any, mixing is necessary by the audio engineer.

Hearings and Conferences

Government hearings; educational, industrial, or civic conferences; and similar types of indoor events present special challenges for the remote production team. Usually television is permitted at these events as long as the equipment and crew remain unobtrusive and avoid interfering with, or upstaging, the very event they have come to cover.

Setup must be carefully planned to allow the crew enough time to prepare the equipment before the event is scheduled to begin. These events rarely accommodate television, and you may be prohibited from moving equipment around after a specified time. The event's coordinator should be contacted for a rough time schedule and sequence of the planned activities. Even so, these events have a tendency to run on their own time, and the television director should be prepared to follow whatever develops.

Lighting can present a serious problem when available light levels are too low for camera operation. The coordinator of the event may be apprehensive about approving the use of television lights, which cause heat, glare, and distractions. The best you can hope for is a compromise in which you use fewer but more efficient units, which are positioned as unobtrusively as possible.

The director must learn where important participants will be located (on the panel, at the podium, in the audience) and plan to cover them for both sound and picture. If audience participation is expected, have a floor assistant on hand with a hand-held or shotgun mike to cover audience speakers. Panel speakers can be miked with either separate desk mikes or by patching into the house PA system. At least one camera should be positioned where it can photograph the audience whenever necessary.

Figure 21-13 shows a three-camera setup for a typical indoor conference of the type just described. These events frequently occur within a small, crowded room, which greatly restricts camera and crew movement. The more you learn about what is to happen, the better you can decide where to position your cameras to best cover the event's various activities. Sometimes an advance text of a speaker's address or comments

can be helpful in planning the integration of graphics, visuals, and close-up shots.

Magazine-Format Feature Story

One of the most common EFP production situations is the "feature story" which is used on magazine-format programs, on news shows, and as insert material for a variety of studio productions. Of course, every story has its own "angle" which makes it interesting and different, but there are a number of production considerations which are common to such productions. An illustration of how this type of EFP production can be approached is this case study of a feature story on a well-known chef which is to be produced for a local station's weekly magazine program.

The story angle concerns the chef's "no salt gourmet diet." A few days before the production is scheduled to tape, the segment's producer visits the chef for a preinterview and to survey the location site which is the chef's home. The producer and the chef decide that the chef will be interviewed and will also show the audience how to prepare a fish dish. This means that two different locations will be used: the living room for the interview and the kitchen for the cooking demonstration. The producer surveys both areas and determines that there is sufficient space for the camera, ample electrical power for the equipment, and no serious lighting problems caused by windows, which can be closed during taping if necessary.

Before leaving the studio, the producer meets briefly with the production crew to run down the story and how it will be shot. When the crew and the reporter arrive at the chef's home, the crew begins to set up in the living room since the producer decided that the interview segment would be taped first. The lighting crew decides to use primarily bounce lighting off the white ceiling and takes care to spread the load of the various instruments across a number of different electrical circuits to avoid overloading. It is decided that rather than attempt to match the daylight color

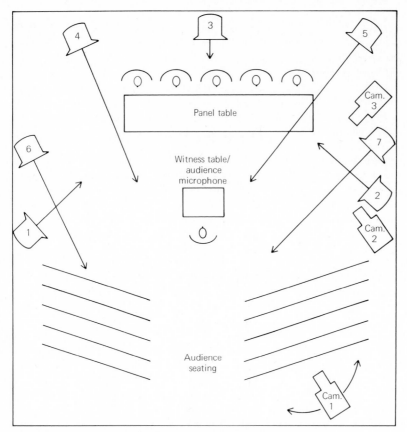

FIGURE 21-13 Camera Positions for Hearing or Conference.
Cameras 1 and 2 are positioned to cover the speakers at the panel table.
Camera 3 shoots the speaker at the witness table or the audience microphone,
depending on the nature of the event. Lighting instruments 1, 2, and 3 are
mounted on floor stands (or if possible, hung from the ceiling with gaffer-grips or
other hanging devices), and cover the panel table. Instruments 4 and 5 illuminate
the witness or audience mike area. Instruments 6 and 7 are used to light the
audience seating area. Lighting must be arranged so it will neither interfere with
the participants by glaring excessively into their eyes nor create flare problems
for any of the cameras.

temperature outside the large livingroom windows, the crew will simply close the window
shutters so that the outdoor illumination will not
interfere with the interior lighting. The audio engineer attaches a lavalier microphone to the chef
and to the reporter, and the interview begins.

During the interview taping, the producer takes
notes on each of the subjects which is discussed
and its approximate running time to help her later
during postproduction editing.

Once the interview segment is taped, the crew

strikes the equipment in the living room and sets
up in the kitchen for the cooking demonstration.
The producer decided earlier simply to follow the
chef as he talks his way through the cooking
demonstration. Once the entire preparation is
complete, the crew will then shoot a number of
close-up inserts of each ingredient and the close-
ups will be edited into the cooking demonstration
so that the audience can clearly see every ingredient before it is mixed into the final dish.

The producer also decided that she would

insert a lower-third graphic caption of each ingredient as it is shown in close-up. The graphic will be inserted by a character generator during postproduction, but the camera operator must be told about this so that he can compose each close-up and leave enough room for the lower-third insert.

The demonstration proceeds smoothly with only one hitch: The camera operator failed to get a shot of the chef as he placed the fish into the skillet, so a pickup shot is taken after the demonstration has ended, and this will be edited into the final tape to eliminate the camera error.

At the end of the shoot, the audio engineer records a few moments of "room noise" to help in postproduction editing, and the producer checks her notes to make certain that all shots necessary for editing were recorded. With nothing more needing to be done, the crew strikes and packs

the equipment, returns any furniture which was moved back to its original place, and stores the production equipment inside the EFP mobile van.

During postproduction, the producer refers to her notes as she screens the raw footage. She decides to use some of the chef's interview comments as a voice-over during the cooking segment, and this is easily accomplished during postproduction. Once the final tape has been completely edited, the lower-third captions are inserted from a character generator, and the finished segment is ready to be inserted into the week's magazine show.

SUMMARY

A remote is any production which takes place outside the studio. Electronic field production (EFP) utilizes a single camera/single VTR production approach, but, unlike ENG, it can be preplanned and permits the production crew to devote more care and attention to production details in order to produce a show with higher production values than is usually possible when covering fast-breaking news events.

Multiple-camera remotes (MCR) utilize a number of cameras which cover the event in real time, enabling the director to cut between cameras from a control room located inside the MCR mobile unit. MCR productions are either covered "live" or videotaped for postproduction editing later. For live coverage, the production unit must arrange for the video and audio feed to be transmitted back to the home station for broadcast.

Planning is indispensable for a successful remote, particularly because all the usual conveniences which are automatically available in the studio must be specifically brought to the location site. A *remote survey* is conducted by the producer, director, and technical supervisor before the actual production to determine, in advance, the project's production and engineering requirements and how they can be met at the remote location.

During preproduction meetings, the survey is reviewed with the key production team members, and the usual program planning considerations are discussed. Of particular importance is whether the remote event is staged or covered. A *staged* event is produced specifically for television and offers maximum production control. A *covered* event would occur with or without the television unit at the scene and offers additional production challenges since the remote crew must adapt to the circumstances of the event as they already exist.

Setup, rehearsal, and production operations are usually determined by the type of remote, the production approach which is employed, and the nature of the event itself. Postproduction involves striking and packing the equipment and the restoration of the location to its original condition before the remote unit returns to the studio. In the case of EFP productions or of an MCR production which is videotaped, postproduction editing is necessary in order to produce the completed program.

GLOSSARY

A

Above-the-Line Budget category which includes the artistic, or creative, elements engaged in a production. Includes primarily nontechnical personnel and activities.

A-B Rolling (1) Preparing film for optical printing so that each new cut is located on either the A or the B roll. (2) Utilizing two or more videotape machines and switching between them to produce a composite master.

AC Alternating current, the electrical energy found in conventional wall outlets.

Academy Leader Film leader with a numerical countdown in seconds. Used to cue film cuts precisely. Also called "SMPTE Leader."

Ace A 1,000-watt lighting instrument; usually refers to a 1,000-watt fresnel.

Acetate Clear, plastic material used in graphics to prepare animated sequences or to superimpose lettering on a background.

A. C. Nielsen A major research company, which provides national and regional television ratings.

Action Axis An imaginary line running through a subject, which is helpful in maintaining consistent screen direction.

AD Assistant director.

Additive Primaries Red, blue, and green colors which, when combined, produce white light. The additive primary colors are used in color reproduction.

Address Code Time code used to distinguish a particular location on a videotape for editing purposes. Address code indicates each frame by hour, minute, second, and frame number. *See SMPTE Time Code.*

Adjustable Hanging Pole An aluminum telescoping pole, which is used to hang lighting instruments at any desired height.

Ad-Lib Dialogue or action which has not been previously rehearsed.

AFM American Federation of Musicians; a union.

AFTRA American Federation of Television and Radio Artists; a major union for television talent.

AGC Automatic gain control, a device which automatically regulates sound or video intensity to maintain the proper technical level.

Alpha Wrap A helical-scan, VTR tape-threading configuration, in which the tape is wrapped completely around the head drum.

Amplitude The strength or range of an electrical signal.

Analog The variation of an electrical signal over a continuous range to represent the original image or sound which is being processed and reproduced.

Animation Combining individual shots, still drawings, or photographs to create the illusion of movement.

Aperture The opening of a lens through which the light passes. The aperture or diaphragm is usually measured in f-stops.

ARBITRON A major supplier of television rating information, mainly for local markets.

Arc Movement of a camera in an arclike, or curving, pattern.

Art Director Individual responsible for designing the setting for a production and for establishing the overall visual "look," including graphics, wardrobe, and sets. Also called "scenic designer."

ASCAP American Society of Composers, Authors, and Publishers; a nonprofit, music-licensing organization.

Aspect Ratio The proportional dimensions of the television screen, which measures three units high and four units wide.

Assemble Edit Addition of new video material following program material already recorded. Does not require existing control track on the tape.

Attenuate To decrease the strength of an electrical signal. A potentiometer, or fader, permits attenuation to be continuously controlled.

Audio The sound portion of a television program.

Audio Layback Process by which multitrack audio mix is re-recorded in sync onto the original videotape.

Audio Level The strength of the audio signal. "Taking an audio level" involves sampling each audio source individually.

Audio Mix Balancing all audio sources to produce the desired composite sound.

Audiospace The audience's only reference of aural reality. Audiospace refers to the way sound can be manipulated to produce a particular audience impression.

Audio Track   The particular area on audio or videotape which carries audio information. On multitrack machines, the tape holds a number of seperate audio tracks.

Audition   (1) A special audio circuit which enables the audio engineer to preview or cue audio sources before sending them through the audio console. (2) A try-out, in which talent is selected for a production.

Auto-Iris   Automatic iris, a device which varies the aperture of a camera lens to adjust for brightness variations in the scene being photographed.

Auto-Transformer Dimmer   A lighting control dimmer, which uses a transformer to regulate the power to the lighting instruments.

B

Background Light   Illumination of the set or background. Also called a "set light."

Background Music   Music used under principal dialogue or program audio to establish a mood.

Backlight   Illumination from behind the subject, used to separate the foreground subject from the background area. The lighting instrument is positioned above and behind the subject, usually at a 45° vertical angle.

Backtime   Figuring the amount of time left in a show by subtracting the present time from the program's end time. *See dead-pot.*

Banding   A series of horizontal lines which occur in the picture when a quadraplex VTR's heads are not properly aligned.

Barndoors   Metal flaps which are mounted on lighting instruments and used to control light distribution.

Barrel Distortion   Optical distortion commonly associated with wide-angle lenses, in which the straight sides of a subject curve outward.

Base   (1) Film base, the material used to make film onto which the light-sensitive emulsion is layered. (2) The metal bottom of a lamp that is inserted into the light socket.

Base Station   Camera control unit for digital cameras.

Batten   The metal pipes from which lighting instruments are hung.

Bayonet Lock Mount   A type of lens mounting which does not use threads and permits rapid changing of camera lenses.

Beepers   A series of low-frequency audio tones which are recorded prior to videotaping program material. They are used for cuing the videotape and for locating particular program segments.

Below-the-Line   Technical and production costs of a program, as indicated in the program budget. Includes production equipment and technical personnel.

Bias Light   Device used to boost the electrical output of television pickup tubes to permit camera operation under low light levels.

Bidirectional A microphone pickup pattern, in which the mike has two equally "live" sides.

Black The darkest portion of the gray scale. In terms of program switching, "black" refers to a video source which provides sync signals but no picture.

Blast Filter A microphone attachment (either external or built-in) which suppresses wind noise and breath popping. Also called a "pop filter" or "wind filter."

Bleeding When the edges of a video key or insert show the background video. Bleeding also occurs when a chroma key insert is improperly set up.

Blocking Working out talent and camera positions for a production.

Blooming Distortion of a television picture caused by an overly high video level or an excessively bright region in the scene being photographed.

Blooping Pencil A magnetic pencil used to erase the audio on a magnetic film track.

BMI Broadcast Music Incorporated; a major music-licensing organization.

Body Mount Device used to mount a portable camera on the operator for steady pictures and flexible movement.

Boom Microphone A microphone suspended from a long arm, which enables the mike to remain out of camera range.

Boom Shadow Shadows created by a boom microphone.

Boom Up; Boom Down (1) To raise or lower the microphone boom arm. (2) To raise or lower a camera mounted on a camera crane.

Border Edges which are electronically produced to visually separate and distinguish wipes, split-screens, or letters.

Box Set A realistic television set, which consists of three walls.

Break (1) Releasing a camera to move to another position or to obtain another shot. (2) Station break; local commercials and identification between programs.

Breakaway Prop A specially designed prop, which shatters harmlessly on impact.

Brightness One of the important attributes of color which determines how it will appear on a black and white gray scale. Sometimes called "value."

Broad A square or rectangular floodlight. Also called a "pan."

B-Roll A film or videotape insert reel which is rolled into the program.

Budget A cost breakdown of all production elements; usually divided into above-the-line and below-the-line sections.

Bulk Eraser A device which produces a strong magnetic field, used to erase quickly audio or video magnetic tapes, cartridges, and cassettes.

Bumper (1) A slide or graphic used at the beginning and end of various program segments, often before and after a commercial break. (2)

Additional tape or film footage following the primary program material as a safeguard.

Bump-up Dubbing a previously recorded videotape from a smaller format to a higher one (e.g., 3/4-inch to 1-inch). Bump-ups are done by running the playback through a time base corrector.

Burn-In Image retention by the camera pickup tube, caused by excessively bright subject, extreme contrast, or photographing a static scene for extended period of time. Also called "sticking" or "lag."

Bus A row of buttons on the video switcher. Also called a "bank."

Bust Shot Framing a subject from midchest to slightly above the head.

C

Calibrating the Zoom Zooming to longest focal length and focusing to ensure a focused picture throughout the entire zoom range. The zoom must be calibrated whenever the camera or subject distance varies significantly.

Call Sheet Schedule indicating talent, production, and technical personnel needed for rehearsal and production.

Cam Head Camera mounting head which produces very smooth camera moves. Utilizes a series of "cams," or cylinders, to control pan and tilt moves.

Cameo Lighting technique in which foreground subjects appear before a completely black background.

Camera Card A graphic which is photographed by a studio camera. Also called "title card," or "flip card."

Camera Control Unit (CCU) Equipment containing the various controls necessary to set up, align, and regulate camera operation. Includes waveform monitor, television monitor, and shading control.

Camera Head The portion of the camera chain which includes the lens system, pickup tubes, and viewfinder.

Cannon Plug A special three-prong audio plug and connector jack, which locks male and female plugs together with a small latching device. Used on all professional audio equipment.

Cans Studio headsets.

Canted Angle A shot in which the subject appears tilted in the screen.

Capacitor Electronic device designed to store an electric charge. Used in condenser microphones as part of the generating element.

Cardioid Microphone Microphone with a heart-shaped, directional pickup pattern.

Cart *See cartridge.*

Cartridge A plastic case which encloses an endless loop of tape. The tape is played in a cartridge player and automatically recues after each use.

Cathode Ray Tube (CRT) Specially designed vacuum tube in which a series of electrons are focused into a beam and strike a phosphor-coated surface, which glows to create the television image.

C-Clamp Device used to fasten lighting instruments to a lighting batten.

Cels *See acetate*.

Cement Splice Joining two pieces of film with cement adhesive.

Character Generator Device which electronically produces lettering and other graphic displays directly on television screen for use in production.

Charge Coupled Device (CCD) A solid-state video transducer which is used in place of a conventional camera pickup tube.

Cheat Repositioning a performer, camera, or object to produce a better shot.

Chroma Key Method of electronically inserting the image from one video source into the picture from another video source. The process utilizes a selected "key color," which, wherever it appears in the foreground shot, is replaced by the background image.

Chroma Key Tracking The ability of digital video manipulators to continually vary the size of the chroma key insert, depending on the movement of the foreground subject camera. This maintains realistic visual perspective.

Chroma Key Window The area on the foreground subject set painted the "key" color, in which the background image will appear on the composite shot. Frequently used on news or instructional programs to insert graphics behind the talent.

Chrominance Channels Red, blue, and green color channels which together produce the color image. Each primary color has a separate chrominance channel in the camera.

Clapstick Identification slate with a hinged top which, when brought down quickly, produces a loud clap, which is used to synchronize double system movie sound.

Clip (1) The control on the video switcher which is used to regulate the intensity of a matte, key, or insert. (2) A short piece of film or videotape which is rolled into a program. (3) To cut off abruptly.

Clock Time Setting the SMPTE time code to reflect the actual time of day. Useful in cataloging tape recordings of real time events.

Close-up A camera shot in which the principal subject is seen as relatively large and dominant in the screen.

C-Mount A threaded, screw-type lens mount.

Color Bars A standard color test signal which is generated electronically by a "color bar generator." Used as a reference in setting up and aligning color video equipment.

Color Correction Filter Filter which changes the color temperature of the scene being photographed.

Color Media *See gel*.

Color Temperature The relative amount of reddish or bluish color quality of light. Color temperature is measured in "degrees Kelvin." Standard indoor, studio lighting instruments are balanced for 3200°K; outdoor color film and electronic cameras are balanced for approximately 5600°K.

Colorizer An electronic device which adds preselected colors to a monochrome image according to the gray scale information present in the picture.

Comet Tailing A smear of light created by an excessively bright portion of the picture. *See also burn-in, lag*.

Complementary Colors Two colors which, when added together, will produce white light. The complementary color is produced by adding two primary colors together in the proper proportions. Yellow, cyan, and magenta are the three complementary colors.

Composite Signal A complete video signal including sync pulse.

Condenser Microphone A microphone which uses an electronic condenser as a part of the generating element. *See also electret condenser microphone*.

Conforming Matching original film or videotaped footage to an edited workprint to produce the final, edited master.

Contrast Range (Contrast Ratio) The difference between the brightest and darkest portions of a picture. In television a contrast ratio of 20:1 to 30:1 is the widest possible brightness range which still permits accurate picture reproduction.

Control Room The area where the program's director and production personnel control the audio and video for a program.

Control Track The area of the videotape which contains information used to control and synchronize the playback and videotape editing operations.

Control Track Time Code A variation on SMPTE time code, which counts control track pulses to produce an eight-digit edit code. Does not identify frames independently, as SMPTE code does, however, but simply counts pulses from any predetermined starting point.

Cookie *See cucalorus*.

Coord Short for *coordination*, in which two or more videotape machines are played through a switcher to produce a composite signal, which is recorded by another VTR.

Core Plastic hub on which audio tape or film is wound.

Countdown (1) The numbers which appear on film or videotape before the actual program material to facilitate accurate cuing. (2) Silent hand signals delivered by the floor manager to talent to smooth the transition into, and out of, program segments.

Counterweight Pedestal Camera pedestal which uses a counterweight system to control camera height and permit smooth on-air movement of camera pedestal.

Crabbing The parallel movement of all wheels on a dolly or pedestal to permit smooth and accurate dollies, trucks, and arcs.

Cradle Head A camera mounting head designed to counteract the weight of the camera during tilts, and to produce smooth pans and tilts on air.

Crane Large camera dolly in which camera and operator are mounted at the end of a long arm which permits extremely high and low camera angles, as well as very fluid on-air movement.

Crawl Device using a large drum, or paper roll, which moves credits or other graphic material horizontally or vertically past the camera.

Crawling Imperfections along the edges of a chroma key, in which the sides of the insert seem to move up and down.

Cross-Fade The simultaneous fade-in of one audio source as another is faded out. Can also be used to vary two lighting effects. In video, the transition is called a "dissolve."

Crush To compress all blacks and dark grays in a picture in order to include white levels. Necessary when the contrast range exceeds the acceptable contrast ratio.

Crystal Black A videotape which has recorded only video black. Used to lay down a continuous control track, which is necessary for "insert" editing.

Cucalorus A metal pattern which is inserted into an ellipsoidal spotlight, used to produce a shadow pattern against a set wall, curtain, or cyclorama.

Cue (1) The signal to begin a program, an action, dialogue, or other production activity. (2) To preset film, videotape, audiotape, or records, so they will be available immediately when called for by the program's director.

Cue Card Large card containing dialogue which talent must deliver during the production. Held next to the camera lens by the floor manager so talent can refer to the card without making it obvious that he or she is reading. Sometimes called "idiot cards."

Cue Track An area on videotape reserved for additional audio information or to carry SMPTE time code information for tape editing.

Cut (1) An instantaneous change from one camera shot to another. *See also take.* (2) A particular segment in a record or tape, usually indicated by number as in "Cut #2." (3) Command which means to immediately stop talking, stop action, or stop production.

Cutaway A shot which focuses on a view other than the principal action. Used frequently to provide transitional footage or to avoid a jump cut.

Cut-in An insert from another source which is introduced into the program.

Cycles Per Second The number of complete cycles of an electronic signal in one second. Expressed in Hertz, abbreviated as Hz.

Cyclorama (Cyc) A continuous piece of canvas fabric which runs around the edges of a studio and is used to produce the illusion of infinite depth. The cyc is often "painted" with colored light for a variety of effects.

D

db Abbreviation for "decibel."

DC Direct current, usually provided from special electrical sources or power generators.

Dead-Pot To play a film, tape, or record with the pot, or fader, off and to bring up the sound or take the video at the proper time on the director's cue. Permits prerecorded material to end precisely at a predetermined time.

Decibel A standard measure of relative intensity or power which is expressed on a logarithmic scale.

Demographics Breaking down a viewing audience by various social and economic characteristics, such as age, sex, income, and educa-- tion.

Depth of Field Area in which all objects photographed by a lens and camera appear in focus. Depends on subject-to-camera distance, focal length of lens, and f-stop.

Deuce A 2,000-watt lighting instrument; usually refers to 2-kw fresnel spot.

DGA Director's Guild of America; union for directors, ADs, and stage managers.

Diaphragm (1) The adjustable opening which varies the aperture size of a lens. (2) The element in a microphone which vibrates according to the pressure variations in the air created by the sound source.

Dichroic Mirror The filter in a color television camera that dissects white light into the three primary colors.

Diffusion Filter A special filter which produces a fuzzy, foglike effect on the image photographed.

Digital Refers to any device in which an electronic signal is represented by computer-type binary numbers.

Digital Camera A camera which operates utilizing digital circuits and technology. Offers extremely stable pictures, automatic setup and picture correction, and flexibility in transmitting the picture from the camera head to the CCU (base station).

Digital Video Manipulator A series of sophisticated special effects generators which utilize digital technology to manipulate and control the video image. Produces effects which are impossible utilizing analog technology.

Digitize To convert an audio or video signal from its "analog" form into computerlike digital code numbers.

Dimmer Device used to control the amount of electric power reaching a lighting instrument and, therefore, the light output of the instrument.

Director The production team member responsible for creating the sound and picture of a program.

Dissolve A simultaneous fade-in of one video source and fade-out of another. Analogous to the audio cross-fade.

Diversity Receiving System A special wireless microphone receiver which eliminates spurious interference or signal dropout, making RF mikes very reliable.

Dolby Noise Reduction An electronic device which, when used during audio recording and playback reduces the background "noise" and produces a better quality audio signal.

Dolly (1) Camera support which permits the camera to move smoothly across the studio floor. (2) Movement of the camera on its pedestal closer to, or farther from, the subject.

Double-Chaining The simultaneous use of two film chains to permit cross-fading of audio and video, and the elimination of lip flap.

Double Re-entry A sophisticated switcher which permits the output of a mix/effects system to be re-entered for further video manipulation.

Double Zoom A zoom lens with a continuously variable "extender" to increase the lens' focal length.

Downstream Keyer A special effects generator which enables the TD to insert or key over a composite video signal just before the video signal leaves the switcher to go over the air.

Dress Rehearsal The final rehearsal of a production which is an exact duplicate of the air show.

Drop A large piece of canvas or other material used as scenery backing.

Drop-Out The loss of a part of the video signal during VTR playback. A *drop-out compensator* is a device which eliminates most drop-outs by reinserting previous video information to hide the drop-out.

Dry Mount Press A piece of graphics equipment that uses pressure and heat to mount pictures or other graphic materials.

Dry Rehearsal Rehearsal outside the studio, in which initial performer blocking is planned and run through without production facilities.

Dual Redundancy Using two identical microphones to cover a sound source. In the event of mike failure, the microphone with the control pot turned off is immediately opened to continue audio without interruption.

Dub A copy of a videotape or audio tape made by recording the output of one machine on another.

Dub-Down Copying a videotape from a larger tape format to a smaller one, e.g., from 2-inch quad to ¾-inch helical.

Dub-Up See *Bump-up*.

DVE Digital video effects.

Dynamic Microphone See *moving coil microphone*.

E

Easel Card See *camera card*.

Echo Strictly speaking, a wave which has been reflected along its transmission with sufficient difference in time and magnitude to be perceived as distinct from the original. Commonly, echo is used interchangeably with "reverberation" to indicate the controlled time delay and repetition of a sound.

Edge Wipe; Edge Key   A wipe or key in which edges are electronically produced to make the key material appear more prominent on screen.

Edit Programmer   Device which automatically produces a videotape edit at the predetermined edit point as it controls the operation of both the playback and edit/record VTRs.

Edit Room   Having enough space before or after the edit point to permit a clean video or audio edit.

Editing on the Fly   (1) Directing a multicamera production, with the director calling shots as the show progresses. (2) Editing videotape without an edit programmer, which requires the edit to be timed perfectly and made as both playback and edit/record VTRs are in operation.

Effects Bus   The various banks on the switcher which control such electronic effects as wipes, keys, mattes, and inserts.

EFP   *See electronic field production.*

Electret Condenser Microphone   A capacitor or condenser microphone which utilizes a precharged element, thus eliminating the need for bulky power sources.

Electron Gun   Device which creates the electron beam used to scan across the photosensitive element of the camera pickup tube and the phosphor-coated surface of the CRT.

Electron Scanning Beam   The electron beam created and controlled by the electron gun and its auxiliary circuitry.

Electronic Editing   A postproduction operation, in which videotape material is edited by dubbing from a playback machine to an edit/record VTR at the precise point where the edit is to be made.

Electronic Field Production   The use of a single camera with a single VTR to shoot on location in the field. EFP is distinguished from ENG by generally higher production values.

Ellipsoidal Spotlight   Spotlight which produces hard, directional light. Internal shutters permit distribution control, and a pattern slot enables the ellipsoidal to project shadow patterns. Also called a "Leko."

Emulsion   The light-sensitive coating of film which produces the image.

Encoder   Electronic device which transforms the color camera's red, blue, and green video signals into a luminance signal and a chrominance signal.

ENG   Electronic news gathering; the use of portable video cameras and portable VTRs to cover news events quickly. Most ENG units also contain a microwave, so live sound and picture can be relayed to the station for immediate broadcast.

Equalizer   Audio device which permits a sound signal to be manipulated by varying specific frequencies to produce a particular sound quality.

Essential Area   The area of the television picture which is sure to be received by all television sets. In graphics, the area where important visual and lettering information must be positioned to ensure its reception by all viewers.

Establishing Shot   The opening shot of a show or scene, which orients the viewer to the surroundings; usually a wide shot.

Exposure   The amount of light which is allowed to enter the camera and affect the light-sensitive surface of the pickup tube or the emulsion of the film.

F

Fade   A gradual increase or decrease of the video or audio signal. In video it is the gradual appearance of a picture from black or vice versa. In audio, it is the gradual increase in sound or vice versa.

Fader   A device used to control sound, video, and lighting intensities.

Fader Bar   Two ganged levers on the switcher which control the output of a double bank of video sources. Fader bars are used to produce fades, dissolves, supers, split screens, and wipes.

Falloff   The degree with which light goes from full intensity on a subject to black. "Rapid falloff" means there is a quick and obvious distinction between an illuminated area and a nonilluminated area. "Slow falloff" means the light intensity gradually reduces from full to none.

Fast Lens   A lens with a low f-stop number which permits photographing with a wide aperture and, consequently, use under low-light conditions.

Fax Sheet   Short for "Facilities Sheet," a listing of all production equipment necessary for a show.

Feed   The transmission of a signal or a show from one point to another.

Feedback   (1) Video: sending the video signal back on itself to produce a series of random streaks or patterns on the television screen. (2) Audio: sending the audio back on itself to produce an echo effect at low levels, or a loud howl when uncontrolled.

Field   Half the television picture composed of either all odd or all even scanning lines. Two fields are interlaced together to produce a *frame*, or complete video picture.

Field of View   The area covered by a lens.

Fill Light   Light used to lighten or eliminate shadows created by the key light.

Film Chain   Film projectors, 35-mm slide projector, multiplexer, and television film camera used to reproduce film and slides for television. Also called "telecine," "film island."

Film Speed   The degree of sensitivity that a film emulsion has to light. "Fast film" means high sensitivity; "slow film" means reduced sensitivity.

Filter   (1) Audio: device which allows some sound frequencies to pass and blocks others to manipulate sound quality. Used to simulate telephone voice, radio or TV audio, etc. (2) A glass or gelatin lens cover, which is used to change the quality of the light which enters the camera.

Fisheye Lens An extremely wide angle lens which produces a 180° field of view.

Fishpole Boom Hand-held boom, used mainly on location or where a larger boom is too unwieldy.

Fixed Focal Length Lens A lens with a fixed focal length, as opposed to a zoom lens, which has a variable focal length. Also called a "primary lens."

Flagging Distortion seen on some helical VTR playbacks where straight lines at the top of the picture become wavy and fluctuate back and forth.

Flare Dark or colored streaks in the picture, caused by a very bright or highly specular reflection in the scene being photographed.

Flashback; Flashforward Varying the temporal order of a show by going backward or forward in time.

Flat (1) A piece of scenery used as a background set. (2) A picture lacking contrast.

Flat-bed Editing Machine *See horizontal editing machine.*

Flat Lighting Lighting characterized by even, diffused light without shadows or contrast.

Flip Card *See camera card.*

Flooding the Beam Focusing a fresnel spotlight to "flood" position to widen the beam distribution and reduce intensity. *See spread.*

Floodlight Wide-aperture light source which produces flat, diffused illumination over a wide area.

Floor Manager Individual responsible for all activities on the studio floor and for relaying director's signals to talent during rehearsal and production. Also called "stage manager."

Floorplan A scale drawing of the studio used in planning scenery design and construction, lighting, and camera and subject blocking.

Floorstand (1) A mounting device for lighting instruments which is positioned on the studio floor. (2) An easel for holding title cards and other camera graphics.

Fluff A mistake, or error.

Focal Length The distance from the optical center of a lens to its focal point. Expressed in millimeters (mm), it indicates the lens's horizontal field of view.

Focus The point where the light rays converge to produce a clear, sharp, and defined image.

Foldback Returning selected portions of an audio mix to the studio so talent can monitor it without creating a feedback loop; e.g., to foldback all musical instruments to a singer but eliminate his voice to prevent creating feedback squeal.

Footcandle   Measure of light intensity.

Foreground Treatment   Photographing certain subjects or objects in the foreground of the shot to induce depth in the picture.

Frame   (1) Video: a complete television picture consisting of two interlaced fields. There are thirty frames produced per second in television reproduction. (2) Film: a single picture in a series of pictures on motion picture film; twenty-four frames are produced each second in film. (3) The outline of the television screen which the director uses to determine which visual elements to include and which to exclude in each camera shot.

Framestore Synchronizer   A digital device which processes each television frame to correct for problems in synchronization before the video signal enters the studio system. This permits the use of wild feeds from remote sources which may not be in synchronization with the home studio.

Freeze-Frame   To stop the action on a single video frame. Can be produced with a slow-motion disc, a still-frame storage unit, a framestore synchronizer, or some helical VTRs.

Frequency   The number of complete cycles per second of an electrical signal expressed in Hertz (Hz).

Fresnel Spotlight   A lighting instrument which uses a fresnel lens to produce a beam of hard, directional light, which can be varied from "spot" to "flood." The most commonly used lighting instrument in television production.

F-Stop   The numerical setting on a camera lens which indicates the size of the aperture opening. The higher the f-stop number, the smaller the opening.

Full-Coat   Magnetic recording tape produced in film format to permit accurate sound and picture editing when utilizing double system sound.

Full Track Recorder   An audio recorder which records a monaural signal scross the entire width of the audio tape.

G

Gaffer-Grip   A strong clamp used to attach lightweight lighting instruments to scenery, doors, and other locations.

Gaffer Tape   Strong, all-purpose tape which is used to dress audio and video cable, secure equipment and set pieces, and for a variety of jobs around the studio and location.

Gain   The amount of signal amplification for audio and video signals. "Riding gain" means varying controls to produce the proper sound level.

Gel   Colored plastic or gelatin material which is mounted in front of lighting instruments to produce colored light.

Generating Element The part of a microphone which transforms sound waves into electrical energy.

Generation The number of dubs away from the master original tape. "First generation" is a tape dubbed directly from the master. "Second generation" is a tape dubbed from the first generation, and so on. As generations increase, technical quality usually decreases.

Giraffe Boom Tripod boom which is smaller than a large perambulator boom but provides more operating flexibility than a fishpole boom.

Glitch Picture interference or distortion which occurs momentarily.

Gobo A foreground set piece designed for the camera to shoot through.

Grain Refers to the degree that the minute granules of silver, which comprise the film emulsion, appear on screen.

Graphics All visuals prepared for a production. Includes camera cards, slides, electronically generated letters and symbols, and special graphic set pieces.

Gray Scale A test pattern or chart progressing in steps from TV white to TV black. Most gray scales use either seven or ten gray scale steps.

Ground Row A curved set piece positioned in front of a cyclorama to increase the perception of depth and to hide cyc strip lights from camera view.

Guillotine Tape Splicer A film splicer which automatically trims splicing tape, producing very fast splices.

H

Hand Card A graphic held by talent.

Hand Cue Silent hand signals given by floor manager to talent during production.

Hand Prop Objects or props which are handled by performers during production.

Hard Key A key or insert with distinctive borders around the edges of the key.

Hard Light Light quality characterized by a strong, directional beam, which produces dark shadows. Hard light is produced by spotlights.

Hardwall Flat A flat with wooden surface.

Headroom The space left between the subject's head and the top of the screen.

Helical Scan (Helical VTR) A method of videotape recording which uses one or two video heads to scan the tape in a slanted track pattern. Also called "slant track" recording.

Hertz (Hz) A unit of frequency indicating the number of cycles per second.

High Band A videotape recording technique which uses high-frequency signals to produce higher-quality pictures and permit a greater number of dubs without significant reduction in picture quality.

High Definition Television (HDTV) The use of specially designed video equipment capable of reproducing pictures with significantly greater resolution than is possible with conventional video equipment. HDTV uses over 1,000 video lines to create the video image as opposed to the 525-line or 625-line systems which are currently in use.

High Key   A lighting approach characterized by light shadows and relatively even illumination.

Highlight   The brightest portion of a picture or area.

High-Z   High impedance microphone signal; rarely used in professional audio equipment.

HMI Lamp   A gas-filled lamp which produces a high-efficiency light output with a marked reduction in heat. The color quality of the light is daylight balanced.

Horizontal Editing Machine   A film editing table which permits multiple picture and sound tracks to be manipulated and edited quickly.

Hot   Equipment is on: current is being supplied to equipment.

Hotspot   An extremely bright concentration of light in one place relative to the surrounding illumination. Can create camera shading problems.

Hue   The color itself of light, paint, etc. The hue is the actual color.

I

IATSE   International Alliance of Theatrical Stage Employees; technician's union.

IBEW   International Brotherhood of Electrical Workers; technician's union.

ID   Short for "station identification."

Image Compression   The special effect available with a digital video manipulator in which a full frame image can be compressed to any desired size and located at any point on the television screen.

Image Enhancer   An electronic device designed to increase picture resolution.

Image-Orthicon (I-O)   A camera pickup tube used in some monochrome and early model color cameras. I-O cameras utilize lenses in 35-mm format.

Image Retention   *See burn-in, comet tailing.*

Impedance   Resistance to current flow, which is especially important in matching microphones and audio equipment. Almost all professional audio microphones are low impedance (low-Z), which permits longer cable runs without a reduction in audio quality.

Incandescent Lamp   A lamp which produces light by heating an internal filament inside a conventional glass globe.

Incident Light Reading   A light meter reading taken from the position of the subject with the meter facing toward the camera and light source. Incident light refers to the illumination which falls directly on the subject.

Insert Edit   Electronic videotape editing in which new video, audio, or both can be inserted into a previously recorded tape without disturbing material before and after the insert. Insert editing utilizes the existing control track and produces very stable edits.

Insert Key   The insertion of one video image into another. *See chroma key.*

Insert Reel   *See B-roll.*

Intercom   Short for "intercommunication system;" the internal communication system, using telephone-type headsets, between production personnel in the control room and crew members on the studio floor. Also called "PL system."

Interlace Scanning   Process in which the odd and even numbered lines of the television raster are transmitted consecutively. Together they make up a single video frame which is a complete picture. Interlace scanning is used to eliminate picture flicker.

Interrupted Feedback (IFB)   A special intercom between the control room and talent which interrupts program audio whenever the director or producer must communicate with talent while the program is on the air.

In The Can   A finished production which has been edited and is ready for film or videotape playback.

I-O   *See image-orthicon.*

i.p.s.   Abbreviation for *inches per second.* Indicates the speed of audio or videotape during recording and playback.

Iris   Same as aperture.

Isolated Camera   A camera which feeds its own videotape machine, as well as being used in the multiple-camera video mix. Started in sports coverage but is now used for many entertainment and dramatic productions to provide additional editing footage.

J

Jack   (1) A female socket or receptacle. (2) A stage brace used to support scenery.

Jog   To slowly move helical videotape back and forth in order to locate a precise edit point.

Joystick   A hand-operated control which offers 360° positioning of the joystick. Used in switchers, to position a cutout insert, and, in some videotape edit programmers, to control the operation of the VTRs. Also used on some audio consoles to pan a sound channel when mixing in stereo.

Jump Cut   An unnatural or jarring transition between two camera shots.

K

Kelvin Scale   A unit of measurement used to indicate the color temperature of a light source.

Key   An electronic effect in which an image is electronically cut into another background picture. *See matte key, insert key, chroma key.*

Key Light   The principal source of illumination on a subject or scene.

Keystone   (1) Picture distortion caused by a camera that is not at a perfect right angle to the surface of the object or graphic being photographed. (2) A piece of plywood used to reinforce the joints of a flat frame.

Kicker Light   A light positioned at the rear and side of the subject.

Kill   To turn off a light, sound, or video feed; to cut or delete a line or portion of a program.

Kinescope   Recording a television program from a monitor using 16-mm motion picture film.

L

Lag   Persistence of an image on the face of the television camera pickup tube. *See also comet tailing, burn-in, image retention.*

Lavalier Microphone   A small microphone designed to be worn by talent either hung from a cord or clipped to tie, lapel, or blouse.

Leko   Ellipsoidal spotlight.

Lens Format   Indicates the size of the television tube's faceplate with which the lens can be used. There are three basic lens formats: 35 mm for I-O, a 16-mm format for 1¼-and 1-inch tubes, and a 16-mm format for ⅔-inch tubes.

Lens Speed   Refers to the maximum aperture of the lens; the lowest numerical f-stop. The faster the lens, the wider the maximum aperture, and the more light-gathering capability.

Lensless Spotlight   Instrument designed to produce hard, directional light without a spotlight. Lightweight and portable design makes them useful for remote production.

Level   The signal strength, or volume, of a video or audio signal.

Light Meter   Meter designed to read light intensity using either *reflected* or *incident* light. Most light meters used in TV have footcandle scales.

Light Plot   A floorplan showing the studio set drawn to scale, with the lighting instruments that are to be used superimposed to indicate instrument, location, and function.

Light Ratio   The relative intensities of various light sources; intensity of light between key, back, and fill lights.

Limbo   (1) An area with a completely neutral background which gives the illusion of endless distance. (2) A lighting approach in which background is light gray with foreground subject prominent on screen. *See also cameo.*

Line   (1) Line monitor: the master video monitor which shows the picture being recorded or broadcast. Also called "air monitor." (2) One of 525 traces of the television pickup tube scanning beam.

Lip Sync   (1) The synchronization of sound and picture. (2) Having a performer mouth words to a prerecorded sound track.

Live   (1) A program broadcast as it happens, in real time. (2) A device or piece of equipment which is turned on, e.g., "live microphone."

Live on Tape   A videotape production approach in which the show is produced with multiple cameras as though it were broadcast live. No postproduction editing is used.

Location   A production site outside the normal studio.

Logo   Symbol used to identify a station, program, sponsor, company, etc.

Long Shot   Camera shot of a set or subject which usually includes a wide-angle field of view. *See establishing shot.*

Look Space   Composing a subject so that the picture compensates for the psychological impression given of looking toward the edges of the television frame.

Low Band   Videotape recording process which uses relatively low-frequency signals to produce a video image. Provides lower-quality picture than high band VTRs.

Low Key   A lighting approach characterized by deep shadows, high contrast ratio, and strong use of highlight and dark areas.

Lower Third   Super graphic designed to appear beneath a subject in lower third of the screen. Used to identify subject, object, place.

Low-Z   Low impedance; used primarily when referring to microphones. All professional mikes are low-Z because it permits longer cable runs.

Luminance Channel   In color television, the channel which carries the monochrome signal and provides brightness information for color receivers.

Luminance Chroma Key   *See soft key.*

Luminaire   A lighting instrument.

M

Magnetic Film   Sound tape produced in the same format as motion picture film with sprocket holes. Used for double system sound editing.

Magnetic Sound   Motion picture film with a magnetic sound strip running along one edge. Small audio recording head in camera lays down the track and a tiny playback head in projector reproduces the sound during projection. Also called "mag sound."

Mark   Tape placed on the studio floor to indicate to talent or camera operators where they should be positioned.

Master   An original audiotape, videotape, or film. Used for broadcast or to produce copies.

Master Control   The room where all video and audio outputs of various production studios are fed for distribution and broadcast or recording.

Master Control Fader   Fader on audio console which regulates the entire output of the console.

Master Shot A single shot of a scene, usually a wide or long shot, which is used as the reference, or master, in editing a sequence.

Matched Dissolve Dissolve from one picture to another which is closely related in appearance or shot size.

Matte Key Keying a graphic or symbol over a background picture. The cut-out lettering can be electronically filled in with any desired color shade or with gray.

Mechanical Editing Videotape editing in which the tape is physically cut and spliced together.

Mechanical Sound Effect A sound effect which is produced through mechanical means, such as crumpling cellophane, opening and closing doors, etc. Requires a live microphone pickup to cover sound.

Microwave The line-of-sight and point-to-point transmission of video and audio signals. Commonly used to feed live and taped signals from a remote production back to the studio for taping or broadcast.

Mix Bus A pair of buses with a fader bar control to permit the production of fades, dissolves, and supers.

Mix/Effects Bus A series of buses which are connected to a fader bar and a special effects generator to produce keys, inserts, wipes, split screens.

mm (millimeter) One-thousandth of a meter; 25.4 mm = 1 inch. Measurement used to express the focal length of a lens.

Modulation Varying the frequency, amplitude, or phase of a wave or electrical current by superimposing another wave on it.

Moiré Effect Spurious color patterns which appear in a television picture when certain stripes, checks, or other complex designs on fabrics or graphics interfere with the television system's scanning operation.

Monaural Single-channel audio.

Monochrome Black and white television.

Montage A rapid sequence of shots used to produce a particular image or mood.

Motivated Key Positioning the key light in a place logically determined by set elements such as a window, lamp, etc.

Moving Coil Microphone Microphone designed with diaphragm connected to a moving coil which creates an electrical current by its motion within a magnetic field. Also called "dynamic microphone."

Multiple Camera/Multiple VTR A videotape production approach in which each camera continuously feeds its own VTR. All editing is done in postproduction.

Multiplexer A device utilizing mirrors and prisms to enable a number of film and slide projectors to feed a single television film camera.

Multiplexing The process of "piggybacking" multiple signals on one carrier.

Multiplier Same as *range extender*.

Multitrack Recorder An audio recorder which permits recording signals on a number of separate tracks or channels for greater creative control in recording and mixing.

Munsell System A color scale which identifies various colors according to hue, saturation, and brightness.

N

NABET National Association of Broadcast Employees and Technicians; primarily a technicians' union.

Narrow-Angle Lens Telephoto lens.

Negative Film Film which produces a reverse image, light areas appearing dark and dark areas appearing light. Used to produce positive release prints.

Neutral Density Filter Filter designed to reduce intensity of light without affecting its color quality.

Neutral Set Completely bare setting which emphasizes foreground subjects.

Noise Unwanted audio or video signals which interfere with program information.

Nonsegmented Helical recording technique using a single video head which lays down a complete video track on each pass. Permits stable freeze-frame and slow-motion operation.

Normal Lens A lens with a focal length that produces the normal field of view and spatial perspective which our eyes produce.

Normal Wiring A circuit which is permanently wired into an audio console. It is broken only if a patch cord is inserted into the appropriate jack on the patch panel.

Notes Suggestions, criticisms, and revisions given to cast and crew by director and producer after each rehearsal.

O

Off-Book When talent must have dialogue memorized and not rely on script.

Off-Line Editing with helical workprint of original master footage. Usually SMPTE time code is used to catalogue edits, and a computer-assisted edit system is used to transform workprint time codes into the final, edited master.

Ohm A measure of electrical resistance. Often expressed with symbol Ω.

Omega Wrap Videotape threading configuration in which the tape on a helical VTR is wrapped around the head drum in a 180° or omega pattern.

Omnidirectional Audio pickup pattern in which microphone is designed to be equally sensitive to sounds emanating from all directions.

On-Line Editing videotape using top-quality recorders, as opposed to working with inexpensive helical workprints of the original master tape. On-line also refers to a computer-assisted edit system that uses SMPTE time code from the workprint to assemble the final master on large format VTRs.

Open Face Spotlight   Lensless spotlight.

Open Set   A set design approach which uses a minimum of flats, and then only to suggest the actual environment.

Optical Sound   Film sound track which uses variations of light and dark patterns to produce the sound. The optical track runs along one edge of the film.

Oscilloscope   A device utilizing a cathode ray tube to visually display electronic signals; used for equipment testing and setup.

Outcue   The final cue of a tape or film cut; the final cue of a program.

Overcut   (1) Changing the inserted image in a key or matte picture without affecting the background picture. (2) Too much cutting between shots within a program or sequence.

Over-the-Shoulder Shot (O.S. Shot)   Camera shot in which a subject is photographed framed by another subject's shoulder in the foreground. Induces depth in the shot.

P

PA   (1) Production assistant. (2) Public address. In audio, the amplification system used to feed the audience area program sound.

Pacing   The overall rhythm of the program, which is determined by the cutting, performance, and other creative aspects. Determines how the audience perceives the time of segments and the overall show.

Pad   (1) A resistance placed in the audio circuit to match impedances or to cut down excessive electrical power to provide greater control to the audio engineer. (2) A flexible segment in a program which can be used to stretch or shorten the running time of the show. (3) Additional video or audio on film or tape to provide protection against running out of program material.

Pan   Horizontal movement of the camera on a stationary pedestal.

Pan Light   Same as *broad*.

Pantograph   A scissor-type hanging device used to vary the height of a lighting instrument.

PAR Light   Short for *parabolic aluminized reflector*; a lamp with built-in reflector unit.

Parabolic Reflector   A large dish with a microphone mounted in the center. Used to pick up audio from large distances.

Parallel Action   Two or more actions or events occurring simultaneously and shown through the use of cross-cutting, flashbacks, or the use of complementary or opposing audio and video tracks.

Patch Board   Refers to a distribution system which enables lighting instruments, audio sources, or video sources to be individually assigned to various control circuits.

Patch Cord   The cord on a patch board which is used to complete a connection assigning control of audio, video, or lighting to a particular circuit.

Patch Panel   Refers to the audio patch board.

Pea Light   A tiny Christmas tree-type light which is often used for stars in conjunction with a cyclorama.

Pedestal   (1) Camera mounting device. (2) Camera operation command meaning to raise or lower the height of the camera by adjusting the pedestal column control.

Perambulator Boom   A large boom arm on a wheel-mounted base. A series of controls permit the microphone to be rotated and extended or retracted to cover the audio.

Performer   Anyone who appears in front of the television camera.

Perspective   (1) Spatial relationships as they appear in a camera shot. (2) Sound perspective: the relationship of the sound quality of the visual image of the sound source.

Pickup Pattern   The pattern or direction with which a microphone is designed to cover sound.

Pickup Tube   The vacuum tube inside the television camera head which converts light into electrical energy.

Pigtail   Cable and connectors running from the lighting power strip which are used to supply power to individual lighting instruments.

Pin   (1) To focus the beam of light on a fresnel down to a highly directional spot. (2) Overloading the audio level to the extent that the VU needle is driven past its maximum operating limit.

Pipe Grid   Crisscrossing metal pipes suspended over the studio floor, from which lighting instruments are hung.

Pixel   The smallest element of the video image. Pixels are the phosphorescent "dots" which run along each of the video lines that make up the video image.

PL   Short for *private line*. (1) Refers to communication intercom system in a studio which enables production and crew members to talk to each other. (2) A special telephone line installed specifically for a production.

Plastics   A very general term which refers to such design elements as graphics, sets, and lighting.

Plumbicon   Camera pickup tube which utilizes a lead oxide compound as the light-sensitive element.

Points   The intensity scale used to indicate the position of a lighting dimmer fader control. Points run from zero (no power) to ten (full power).

Polar Diagram   Illustration of the pickup pattern of a microphone.

Polarity Reversal   Electronically reversing the gray scale of an image, so the camera's picture shows a negative image.

Polecat   Spring-loaded, aluminum pole which is used to hang graphics, set pieces, lighting instruments.

Polish   A script revision or rewrite.

Polystyrene   A plastic material used in set construction and for special props. Also called "Styrofoam."

Pop Filter   *See blast filter.*

Posting   A term for *postproduction*.

Postproduction   The final stage of the production process in which videotape is edited and audio is added or "sweetened" after the actual production is completed.

Pot   Short for potentiometer; a control operated in a clockwise or counterclockwise direction to vary the intensity of an electrical signal. Commonly refers to knob controls on audio console.

Power Rail   The cable trough in which power cables for lighting instruments are run along pipe grids.

Practical Lights   Set pieces, such as table lamps and chandeliers, which must actually operate during the production.

Preproduction Planning   The first production stage in which the program is planned and coordinated.

Preroll   Starting a film or tape earlier than it is needed on the air to permit it to attain the proper operating speed and to stabilize. The time necessary for a preroll depends on the film or tape machines, as well as the program's script, and director and talent preference.

Preview Bus   A row of video source buttons which enable the TD to look at any video picture before actually putting it on the air.

Preview Monitor   The control room monitor which displays the output of the preview bus.

Primary Lens   Fixed focal length lens.

Primary Video Source   Any video source which produces its own picture image; e.g., cameras, VTRs, telecine, and character generator.

Prism Beam Splitter   Optical device in a color camera which dissects the reflected light into the three primary colors *See dichroic mirror.*

Producer   The production team member responsible for the entire production.

Production   The third stage in the production process, when sound and picture are broadcast or recorded on tape.

Production Console   The long table in the control room which faces the monitor bank in which the video switcher is installed. Seated at the production console are the director, TD, AD, and other production staff members.

Production Music   Specially written and recorded music, designed to be used as background, theme, or transitional music on a show.

Production Switcher   Video switcher which enables the TD to put any video source or composite picture on the air.

Program Bus   The master bus on a video switcher which controls the output signal of the switcher.

Program Intercom   PL system used to communicate directions and cues between control room and studio floor during rehearsal and production.

Program Monitor   Large control room monitor which displays the video output of the switcher which is being broadcast or recorded. Also called "line monitor" or "air monitor."

Prop   Short for *property*; any scenic element used to dress the set but which is not structurally a part of the background. Includes furniture, pictures, and various items used by performers. *See hand prop.*

Protection   A duplicate made of a film or tape and run simultaneously with the master machine to safeguard against technical problems during air.

Punch Up   To "take" or "cut" to a video source. Often used as a director command; e.g. "punch up" a video source, or put it on the air.

Push Film   Force the development of film beyond its normal exposure rating to permit filming under low-light conditions.

Push-Off   Special effect possible with a digital video manipulator unit. Effect appears as though one image literally pushes the other off the screen.

Q

Quad   Short for quadraplex.

Quad Split   Special video effect in which four different images appear on screen simultaneously.

Quadraplex   A large-format videotape recording process which utilizes four video heads mounted on a rapidly rotating head drum. All quad VTRs use 2-inch videotape.

Quantitize   To convert an analog signal into a digital signal.

Quartz Lamp   Lamp which provides a high-intensity illumination with a constant color temperature. Also called "tungsten-halogen" lamps.

R

Rack Focus   Varying the focus of a lens to change the areas which appear in and out of focus.

Radio Frequency (RF)   Wireless transmission of video and audio signals via various broadcast channel frequencies.

Range Extender   Optical device which increases the focal length of a lens.

Raster   The illuminated area of the television screen which is produced by the scanning lines.

Rating A statistical estimate of a program's popularity. Expressed as a percentage of the number of households watching among all television households.

Reaction Shot A cutaway to a shot which shows the reaction of another subject. *See also cutaway shot.*

Real Time The actual time in which an event or program takes place. Used to distinguish between cutting or editing while a show's production progresses or afterwards in postproduction.

Realistic/Representational Set Set design approach in which the setting is meant to appear as realistic as possible.

Rear Screen Projection Projecting slides or film on a translucent screen with the projector positioned behind the screen. Used as a background set.

Recording Head The electromagnetic device in audio and video tape recorders which produces audio or video information on the tape.

Recording in Segments A videotape production technique in which multiple cameras are used to tape a portion of a larger program. Once all segments are recorded, they are assembled together in postproduction.

Reflected Reading A light meter reading taken with the meter facing the subject and measuring the light reflected from the subject into the camera lens.

Registration The accurate alignment of the three electronic pickup tube images in a camera to produce the composite color picture.

Rehearsal The second production stage, which is often divided into two parts: (1) dry rehearsal and (2) studio rehearsal.

Release Form A standard form signed by all performers before appearing on a television production.

Remote A television production produced outside the studio.

Remote Survey Preliminary visit to the location site to determine all technical and creative production requirements for the show.

Resolution The degree of clarity or definition, of a picture. The higher the resolution, the sharper the image.

Reverberation The persistence of a sound which is produced either acoustically or electronically. Often referred to interchangeably as "reverb" and "echo."

RF Microphone A wireless microphone.

Ribbon Microphone Microphone which utilizes a sensitive ribbon suspended in a magnetic field as the generating element.

Riding Gain Controlling the level of a video or audio signal.

Riding Levels Same as riding gain.

Riser A platform, usually made from plywood, which is used to elevate talent or to produce multiple levels on a studio set.

Roll Cue The command to start a film, videotape, or audio tape.

Rolling Means that a film, videotape, or audio tape has been started and is ready for playback or recording.

Room Noise Recording the ambient sound of a production location (either interior or exterior). The room noise is useful later during postproduction editing in smoothing over audio edits on audio or video tape.

R.P. Short for rear screen projection.

Rub-on Lettering *See transfer lettering.*

Rundown Sheet Indicates each sequence within a program, its segment time, and the program's overall running time. Some rundown sheets also include talent and set information which may be needed during production.

Run-through Rehearsal, usually with full facilities.

S

SAG Screen Actor's Guild; performers' union.

Saticon A vidicon-type color pickup tube with improved operational characteristics over standard vidicons.

Saturation The strength of a color determined by its relative purity; i.e., a fully saturated color has no white light mixed in, an unsaturated color appears washed out with the addition of white light.

Scanning The television reproduction process which involves moving an electron beam across a camera pickup tube (to convert light energy into electrical impulses) and across a CRT (to reproduce the image).

Scanning Area The total area scanned by the television camera and reproduced on the studio monitor.

Scene Breakdown Dissecting each scene into its component parts; i.e., location, performers, sets, and props, in order to develop an efficient videotape shooting schedule.

Scenic Designer Production team member responsible for designing sets and supervising set construction.

Scoop A television floodlight.

SCR Dimmer Short for *silicon-controlled rectifier*. A dimmer control which uses a low-level pilot voltage to control power to lighting instruments.

Scrim (1) A translucent gauze or fiberglass material used to soften and diffuse illumination from a lighting instrument. (2) A gauzelike curtain often used in conjunction with a studio cyclorama.

Secondary Video Source The output of a mix or mix/effects bus on the video switcher, which can be re-entered for additional video manipulation.

Segmented Scanning A VTR recording/reproduction process in which two or more video heads are used to divide or "segment" the video track information as it is written on the videotape.

Segue To begin next sound source immediately after preceding sound without interruption.

Selective Focus Utilizing depth of field to direct the viewer's attention to certain areas of the scene by varying those elements in and out of focus.

Servozoom   A zoom lens which is operated via electronically controlled motors.

Setup and Rehearsal   The second stage of the production process, when the studio is prepared for the production, and performers and technical personnel run through their various responsibilities in order to coordinate their activities for production.

Shading   Varying the video controls on a CCU to compensate for changes in scene brightness.

Shadow Keyer   Chroma keyer which can produce shadows, along with the insert image, for a more realistic composite effect.

Share   Audience measurement estimate expressed as the percentage of viewers watching a particular show among all those watching TV at the time.

Shot Box   Zoom lens control mounted at rear of camera, which permits the camera operator to preselect a number of specific lens focal lengths.

Shotgun Microphone   Highly directional microphone designed to pick up audio from large mike-to-subject distances.

Shot Sheets   Listings of each camera's individual shots which are given to camera operators prior to rehearsal and production.

Shot Shopper   Slang term for program directors who permit their camera operators freedom in selecting shots.

Show Reel   The edited film or videotape reel which is to be used for a production. Usually includes all cuts or segments, separated by cuing leader.

Side Light   Lighting instrument positioned to side of subject to emphasize body shape and form.

Signal-to-Noise Ratio   The relationship of the strength of the video or audio signal to interfering noise. The higher the SN ratio, the better, since this means the signal is masking the interfering noise.

Simulcast   Refers to the playing of synchronized FM stereo audio along with the conventional video and audio to provide high-quality, stereo sound.

Single Camera/Single VTR   A videotape production approach utilizing one camera and a single VTR. The camera shoots a scene from a variety of different angles, or setups, and all editing is done in postproduction.

Single System Sound   The process of recording both sound and picture on the same film or videotape.

16 mm   Most commonly used television film format.

Slant Track Scanning   Helical VTR recording/reproduction process in which video information is written on the tape in a slanted pattern.

Slate   A blackboard or chart used to visually identify a film or videotape take according to take number, recording date, and additional identification information.

Slave Synchronizing a video or audio tape machine to a "master" machine for recording, editing, and playback operations. Commonly, a multitrack audio tape recorder is slaved to a VTR via SMPTE time code.

Slide Chain 35-mm slide projector; part of the telecine.

Slip Cue Method of cuing records to permit instantaneous program sound when called for by director.

Slow Lens A lens with a relatively small maximum aperture; a lens with a relatively high minimum numerical f-stop which does not permit it to gather much light.

Slug A section of blank film or videotape which is inserted into the film or tape to represent additional program information which is forthcoming. The slug usually runs the exact length of the intended program material.

SMPTE Society of Motion Picture and Television Engineers.

SMPTE Time Code Eight-digit address code used to identify each videotape frame by hour, minute, second and frame number for precision editing.

Sneak Slowly fade-in audio source.

SOF; SOT Short for *sound on film* or *sound on tape*.

Soft Key A chroma key in which the edges of the insert image gradually blend into the background. The soft key permits keying transparent objects and shadows into a background picture creating a more realistic key. Also called "luminance keying."

Soft Light Wide aperture floodlight which produces a very diffused illumination.

Soft Wipe A wipe with edges similar to those seen in a soft key.

Solarization Video special effect in which insert key is used to produce a high-contrast image which can be colorized using the switcher SEG.

Sound Bite A film or tape clip containing lip sync sound from a news story. Usually refers to a piece of an interview or a statement which is used in the news story.

Sound Intensity The volume of the sound.

Sound Perspective The relationship of the sound quality with the visual image of the sound source.

Sound Presence A subjective impression of sound quality which depends upon the microphone, mike-to-subject distance, and room acoustics. More presence produces a fuller, richer, closer sounding audio.

Special Effects Generator (SEG) Electronic device, usually installed in the video switcher, which is used to produce wipes, split screens, inserts, keys, and mattes.

Specular Reflection Hard, intense reflection from a shiny surface.

Splice (1) The point at which film or tape is physically joined. (2) The process
of joining two pieces of film.

Split Reel Motion picture film reel which can be unscrewed to remove
film wrapped around a central core.

Split Screen A wipe carried only partway through; thus, two or more
images appear on screen simultaneously.

Spot Down Focusing the fresnel spotlight into its spot beam; narrowing
the beam distribution and increasing light intensity.

Spot Meter A reflected light meter with an extremely narrow field of view
to permit highly specific and accurate light readings over a limited
area.

Spotlight Lighting instrument which produces a hard, directional, intense beam of light.

Spotlight Effect Special video effect on many switchers which enables
a portion of the screen to appear brighter in order to highlight a
particular element within the shot.

Spotter A person who helps the producer, director, or talent at a remote
event (usually sports) to identify participants and important action.

Spread Focusing the fresnel spotlight into its flooded beam; widening
the light distribution and reducing the light intensity.

Stage Manager Same as floor manager.

Standupper A news story delivered by a reporter on location talking
directly into the camera. Also refers to the open and closing sections of
a news story in which reporter talks directly to audience.

Starlight Filter Special filter designed to produce a starburst effect
whenever the camera photographs a high-intensity light source.

Still-Frame-Storage Unit A digital device which is used to store
individual video frames, any of which can be instantly recalled by
entering its code address number. Eliminates the need for slides and
camera cards.

Stop Down To close the aperture of a lens to permit less light to enter the
camera.

Storyboard A script which includes sketches of important visual sequences to more precisely illustrate the author's concept.

Straight-up Refers to the clock time when the seconds and minutes are
both at 12, or straight-up.

Stretch (1) Director command to talent to slow dialogue or action. (2)
Expanding either the white or black levels of the television camera
video signal as in "stretching the whites," or "stretching the blacks."

Strike (1) To dismantle a set after the production is over. (2) To remove a
prop, set piece, or item of equipment after it is no longer needed.

Striplight A row of boards attached into a strip.

Studio Address (SA) Public address system which enables the director in the control room to talk to everyone on the studio floor via a
loudspeaker system.

Studio Plot Scale drawing of studio which is used to produce floorplan and lighting plot.

Stylized/Abstract Set (1) Set design approach in which the setting is designed to suggest a particular environment without actually portraying it. (2) A fantasy or unreal setting.

Subjective Angle Camera angle in which the camera is positioned to show the scene from a participant's point of view.

Submaster A single control which enables the operator to group a number of different audio or lighting sources and to control them with one fader.

Super Short for *superimposition*.

Super Card White lettering mounted against a black card for use in supering.

Super Slide 35-mm slide with white lettering against black background for use in supers.

Supercardioid Pickup An extremely directional microphone pickup pattern.

Superimposition Combining two of more complete video images simultaneously using mix or mix/effects buses.

Sweep Reversal Electronically reversing the operation of the scanning beam to reverse horizontal or vertical placement of the image.

Sweetening Postproduction audio production to add, modify, and enhance the program audio.

Swishpan An extremely rapid pan which appears as a blur on screen.

Switcher (1) Electronic device used to select the image or composite images which are either broadcast or recorded. (2) The production crew member who operates the video switcher. Also called a "technical director."

Sync Generator An electronic component which produces various synchronizing pulses necessary for the operation of the television system.

T

Take (1) To cut to a video source. Usually a director's command, as in "Take 1." (2) Individual scenes, segments, or shots recorded on film or videotape. Each is assigned a "take number" which is used to locate and identify the segments for screening and editing.

Take Sheet A form used to keep track of each take. Includes take number, time, and whether it is "good" or "bad."

Talent Anyone who appears on camera or before the microphone.

Tally Light Lights atop the camera, inside the viewfinder, and on the control room monitors which automatically light each time the camera is punched up on the air.

Tape Degausser Same as *bulk eraser*.

Target Audience The intended audience for a program. Characteristics of the target audience must be determined by the producer in preproduction.

TD   Technical director.

Technical Director   Individual responsible for the technical aspects of a production. Often operates the video switcher during rehearsal and production.

Telecine   *See film chain.*

Telephoto Lens   A long focal length lens which produces a magnified image with a narrow horizontal field of view.

Teleprompter   A prompting device which uses either a long roll of paper or a closed-circuit television feed to display script copy to talent. It is usually mounted atop the camera.

Television Black   Darkest gray scale element which measures approximately 3 percent reflectance but appears black on screen.

Television White   Brightest element of TV gray scale which measures between 60 and 70 percent white reflectance but appears natural white on screen.

Test Tone   Audio tone produced by a tone generator and used as a reference in adjusting sound levels during recording and playback.

35 mm   (1) 2 × 2 inch slides used in slide chain. (2) Motion picture film format which is used only by largest stations and networks.

Threefold   Three set flats hinged together into a single unit.

Three-Point Lighting   Basic lighting approach which uses a key, back, and fill light to illuminate subject and create depth and texture.

Three-Shot   Camera shot which includes three subjects.

Throw Focus   Same as *rack focus*.

Tilt   Vertical movement of the camera on a stationary base.

Time Base Corrector (TBC)   A device which corrects for technical errors in helical scan VTR formats and permits the tape to be broadcast or dubbed-up to larger tape formats.

Time Code   *See SMPTE time code.*

Title Card   *See camera card.*

Track   (1) A special area of audio or videotape which contains program information or technical control information. (2) The sound portion of a motion picture film. (3) The accuracy with which a recorded videotape plays back.

Tracking Shot   Same as a *truck*.

Transducer   A device used to convert energy from one form to another.

Transfer Lettering   Graphic process in which specially produced lettering is transferred from a carrier sheet to a graphic card by rubbing over each letter.

Transverse Scanning   The videotape record/reproduce approach utilized in all quadraplex machines in which video information is written in an up-down or transverse configuration.

Treatment   A brief proposal outlining a television production.

Triaxial Cable   Thin, lightweight cable which is used on digital cameras. Weighs about one-fifth the amount of conventional cable. Also called "triax."

Trim   To adjust lighting barndoors or shutters to control extraneous light distribution falling on a subject or area.

Tripod   A three-legged camera mount, sometimes attached to a dolly for maneuverability. Most tripods are lightweight and are used for remote productions.

Trombone   A special lighting hanging device which attaches over a set wall to position the instrument where necessary.

Truck   Horizontal movement of the camera on its pedestal. The movement can also follow a moving subject in which case it is sometimes called a "tracking" shot.

Tungsten Film   Motion picture film balanced for indoor, 3200°K color temperature.

Tungsten-Halogen Lamp   Specially designed lamp which utilizes a tungsten filament surrounded by halogen gas to provide constant color temperature. Sometimes referred to as a "quartz-iodine lamp."

Turret   A number of primary lenses mounted on a movable device which enables the camera operator to position any one lens in front of the camera pickup tube.

Twofold   Two set flats hinged together into a single unit.
    Two-Shot   Camera shot including two subjects.

Type B Format   Helical VTR record/reproduction standard used on top-quality 1-inch production VTRs, characterized by a segmented scanning process.

Type C Format   Helical VTR record/reproduction standard used on top-quality 1-inch VTRs, characterized by a nonsegmented scanning process.

U

Under Five   Refers to a category of the SAG and AFTRA contracts, which stipulate a wage difference between performers with more or less than five lines; synonym for "extra" player in a casting sheet.

Undercut   Changing the background picture of a composite image key or insert without affecting the foreground picture.

Unidirectional   Microphone pickup pattern in which the microphone is designed for increased sensitivity to sounds emanating from a particular direction.

Universal Zoom   Generally describes a lens with a wide horizontal field of view (to permit great indoor production flexibility) and a very long telephoto focal length (for remote production coverage, especially sports). The wide zoom range permits the lens to be used in virtually every production situation.

V

Variable Focal Length Lens Zoom lens.

Vectorscope A specially designed oscilloscope which is used to set up and align color equipment.

Velocity Microphone Same as *ribbon microphone*.

Vertical Interval Switcher A television switcher which produces glitchfree transitions by switching between picture sources during the vertical interval between video frames.

Video The picture portion of the television signal.

Video Cartridge A plastic container which holds videotape; used in automatic-threading VTR cartridge machines.

Video Feedback Rephotographing the output of the video switcher off a video monitor and feeding the camera's output back into the switcher to create a continuous feedback loop, which produces multiple images.

Video Level The strength of the video signal; controlled by the video engineer at the CCU.

Videofilm Motion picture film shot expressly for use on television.

Videospace Refers to the fact that the audience's only measure of video reality is what appears on the television screen. The videospace is the sum total of all visual elements which interact to create the video; the visual counterpart of audiospace.

Videotape Recorder Electronic recording device which stores video and audio signals on magnetic tape for playback and postproduction editing.

Vidicon A type of camera pickup tube which utilizes lenses in 16-mm format.

Viewfinder The device through which the camera operator views the scene being photographed. The camera viewfinder is a miniature television monitor.

Voice Coil Electromagnetic device used in dynamic microphones, as well as in loudspeaker systems.

Voice-Over (V.O.) Using an announcer or performer's voice over visual material, so that the speaker is not shown on camera.

VTR Short for videotape recorder.

VU Meter Audio meter which measures the intensity of sound in volume units.

W

Wall Sled Lighting instrument mounting device which braces the weight of equipment against a set wall.

Watt A unit of electrical power; used to describe the relative light output of various lamps and lighting instruments.

Waveform Monitor   A specially designed oscilloscope used to graphically display a video signal.

Wedge Mount   Camera mounting device which enables cameras to be quickly mounted and dismounted from studio pedestals and tripods.

WGA   Writer's Guild of America; writers' union.

Whip Pan   Same as *swish pan.*

Wide-Angle Lens   A camera lens with a short focal length and a wide horizontal field of view; tends to increase perception of depth and force perspective.

Wild Feed   A nonsynchronous video feed.

Wild Sound   Recording nonsynchronous sound for either film or videotape.

Wing It   Slang term refers to improvising or ad-libbing the production without prior rehearsal.

Windscreen   Same as *blast filter.*

Wipe   Video transition in which one image wipes across another to replace it.

Wireless Microphone   A microphone that transmits a low-power radio signal which permits cable-free operation.

Workprint   (1) Film: dub of original film which is used during editing to protect the master. (2) Video: small-format, helical dub of a 2-inch or 1-inch master tape which is used for viewing and editing off-line.

Wow and Flutter   Speed variations in record, tape, or film playback which distort the sound, picture, or both.

Wow In   When a tape, record, or film is put on the air before it is completely up to speed. Produces a distorted sound and/or picture.

Writer's Bible   Information and script direction prepared by the producer about the program, objectives, characters, and other production elements which are used by the writers.

X

XLR Connector   Cannon-type audio connector used on all professional microphones and audio equipment.

Z

Zero Start   Setting SMPTE time code digits for minutes, seconds, and frames to zero before recording each new tape, with the hours set to distinguish among different tape reels. Opposite of *clock time.*

Zoom Lens   A variable focal length lens which provides a continuously changing field of view from wide angle to telephoto.

Zoom Range (Zoom Ratio)   The range of the zoom from widest possible angle to narrowest possible angle. Often expressed as a ratio such as 10:1, 15:1, etc., the 1 referring to the shortest possible focal length.

ADDITIONAL READINGS

Chapter 1 Introduction to Television Production

Schwartz, Tony: *The Responsive Chord,* Anchor Books, New York, 1974.
Shanks, Bob: *The Cool Fire,* Norton, New York, 1976.

Chapter 2 The Television Camera

Ennes, Harold E.: *Television Broadcasting,* 2d ed., Howard W. Sams, Indianapolis, 1979.
Marsh, Ken: *Independent Video,* Straight Arrow, San Francisco, 1974.

Chapter 3 Television Lenses

Ray, Sidney: *The Lens in Action,* Hastings House, New York, 1976.

Chapter 4 Camera Operation

Jones, Peter: *The Technique of the Television Cameraman,* 3d ed., Focal Press, London and Woburn, Mass., 1974.
Millerson, Gerald: *Television Camera Operation,* Focal Press, London and Woburn, Mass., 1974.

Chapter 5 and Chapter 6 Television Lighting

Clark, Charles G., and Walter Strenge, eds.: *American Cinematographer Handbook,* 4th ed., American Society of Cinematographers, Hollywood, 1973.

GTE/Sylvania Lighting Handbook, 6th ed., available from GTE/Sylvania, 1978.

Millerson, Gerald: *The Technique of Lighting for Television and Motion Pictures,* Hastings House, New York, 1972.

Chapter 7 and Chapter 8   Television Audio

Alkin, Glyn: *Television Sound Operations,* Hastings House, New York, 1975.

Alten, Stanley: *Audio in Media,* Wadsworth, Belmont, Calif., 1981.

Clifford, Martin: *Microphones: How They Work & How to Use Them,* Tab Books, Blue Ridge Summit, Pa., 1977.

Eargle, John: *Sound Recording,* 2d ed., Van Nostrand Reinhold, New York, 1980.

Nisbett, Alec: *The Technique of the Sound Studio,* Focal Press, London, 1979.

Oringel, Robert: *Audio Control Handbook,* 4th ed., Hastings House, New York, 1972.

Tremaine, Howard: *Audio Cyclopedia,* 2d ed., Howard Sams, Indianapolis, 1979.

Woram, John: *Recording Studio Handbook,* Sagamore Publishing Co., New York, 1976.

Chapter 9 and Chapter 10   Video Recording

Ennes, Harold: *Television Broadcasting: Tape Recording Systems,* 2d ed., Howard Sams, Indianapolis, 1979.

Kybett, Harry: *Videotape Recording,* 2d ed., Howard Sams, Indianapolis, 1978.

Robinson, J. F., and P. H. Beards: *Using Videotape,* 2d ed., Focal Press, London, 1981.

Chapter 11   The Switcher and Special Electronic Effects

Wilkie, Bernard: *The Technique of Special Effects in Television,* Focal Press, London, 1971.

Chapter 12   Electronic News Gathering (ENG) Production

Garvey, C. Robert, et al.: *BME's ENG/EFP Handbook,* Broadband Information Services, New York, 1981.

Chapter 13   Television Graphics

Hurrell, Ron: *Television Graphics,* Von Nostrand Reinhold, New York, 1973.
Zettl, Herbert: *Sight-Sound-Motion,* Wadsworth, Belmont, Calif., 1973.

Chapter 14   Film for Television

Churchill, Hugh B.: *Film Editing Handbook,* Wadsworth, Belmont, Calif., 1972.
Corbett, D. J.: *Motion Picture and Television Film: Image Control and Processing Techniques,* Focal Press, London, 1968.
Clarke, Charles G., and Walter Strenge, eds.: *American Cinematographer Handbook,* 4th ed., American Society of Cinematographers, Hollywood, 1973.
Reisz, Karel, and Gavin Millar: *The Technique of Film Editing,* Focal Press, London, 1968.
Samuelson, David W.: *Motion Picture Camera Technique,* Focal Press, London, 1978.

Chapter 15   Set and Staging Design

Bellman, Willard F.: *Scenography and Stage Technology,* Crowell, New York, 1977.
Clark, Frank P.: *Special Effects in Motion Pictures,* Society of Motion Picture and Television Engineers, New York, 1966.
Millerson, Gerald: *Basic TV Staging,* Hastings House, New York, 1974.
Wilkie, Bernard: *Creating Special Effects,* Hastings House, New York, 1977.

Chapter 16   Television Scripts

Bronfeld, Stewart: *Writing for Film and Television,* Prentice-Hall, Englewood Cliffs, N.J., 1981.
Coe, Michelle: *How to Write for Television,* Crown Publishers, New York, 1980.
Maloney, Martin, and Paul Max Rubenstein: *Writing for the Media,* Prentice-Hall, Englewood Cliffs, N.J., 1980.
Willis, Edgar, and Camille D'Arienzo: *Writing Scripts for Television,* Radio and Film, Holt, Rinehart and Winston, New York, 1981.

Chapter 17   Television Talent: The Performer and the Actor

Buchman, Herman: *Film and Television Makeup,* Watson-Guptill, New York, 1973.

Dudek, Lee J.: *Professional Broadcast Announcing,* 2d ed., Allyn and Bacon, Boston, 1981.
Hindman, James et al.: *Television Acting,* Hastings House, New York, 1979.
Hyde, Stuart: *Television and Radio Announcing,* 3d ed., Houghton Mifflin, Boston, 1979.

Chapter 18   Producing

Costa, Sylvia Allen: *How to Prepare a Production Budget for Film & Videotape,* TAB Books, Blue Ridge Summit, Pa., 1975.
Gompertz, Rolf: *Promotion and Publicity Handbook for Broadcasters,* TAB Books, Blue Ridge Summit, Pa., 1977.
Shanks, Bob: *The Cool Fire,* Norton, New York, 1976.

Chapter 19 and Chapter 20   Directing

Lewis, Colby: *The TV Director/Interpreter,* Hastings House, New York, 1968.
Reisz, Karel: *The Technique of Film Editing,* Hastings House, New York, 1968.
Rowlands, Avril: *Script Continuity for the Production Secretary,* Focal Press, London, 1977.
Zettl, Herbert: *Sight-Sound-Motion,* Wadsworth, Belmont, Calif., 1973.

Chapter 21   Remote Production

Garvey, C. Robert, et al.: *BME's ENG/EFP Handbook,* Broadband Information Services, New York, 1981.

INDEX